W9-CLC-010

Applied Mathematical Sciences
Volume 15

M. Braun

Differential Equations and Their Applications

An Introduction to Applied Mathematics

3rd Edition

Springer-Verlag New York · Heidelberg · Berlin

Martin Braun

Department of Mathematics
Queens College
City University of New York
Flushing, NY 11367
USA

AMS Subject Classifications: 98A20, 98A35, 34-01

Library of Congress Cataloging in Publication Data

Braun, Martin, 1941–
Differential equations and their applications.
(Applied mathematical sciences; v. 15)
Includes index.
1. Differential equations. I. Title. II. Series:
Applied mathematical sciences (Springer-Verlag
New York Inc.); v. 15.
QA1.A647 vol. 15 1983 [QA371] 510s 82-19611
[515.3′5]

Printed in the United States of America.

9 8 7 6 5 4 3 (Third printing, 1986)

ISBN 0-387-90806-4 Springer-Verlag New York Heidelberg Berlin
ISBN 3-540-90806-4 Springer-Verlag Berlin Heidelberg New York

To four beautiful people:

Zelda Lee

Adeena Rachelle, I. Nasanayl, and Shulamit

Preface to the Third Edition

There are three major changes in the Third Edition of *Differential Equations and Their Applications*. First, we have completely rewritten the section on singular solutions of differential equations. A new section, 2.8.1, dealing with Euler equations has been added, and this section is used to motivate a greatly expanded treatment of singular equations in sections 2.8.2 and 2.8.3.

Our second major change is the addition of a new section, 4.9, dealing with bifurcation theory, a subject of much current interest. We felt it desirable to give the reader a brief but nontrivial introduction to this important topic.

Our third major change is in Section 2.6, where we have switched to the metric system of units. This change was requested by many of our readers.

In addition to the above changes, we have updated the material on population models, and have revised the exercises in this section. Minor editorial changes have also been made throughout the text.

New York City
November, 1982

Martin Braun

Preface to the First Edition

This textbook is a unique blend of the theory of differential equations and their exciting application to "real world" problems. First, and foremost, it is a rigorous study of ordinary differential equations and can be fully understood by anyone who has completed one year of calculus. However, in addition to the traditional applications, it also contains many exciting "real life" problems. These applications are completely self contained. First, the problem to be solved is outlined clearly, and one or more differential equations are derived as a model for this problem. These equations are then solved, and the results are compared with real world data. The following applications are covered in this text.

1. In Section 1.3 we prove that the beautiful painting "Disciples of Emmaus" which was bought by the Rembrandt Society of Belgium for $170,000 was a modern forgery.

2. In Section 1.5 we derive differential equations which govern the population growth of various species, and compare the results predicted by our models with the known values of the populations.

3. In Section 1.6 we derive differential equations which govern the rate at which farmers adopt new innovations. Surprisingly, these same differential equations govern the rate at which technological innovations are adopted in such diverse industries as coal, iron and steel, brewing, and railroads.

4. In Section 1.7 we try to determine whether tightly sealed drums filled with concentrated waste material will crack upon impact with the ocean floor. In this section we also describe several tricks for obtaining information about solutions of a differential equation that cannot be solved explicitly.

5. In Section 2.7 we derive a very simple model of the blood glucose regulatory system and obtain a fairly reliable criterion for the diagnosis of diabetes.

6. Section 4.5 describes two applications of differential equations to arms races and actual combat. In Section 4.5.1 we discuss L. F. Richardson's theory of the escalation of arms races and fit his model to the arms race which led eventually to World War I. This section also provides the reader with a concrete feeling for the concept of stability. In Section 4.5.2 we derive two Lanchestrian combat models, and fit one of these models, with astonishing accuracy, to the battle of Iwo Jima in World War II.

7. In Section 4.10 we show why the predator portion (sharks, skates, rays, etc.) of all fish caught in the port of Fiume, Italy rose dramatically during the years of World War I. The theory we develop here also has a spectacular application to the spraying of insecticides.

8. In Section 4.11 we derive the "principle of competitive exclusion," which states, essentially, that no two species can earn their living in an identical manner.

9. In Section 4.12 we study a system of differential equations which govern the spread of epidemics in a population. This model enables us to prove the famous "threshold theorem of epidemiology," which states that an epidemic will occur only if the number of people susceptible to the disease exceeds a certain threshold value. We also compare the predictions of our model with data from an actual plague in Bombay.

10. In Section 4.13 we derive a model for the spread of gonorrhea and prove that either this disease dies out, or else the number of people who have gonorrhea will ultimately approach a fixed value.

This textbook also contains the following important, and often unique features.

1. In Section 1.10 we give a complete proof of the existence–uniqueness theorem for solutions of first-order equations. Our proof is based on the method of Picard iterates, and can be fully understood by anyone who has completed one year of calculus.

2. In Section 1.11 we show how to solve equations by iteration. This section has the added advantage of reinforcing the reader's understanding of the proof of the existence–uniqueness theorem.

3. Complete Fortran and APL programs are given for every computer example in the text. Computer problems appear in Sections 1.13–1.17, which deal with numerical approximations of solutions of differential equations; in Section 1.11, which deals with solving the equations $x=f(x)$ and $g(x)=0$; and in Section 2.8, where we show how to obtain a power-series solution of a differential equation even though we cannot explicitly solve the recurrence formula for the coefficients.

4. A self-contained introduction to the computing language APL is presented in Appendix C. Using this appendix we have been able to teach our students APL in just two lectures.

5. Modesty aside, Section 2.12 contains an absolutely super and unique treatment of the Dirac delta function. We are very proud of this section because it eliminates all the ambiguities which are inherent in the traditional exposition of this topic.

6. All the linear algebra pertinent to the study of systems of equations is presented in Sections 3.1–3.7. One advantage of our approach is that the reader gets a concrete feeling for the very important but extremely abstract properties of linear independence, spanning, and dimension. Indeed, many linear algebra students sit in on our course to find out what's really going on in their course.

Differential Equations and Their Applications can be used for a one- or two-semester course in ordinary differential equations. It is geared to the student who has completed two semesters of calculus. Traditionally, most authors present a "suggested syllabus" for their textbook. We will not do so here, though, since there are already more than twenty different syllabi in use. Suffice it to say that this text can be used for a wide variety of courses in ordinary differential equations.

I greatly appreciate the help of the following people in the preparation of this manuscript: Douglas Reber who wrote the Fortran programs, Eleanor Addison who drew the original figures, and Kate MacDougall, Sandra Spinacci, and Miriam Green who typed portions of this manuscript.

I am grateful to Walter Kaufmann-Bühler, the mathematics editor at Springer-Verlag, and Elizabeth Kaplan, the production editor, for their extensive assistance and courtesy during the preparation of this manuscript. It is a pleasure to work with these true professionals.

Finally, I am especially grateful to Joseph P. LaSalle for the encouragement and help he gave me. Thanks again, Joe.

New York City
July, 1976

Martin Braun

Contents

Chapter 2
Second-order linear differential equations 125

Chapter 3
Systems of differential equations 262

Chapter 4
Qualitative theory of differential equations 370

First-order differential equations 1

1.1 Introduction

This book is a study of differential equations and their applications. A differential equation is a relationship between a function of time and its derivatives. The equations

$$\frac{dy}{dt}=3y^2\sin(t+y) \tag{i}$$

and

$$\frac{d^3y}{dt^3}=e^{-y}+t+\frac{d^2y}{dt^2} \tag{ii}$$

are both examples of differential equations. The order of a differential equation is the order of the highest derivative of the function y that appears in the equation. Thus (i) is a first-order differential equation and (ii) is a third-order differential equation. By a solution of a differential equation we will mean a continuous function $y(t)$ which together with its derivatives satisfies the relationship. For example, the function

$$y(t)=2\sin t-\tfrac{1}{3}\cos 2t$$

is a solution of the second-order differential equation

$$\frac{d^2y}{dt^2}+y=\cos 2t$$

since

$$\frac{d^2}{dt^2}\left(2\sin t-\tfrac{1}{3}\cos 2t\right)+\left(2\sin t-\tfrac{1}{3}\cos 2t\right)$$

$$=\left(-2\sin t+\tfrac{4}{3}\cos 2t\right)+2\sin t-\tfrac{1}{3}\cos 2t=\cos 2t.$$

Differential equations appear naturally in many areas of science and the humanities. In this book, we will present serious discussions of the applications of differential equations to such diverse and fascinating problems as the detection of art forgeries, the diagnosis of diabetes, the increase in the percentage of sharks present in the Mediterranean Sea during World War I, and the spread of gonorrhea. Our purpose is to show how researchers have used differential equations to solve, or try to solve, *real life* problems. And while we will discuss some of the great success stories of differential equations, we will also point out their limitations and document some of their failures.

1.2 First-order linear differential equations

We begin by studying first-order differential equations and we will assume that our equation is, or can be put, in the form

$$\frac{dy}{dt}=f(t,y). \tag{1}$$

The problem before us is this: Given $f(t,y)$ find all functions $y(t)$ which satisfy the differential equation (1). We approach this problem in the following manner. A fundamental principle of mathematics is that the way to solve a new problem is to reduce it, in some manner, to a problem that we have already solved. In practice this usually entails successively simplifying the problem until it resembles one we have already solved. Since we are presently in the business of solving differential equations, it is advisable for us to take inventory and list all the differential equations we can solve. If we assume that our mathematical background consists of just elementary calculus then the very sad fact is that the only first-order differential equation we can solve at present is

$$\frac{dy}{dt}=g(t) \tag{2}$$

where g is any integrable function of time. To solve Equation (2) simply integrate both sides with respect to t, which yields

$$y(t)=\int g(t)\,dt+c.$$

Here c is an arbitrary constant of integration, and by $\int g(t)\,dt$ we mean an anti-derivative of g, that is, a function whose derivative is g. Thus, to solve any other differential equation we must somehow reduce it to the form (2). As we will see in Section 1.9, this is impossible to do in most cases. Hence, we will not be able, without the aid of a computer, to solve most differential equations. It stands to reason, therefore, that to find those differential equations that we *can* solve, we should start with very simple equations and not ones like

$$\frac{dy}{dt} = e^{\sin(t - 37\sqrt{|y|}\,)}$$

(which incidentally, cannot be solved exactly). Experience has taught us that the "simplest" equations are those which are *linear* in the dependent variable y.

Definition. The general first-order linear differential equation is

$$\frac{dy}{dt} + a(t)y = b(t). \tag{3}$$

Unless otherwise stated, the functions $a(t)$ and $b(t)$ are assumed to be continuous functions of time. We single out this equation and call it linear because the dependent variable y appears by itself, that is, no terms such as e^{-y}, y^3 or $\sin y$ etc. appear in the equation. For example $dy/dt = y^2 + \sin t$ and $dy/dt = \cos y + t$ are both *nonlinear* equations because of the y^2 and $\cos y$ terms respectively.

Now it is not immediately apparent how to solve Equation (3). Thus, we simplify it even further by setting $b(t) = 0$.

Definition. The equation

$$\frac{dy}{dt} + a(t)y = 0 \tag{4}$$

is called the ***homogeneous*** first-order linear differential equation, and Equation (3) is called the ***nonhomogeneous*** first-order linear differential equation for $b(t)$ not identically zero.

Fortunately, the homogeneous equation (4) can be solved quite easily. First, divide both sides of the equation by y and rewrite it in the form

$$\frac{\frac{dy}{dt}}{y} = -a(t).$$

Second, observe that

$$\frac{\frac{dy}{dt}}{y} \equiv \frac{d}{dt}\ln|y(t)|$$

where by $\ln|y(t)|$ we mean the natural logarithm of $|y(t)|$. Hence Equation (4) can be written in the form

$$\frac{d}{dt}\ln|y(t)| = -a(t). \tag{5}$$

But this is Equation (2) "essentially" since we can integrate both sides of (5) to obtain that

$$\ln|y(t)| = -\int a(t)\,dt + c_1$$

where c_1 is an arbitrary constant of integration. Taking exponentials of both sides yields

$$|y(t)| = \exp\left(-\int a(t)\,dt + c_1\right) = c\exp\left(-\int a(t)\,dt\right)$$

or

$$\left|y(t)\exp\left(\int a(t)\,dt\right)\right| = c. \tag{6}$$

Now, $y(t)\exp\left(\int a(t)\,dt\right)$ is a continuous function of time and Equation (6) states that its absolute value is constant. But if the absolute value of a continuous function $g(t)$ is constant then g itself must be constant. To prove this observe that if g is not constant, then there exist two different times t_1 and t_2 for which $g(t_1) = c$ and $g(t_2) = -c$. By the intermediate value theorem of calculus g must achieve all values between $-c$ and $+c$ which is impossible if $|g(t)| = c$. Hence, we obtain the equation $y(t)\exp\left(\int a(t)\,dt\right) = c$ or

$$y(t) = c\exp\left(-\int a(t)\,dt\right). \tag{7}$$

Equation (7) is said to be the *general solution* of the homogeneous equation since every solution of (4) must be of this form. Observe that an arbitrary constant c appears in (7). This should not be too surprising. Indeed, we will always expect an arbitrary constant to appear in the general solution of any first-order differential equation. To wit, if we are given dy/dt and we want to recover $y(t)$, then we must perform an integration, and this, of necessity, yields an arbitrary constant. Observe also that Equation (4) has infinitely many solutions; for each value of c we obtain a distinct solution $y(t)$.

Example 1. Find the general solution of the equation $(dy/dt) + 2ty = 0$.

Solution. Here $a(t) = 2t$ so that $y(t) = c\exp\left(-\int 2t\,dt\right) = ce^{-t^2}$.

Example 2. Determine the behavior, as $t \to \infty$, of all solutions of the equation $(dy/dt) + ay = 0$, a constant.

Solution. The general solution is $y(t) = c\exp\left(-\int a\,dt\right) = ce^{-at}$. Hence if $a < 0$, all solutions, with the exception of $y = 0$, approach infinity, and if $a > 0$, all solutions approach zero as $t \to \infty$.

In applications, we are usually not interested in all solutions of (4). Rather, we are looking for the *specific* solution $y(t)$ which at some initial time t_0 has the value y_0. Thus, we want to determine a function $y(t)$ such that

$$\frac{dy}{dt}+a(t)y=0, \qquad y(t_0)=y_0. \tag{8}$$

Equation (8) is referred to as an initial-value problem for the obvious reason that of the totality of all solutions of the differential equation, we are looking for the one solution which initially (at time t_0) has the value y_0. To find this solution we integrate both sides of (5) between t_0 and t. Thus

$$\int_{t_0}^{t}\frac{d}{ds}\ln|y(s)|\,ds=-\int_{t_0}^{t}a(s)\,ds$$

and, therefore

$$\ln|y(t)|-\ln|y(t_0)|=\ln\left|\frac{y(t)}{y(t_0)}\right|=-\int_{t_0}^{t}a(s)\,ds.$$

Taking exponentials of both sides of this equation we obtain that

$$\left|\frac{y(t)}{y(t_0)}\right|=\exp\left(-\int_{t_0}^{t}a(s)\,ds\right)$$

or

$$\left|\frac{y(t)}{y(t_0)}\exp\left(\int_{t_0}^{t}a(s)\,ds\right)\right|=1.$$

The function inside the absolute value sign is a continuous function of time. Thus, by the argument given previously, it is either identically $+1$ or identically -1. To determine which one it is, evaluate it at the point t_0; since

$$\frac{y(t_0)}{y(t_0)}\exp\left(\int_{t_0}^{t_0}a(s)\,ds\right)=1$$

we see that

$$\frac{y(t)}{y(t_0)}\exp\left(\int_{t_0}^{t}a(s)\,ds\right)=1.$$

Hence

$$y(t)=y(t_0)\exp\left(-\int_{t_0}^{t}a(s)\,ds\right)=y_0\exp\left(-\int_{t_0}^{t}a(s)\,ds\right).$$

Example 3. Find the solution of the initial-value problem

$$\frac{dy}{dt}+(\sin t)y=0, \qquad y(0)=\tfrac{3}{2}.$$

Solution. Here $a(t)=\sin t$ so that

$$y(t)=\tfrac{3}{2}\exp\left(-\int_0^t \sin s\, ds\right)=\tfrac{3}{2}e^{(\cos t)-1}.$$

Example 4. Find the solution of the initial-value problem

$$\frac{dy}{dt}+e^{t^2}y=0, \qquad y(1)=2.$$

Solution. Here $a(t)=e^{t^2}$ so that

$$y(t)=2\exp\left(-\int_1^t e^{s^2}\, ds\right).$$

Now, at first glance this problem would seem to present a very serious difficulty in that we cannot integrate the function e^{s^2} directly. However, this solution is equally as valid and equally as useful as the solution to Example 3. The reason for this is twofold. First, there are very simple numerical schemes to evaluate the above integral to any degree of accuracy with the aid of a computer. Second, even though the solution to Example 3 is given explicitly, we still cannot evaluate it at any time t without the aid of a table of trigonometric functions and some sort of calculating aid, such as a slide rule, electronic calculator or digital computer.

We return now to the nonhomogeneous equation

$$\frac{dy}{dt}+a(t)y=b(t).$$

It should be clear from our analysis of the homogeneous equation that the way to solve the nonhomogeneous equation is to express it in the form

$$\frac{d}{dt}(\text{“something”})=b(t)$$

and then to integrate both sides to solve for “something”. However, the expression $(dy/dt)+a(t)y$ does not appear to be the derivative of some simple expression. The next logical step in our analysis therefore should be the following: Can we make the left hand side of the equation to be d/dt of “something”? More precisely, we can multiply both sides of (3) by any continuous function $\mu(t)$ to obtain the equivalent equation

$$\mu(t)\frac{dy}{dt}+a(t)\mu(t)y=\mu(t)b(t). \tag{9}$$

(By equivalent equations we mean that every solution of (9) is a solution of (3) and vice-versa.) Thus, can we *choose* $\mu(t)$ so that $\mu(t)(dy/dt)+a(t)\mu(t)y$ is the derivative of some simple expression? The answer to this question is yes, and is obtained by observing that

$$\frac{d}{dt}\mu(t)y=\mu(t)\frac{dy}{dt}+\frac{d\mu}{dt}y.$$

Hence, $\mu(t)(dy/dt)+a(t)\mu(t)y$ will be equal to the derivative of $\mu(t)y$ if and only if $d\mu(t)/dt=a(t)\mu(t)$. But this is a first-order linear homogeneous equation for $\mu(t)$, i.e. $(d\mu/dt)-a(t)\mu=0$ which we already know how to solve, and since we only need one such function $\mu(t)$ we set the constant c in (7) equal to one and take

$$\mu(t)=\exp\left(\int a(t)\,dt\right).$$

For this $\mu(t)$, Equation (9) can be written as

$$\frac{d}{dt}\mu(t)y=\mu(t)b(t). \tag{10}$$

To obtain the general solution of the nonhomogeneous equation (3), that is, to find all solutions of the nonhomogeneous equation, we take the indefinite integral (anti-derivative) of both sides of (10) which yields

$$\mu(t)y=\int\mu(t)b(t)\,dt+c$$

or

$$y=\frac{1}{\mu(t)}\left(\int\mu(t)b(t)\,dt+c\right)=\exp\left(-\int a(t)\,dt\right)\left(\int\mu(t)b(t)\,dt+c\right). \tag{11}$$

Alternately, if we are interested in the specific solution of (3) satisfying the initial condition $y(t_0)=y_0$, that is, if we want to solve the initial-value problem

$$\frac{dy}{dt}+a(t)y=b(t), \qquad y(t_0)=y_0$$

then we can take the definite integral of both sides of (10) between t_0 and t to obtain that

$$\mu(t)y-\mu(t_0)y_0=\int_{t_0}^{t}\mu(s)b(s)\,ds$$

or

$$y=\frac{1}{\mu(t)}\left(\mu(t_0)y_0+\int_{t_0}^{t}\mu(s)b(s)\,ds\right). \tag{12}$$

Remark 1. Notice how we used our knowledge of the solution of the homogeneous equation to find the function $\mu(t)$ which enables us to solve the nonhomogeneous equation. This is an excellent illustration of how we use our knowledge of the solution of a simpler problem to solve a harder problem.

Remark 2. The function $\mu(t)=\exp\left(\int a(t)\,dt\right)$ is called an *integrating factor* for the nonhomogeneous equation since after multiplying both sides by $\mu(t)$ we can immediately integrate the equation to find all solutions.

Remark 3. The reader should not memorize formulae (11) and (12). Rather, we will solve all nonhomogeneous equations by first multiplying both sides by $\mu(t)$, by writing the new left-hand side as the derivative of $\mu(t)y(t)$, and then by integrating both sides of the equation.

Remark 4. An alternative way of solving the initial-value problem $(dy/dt)+a(t)y=b(t)$, $y(t_0)=y_0$ is to find the general solution (11) of (3) and then use the initial condition $y(t_0)=y_0$ to evaluate the constant c. If the function $\mu(t)b(t)$ cannot be integrated directly, though, then we must take the definite integral of (10) to obtain (12), and this equation is then approximated numerically.

Example 5. Find the general solution of the equation $(dy/dt)-2ty=t$.
Solution. Here $a(t)=-2t$ so that

$$\mu(t)=\exp\left(\int a(t)\,dt\right)=\exp\left(-\int 2t\,dt\right)=e^{-t^2}.$$

Multiplying both sides of the equation by $\mu(t)$ we obtain the equivalent equation

$$e^{-t^2}\left(\frac{dy}{dt}-2ty\right)=te^{-t^2} \quad\text{or}\quad \frac{d}{dt}e^{-t^2}y=te^{-t^2}.$$

Hence,

$$e^{-t^2}y=\int te^{-t^2}dt+c=\frac{-e^{-t^2}}{2}+c$$

and

$$y(t)=-\tfrac{1}{2}+ce^{t^2}.$$

Example 6. Find the solution of the initial-value problem

$$\frac{dy}{dt}+2ty=t, \qquad y(1)=2.$$

Solution. Here $a(t)=2t$ so that

$$\mu(t)=\exp\left(\int a(t)\,dt\right)=\exp\left(\int 2t\,dt\right)=e^{t^2}.$$

Multiplying both sides of the equation by $\mu(t)$ we obtain that

$$e^{t^2}\left(\frac{dy}{dt}+2ty\right)=te^{t^2} \quad\text{or}\quad \frac{d}{dt}(e^{t^2}y)=te^{t^2}.$$

Hence,

$$\int_1^t \frac{d}{ds} e^{s^2} y(s)\,ds = \int_1^t s e^{s^2}\,ds$$

so that

$$e^{s^2} y(s)\Big|_1^t = \frac{e^{s^2}}{2}\Big|_1^t.$$

Consequently,

$$e^{t^2} y - 2e = \frac{e^{t^2}}{2} - \frac{e}{2}$$

and

$$y = \frac{1}{2} + \frac{3e}{2} e^{-t^2} = \tfrac{1}{2} + \tfrac{3}{2} e^{1-t^2}.$$

Example 7. Find the solution of the initial-value problem

$$\frac{dy}{dt} + y = \frac{1}{1+t^2}, \quad y(2) = 3.$$

Solution. Here $a(t) = 1$, so that

$$\mu(t) = \exp\left(\int a(t)\,dt\right) = \exp\left(\int 1\,dt\right) = e^t.$$

Multiplying both sides of the equation by $\mu(t)$ we obtain that

$$e^t\left(\frac{dy}{dt} + y\right) = \frac{e^t}{1+t^2} \quad \text{or} \quad \frac{d}{dt} e^t y = \frac{e^t}{1+t^2}.$$

Hence

$$\int_2^t \frac{d}{ds} e^s y(s)\,ds = \int_2^t \frac{e^s}{1+s^2}\,ds,$$

so that

$$e^t y - 3e^2 = \int_2^t \frac{e^s}{1+s^2}\,ds$$

and

$$y = e^{-t}\left[3e^2 + \int_2^t \frac{e^s}{1+s^2}\,ds\right].$$

EXERCISES

In each of Problems 1–7 find the general solution of the given differential equation.

1. $\dfrac{dy}{dt} + y\cos t = 0$

2. $\dfrac{dy}{dt} + y\sqrt{t}\,\sin t = 0$

3. $\frac{dy}{dt}+\frac{2t}{1+t^2}y=\frac{1}{1+t^2}$

4. $\frac{dy}{dt}+y=te^t$

5. $\frac{dy}{dt}+t^2y=1$

6. $\frac{dy}{dt}+t^2y=t^2$

7. $\frac{dy}{dt}+\frac{t}{1+t^2}y=1-\frac{t^3}{1+t^4}y$

In each of Problems 8–14, find the solution of the given initial-value problem.

8. $\frac{dy}{dt}+\sqrt{1+t^2}\,y=0, \qquad y(0)=\sqrt{5}$

9. $\frac{dy}{dt}+\sqrt{1+t^2}\,e^{-t}y=0, \qquad y(0)=1$

10. $\frac{dy}{dt}+\sqrt{1+t^2}\,e^{-t}y=0, \qquad y(0)=0$

11. $\frac{dy}{dt}-2ty=t, \qquad y(0)=1$

12. $\frac{dy}{dt}+ty=1+t, \qquad y(\frac{3}{2})=0$

13. $\frac{dy}{dt}+y=\frac{1}{1+t^2}, \qquad y(1)=2$

14. $\frac{dy}{dt}-2ty=1, \qquad y(0)=1$

15. Find the general solution of the equation

$$(1+t^2)\frac{dy}{dt}+ty=(1+t^2)^{5/2}.$$

(*Hint*: Divide both sides of the equation by $1+t^2$.)

16. Find the solution of the initial-value problem

$$(1+t^2)\frac{dy}{dt}+4ty=t, \qquad y(1)=\tfrac{1}{4}.$$

17. Find a continuous solution of the initial-value problem

$$y'+y=g(t), \qquad y(0)=0$$

where

$$g(t)=\begin{cases} 2, & 0\leqslant t\leqslant 1 \\ 0, & t>1 \end{cases}.$$

18. Show that every solution of the equation $(dy/dt)+ay=be^{-ct}$ where a and c are positive constants and b is any real number approaches zero as t approaches infinity.

19. Given the differential equation $(dy/dt)+a(t)y=f(t)$ with $a(t)$ and $f(t)$ continuous for $-\infty<t<\infty$, $a(t)\geqslant c>0$, and $\lim_{t\to\infty}f(t)=0$, show that every solution tends to zero as t approaches infinity.

When we derived the solution of the nonhomogeneous equation we tacitly assumed that the functions $a(t)$ and $b(t)$ were continuous so that we could perform the necessary integrations. If either of these functions was discontinuous at a point t_1, then we would expect that our solutions might be discontinuous at $t=t_1$. Problems 20–23 illustrate the variety of things that may happen. In Problems 20–22 determine the behavior of all solutions of the given differential equation as $t \to 0$, and in Problem 23 determine the behavior of all solutions as $t \to \pi/2$.

20. $\dfrac{dy}{dt}+\dfrac{1}{t}y=\dfrac{1}{t^2}$

21. $\dfrac{dy}{dt}+\dfrac{1}{\sqrt{t}}y=e^{\sqrt{t}/2}$

22. $\dfrac{dy}{dt}+\dfrac{1}{t}y=\cos t+\dfrac{\sin t}{t}$

23. $\dfrac{dy}{dt}+y\tan t=\sin t\cos t.$

1.3 The Van Meegeren art forgeries

After the liberation of Belgium in World War II, the Dutch Field Security began its hunt for Nazi collaborators. They discovered, in the records of a firm which had sold numerous works of art to the Germans, the name of a banker who had acted as an intermediary in the sale to Goering of the painting "Woman Taken in Adultery" by the famed 17th century Dutch painter Jan Vermeer. The banker in turn revealed that he was acting on behalf of a third rate Dutch painter H. A. Van Meegeren, and on May 29, 1945 Van Meegeren was arrested on the charge of collaborating with the enemy. On July 12, 1945 Van Meegeren startled the world by announcing from his prison cell that he had never sold "Woman Taken in Adultery" to Goering. Moreover, he stated that this painting and the very famous and beautiful "Disciples at Emmaus", as well as four other presumed Vermeers and two de Hooghs (a 17th century Dutch painter) were his own work. Many people, however, thought that Van Meegeren was only lying to save himself from the charge of treason. To prove his point, Van Meegeren began, while in prison, to forge the Vermeer painting "Jesus Amongst the Doctors" to demonstrate to the skeptics just how good a forger of Vermeer he was. The work was nearly completed when Van Meegeren learned that a charge of forgery had been substituted for that of collaboration. He, therefore, refused to finish and age the painting so that hopefully investigators would not uncover his secret of aging his forgeries. To settle the question an international panel of distinguished chemists, physicists and art historians was appointed to investigate the matter. The panel took x-rays of the paintings to determine whether other paintings were underneath them. In addition, they analyzed the pigments (coloring materials) used in the paint, and examined the paintings for certain signs of old age.

Now, Van Meegeren was well aware of these methods. To avoid detection, he scraped the paint from old paintings that were not worth much, just to get the canvas, and he tried to use pigments that Vermeer would have used. Van Meegeren also knew that old paint was extremely hard, and impossible to dissolve. Therefore, he very cleverly mixed a chemical, phenoformaldehyde, into the paint, and this hardened into bakelite when the finished painting was heated in an oven.

However, Van Meegeren was careless with several of his forgeries, and the panel of experts found traces of the modern pigment cobalt blue. In addition, they also detected the phenoformaldehyde, which was not discovered until the turn of the 19th century, in several of the paintings. On the basis of this evidence Van Meegeren was convicted, of forgery, on October 12, 1947 and sentenced to one year in prison. While in prison he suffered a heart attack and died on December 30, 1947.

However, even following the evidence gathered by the panel of experts, many people still refused to believe that the famed "Disciples at Emmaus" was forged by Van Meegeren. Their contention was based on the fact that the other alleged forgeries and Van Meegeren's nearly completed "Jesus Amongst the Doctors" were of a very inferior quality. Surely, they said, the creator of the beautiful "Disciples at Emmaus" could not produce such inferior pictures. Indeed, the "Disciples at Emmaus" was certified as an authentic Vermeer by the noted art historian A. Bredius and was bought by the Rembrandt Society for $170,000. The answer of the panel to these skeptics was that because Van Meegeren was keenly disappointed by his lack of status in the art world, he worked on the "Disciples at Emmaus" with the fierce determination of proving that he was better than a third rate painter. After producing such a masterpiece his determination was gone. Moreover, after seeing how easy it was to dispose of the "Disciples at Emmaus" he devoted less effort to his subsequent forgeries. This explanation failed to satisfy the skeptics. They demanded a thoroughly scientific and conclusive proof that the "Disciples at Emmaus" was indeed a forgery. This was done recently in 1967 by scientists at Carnegie Mellon University, and we would now like to describe their work.

The key to the dating of paintings and other materials such as rocks and fossils lies in the phenomenon of radioactivity discovered at the turn of the century. The physicist Rutherford and his colleagues showed that the atoms of certain "radioactive" elements are unstable and that within a given time period a fixed proportion of the atoms spontaneously disintegrates to form atoms of a new element. Because radioactivity is a property of the atom, Rutherford showed that the radioactivity of a substance is directly proportional to the number of atoms of the substance present. Thus, if $N(t)$ denotes the number of atoms present at time t, then dN/dt, the number of atoms that disintegrate per unit time is proportional to N, that is,

$$\frac{dN}{dt} = -\lambda N. \tag{1}$$

The constant λ which is positive, is known as the decay constant of the substance. The larger λ is, of course, the faster the substance decays. One measure of the rate of disintegration of a substance is its *half-life* which is defined as the time required for half of a given quantity of radioactive atoms to decay. To compute the half-life of a substance in terms of λ, assume that at time t_0, $N(t_0) = N_0$. Then, the solution of the initial-value problem $dN/dt = -\lambda N$, $N(t_0) = N_0$ is

$$N(t) = N_0 \exp\left(-\lambda \int_{t_0}^{t} ds\right) = N_0 e^{-\lambda(t-t_0)}$$

or $N/N_0 = \exp(-\lambda(t-t_0))$. Taking logarithms of both sides we obtain that

$$-\lambda(t-t_0) = \ln\frac{N}{N_0}. \tag{2}$$

Now, if $N/N_0 = \frac{1}{2}$ then $-\lambda(t-t_0) = \ln\frac{1}{2}$ so that

$$(t-t_0) = \frac{\ln 2}{\lambda} \doteq \frac{0.6931}{\lambda}. \tag{3}$$

Thus, the half-life of a substance is ln 2 divided by the decay constant λ. The dimension of λ, which we suppress for simplicity of writing, is reciprocal time. If t is measured in years then λ has the dimension of reciprocal years, and if t is measured in minutes, then λ has the dimension of reciprocal minutes. The half-lives of many substances have been determined and recorded. For example, the half-life of carbon-14 is 5568 years and the half-life of uranium-238 is 4.5 billion years.

Now the basis of "radioactive dating" is essentially the following. From Equation (2) we can solve for $t-t_0 = 1/\lambda \ln(N_0/N)$. If t_0 is the time the substance was initially formed or manufactured, then the age of the substance is $1/\lambda \ln(N_0/N)$. The decay constant λ is known or can be computed, in most instances. Moreover, we can usually evaluate N quite easily. Thus, if we knew N_0 we could determine the age of the substance. But this is the real difficulty of course, since we usually do not know N_0. In some instances though, we can either determine N_0 indirectly, or else determine certain suitable ranges for N_0, and such is the case for the forgeries of Van Meegeren.

We begin with the following well-known facts of elementary chemistry. Almost all rocks in the earth's crust contain a small quantity of uranium. The uranium in the rock decays to another radioactive element, and that one decays to another and another, and so forth (see Figure 1) in a series of elements that results in lead, which is not radioactive. The uranium (whose half-life is over four billion years) keeps feeding the elements following it in the series, so that as fast as they decay, they are replaced by the elements before them.

Now, all paintings contain a small amount of the radioactive element lead-210 (^{210}Pb), and an even smaller amount of radium-226 (^{226}Ra), since these elements are contained in white lead (lead oxide), which is a pigment that artists have used for over 2000 years. For the analysis which follows, it is important to note that white lead is made from lead metal, which, in turn, is extracted from a rock called lead ore, in a process called smelting. In this process, the lead-210 in the ore goes along with the lead metal. However, 90–95% of the radium and its descendants are removed with

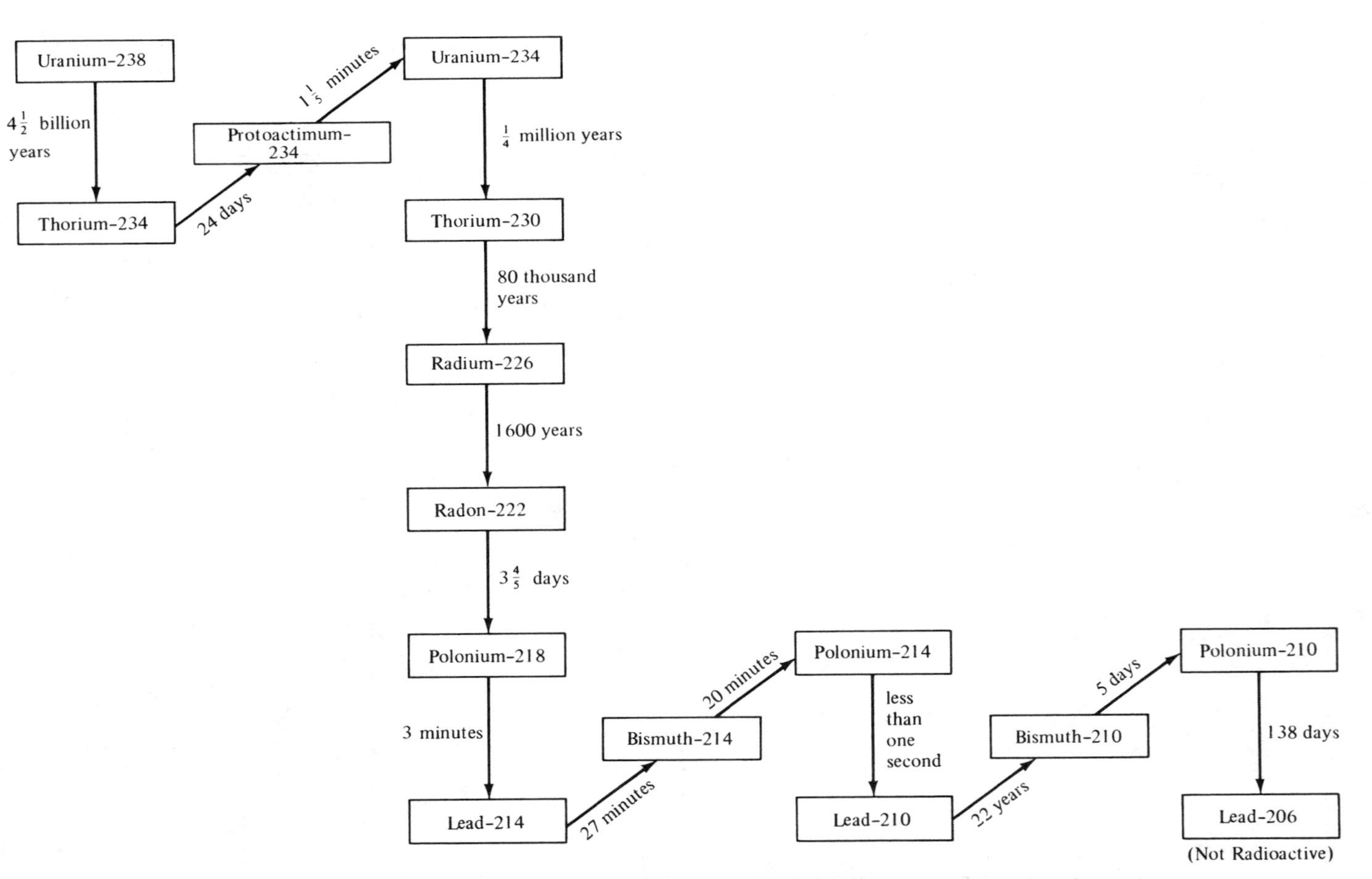

Figure 1. The Uranium series. (The times shown on the arrows are the half-lives of each step.)

other waste products in a material called slag. Thus, most of the supply of lead-210 is cut off and it begins to decay very rapidly, with a half-life of 22 years. This process continues until the lead-210 in the white lead is once more in radioactive equilibrium with the small amount of radium present, i.e. the disintegration of the lead-210 is exactly balanced by the disintegration of the radium.

Let us now use this information to compute the amount of lead-210 present in a sample in terms of the amount originally present at the time of manufacture. Let $y(t)$ be the amount of lead-210 per gram of white lead at time t, y_0 the amount of lead-210 per gram of white lead present at the time of manufacture t_0, and $r(t)$ the number of disintegrations of radium-226 per minute per gram of white lead, at time t. If λ is the decay constant for lead-210, then

$$\frac{dy}{dt} = -\lambda y + r(t), \qquad y(t_0) = y_0. \tag{4}$$

Since we are only interested in a time period of at most 300 years we may assume that the radium-226, whose half-life is 1600 years, remains constant, so that $r(t)$ is a constant r. Multiplying both sides of the differential equation by the integrating factor $\mu(t) = e^{\lambda t}$ we obtain that

$$\frac{d}{dt} e^{\lambda t} y = r e^{\lambda t}.$$

Hence

$$e^{\lambda t} y(t) - e^{\lambda t_0} y_0 = \frac{r}{\lambda}\left(e^{\lambda t} - e^{\lambda t_0}\right)$$

or

$$y(t) = \frac{r}{\lambda}\left(1 - e^{-\lambda(t-t_0)}\right) + y_0 e^{-\lambda(t-t_0)}. \tag{5}$$

Now $y(t)$ and r can be easily measured. Thus, if we knew y_0 we could use Equation (5) to compute $(t-t_0)$ and consequently, we could determine the age of the painting. As we pointed out, though, we cannot measure y_0 directly. One possible way out of this difficulty is to use the fact that the original quantity of lead-210 was in radioactive equilibrium with the larger amount of radium-226 in the ore from which the metal was extracted. Let us, therefore, take samples of different ores and count the number of disintegrations of the radium-226 in the ores. This was done for a variety of ores and the results are given in Table 1 below. These numbers vary from 0.18 to 140. Consequently, the number of disintegrations of the lead-210 per minute per gram of white lead at the time of manufacture will vary from 0.18 to 140. This implies that y_0 will also vary over a very large interval, since the number of disintegrations of lead-210 is proportional to the amount present. Thus, we cannot use Equation (5) to obtain an accurate, or even a crude estimate, of the age of a painting.

Table 1. Ore and ore concentrate samples. All disintegration rates are per gram of white lead.

Description and Source		Disintegrations per minute of ^{226}Ra
Ore concentrate	(Oklahoma–Kansas)	4.5
Crushed raw ore	(S.E. Missouri)	2.4
Ore concentrate	(S.E. Missouri)	0.7
Ore concentrate	(Idaho)	2.2
Ore concentrate	(Idaho)	0.18
Ore concentrate	(Washington)	140.0
Ore concentrate	(British Columbia)	1.9
Ore concentrate	(British Columbia)	0.4
Ore concentrate	(Bolivia)	1.6
Ore concentrate	(Australia)	1.1

However, we can still use Equation (5) to distinguish between a 17th century painting and a modern forgery. The basis for this statement is the simple observation that if the paint is very old compared to the 22 year half-life of lead, then the amount of radioactivity from the lead-210 in the paint will be nearly equal to the amount of radioactivity from the radium in the paint. On the other hand, if the painting is modern (approximately 20 years old, or so) then the amount of radioactivity from the lead-210 will be much greater than the amount of radioactivity from the radium.

We make this argument precise in the following manner. Let us assume that the painting in question is either very new or about 300 years old. Set $t-t_0=300$ years in (5). Then, after some simple algebra, we see that

$$\lambda y_0=\lambda y(t)e^{300\lambda}-r(e^{300\lambda}-1). \tag{6}$$

If the painting is indeed a modern forgery, then λy_0 will be absurdly large. To determine what is an absurdly high disintegration rate we observe (see Exercise 1) that if the lead-210 decayed originally (at the time of manufacture) at the rate of 100 disintegrations per minute per gram of white lead, then the ore from which it was extracted had a uranium content of approximately 0.014 per cent. This is a fairly high concentration of uranium since the average amount of uranium in rocks of the earth's crust is about 2.7 parts per million. On the other hand, there are some very rare ores in the Western Hemisphere whose uranium content is 2-3 per cent. To be on the safe side, we will say that a disintegration rate of lead-210 is certainly absurd if it exceeds 30,000 disintegrations per minute per gram of white lead.

To evaluate λy_0, we must evaluate the present disintegration rate, $\lambda y(t)$, of the lead-210, the disintegration rate r of the radium-226, and $e^{300\lambda}$. Since the disintegration rate of polonium-210 (^{210}Po) equals that of lead-210 after several years, and since it is easier to measure the disintegration rate of polonium-210, we substitute these values for those of lead-210. To compute

$e^{300\lambda}$, we observe from (3) that $\lambda=(\ln 2/22)$. Hence

$$e^{300\lambda}=e^{(300/22)\ln 2}=2^{(150/11)}.$$

The disintegration rates of polonium-210 and radium-226 were measured for the "Disciples at Emmaus" and various other alleged forgeries and are given in Table 2 below.

Table 2. Paintings of questioned authorship. All disintegration rates are per minute, per gram of white lead.

Description	^{210}Po disintegration	^{226}Ra disintegration
"Disciples at Emmaus"	8.5	0.8
"Washing of Feet"	12.6	0.26
"Woman Reading Music"	10.3	0.3
"Woman Playing Mandolin"	8.2	0.17
"Lace Maker"	1.5	1.4
"Laughing Girl"	5.2	6.0

If we now evaluate λy_0 from (6) for the white lead in the painting "Disciples at Emmaus" we obtain that

$$\lambda y_0=(8.5)2^{150/11}-0.8(2^{150/11}-1)$$
$$=98{,}050$$

which is unacceptably large. Thus, this painting must be a modern forgery. By a similar analysis, (see Exercises 2–4) the paintings "Washing of Feet", "Woman Reading Music" and "Woman Playing Mandolin" were indisputably shown to be faked Vermeers. On the other hand, the paintings "Lace Maker" and "Laughing Girl" cannot be recently forged Vermeers, as claimed by some experts, since for these two paintings, the polonium-210 is very nearly in radioactive equilibrium with the radium-226, and no such equilibrium has been observed in any samples from 19th or 20th century paintings.

References

Coremans, P., Van Meegeren's Faked Vermeers and De Hooghs, Meulenhoff, Amsterdam, 1949.

Keisch, B., Feller, R. L., Levine, A. S., Edwards, P. R., Dating and Authenticating Works of Art by Measurement of Natural Alpha Emitters, Science (155), 1238–1241, March 1967.

Keisch, B., Dating Works of Art through Their Natural Radioactivity: Improvements and Applications, Science, 160, 413–415, April 1968.

Exercises

1. In this exercise we show how to compute the concentration of uranium in an ore from the dpm/(g of Pb) of the lead-210 in the ore.
(a) The half-life of uranium-238 is 4.51×10^9 years. Since this half-life is so large, we may assume that the amount of uranium in the ore is constant over a period of two to three hundred years. Let $N(t)$ denote the number of atoms of ^{238}U per gram of ordinary lead in the ore at time t. Since the lead-210 is in radioactive equilibrium with the uranium-238 in the ore, we know that $dN/dt=-\lambda N=-100$ dpm/g of Pb at time t_0. Show that there are 3.42×10^{17} atoms of uranium-238 per gram of ordinary lead in the ore at time t_0. (Hint: 1 year = 525,600 minutes.)
(b) Using the fact that one mole of uranium-238 weighs 238 grams, and that there are 6.02×10^{23} atoms in a mole, show that the concentration of uranium in the ore is approximately 0.014 percent.

For each of the paintings 2, 3, and 4 use the data in Table 2 to compute the disintegrations per minute of the original amount of lead-210 per gram of white lead, and conclude that each of these paintings is a forged Vermeer.

2. "Washing of Feet"

3. "Woman Reading Music"

4. "Woman Playing Mandolin"

5. The following problem describes a very accurate derivation of the age of uranium.
(a) Let $N_{238}(t)$ and $N_{235}(t)$ denote the number of atoms of ^{238}U and ^{235}U at time t in a given sample of uranium, and let $t=0$ be the time this sample was created. By the radioactive decay law,

$$\frac{d}{dt}N_{238}(t)=\frac{-\ln 2}{(4.5)10^9}N_{238}(t),$$

$$\frac{d}{dt}N_{235}(t)=\frac{-\ln 2}{0.707(10)^9}N_{235}(t).$$

Solve these equations for $N_{238}(t)$ and $N_{235}(t)$ in terms of their original numbers $N_{238}(0)$ and $N_{235}(0)$.
(b) In 1946 the ratio of $^{238}U/^{235}U$ in any sample was 137.8. Assuming that equal amounts of ^{238}U and ^{235}U appeared in any sample at the time of its creation, show that the age of uranium is 5.96×10^9 years. This figure is universally accepted as the age of uranium.

6. In a samarskite sample discovered recently, there was 3 grams of Thorium (^{232}TH). Thorium decays to lead-208 (^{208}Pb) through the reaction $^{232}Th\rightarrow{}^{208}Pb+6(4\,^4He)$. It was determined that 0.0376 of a gram of lead-208 was produced by the disintegration of the original Thorium in the sample. Given that the half-life of Thorium is 13.9 billion years, derive the age of this samarskite sample. (Hint: 0.0376 grams of ^{208}Pb is the product of the decay of $(232/208)\times 0.0376$ grams of Thorium.)

One of the most accurate ways of dating archaeological finds is the method of carbon-14 (^{14}C) dating discovered by Willard Libby around 1949. The basis of this method is delightfully simple: The atmosphere of the earth is continuously bombarded by cosmic rays. These cosmic rays produce neutrons in the earth's atmosphere, and these neutrons combine with nitrogen to produce ^{14}C, which is usually called radiocarbon, since it decays radioactively. Now, this radiocarbon is incorporated in carbon dioxide and thus moves through the atmosphere to be absorbed by plants. Animals, in turn, build radiocarbon into their tissues by eating the plants. In living tissue, the rate of ingestion of ^{14}C exactly balances the rate of disintegration of ^{14}C. When an organism dies, though, it ceases to ingest carbon-14 and thus its ^{14}C concentration begins to decrease through disintegration of the ^{14}C present. Now, it is a fundamental assumption of physics that the rate of bombardment of the earth's atmosphere by cosmic rays has always been constant. This implies that the original rate of disintegration of the ^{14}C in a sample such as charcoal is the same as the rate measured today.* This assumption enables us to determine the age of a sample of charcoal. Let $N(t)$ denote the amount of carbon-14 present in a sample at time t, and N_0 the amount present at time $t=0$ when the sample was formed. If λ denotes the decay constant of ^{14}C (the half-life of carbon-14 is 5568 years) then $dN(t)/dt=-\lambda N(t)$, $N(0)=N_0$. Consequently, $N(t)=N_0e^{-\lambda t}$. Now the present rate $R(t)$ of disintegration of the ^{14}C in the sample is given by $R(t)=\lambda N(t)=\lambda N_0e^{-\lambda t}$ and the original rate of disintegration is $R(0)=\lambda N_0$. Thus, $R(t)/R(0)=e^{-\lambda t}$ so that $t=(1/\lambda)\ln[R(0)/R(t)]$. Hence if we measure $R(t)$, the present rate of disintegration of the ^{14}C in the charcoal, and observe that $R(0)$ must equal the rate of disintegration of the ^{14}C in a comparable amount of living wood, then we can compute the age t of the charcoal. The following two problems are real life illustrations of this method.

7. Charcoal from the occupation level of the famous Lascaux Cave in France gave an average count in 1950 of 0.97 disintegrations per minute per gram. Living wood gave 6.68 disintegrations. Estimate the date of occupation and hence the probable date of the remarkable paintings in the Lascaux Cave.

8. In the 1950 excavation at Nippur, a city of Babylonia, charcoal from a roof beam gave a count of 4.09 disintegrations per minute per gram. Living wood gave 6.68 disintegrations. Assuming that this charcoal was formed during the time of Hammurabi's reign, find an estimate for the likely time of Hamurabi's succession.

*Since the mid 1950's the testing of nuclear weapons has significantly increased the amount of radioactive carbon in our atmosphere. Ironically this unfortunate state of affairs provides us with yet another extremely powerful method of detecting art forgeries. To wit, many artists' materials, such as linseed oil and canvas paper, come from plants and animals, and so will contain the same concentration of carbon-14 as the atmosphere at the time the plant or animal dies. Thus linseed oil (which is derived from the flax plant) that was produced during the last few years will contain a much greater concentration of carbon-14 than linseed oil produced before 1950.

1.4 Separable equations

We solved the first-order linear homogeneous equation

$$\frac{dy}{dt}+a(t)y=0 \tag{1}$$

by dividing both sides of the equation by $y(t)$ to obtain the equivalent equation

$$\frac{1}{y(t)}\frac{dy(t)}{dt}=-a(t) \tag{2}$$

and observing that Equation (2) can be written in the form

$$\frac{d}{dt}\ln|y(t)|=-a(t). \tag{3}$$

We then found $\ln|y(t)|$, and consequently $y(t)$, by integrating both sides of (3). In an exactly analogous manner, we can solve the more general differential equation

$$\frac{dy}{dt}=\frac{g(t)}{f(y)} \tag{4}$$

where f and g are continuous functions of y and t. This equation, and any other equation which can be put into this form, is said to be separable. To solve (4), we first multiply both sides by $f(y)$ to obtain the equivalent equation

$$f(y)\frac{dy}{dt}=g(t). \tag{5}$$

Then, we observe that (5) can be written in the form

$$\frac{d}{dt}F(y(t))=g(t) \tag{6}$$

where $F(y)$ is any anti-derivative of $f(y)$; i.e., $F(y)=\int f(y)dy$. Consequently,

$$F(y(t))=\int g(t)dt+c \tag{7}$$

where c is an arbitrary constant of integration, and we solve for $y=y(t)$ from (7) to find the general solution of (4).

Example 1. Find the general solution of the equation $dy/dt=t^2/y^2$.

Solution. Multiplying both sides of this equation by y^2 gives

$$y^2\frac{dy}{dt}=t^2, \quad \text{or} \quad \frac{d}{dt}\frac{y^3(t)}{3}=t^2.$$

Hence, $y^3(t)=t^3+c$ where c is an arbitrary constant, and $y(t)=(t^3+c)^{1/3}$.

Example 2. Find the general solution of the equation

$$e^{y}\frac{dy}{dt}-t-t^{3}=0.$$

Solution. This equation can be written in the form

$$\frac{d}{dt}e^{y(t)}=t+t^{3}$$

and thus $e^{y(t)}=t^2/2+t^4/4+c$. Taking logarithms of both sides of this equation gives $y(t)=\ln(t^2/2+t^4/4+c)$.

In addition to the differential equation (4), we will often impose an initial condition on $y(t)$ of the form $y(t_0)=y_0$. The differential equation (4) together with the initial condition $y(t_0)=y_0$ is called an initial-value problem. We can solve an initial-value problem two different ways. Either we use the initial condition $y(t_0)=y_0$ to solve for the constant c in (7), or else we integrate both sides of (6) between t_0 and t to obtain that

$$F(y(t))-F(y_0)=\int_{t_0}^{t}g(s)\,ds. \tag{8}$$

If we now observe that

$$F(y)-F(y_0)=\int_{y_0}^{y}f(r)\,dr, \tag{9}$$

then we can rewrite (8) in the simpler form

$$\int_{y_0}^{y}f(r)\,dr=\int_{t_0}^{t}g(s)\,ds. \tag{10}$$

Example 3. Find the solution $y(t)$ of the initial-value problem

$$e^{y}\frac{dy}{dt}-(t+t^{3})=0, \qquad y(1)=1.$$

Solution. Method (i). From Example 2, we know that the general solution of this equation is $y=\ln(t^2/2+t^4/4+c)$. Setting $t=1$ and $y=1$ gives $1=\ln(3/4+c)$, or $c=e-3/4$. Hence, $y(t)=\ln(e-3/4+t^2/2+t^4/4)$.
Method (ii). From (10),

$$\int_{1}^{y}e^{r}\,dr=\int_{1}^{t}(s+s^{3})\,ds.$$

Consequently,

$$e^{y}-e=\frac{t^{2}}{2}+\frac{t^{4}}{4}-\frac{1}{2}-\frac{1}{4}, \quad \text{and} \quad y(t)=\ln(e-3/4+t^{2}/2+t^{4}/4).$$

Example 4. Solve the initial-value problem $dy/dt=1+y^2$, $y(0)=0$.
Solution. Divide both sides of the differential equation by $1+y^2$ to obtain

the equivalent equation $1/(1+y^2)dy/dt=1$. Then, from (10)

$$\int_0^y \frac{dr}{1+r^2}=\int_0^t ds.$$

Consequently, $\arctan y=t$, and $y=\tan t$.

The solution $y=\tan t$ of the above problem has the disturbing property that it goes to $\pm\infty$ at $t=\pm\pi/2$. And what's even more disturbing is the fact that there is nothing at all in this initial-value problem which even hints to us that there is any trouble at $t=\pm\pi/2$. The sad fact of life is that solutions of perfectly nice differential equations can go to infinity in finite time. Thus, solutions will usually exist only on a finite open interval $a<t<b$, rather than for all time. Moreover, as the following example shows, different solutions of the same differential equation usually go to infinity at different times.

Example 5. Solve the initial-value problem $dy/dt=1+y^2$, $y(0)=1$.
Solution. From (10)

$$\int_1^y \frac{dr}{1+r^2}=\int_0^t ds.$$

Consequently, $\arctan y-\arctan 1=t$, and $y=\tan(t+\pi/4)$. This solution exists on the open interval $-3\pi/4<t<\pi/4$.

Example 6. Find the solution $y(t)$ of the initial-value problem

$$y\frac{dy}{dt}+(1+y^2)\sin t=0, \qquad y(0)=1.$$

Solution. Dividing both sides of the differential equation by $1+y^2$ gives

$$\frac{y}{1+y^2}\frac{dy}{dt}=-\sin t.$$

Consequently,

$$\int_1^y \frac{r\,dr}{1+r^2}=\int_0^t -\sin s\,ds,$$

so that

$$\tfrac{1}{2}\ln(1+y^2)-\tfrac{1}{2}\ln 2=\cos t-1.$$

Solving this equation for $y(t)$ gives

$$y(t)=\pm\left(2e^{-4\sin^2 t/2}-1\right)^{1/2}.$$

To determine whether we take the plus or minus branch of the square root, we note that $y(0)$ is positive. Hence,

$$y(t)=\left(2e^{-4\sin^2 t/2}-1\right)^{1/2}$$

This solution is only defined when

$$2e^{-4\sin^2 t/2} \geqslant 1$$

or

$$e^{4\sin^2 t/2} \leqslant 2. \tag{11}$$

Since the logarithm function is monotonic increasing, we may take logarithms of both sides of (11) and still preserve the inequality. Thus, $4\sin^2 t/2 \leqslant \ln 2$, which implies that

$$\left|\frac{t}{2}\right| \leqslant \arcsin\frac{\sqrt{\ln 2}}{2}$$

Therefore, $y(t)$ only exists on the open interval $(-a,a)$ where

$$a = 2\arcsin\left[\sqrt{\ln 2}\,/2\right].$$

Now, this appears to be a new difficulty associated with nonlinear equations, since $y(t)$ just "disappears" at $t = \pm a$, without going to infinity. However, this apparent difficulty can be explained quite easily, and moreover, can even be anticipated, if we rewrite the differential equation above in the standard form

$$\frac{dy}{dt} = -\frac{(1+y^2)\sin t}{y}.$$

Notice that this differential equation is not defined when $y=0$. Therefore, if a solution $y(t)$ achieves the value zero at some time $t = t^*$, then we cannot expect it to be defined for $t > t^*$. This is exactly what happens here, since $y(\pm a) = 0$.

Example 7. Solve the initial-value problem $dy/dt = (1+y)t$, $y(0) = -1$.
Solution. In this case, we cannot divide both sides of the differential equation by $1+y$, since $y(0) = -1$. However, it is easily seen that $y(t) = -1$ is one solution of this initial-value problem, and in Section 1.10 we show that it is the only solution. More generally, consider the initial-value problem $dy/dt = f(y)g(t)$, $y(t_0) = y_0$, where $f(y_0) = 0$. Certainly, $y(t) = y_0$ is one solution of this initial-value problem, and in Section 1.10 we show that it is the only solution if $\partial f/\partial y$ exists and is continuous.

Example 8. Solve the initial-value problem

$$(1+e^y)\,dy/dt = \cos t, \qquad y(\pi/2) = 3.$$

Solution. From (10),

$$\int_3^y (1+e^r)\,dr = \int_{\pi/2}^t \cos s\,ds$$

so that $y + e^y = 2 + e^3 + \sin t$. This equation cannot be solved explicitly for y

as a function of t. Indeed, most separable equations cannot be solved explicitly for y as a function of t. Thus, when we say that

$$y+e^{y}=2+e^{3}+\sin t$$

is the solution of this initial-value problem, we really mean that it is an implicit, rather than an explicit solution. This does not present us with any difficulties in applications, since we can always find $y(t)$ numerically with the aid of a digital computer (see Section 1.11).

Example 9. Find all solutions of the differential equation $dy/dt=-t/y$.
Solution. Multiplying both sides of the differential equation by y gives $y\,dy/dt=-t$. Hence

$$y^{2}+t^{2}=c^{2}. \tag{12}$$

Now, the curves (12) are *closed*, and we cannot solve for y as a *single-valued* function of t. The reason for this difficulty, of course, is that the differential equation is not defined when $y=0$. Nevertheless, the circles $t^2+y^2=c^2$ are perfectly well defined, even when $y=0$. Thus, we will call the circles $t^2+y^2=c^2$ *solution curves* of the differential equation

$$dy/dt=-t/y.$$

More generally, we will say that any curve defined by (7) is a solution curve of (4).

Exercises

In each of Problems 1–5, find the general solution of the given differential equation.

1. $(1+t^2)\dfrac{dy}{dt}=1+y^2$. *Hint*: $\tan(x+y)=\dfrac{\tan x+\tan y}{1-\tan x\tan y}$.

2. $\dfrac{dy}{dt}=(1+t)(1+y)$

3. $\dfrac{dy}{dt}=1-t+y^2-ty^2$

4. $\dfrac{dy}{dt}=e^{t+y+3}$

5. $\cos y\sin t\dfrac{dy}{dt}=\sin y\cos t$

In each of Problems 6–12, solve the given initial-value problem, and determine the interval of existence of each solution.

6. $t^2(1+y^2)+2y\dfrac{dy}{dt}=0,\quad y(0)=1$

7. $\dfrac{dy}{dt}=\dfrac{2t}{y+yt^2},\quad y(2)=3$

8. $(1+t^2)^{1/2}\dfrac{dy}{dt}=ty^3(1+t^2)^{-1/2},\quad y(0)=1$

9. $\dfrac{dy}{dt}=\dfrac{3t^2+4t+2}{2(y-1)},\quad y(0)=-1$

10. $\cos y\dfrac{dy}{dt}=\dfrac{-t\sin y}{1+t^2},\quad y(1)=\pi/2$

11. $\dfrac{dy}{dt}=k(a-y)(b-y),\quad y(0)=0,\ a,b>0$

12. $3t\dfrac{dy}{dt}=y\cos t,\quad y(1)=0$

13. Any differential equation of the form $dy/dt=f(y)$ is separable. Thus, we can solve all those first-order differential equations in which time does not appear explicitly. Now, suppose we have a differential equation of the form $dy/dt=f(y/t)$, such as, for example, the equation $dy/dt=\sin(y/t)$. Differential equations of this form are called homogeneous equations. Since the right-hand side only depends on the single variable y/t, it suggests itself to make the substitution $y/t=v$ or $y=tv$.
(a) Show that this substitution replaces the equation $dy/dt=f(y/t)$ by the equivalent equation $t\,dv/dt+v=f(v)$, which is separable.
(b) Find the general solution of the equation $dy/dt=2(y/t)+(y/t)^2$.

14. Determine whether each of the following functions of t and y can be expressed as a function of the single variable y/t.

(a) $\dfrac{y^2+2ty}{y^2}$ (b) $\dfrac{y^3+t^3}{yt^2+y^3}$

(c) $\dfrac{y^3+t^3}{t^2+y^3}$ (d) $\ln y-\ln t+\dfrac{t+y}{t-y}$

(e) $\dfrac{e^{t+y}}{e^{t-y}}$ (f) $\ln\sqrt{t+y}-\ln\sqrt{t-y}$

(g) $\sin\dfrac{t+y}{t-y}$ (h) $\dfrac{(t^2+7ty+9y^2)^{1/2}}{3t+5y}$.

15. Solve the initial-value problem $t(dy/dt)=y+\sqrt{t^2+y^2}\,,\ y(1)=0$.

Find the general solution of the following differential equations.

16. $2ty\dfrac{dy}{dt}=3y^2-t^2$

17. $(t-\sqrt{ty}\,)\dfrac{dy}{dt}=y$

18. $\dfrac{dy}{dt}=\dfrac{t+y}{t-y}$.

19. $e^{t/y}(y-t)\dfrac{dy}{dt}+y(1+e^{t/y})=0$

$$\left[\textit{Hint}: \int\frac{v-1}{ve^{-1/v}+v^2}\,dv=\ln(1+ve^{1/v})\right]$$

20. Consider the differential equation

$$\frac{dy}{dt}=\frac{t+y+1}{t-y+3}. \qquad (*)$$

We could solve this equation if the constants 1 and 3 were not present. To eliminate these constants, we make the substitution $t=T+h$, $y=Y+k$.

(a) Determine h and k so that (*) can be written in the form $dY/dT=(T+Y)/(T-Y)$.

(b) Find the general solution of (*). (See Exercise 18).

21. (a) Prove that the differential equation

$$\frac{dy}{dt}=\frac{at+by+m}{ct+dy+n}$$

where a, b, c, d, m, and n are constants, can always be reduced to $dy/dt=(at+by)/(ct+dy)$ if $ad-bc\neq 0$.

(b) Solve the above equation in the special case that $ad=bc$.

Find the general solution of the following equations.

22. $(1+t-2y)+(4t-3y-6)dy/dt=0$

23. $(t+2y+3)+(2t+4y-1)dy/dt=0$

1.5 Population models

In this section we will study first-order differential equations which govern the growth of various species. At first glance it would seem impossible to model the growth of a species by a differential equation since the population of any species always changes by integer amounts. Hence the population of any species can never be a differentiable function of time. However, if a given population is very large and it is suddenly increased by one, then the change is very small compared to the given population. Thus, we make the approximation that large populations change continuously and even differentiably with time.

Let $p(t)$ denote the population of a given species at time t and let $r(t,p)$ denote the difference between its birth rate and its death rate. If this population is isolated, that is, there is no net immigration or emigration, then dp/dt, the rate of change of the population, equals $rp(t)$. In the most simplistic model we assume that r is constant, that is, it does not change with either time or population. Then, we can write down the following differential equation governing population growth:

$$\frac{dp(t)}{dt}=ap(t), \qquad a=\text{constant}.$$

This is a linear equation and is known as the Malthusian law of population growth. If the population of the given species is p_0 at time t_0, then $p(t)$ satisfies the initial-value problem $dp(t)/dt = ap(t)$, $p(t_0)=p_0$. The solution of this initial-value problem is $p(t)=p_0e^{a(t-t_0)}$. Hence any species satisfying the Malthusian law of population growth grows exponentially with time.

Now, we have just formulated a very simple model for population growth; so simple, in fact, that we have been able to solve it completely in a few lines. It is important, therefore, to see if this model, with its simplicity, has any relationship at all with reality. Let $p(t)$ denote the human population of the earth at time t. It was estimated that the earth's human population was increasing at an average rate of 2% per year during the period 1960–1970. Let us start in the middle of this decade on January 1, 1965, at which time the U.S. Department of Commerce estimated the earth's population to be 3.34 billion people. Then, $t_0 = 1965$, $p_0 = 3.34 \times 10^9$ and $a = .02$, so that

$$p(t) = (3.34)10^9 e^{.02(t-1965)}.$$

One way of checking the accuracy of this formula is to compute the time required for the population of the earth to double, and then compare it to the observed value of 35 years. Our formula predicts that the population of the earth will double every T years, where

$$e^{.02T} = 2.$$

Taking logarithms of both sides of this equation gives $.02T = \ln 2$, so that

$$T = 50 \ln 2 \simeq 34.6 \text{ years}.$$

This is in excellent agreement with the observed value. On the other hand, though, let us look into the distant future. Our equation predicts that the earth's population will be 200,000 billion in the year 2515, 1,800,000 billion in the year 2625, and 3,600,000 billion in the year 2660. These are astronomical numbers whose significance is difficult to gauge. The total surface of this planet is approximately 1,860,000 billion square feet. Eighty percent of this surface is covered by water. Assuming that we are willing to live on boats as well as land, it is easy to see that by the year 2515 there will be only 9.3 square feet per person; by 2625 each person will have only one square foot on which to stand; and by 2660 we will be standing two deep on each other's shoulders.

It would seem therefore, that this model is unreasonable and should be thrown out. However, we cannot ignore the fact that it offers exceptional agreement in the past. Moreover, we have additional evidence that populations do grow exponentially. Consider the Microtus Arvallis Pall, a small rodent which reproduces very rapidly. We take the unit of time to be a month, and assume that the population is increasing at the rate of 40% per month. If there are two rodents present initially at time $t=0$, then $p(t)$, the number of rodents at time t, satisfies the initial-value problem

$$dp(t)/dt = 0.4p, \qquad p(0) = 2.$$

Consequently,

$$p(t) = 2e^{0.4t}. \tag{1}$$

Table 1 compares the observed population with the population calculated from Equation (1).

Table 1. The growth of Microtus Arvallis Pall.

Months	0	2	6	10
p Observed	2	5	20	109
p Calculated	2	4.5	22	109.1

As one can see, there is excellent agreement.

Remark. In the case of the Microtus Arvallis Pall, p observed is very accurate since the pregnancy period is three weeks and the time required for the census taking is considerably less. If the pregnancy period were very short then p observed could not be accurate since many of the pregnant rodents would have given birth before the census was completed.

The way out of our dilemma is to observe that linear models for population growth are satisfactory *as long as* the population is not too large. When the population gets extremely large though, these models cannot be very accurate, since they do not reflect the fact that individual members are now competing with each other for the limited living space, natural resources and food available. Thus, we must add a competition term to our linear differential equation. A suitable choice of a competition term is $-bp^2$, where b is a constant, since the statistical average of the number of encounters of two members per unit time is proportional to p^2. We consider, therefore, the modified equation

$$\frac{dp}{dt} = ap - bp^2.$$

This equation is known as the logistic law of population growth and the numbers a,b are called the vital coefficients of the population. It was first introduced in 1837 by the Dutch mathematical-biologist Verhulst. Now, the constant b, in general, will be very small compared to a, so that if p is not too large then the term $-bp^2$ will be negligible compared to ap and the population will grow exponentially. However, when p is very large, the term $-bp^2$ is no longer negligible, and thus serves to slow down the rapid rate of increase of the population. Needless to say, the more industrialized a nation is, the more living space it has, and the more food it has, the smaller the coefficient b is.

Let us now use the logistic equation to predict the future growth of an isolated population. If p_0 is the population at time t_0, then $p(t)$, the population at time t, satisfies the initial-value problem

$$\frac{dp}{dt} = ap - bp^2, \qquad p(t_0) = p_0.$$

This is a separable differential equation, and from Equation (10), Section 1.4,

$$\int_{p_0}^{p} \frac{dr}{ar - br^2} = \int_{t_0}^{t} ds = t - t_0.$$

To integrate the function $1/(ar - br^2)$ we resort to partial fractions. Let

$$\frac{1}{ar - br^2} \equiv \frac{1}{r(a - br)} = \frac{A}{r} + \frac{B}{a - br}.$$

To find A and B, observe that

$$\frac{A}{r} + \frac{B}{a - br} = \frac{A(a - br) + Br}{r(a - br)} = \frac{Aa + (B - bA)r}{r(a - br)}.$$

Therefore, $Aa + (B - bA)r = 1$. Since this equation is true for all values of r, we see that $Aa = 1$ and $B - bA = 0$. Consequently, $A = 1/a$, $B = b/a$, and

$$\int_{p_0}^{p} \frac{dr}{r(a - br)} = \frac{1}{a} \int_{p_0}^{p} \left(\frac{1}{r} + \frac{b}{a - br} \right) dr$$

$$= \frac{1}{a} \left[\ln \frac{p}{p_0} + \ln \left| \frac{a - bp_0}{a - bp} \right| \right] = \frac{1}{a} \ln \frac{p}{p_0} \left| \frac{a - bp_0}{a - bp} \right|.$$

Thus,

$$a(t - t_0) = \ln \frac{p}{p_0} \left| \frac{a - bp_0}{a - bp} \right|. \tag{2}$$

Now, it is a simple matter to show (see Exercise 1) that

$$\frac{a - bp_0}{a - bp(t)}$$

is always positive. Hence,

$$a(t - t_0) = \ln \frac{p}{p_0} \frac{a - bp_0}{a - bp}.$$

Taking exponentials of both sides of this equation gives

$$e^{a(t-t_0)} = \frac{p}{p_0}\frac{a-bp_0}{a-bp},$$

or

$$p_0(a-bp)e^{a(t-t_0)} = (a-bp_0)p.$$

Bringing all terms involving p to the left-hand side of this equation, we see that

$$\left[a-bp_0+bp_0e^{a(t-t_0)}\right]p(t) = ap_0e^{a(t-t_0)}.$$

Consequently,

$$p(t) = \frac{ap_0e^{a(t-t_0)}}{a-bp_0+bp_0e^{a(t-t_0)}} = \frac{ap_0}{bp_0+(a-bp_0)e^{-a(t-t_0)}}. \tag{3}$$

Let us now examine Equation (3) to see what kind of population it predicts. Observe that as $t\to\infty$,

$$p(t)\to\frac{ap_0}{bp_0} = \frac{a}{b}.$$

Thus, *regardless of its initial value, the population always approaches the limiting value* a/b. Next, observe that $p(t)$ is a monotonically increasing function of time if $0<p_0<a/b$. Moreover, since

$$\frac{d^2p}{dt^2} = a\frac{dp}{dt} - 2bp\frac{dp}{dt} = (a-2bp)p(a-bp),$$

we see that dp/dt is increasing if $p(t)<a/2b$, and that dp/dt is decreasing if $p(t)>a/2b$. Hence, if $p_0<a/2b$, the graph of $p(t)$ must have the form given in Figure 1. Such a curve is called a logistic, or S-shaped curve. From its shape we conclude that the time period before the population reaches half its limiting value is a period of accelerated growth. After this point, the rate of growth decreases and in time reaches zero. This is a period of diminishing growth.

These predictions are borne out by an experiment on the protozoa Paramecium caudatum performed by the mathematical biologist G. F. Gause. Five individuals of Paramecium were placed in a small test tube containing 0.5 cm^3 of a nutritive medium, and for six days the number of individuals in every tube was counted daily. The Paramecium were found to increase at a rate of 230.9% per day when their numbers were low. The number of individuals increased rapidly at first, and then more slowly, until towards the fourth day it attained a maximum level of 375, saturating the test tube. From this data we conclude that if the Paramecium caudatum grow according to the logistic law $dp/dt = ap - bp^2$, then $a = 2.309$ and

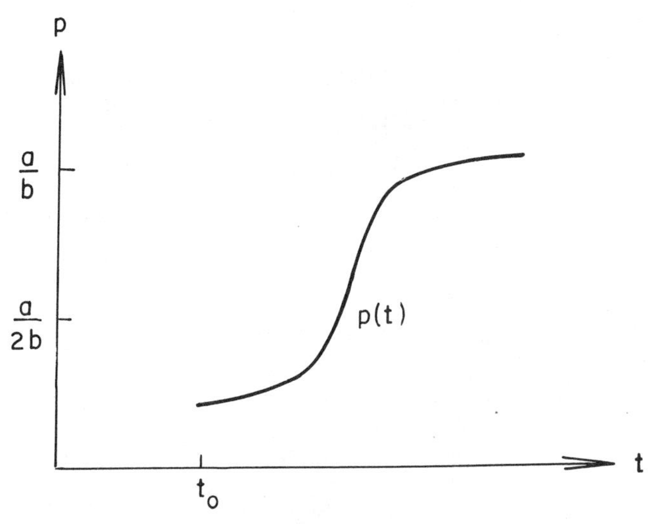

Figure 1. Graph of $p(t)$

$b=2.309/375$. Consequently, the logistic law predicts that

$$p(t)=\frac{(2.309)5}{\frac{(2.309)5}{375}+\left(2.309-\frac{(2.309)5}{375}\right)e^{-2.309t}}$$

$$=\frac{375}{1+74e^{-2.309t}}. \tag{4}$$

(We have taken the initial time t_0 to be 0.) Figure 2 compares the graph of $p(t)$ predicted by Equation (4) with the actual measurements, which are denoted by o. As can be seen, the agreement is remarkably good.

In order to apply our results to predict the future human population of the earth, we must estimate the vital coefficients a and b in the logistic equation governing its growth. Some ecologists have estimated that the natural value of a is 0.029. We also know that the human population was increasing at the rate of 2% per year when the population was $(3.34)10^9$. Since $(1/p)(dp/dt)=a-bp$, we see that

$$0.02=a-b\,(3.34)10^9.$$

Consequently, $b=2.695\times10^{-12}$. Thus, according to the logistic law of population growth, the human population of the earth will tend to the limiting value of

$$\frac{a}{b}=\frac{0.029}{2.695\times10^{-12}}=10.76\text{ billion people}$$

Note that according to this prediction, we were still on the accelerated

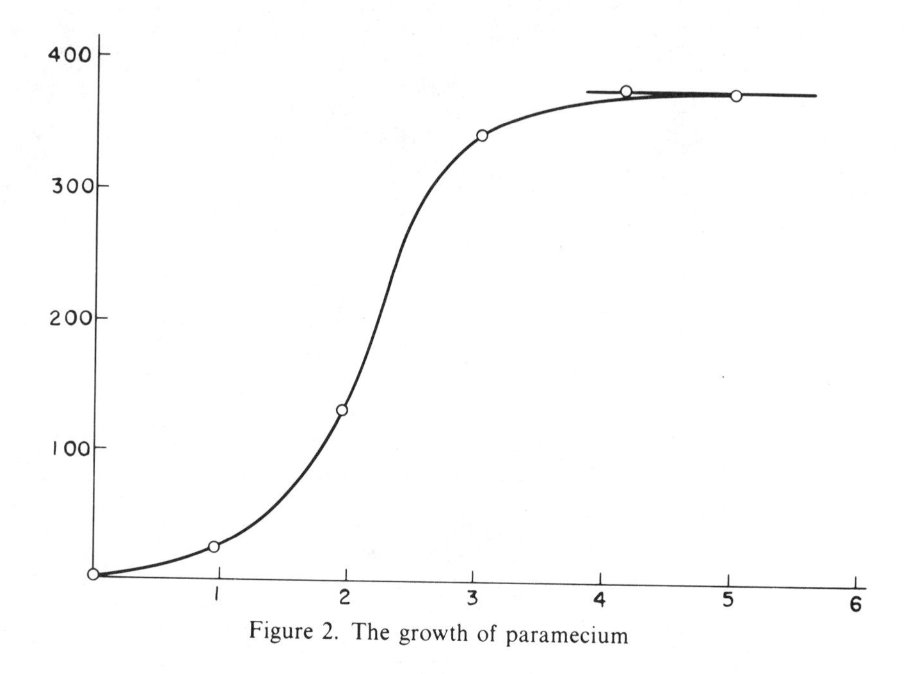

Figure 2. The growth of paramecium

growth portion of the logistic curve in 1965, since we had not yet attained half the limiting population predicted for us.

Remark. A student of mine once suggested that we use Equation (3) to find the time t at which $p(t)=2$, and then we can deduce how long ago mankind appeared on earth. On the surface this seems like a fantastic idea. However, we cannot travel that far backwards into the past, since our model is no longer accurate when the population is small.

As another verification of the validity of the logistic law of population growth, we consider the equation

$$p(t)=\frac{197{,}273{,}000}{1+e^{-0.03134(t-1913.25)}} \tag{5}$$

which was introduced by Pearl and Reed as a model of the population growth of the United States. First, using the census figures for the years 1790, 1850, and 1910, Pearl and Reed found from (3) (see Exercise 2a) that $a=0.03134$ and $b=(1.5887)10^{-10}$. Then (see Exercise 2b), Pearl and Reed calculated that the population of the United States reached half its limiting population of $a/b=197{,}273{,}000$ in April 1913. Consequently (see Exercise 2c), we can rewrite (3) in the simpler form (5).

Table 2 below compares Pearl and Reed's predictions with the observed values of the population of the United States. These results are remarkable,

Table 2. Population of the U.S. from 1790–1950. (The last 4 entries were added by the Dartmouth College Writing Group.)

	Actual	Predicted	Error	%
1790	3,929,000	3,929,000	0	0.0
1800	5,308,000	5,336,000	28,000	0.5
1810	7,240,000	7,228,000	−12,000	−0.2
1820	9,638,000	9,757,000	119,000	1.2
1830	12,866,000	13,109,000	243,000	1.9
1840	17,069,000	17,506,000	437,000	2.6
1850	23,192,000	23,192,000	0	0.0
1860	31,443,000	30,412,000	−1,031,000	−3.3
1870	38,558,000	39,372,000	814,000	2.1
1880	50,156,000	50,177,000	21,000	0.0
1890	62,948,000	62,769,000	−179,000	−0.3
1900	75,995,000	76,870,000	875,000	1.2
1910	91,972,000	91,972,000	0	0.0
1920	105,711,000	107,559,000	1,848,000	1.7
1930	122,775,000	123,124,000	349,000	0.3
1940	131,669,000	136,653,000	4,984,000	3.8
1950	150,697,000	149,053,000	−1,644,000	−1.1

especially since we have not taken into account the large waves of immigration into the United States, and the fact that the United States was involved in five wars during this period.

In 1845 Verhulst prophesied a maximum population for Belgium of 6,600,000, and a maximum population for France of 40,000,000. Now, the population of Belgium in 1930 was already 8,092,000. This large discrepancy would seem to indicate that the logistic law of population growth is very inaccurate, at least as far as the population of Belgium is concerned. However, this discrepancy can be explained by the astonishing rise of industry in Belgium, and by the acquisition of the Congo which secured for the country sufficient additional wealth to support the extra population. Thus, after the acquisition of the Congo, and the astonishing rise of industry in Belgium, Verhulst should have lowered the vital coefficient b.

On the other hand, the population of France in 1930 was in remarkable agreement with Verhulst's forecast. Indeed, we can now answer the following tantalizing paradox: Why was the population of France increasing extremely slowly in 1930 while the French population of Canada was increasing very rapidly? After all, they are the same people! The answer to this paradox, of course, is that the population of France in 1930 was very near its limiting value and thus was far into the period of diminishing growth, while the population of Canada in 1930 was still in the period of accelerated growth.

Remark 1. It is clear that technological developments, pollution considerations and sociological trends have significant influence on the vital coefficients a and b. Therefore, they must be re-evaluated every few years.

Remark 2. To derive more accurate models of population growth, we should not consider the population as made up of one homogeneous group of individuals. Rather, we should subdivide it into different age groups. We should also subdivide the population into males and females, since the reproduction rate in a population usually depends more on the number of females than on the number of males.

Remark 3. Perhaps the severest criticism leveled at the logistic law of population growth is that some populations have been observed to fluctuate periodically between two values, and any type of fluctuation is ruled out in a logistic curve. However, some of these fluctuations can be explained by the fact that when certain populations reach a sufficiently high density, they become susceptible to epidemics. The epidemic brings the population down to a lower value where it again begins to increase, until when it is large enough, the epidemic strikes again. In Exercise 10 we derive a model to describe this phenomenon, and we apply this model in Exercise 11 to explain the sudden appearance and disappearance of hordes of small rodents.

Epilog. The following article appeared in the New York Times on March 26, 1978, and was authored by Nick Eberstadt.

The gist of the following article is that it is very difficult, using statistical methods alone, to make accurate population projections even 30 years into the future. In 1970, demographers at the United Nations projected the population of the earth to be 6.5 billion people by the year 2000. Only six years later, this forecast was revised downward to 5.9 billion people.

Let us now use Equation (3) to predict the population of the earth in the year 2000. Setting $a = .029$, $b = 2.695\times 10^{-12}$, $p_0 = 3.34\times 10^9$, $t_0 = 1965$, and $t = 2{,}000$ gives

$$\begin{aligned} p(2000) &= \frac{(.029)(3.34)10^9}{.009+(.02)e^{-(.029)35}} \\ &= \frac{29(3.34)}{9+20e^{-1.015}}10^9 \\ &= 5.96 \text{ billion people!} \end{aligned}$$

This is another spectacular application of the logistic equation.

World Population Figures Are Misleading

The rate of world population growth has risen fairly steadily for most of man's history, but within the last decade it has peaked, and now appears to be declining. How has this happened, and why?

The "how" is fairly easy. It is not because famines and ecological catastrophes have elevated the death rates. Rather, a large and generally unexpected decrease in fertility in the less developed countries has taken place. From 1970 to 1977 birth rates in the less developed world (excluding China) fell from about 42 to nearly 36 per thousand. This is still higher than the 17 per thousand in developed countries, but the rate of fertility decline appears to be accelerating: the six-point drop of the past seven years compares with a two-point decline for the previous twenty.

This fertility decline has been a very uneven process. The average birth-rate drop of about 13 percent since 1970 in poor world birth rates reflects a very rapid decline in certain countries, while a great many others have remained almost totally unaffected.

Why fertility has dropped so rapidly in the past decade—and why it has dropped so dramatically in some places, but not in others—is far more difficult to explain. Demographers and sociologists offer explanations having to do with social change in the poor world. Unfortunately, these partial explanations are more often theory than tested fact, and there seems to be an exception for almost every rule.

The debate over family planning is characteristic. Surveys show that in some nations as many as a fifth of the children were "mistakes" who presumably would not have been born if parents had had better contraceptives. Family planning experts such as Parker Mauldin of the Population Council have pointed out that no poor nation without an active family planning program has significantly lowered its fertility. On the other hand, such sociologists as William Petersen of Ohio State University attribute the population decline in these nations to social and economic development rather than increased contraceptive use, arguing that international "population control" programs have usually been clumsy and insensitive (or worse), and that in any event even a well-received change in contraceptive "technology" does not necessarily influence parents to want fewer children.

The effects of income distribution are less vociferously debated, but are almost as mysterious. James Kocher and Robert Repetto, both of Harvard, have argued that more equitable income distribution in less developed countries contributes to fertility decline. They have pointed out that such countries as Sri Lanka, South Korea, Cuba and China have seen their fertility rates fall as their income distribution became more nearly equal. Improving a nation's income distribution, however, appears to be neither a necessary nor a sufficient condition for inducing a fertility drop. Income distribution in Burma, for example, has presumably equalized somewhat under 30 years of homemade socialism, but birth rates have hardly fallen at all, while Mexico and Colombia, with highly unequal income distributions, have found their birth rates plummeting in recent years.

One key to changes in fertility levels may be the economic costs and benefits from children. In peasant societies, where children are afforded few amenities and start work young, they may become economic assets early: A recent study in Bangladesh by Mead Cain of the Population Council put the age for boys at 12. Furthermore, children (or more precisely, sons) may also serve as social security and unemployment insurance when parents become too old and weak to work, or when work is unavailable. Where birth rates in poor countries are dropping, social and economic development may be making children less necessary as sources of income and security, but so little work has been done in this area that this is still just a reasonable speculation.

Some of the many other factors whose effects on fertility have been studied are urbanization, education, occupational structure, public health and the status of women. One area which population experts seem to have shied away from, however, is the non-quantifiable realm of attitudes, beliefs and values which may have had much to do with the recent changes in the decisions of hundreds of millions of couples. Cultural differences, ethnic conflicts, psychological, ideological and even political changes could clearly have effects on fertility. As Maris Vonovskis of the House Select Committee on Population has said, just because you can't measure something doesn't mean it isn't important.

What does the decline in fertility mean about future levels of population? Obviously, if the drop continues, population growth will be slower than previously anticipated, and world population will eventually stabilize at a lower level. Only five years ago the United Nations "medium variant" projection for world population in the year 2000 was 6.5 billion; last year this was dropped more than 200 million, and recent work by Gary Littman and Nathan Keyfitz at the Harvard Center for Population Studies shows that in the light of recent changes, one might easily drop it 400 million more.

Population projections, however, are a very tricky business. To begin with, the figures for today's population, upon which tomorrow's projections must be based, contain large margins of error. For example, estimates for China's population run from 750 million to over 950 million. By the account of John Durand of the University of Pennsylvania, the margins of error for world population add up to over 200 million; historian Fernand Braudel puts the margin of error at 10 percent, which, given the world's approximate present population, means about 400 million people.

Population projections inspire even less confidence than population estimates, for they hinge on predicting birth and death rates for the future. These can change rapidly and unexpectedly: two extreme examples are Sri Lanka's 34 percent drop in the death rate in just two years, and Japan's 50 percent drop in the birth rate in 10. "Medium variant" U.N. projections computed just 17 years before 1975 overestimate Russia's population by 10 to 20 million, and underestimate India's by 50 million. Even projections for the United States done in 1966 overestimate its population only nine years later by over 10 million. Enormous as that gap may sound, it seems quite small next to those of the 1930's estimates which extrapolated low Depression era birth rates into an American population peaking at 170 million in the late 1970's (the population now is over 220 million), and then declining!

Could birth rates in the less developed world, which now appear to be declining at an accelerating pace, suddenly stabilize, or even rise again? This could theoretically happen. Here are four of the many reasons: 1) The many countries where fertility has as yet been unaffected by the decline might simply continue to be unaffected far into the future. 2) Since sterility and infertility are widespread in many of the poorest and most disease-ridden areas of the world, improvements in health and nutrition there could raise birth rates. 3) The Gandhi regime's cold-hearted and arbitrary mass sterilization regimen may have hardened that sixth of the world against future family limitation messages. 4) If John Aird of the Department of Commerce and others are correct that China's techniques of political mobilization and social persuasion have induced many parents to have fewer children than they actually want, a relaxation of these rules for whatever reasons might make the birth rate of China's enormous population rise. One of the only long-term rules about population projections which has held up is that within their limits of accuracy (about five years in the future) they can tell nothing interesting, and when they start giving interesting results, they are no longer accurate.

Nick Eberstadt is an affiliate of the Harvard Center for Population Studies.

References

1. Gause, G. F., *The Struggle for Existence*, Dover Publications, New York, 1964.
2. Pearl and Reed, *Proceedings of the National Academy of Sciences*, 1920, p. 275.

Exercises

1. Prove that $(a-bp_0)/(a-bp(t))$ is positive for $t_0<t<\infty$. *Hint*: Use Equation (2) to show that $p(t)$ can never equal a/b if $p_0 \neq a/b$.

2. (a) Choose 3 times t_0, t_1, and t_2, with $t_1-t_0=t_2-t_1$. Show that (3) determines a and b uniquely in terms of t_0, $p(t_0)$, t_1, $p(t_1)$, t_2, and $p(t_2)$.
(b) Show that the period of accelerated growth for the United States ended in April, 1913.
(c) Let a population $p(t)$ grow according to the logistic law (3), and let $\bar{t}$ be the time at which half the limiting population is achieved. Show that

$$p(t)=\frac{a/b}{1+e^{-a(t-\bar{t})}}.$$

3. In 1879 and 1881 a number of yearling bass were seined in New Jersey, taken across the continent in tanks by train, and planted in San Francisco Bay. A total of only 435 Striped Bass survived the rigors of these two trips. Yet, in 1899, the commercial net catch alone was 1,234,000 pounds. Since the growth of this population was so fast, it is reasonable to assume that it obeyed the Malthusian law $dp/dt=ap$. Assuming that the average weight of a bass fish is three pounds, and that in 1899 every tenth bass fish was caught, find a lower bound for a.

4. Suppose that a population doubles its original size in 100 years, and triples it in 200 years. Show that this population cannot satisfy the Malthusian law of population growth.

5. Assume that $p(t)$ satisfies the Malthusian law of population growth. Show that the increases in p in successive time intervals of equal duration form the terms of a geometric progression. This is the source of Thomas Malthus' famous dictum "Population when unchecked increases in a geometrical ratio. Subsistence increases only in an arithmetic ratio. A slight acquaintance with numbers will show the immensity of the first power in comparison of the second."

6. A population grows according to the logistic law, with a limiting population of 5×10^8 individuals. When the population is low it doubles every 40 minutes. What will the population be after two hours if initially it is (a) 10^8, (b) 10^9?

7. A family of salmon fish living off the Alaskan Coast obeys the Malthusian law of population growth $dp(t)/dt=0.003p(t)$, where t is measured in minutes. At time $t=0$ a group of sharks establishes residence in these waters and begins attacking the salmon. The rate at which salmon are killed by the sharks is $0.001p^2(t)$, where $p(t)$ is the population of salmon at time t. Moreover, since an undesirable element has moved into their neighborhood, 0.002 salmon per minute leave the Alaskan waters.
(a) Modify the Malthusian law of population growth to take these two factors into account.

(b) Assume that at time $t=0$ there are one million salmon. Find the population $p(t)$. What happens as $t \to \infty$?

(c) Show that the above model is really absurd. Hint: Show, according to this model, that the salmon population decreases from one million to about one thousand in one minute.

8. The population of New York City would satisfy the logistic law

$$\frac{dp}{dt} = \frac{1}{25}p - \frac{1}{(25)10^6}p^2,$$

where t is measured in years, if we neglected the high emigration and homicide rates.

(a) Modify this equation to take into account the fact that 9,000 people per year move from the city, and 1,000 people per year are murdered.

(b) Assume that the population of New York City was 8,000,000 in 1970. Find the population for all future time. What happens as $t \to \infty$?

9. An initial population of 50,000 inhabits a microcosm with a carrying capacity of 100,000. After five years, the population has increased to 60,000. Show that the natural growth rate a for this population is $(1/5)\ln 3/2$.

10. We can model a population which becomes susceptible to epidemics in the following manner. Assume that our population is originally governed by the logistic law

$$\frac{dp}{dt} = ap - bp^2 \qquad \text{(i)}$$

and that an epidemic strikes as soon as p reaches a certain value Q, with Q less than the limiting population a/b. At this stage the vital coefficients become $A<a$, $B<b$, and Equation (i) is replaced by

$$\frac{dp}{dt} = Ap - Bp^2. \qquad \text{(ii)}$$

Suppose that $Q > A/B$. The population then starts decreasing. A point is reached when the population falls below a certain value $q > A/B$. At this moment the epidemic ceases and the population again begins to grow following Equation (i), until the incidence of a fresh epidemic. In this way there are periodic fluctuations of p between q and Q. We now indicate how to calculate the period T of these fluctuations.

(a) Show that the time T_1 taken by the first part of the cycle, when p increases from q to Q is given by

$$T_1 = \frac{1}{a}\ln\frac{Q(a-bq)}{q(a-bQ)}.$$

(b) Show that the time T_2 taken by the second part of the cycle, when p decreases from Q to q is given by

$$T_2 = \frac{1}{A}\ln\frac{q(QB-A)}{Q(qB-A)}.$$

Thus, the time for the entire cycle is $T_1 + T_2$.

11. It has been observed that plagues appear in mice populations whenever the population becomes too large. Further, a local increase of density attracts predators in large numbers. These two factors will succeed in destroying 97–98% of a population of small rodents in two or three weeks, and the density then falls to a level at which the disease cannot spread. The population, reduced to 2% of its maximum, finds its refuges from the predators sufficient, and its food abundant. The population therefore begins to grow again until it reaches a level favorable to another wave of disease and predation. Now, the speed of reproduction in mice is so great that we may set $b=0$ in Equation (i) of Exercise 7. In the second part of the cycle, on the contrary, A is very small in comparison with B, and it may be neglected therefore in Equation (ii).

(a) Under these assumptions, show that

$$T_1=\frac{1}{a}\ln\frac{Q}{q} \quad \text{and} \quad T_2=\frac{Q-q}{qQB}.$$

(b) Assuming that T_1 is approximately four years, and Q/q is approximately fifty, show that a is approximately one. This value of a, incidentally, corresponds very well with the rate of multiplication of mice in natural circumstances.

12. There are many important classes of organisms whose birth rate is *not* proportional to the population size. Suppose, for example, that each member of the population requires a partner for reproduction, and that each member relies on chance encounters for meeting a mate. If the expected number of encounters is proportional to the product of the numbers of males and females, and if these are equally distributed in the population, then the number of encounters, and hence the birthrate too, is proportional to p^2. The death rate is still proportional to p. Consequently, the population size $p(t)$ satisfies the differential equation

$$\frac{dp}{dt}=bp^2-ap, \quad a,b>0.$$

Show that $p(t)$ approaches 0 as $t\to\infty$ if $p_0<a/b$. Thus, once the population size drops below the critical size a/b, the population tends to extinction. Thus, a species is classified endangered if its current size is perilously close to its critical size.

1.6 The spread of technological innovations

Economists and sociologists have long been concerned with how a technological change, or innovation, spreads in an industry. Once an innovation is introduced by one firm, how soon do others in the industry come to adopt it, and what factors determine how rapidly they follow? In this section we construct a model of the spread of innovations among farmers, and then show that this same model also describes the spread of innovations in such diverse industries as bituminous coal, iron and steel, brewing, and railroads.

Assume that a new innovation is introduced into a fixed community of N farmers at time $t=0$. Let $p(t)$ denote the number of farmers who have adopted at time t. As in the previous section, we make the approximation that $p(t)$ is a continuous function of time, even though it obviously changes by integer amounts. The simplest realistic assumption that we can make concerning the spread of this innovation is that a farmer adopts the innovation only after he has been told of it by a farmer who has already adopted. Then, the number of farmers Δp who adopt the innovation in a small time interval Δt is directly proportional to the number of farmers p who have already adopted, and the number of farmers $N-p$ who are as yet unaware. Hence, $\Delta p = cp(N-p)\Delta t$ or $\Delta p/\Delta t = cp(N-p)$ for some positive constant c. Letting $\Delta t \to 0$, we obtain the differential equation

$$\frac{dp}{dt} = cp(N-p). \tag{1}$$

This is the logistic equation of the previous section if we set $a = cN$, $b = c$. Assuming that $p(0) = 1$; i.e., one farmer has adopted the innovation at time $t = 0$, we see that $p(t)$ satisfies the initial-value problem

$$\frac{dp}{dt} = cp(N-p), \qquad p(0) = 1. \tag{2}$$

The solution of (2) is

$$p(t) = \frac{Ne^{cNt}}{N-1+e^{cNt}} \tag{3}$$

which is a logistic function (see Section 1.5). Hence, our model predicts that the adoption process accelerates up to that point at which half the community is aware of the innovation. After this point, the adoption process begins to decelerate until it eventually reaches zero.

Let us compare the predictions of our model with data on the spread of two innovations through American farming communities in the middle 1950's. Figure 1 represents the cumulative number of farmers in Iowa during 1944–1955 who adopted 2,4-D weed spray, and Figure 2 represents the cumulative percentage of corn acreage in hybrid corn in three American states during the years 1934–1958. The circles in these figures are the actual measurements, and the graphs were obtained by connecting these measurements with straight lines. As can be seen, these curves have all the properties of logistic curves, and on the whole, offer very good agreement with our model. However, there are two discrepancies. First, the actual point at which the adoption process ceases to accelerate is not always when fifty per cent of the population has adopted the innovation. As can be seen from Figure 2, the adoption process for hybrid corn began to decelerate in Alabama only after nearly sixty per cent of the farmers had adopted the innovation. Second, the agreement with our model is much better in the later stages of the adoption process than in the earlier stages.

The source of the second discrepancy is our assumption that a farmer only learns of an innovation through contact with another farmer. This is not entirely true. Studies have shown that mass communication media such

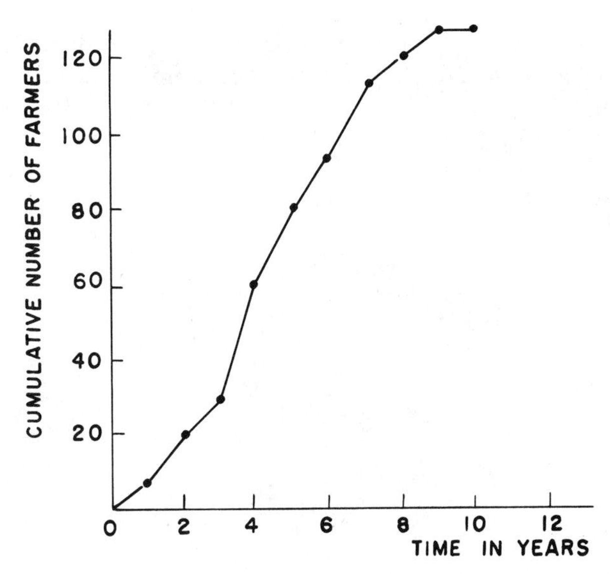

Figure 1. Cumulative number of farmers who adopted 2,4-D weed spray in Iowa

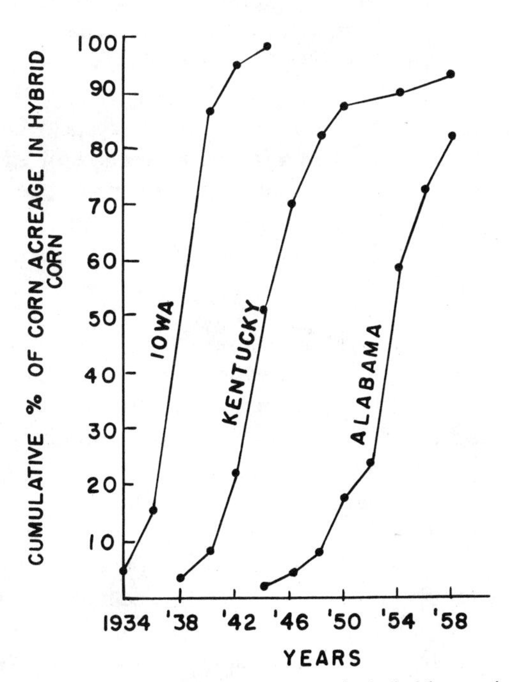

Figure 2. Cumulative percentage of corn acreage in hybrid corn in three American states

as radio, television, newspapers and farmers' magazines play a large role in the early stages of the adoption process. Therefore, we must add a term to the differential equation (1) to take this into account. To compute this term, we assume that the number of farmers Δp who learn of the innovation through the mass communication media in a short period of time Δt is proportional to the number of farmers who do not yet know; i.e.,

$$\Delta p = c'(N-p)\Delta t$$

for some positive constant c'. Letting $\Delta t \to 0$, we see that $c'(N-p)$ farmers, per unit time, learn of the innovation through the mass communication media. Thus, if $p(0)=0$, then $p(t)$ satisfies the initial-value problem

$$\frac{dp}{dt} = cp(N-p) + c'(N-p), \qquad p(0)=0. \tag{4}$$

The solution of (4) is

$$p(t) = \frac{Nc'\left[e^{(c'+cN)t}-1\right]}{cN + c'e^{(c'+cN)t}}, \tag{5}$$

and in Exercises 2 and 3 we indicate how to determine the shape of the curve (5).

The corrected curve (5) now gives remarkably good agreement with Figures 1 and 2, for suitable choices of c and c'. However, (see Exercise 3c) it still doesn't explain why the adoption of hybrid corn in Alabama only began to decelerate after sixty per cent of the farmers had adopted the innovation. This indicates, of course, that other factors, such as the time interval that elapses between when a farmer first learns of an innovation and when he actually adopts it, may play an important role in the adoption process, and must be taken into account in any model.

We would now like to show that the differential equation

$$dp/dt = cp(N-p)$$

also governs the rate at which firms in such diverse industries as bituminous coal, iron and steel, brewing, and railroads adopted several major innovations in the first part of this century. This is rather surprising, since we would expect that the number of firms adopting an innovation in one of these industries certainly depends on the profitability of the innovation and the investment required to implement it, and we haven't mentioned these factors in deriving Equation (1). However, as we shall see shortly, these two factors are incorporated in the constant c.

Let n be the total number of firms in a particular industry who have adopted an innovation at time t. It is clear that the number of firms Δp who adopt the innovation in a short time interval Δt is proportional to the number of firms $n-p$ who have not yet adopted; i.e., $\Delta p = \lambda(n-p)\Delta t$. Letting $\Delta t \to 0$, we see that

$$\frac{dp}{dt} = \lambda(n-p).$$

The proportionality factor λ depends on the profitability π of installing this innovation relative to that of alternative investments, the investment s required to install this innovation as a percentage of the total assets of the firm, and the percentage of firms who have already adopted. Thus,

$$\lambda = f(\pi, s, p/n).$$

Expanding f in a Taylor series, and dropping terms of degree more than two, gives

$$\lambda = a_1 + a_2\pi + a_3 s + a_4\frac{p}{n} + a_5\pi^2 + a_6 s^2 + a_7\pi s + a_8\pi\left(\frac{p}{n}\right) + a_9 s\left(\frac{p}{n}\right) + a_{10}\left(\frac{p}{n}\right)^2.$$

In the late 1950's, Edwin Mansfield of Carnegie Mellon University investigated the spread of twelve innovations in four major industries. From his exhaustive studies, Mansfield concluded that $a_{10} = 0$ and

$$a_1 + a_2\pi + a_3 s + a_5\pi^2 + a_6 s^2 + a_7\pi s = 0.$$

Thus, setting

$$k = a_4 + a_8\pi + a_9 s, \tag{6}$$

we see that

$$\frac{dp}{dt} = k\frac{p}{n}(n-p).$$

(This is the equation obtained previously for the spread of innovations among farmers, if we set $k/n = c$.) We assume that the innovation is first adopted by one firm in the year t_0. Then, $p(t)$ satisfies the initial-value problem

$$\frac{dp}{dt} = \frac{k}{n}p(n-p), \qquad p(t_0) = 1 \tag{7}$$

and this implies that

$$p(t) = \frac{n}{1 + (n-1)e^{-k(t-t_0)}}.$$

Mansfield studied how rapidly the use of twelve innovations spread from enterprise to enterprise in four major industries—bituminous coal, iron and steel, brewing, and railroads. The innovations are the shuttle car, trackless mobile loader, and continuous mining machine (in bituminous coal); the by-product coke oven, continuous wide strip mill, and continuous annealing line for tin plate (in iron and steel); the pallet-loading machine, tin container, and high speed bottle filler (in brewing); and the diesel locomotive, centralized traffic control, and car retarders (in railroads). His results are described graphically in Figure 3. For all but the by-product coke oven and tin container, the percentages given are for every two years from the year of initial introduction. The length of the interval for the by-product coke oven is about six years, and for the tin container, it is six months. Notice how all these curves have the general appearance of a logistic curve.

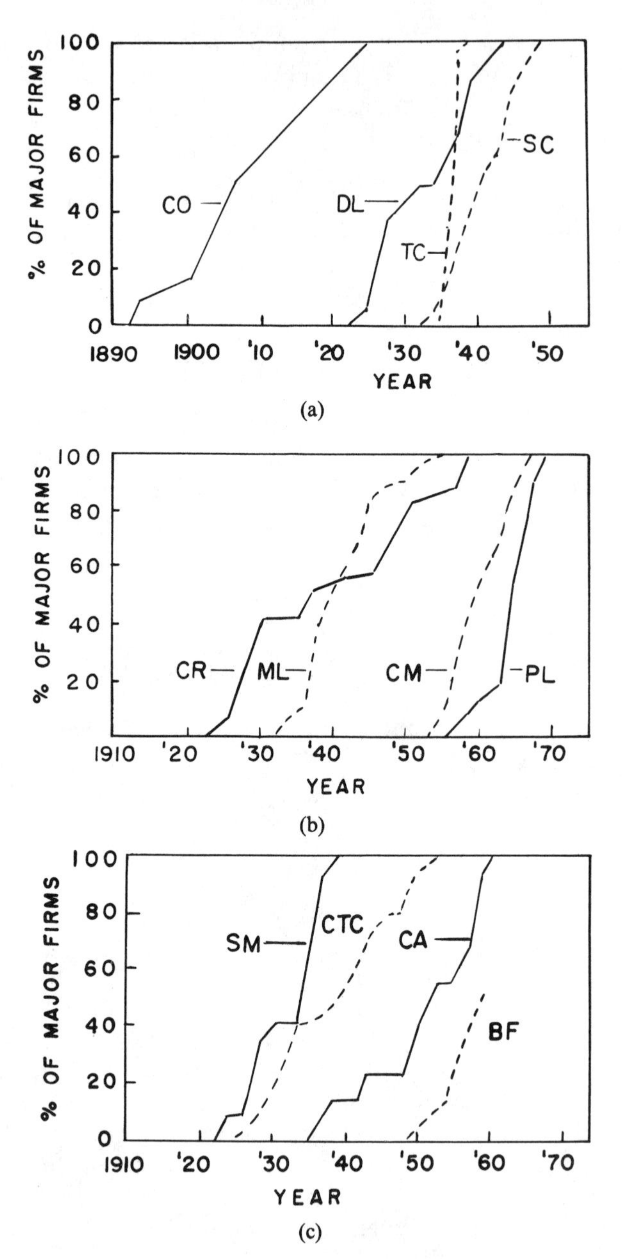

Figure 3. Growth in the percentage of major firms that introduced twelve innovations; bituminous coal, iron and steel, brewing, and railroad industries, 1890–1958; (a) By-product coke oven (CO), diesel locomotive (DL), tin container (TC), and shuttle car (SC); (b) Car retarder (CR), trackless mobile loader (ML), continuous mining machine (CM), and pallet-loading machine (PL); (c) Continuous wide-strip mill (SM), centralized traffic control (CTC), continuous annealing (CA), and highspeed bottle filler (BF).

Table 1.

Innovation	n	t_0	a_4	a_8	a_9	π	s
Diesel locomotive	25	1925	−0.59	0.530	−0.027	1.59	0.015
Centralized traffic control	24	1926	−0.59	0.530	−0.027	1.48	0.024
Car retarders	25	1924	−0.59	0.530	−0.027	1.25	0.785
Continuous wide strip mill	12	1924	−0.52	0.530	−0.027	1.87	4.908
By-product coke oven	12	1894	−0.52	0.530	−0.027	1.47	2.083
Continuous annealing	9	1936	−0.52	0.530	−0.027	1.25	0.554
Shuttle car	15	1937	−0.57	0.530	−0.027	1.74	0.013
Trackless mobile loader	15	1934	−0.57	0.530	−0.027	1.65	0.019
Continuous mining machine	17	1947	−0.57	0.530	−0.027	2.00	0.301
Tin container	22	1935	−0.29	0.530	−0.027	5.07	0.267
High speed bottle filler	16	1951	−0.29	0.530	−0.027	1.20	0.575
Pallet-loading machine	19	1948	−0.29	0.530	−0.027	1.67	0.115

For a more detailed comparison of the predictions of our model (7) with these observed results, we must evaluate the constants n, k, and t_0 for each of the twelve innovations. Table 1 gives the value of n, t_0, a_4, a_5, a_9, π, and s for each of the twelve innovations; the constant k can then be computed from Equation (6). As the answers to Exercises 5 and 6 will indicate, our model (7) predicts with reasonable accuracy the rate of adoption of these twelve innovations.

Reference

Mansfield, E., "Technical change and the rate of imitation," *Econometrica*, Vol. 29, No. 4, Oct. 1961.

Exercises

1. Solve the initial-value problem (2).

2. Let $c=0$ in (5). Show that $p(t)$ increases monotonically from 0 to N, and has no points of inflection.

3. Here is a heuristic argument to determine the behavior of the curve (5). If $c'=0$, then we have a logistic curve, and if $c=0$, then we have the behavior described in Exercise 2. Thus, if c is large relative to c', then we have a logistic curve, and if c is small relative to c' then we have the behavior illustrated in Exercise 2.

(a) Let $p(t)$ satisfy (4). Show that

$$\frac{d^2p}{dt^2}=(N-p)(cp+c')(cN-2cp-c').$$

(b) Show that $p(t)$ has a point of inflection, at which dp/dt achieves a maximum, if, and only if, $c'/c<N$.

(c) Assume that $p(t)$ has a point of inflection at $t=t^*$. Show that $p(t^*)\leqslant N/2$.

4. Solve the initial-value problem (7).

5. It seems reasonable to take the time span between the date when 20% of the firms had introduced the innovation and the date when 80% of the firms had introduced the innovation, as the rate of imitation.

(a) Show from our model that this time span is $4(\ln 2)/k$.

(b) For each of the twelve innovations, compute this time span from the data in Table 1, and compare with the observed value in Figure 3.

6. (a) Show from our model that $(1/k)\ln(n-1)$ years elapse before 50% of the firms introduce an innovation.

(b) Compute this time span for each of the 12 innovations and compare with the observed values in Figure 3.

1.7 An atomic waste disposal problem

For several years the Atomic Energy Commission (now known as the Nuclear Regulatory Commission) had disposed of concentrated radioactive waste material by placing it in tightly sealed drums which were then dumped at sea in fifty fathoms (300 feet) of water. When concerned ecologists and scientists questioned this practice, they were assured by the A.E.C. that the drums would never develop leaks. Exhaustive tests on the drums proved the A.E.C. right. However, several engineers then raised the question of whether the drums could crack from the impact of hitting the ocean floor. "Never," said the A.E.C. "We'll see about that," said the engineers. After performing numerous experiments, the engineers found that the drums could crack on impact if their velocity exceeded forty feet per second. The problem before us, therefore, is to compute the velocity of the drums upon impact with the ocean floor. To this end, we digress briefly to study elementary Newtonian mechanics.

Newtonian mechanics is the study of Newton's famous laws of motion and their consequences. Newton's first law of motion states that an object will remain at rest, or move with constant velocity, if no force is acting on it. A force should be thought of as a push or pull. This push or pull can be exerted directly by something in contact with the object, or it can be exerted indirectly, as the earth's pull of gravity is.

Newton's second law of motion is concerned with describing the motion of an object which is acted upon by several forces. Let $y(t)$ denote the position of the center of gravity of the object. (We assume that the object moves in only one direction.) Those forces acting on the object, which tend to increase y, are considered positive, while those forces tending to decrease y are considered negative. The resultant force F acting on an object is defined to be the sum of all positive forces minus the sum of all negative forces. Newton's second law of motion states that the acceleration d^2y/dt^2 of an object is proportional to the resultant force F acting on it; i.e.,

$$\frac{d^2y}{dt^2} = \frac{1}{m}F. \tag{1}$$

The constant m is the mass of the object. It is related to the weight W of the object by the relation $W = mg$, where g is the acceleration of gravity. Unless otherwise stated, we assume that the weight of an object and the acceleration of gravity are constant. We will also adopt the English system of units, so that t is measured in seconds, y is measured in feet, and F is measured in pounds. The units of m are then slugs, and the gravitational acceleration g equals 32.2 ft/s^2.

Remark. We would prefer to use the mks system of units, where y is measured in meters and F is measured in newtons. The units of m are then kilograms, and the gravitational acceleration equals 9.8 m/s^2. In the third edition of this text, we have changed from the English system of units to the mks system in Section 2.6. However, changing to the mks system in this section would have caused undue confusion to the users of the first and second editions. This is because of the truncation error involved in converting from feet to meters and pounds to newtons.

We return now to our atomic waste disposal problem. As a drum descends through the water, it is acted upon by three forces W, B, and D. The force W is the weight of the drum pulling it down, and in magnitude, $W = 527.436$ lb. The force B is the buoyancy force of the water acting on the drum. This force pushes the drum up, and its magnitude is the weight of the water displaced by the drum. Now, the Atomic Energy Commission used 55 gallon drums, whose volume is 7.35 ft^3. The weight of one cubic foot of salt water is 63.99 lb. Hence $B = (63.99)(7.35) = 470.327$ lb.

The force D is the drag force of the water acting on the drum; it resists the motion of the drum through the water. Experiments have shown that any medium such as water, oil, and air resists the motion of an object through it. This resisting force acts in the direction opposite the motion, and is usually directly proportional to the velocity V of the object. Thus, $D = cV$, for some positive constant c. Notice that the drag force increases as V increases, and decreases as V decreases. To calculate D, the engineers conducted numerous towing experiments. They concluded that the orientation of the drum had little effect on the drag force, and that

$$D = 0.08\,V\frac{(\text{lb})(\text{s})}{\text{ft}}.$$

Now, set $y=0$ at sea level, and let the direction of increasing y be downwards. Then, W is a positive force, and B and D are negative forces. Consequently, from (1),

$$\frac{d^2y}{dt^2}=\frac{1}{m}(W-B-cV)=\frac{g}{W}(W-B-cV).$$

We can rewrite this equation as a first-order linear differential equation for $V=dy/dt$; i.e.,

$$\frac{dV}{dt}+\frac{cg}{W}V=\frac{g}{W}(W-B). \tag{2}$$

Initially, when the drum is released in the ocean, its velocity is zero. Thus, $V(t)$, the velocity of the drum, satisfies the initial-value problem

$$\frac{dV}{dt}+\frac{cg}{W}V=\frac{g}{W}(W-B), \qquad V(0)=0, \tag{3}$$

and this implies that

$$V(t)=\frac{W-B}{c}\left[1-e^{(-cg/W)t}\right]. \tag{4}$$

Equation (4) expresses the velocity of the drum as a function of time. In order to determine the impact velocity of the drum, we must compute the time t at which the drum hits the ocean floor. Unfortunately, though, it is impossible to find t as an explicit function of y (see Exercise 2). Therefore, we cannot use Equation (4) to find the velocity of the drum when it hits the ocean floor. However, the A.E.C. can use this equation to try and prove that the drums do not crack on impact. To wit, observe from (4) that $V(t)$ is a monotonic increasing function of time which approaches the limiting value

$$V_T=\frac{W-B}{c}$$

as t approaches infinity. The quantity V_T is called the terminal velocity of the drum. Clearly, $V(t)\leqslant V_T$, so that the velocity of the drum when it hits the ocean floor is certainly less than $(W-B)/c$. Now, if this terminal velocity is less than 40 ft/s, then the drums could not possibly break on impact. However,

$$\frac{W-B}{c}=\frac{527.436-470.327}{0.08}=713.86 \text{ ft/s},$$

and this is way too large.

It should be clear now that the only way we can resolve the dispute between the A.E.C. and the engineers is to find $v(y)$, the velocity of the drum as a function of position. The function $v(y)$ is very different from the function $V(t)$, which is the velocity of the drum as a function of time. However, these two functions are related through the equation

$$V(t)=v(y(t))$$

if we express y as a function of t. By the chain rule of differentiation, $dV/dt=(dv/dy)(dy/dt)$. Hence

$$\frac{W}{g}\frac{dv}{dy}\frac{dy}{dt}=W-B-cV.$$

But $dy/dt=V(t)=v(y(t))$. Thus, suppressing the dependence of y on t, we see that $v(y)$ satisfies the first-order differential equation

$$\frac{W}{g}v\frac{dv}{dy}=W-B-cv,\quad \text{or}\quad \frac{v}{W-B-cv}\frac{dv}{dy}=\frac{g}{W}.$$

Moreover,

$$v(0)=v(y(0))=V(0)=0.$$

Hence,

$$\int_0^v\frac{r\,dr}{W-B-cr}=\int_0^y\frac{g}{W}ds=\frac{gy}{W}.$$

Now,

$$\begin{aligned}\int_0^v\frac{r\,dr}{W-B-cr}&=\int_0^v\frac{r-(W-B)/c}{W-B-cr}dr+\frac{W-B}{c}\int_0^v\frac{dr}{W-B-cr}\\&=-\frac{1}{c}\int_0^v dr+\frac{W-B}{c}\int_0^v\frac{dr}{W-B-cr}\\&=-\frac{v}{c}-\frac{(W-B)}{c^2}\ln\frac{|W-B-cv|}{W-B}.\end{aligned}$$

We know already that $v<(W-B)/c$. Consequently, $W-B-cv$ is always positive, and

$$\frac{gy}{W}=-\frac{v}{c}-\frac{(W-B)}{c^2}\ln\frac{W-B-cv}{W-B}. \tag{5}$$

At this point, we are ready to scream in despair since we cannot find v as an explicit function of y from (5). This is not an insurmountable difficulty, though. As we show in Section 1.11, it is quite simple, with the aid of a digital computer, to find $v(300)$ from (5). We need only supply the computer with a good approximation of $v(300)$ and this is obtained in the following manner. The velocity $v(y)$ of the drum satisfies the initial-value problem

$$\frac{W}{g}v\frac{dv}{dy}=W-B-cv,\qquad v(0)=0. \tag{6}$$

Let us, for the moment, set $c=0$ in (6) to obtain the new initial-value problem

$$\frac{W}{g}u\frac{du}{dy}=W-B,\qquad u(0)=0. \tag{6'}$$

(We have replaced v by u to avoid confusion later.) We can integrate (6′) immediately to obtain that

$$\frac{W}{g}\frac{u^2}{2}=(W-B)y,\quad\text{or}\quad u(y)=\left[\frac{2g}{W}(W-B)y\right]^{1/2}.$$

In particular,

$$u(300)=\left[\frac{2g}{W}(W-B)300\right]^{1/2}=\left[\frac{2(32.2)(57.109)(300)}{527.436}\right]^{1/2}$$

$$\cong\sqrt{2092}\cong 45.7\text{ ft/s}.$$

We claim, now, that $u(300)$ is a very good approximation of $v(300)$. The proof of this is as follows. First, observe that the velocity of the drum is always greater if there is no drag force opposing the motion. Hence,

$$v(300)<u(300).$$

Second, the velocity v increases as y increases, so that $v(y)\leqslant v(300)$ for $y\leqslant 300$. Consequently, the drag force D of the water acting on the drum is always less than $0.08\times u(300)\cong 3.7$ lb. Now, the resultant force $W-B$ pulling the drum down is approximately 57.1 lb, which is very large compared to D. It stands to reason, therefore, that $u(y)$ should be a very good approximation of $v(y)$. And indeed, this is the case, since we find numerically (see Section 1.11) that $v(300)=45.1$ ft/s. Thus, the drums can break upon impact, and the engineers were right.

Epilog. The rules of the Atomic Energy Commission now expressly forbid the dumping of low level atomic waste at sea. This author is uncertain though, as to whether Western Europe has also forbidden this practice.

Remark. The methods introduced in this section can also be used to find the velocity of any object which is moving through a medium that resists the motion. We just disregard the buoyancy force if the medium is not water. For example, let $V(t)$ denote the velocity of a parachutist falling to earth under the influence of gravity. Then,

$$\frac{W}{g}\frac{dV}{dt}=W-D$$

where W is the weight of the man and the parachute, and D is the drag force exerted by the atmosphere on the falling parachutist. The drag force on a bluff object in air, or in any fluid of small viscosity is usually very nearly proportional to V^2. Proportionality to V is the exceptional case, and occurs only at very low speeds. The criterion as to whether the square or the linear law applies is the "Reynolds number"

$$R=\rho VL/\mu.$$

L is a representative length dimension of the object, and ρ and μ are the density and viscosity of the fluid. If $R<10$, then $D\sim V$, and if $R>10^3$, $D\sim V^2$. For $10<R<10^3$, neither law is accurate.

EXERCISES

1. Solve the initial-value problem (3).

2. Solve for $y=y(t)$ from (4), and then show that the equation $y=y(t)$ cannot be solved explicitly for $t=t(y)$.

3. Show that the drums of atomic waste will not crack upon impact if they are dropped into L feet of water with $(2g(W-B)L/W)^{1/2}<40$.

4. Fat Richie, an enormous underworld hoodlum weighing 400 lb, was pushed out of a penthouse window 2800 feet above the ground in New York City. Neglecting air resistance find (a) the velocity with which Fat Richie hit the ground; (b) the time elapsed before Fat Richie hit the ground.

5. An object weighing 300 lb is dropped into a river 150 feet deep. The volume of the object is 2 ft^3, and the drag force exerted by the water on it is 0.05 times its velocity. The drag force may be considered negligible if it does not exceed 5% of the resultant force pulling the drum down. Prove that the drag force is negligible in this case. (Here $B=2(62.4)=124.8$.)

6. A 400 lb sphere of volume $4\pi/3$ and a 300 lb cylinder of volume π are simultaneously released from rest into a river. The drag force exerted by the water on the falling sphere and cylinder is λV_s and λV_c, respectively, where V_s and V_c are the velocities of the sphere and cylinder, and λ is a positive constant. Determine which object reaches the bottom of the river first.

7. A parachutist falls from rest toward earth. The combined weight of man and parachute is 161 lb. Before the parachute opens, the air resistance equals $V/2$. The parachute opens 5 seconds after the fall begins; and the air resistance is then $V^2/2$. Find the velocity $V(t)$ of the parachutist after the parachute opens.

8. A man wearing a parachute jumps from a great height. The combined weight of man and parachute is 161 lb. Let $V(t)$ denote his speed at time t seconds after the fall begins. During the first 10 seconds, the air resistance is $V/2$. Thereafter, while the parachute is open, the air resistance is $10V$. Find an explicit formula for $V(t)$ at any time t greater than 10 seconds.

9. An object of mass m is projected vertically downward with initial velocity V_0 in a medium offering resistance proportional to the square root of the magnitude of the velocity.
   (a) Find a relation between the velocity V and the time t if the drag force equals $c\sqrt{V}$.
   (b) Find the terminal velocity of the object. *Hint*: You can find the terminal velocity even though you cannot solve for $V(t)$.

10. A body of mass m falls from rest in a medium offering resistance proportional to the square of the velocity; that is, $D=cV^2$. Find $V(t)$ and compute the terminal velocity V_T.

11. A body of mass m is projected upward from the earth's surface with an initial velocity V_0. Take the y-axis to be positive upward, with the origin on the surface of the earth. Assuming there is no air resistance, but taking into

account the variation of the earth's gravitational field with altitude, we obtain that

$$m\frac{dV}{dt} = -\frac{mgR^2}{(y+R)^2}$$

where R is the radius of the earth.

(a) Let $V(t) = v(y(t))$. Find a differential equation satisfied by $v(y)$.

(b) Find the smallest initial velocity V_0 for which the body will not return to earth. This is the so-called escape velocity. *Hint*: The escape velocity is found by requiring that $v(y)$ remain strictly positive.

12. It is not really necessary to find $v(y)$ explicitly in order to prove that $v(300)$ exceeds 40 ft/s. Here is an alternate proof. Observe first that $v(y)$ increases as y increases. This implies that y is a monotonic increasing function of v. Therefore, if y is less than 300 ft when v is 40 ft/s, then v must be greater than 40 ft/s when y is 300 ft. Substitute $v = 40$ ft/s in Equation (5), and show that y is less than 300 ft. Conclude, therefore, that the drums can break upon impact.

1.8 The dynamics of tumor growth, mixing problems and orthogonal trajectories

In this section we present three very simple but extremely useful applications of first-order equations. The first application concerns the growth of solid tumors; the second application is concerned with "mixing problems" or "compartment analysis"; and the third application shows how to find a family of curves which is orthogonal to a given family of curves.

(a) *The dynamics of tumor growth*

It has been observed experimentally, that "free living" dividing cells, such as bacteria cells, grow at a rate proportional to the volume of dividing cells at that moment. Let $V(t)$ denote the volume of dividing cells at time t. Then,

$$\frac{dV}{dt} = \lambda V \tag{1}$$

for some positive constant λ. The solution of (1) is

$$V(t) = V_0 e^{\lambda(t-t_0)} \tag{2}$$

where V_0 is the volume of dividing cells at the initial time t_0. Thus, free living dividing cells grow *exponentially* with time. One important consequence of (2) is that the volume of cells keeps doubling (see Exercise 1) every time interval of length $\ln 2/\lambda$.

On the other hand, solid tumors do not grow exponentially with time. As the tumor becomes larger, the doubling time of the total tumor volume continuously increases. Various researchers have shown that the data for many solid tumors is fitted remarkably well, over almost a 1000 fold increase in tumor volume, by the equation

$$V(t)=V_0\exp\left(\frac{\lambda}{\alpha}(1-\exp(-\alpha t))\right) \tag{3}$$

where $\exp(x)\equiv e^x$, and λ and α are positive constants.

Equation (3) is usually referred to as a Gompertzian relation. It says that the tumor grows more and more slowly with the passage of time, and that it ultimately approaches the limiting volume $V_0e^{\lambda/\alpha}$. Medical researchers have long been concerned with explaining this deviation from simple exponential growth. A great deal of insight into this problem can be gained by finding a differential equation satisfied by $V(t)$. Differentiating (3) gives

$$\begin{aligned}\frac{dV}{dt} &= V_0\lambda\exp(-\alpha t)\exp\left(\frac{\lambda}{\alpha}(1-\exp(-\alpha(t)))\right)\\ &=\lambda e^{-\alpha t}V.\end{aligned} \tag{4}$$

Two conflicting theories have been advanced for the dynamics of tumor growth. They correspond to the two arrangements

$$\frac{dV}{dt}=(\lambda e^{-\alpha t})V \tag{4a}$$

$$\frac{dV}{dt}=\lambda(e^{-\alpha t}V) \tag{4b}$$

of the differential equation (4). According to the first theory, the retarding effect of tumor growth is due to an increase in the mean generation time of the cells, without a change in the proportion of reproducing cells. As time goes on, the reproducing cells mature, or age, and thus divide more slowly. This theory corresponds to the bracketing (a).

The bracketing (b) suggests that the mean generation time of the dividing cells remains constant, and the retardation of growth is due to a loss in reproductive cells in the tumor. One possible explanation for this is that a *necrotic region* develops in the center of the tumor. This necrosis appears at a critical size for a particular type of tumor, and thereafter the necrotic "core" increases rapidly as the total tumor mass increases. According to this theory, a necrotic core develops because in many tumors the supply of blood, and thus of oxygen and nutrients, is almost completely confined to the surface of the tumor and a short distance beneath it. As the tumor grows, the supply of oxygen to the central core by diffusion becomes more and more difficult resulting in the formation of a necrotic core.

(b) *Mixing problems*

Many important problems in biology and engineering can be put into the following framework. A solution containing a fixed concentration of substance x flows into a tank, or compartment, containing the substance x and possibly other substances, at a specified rate. The mixture is stirred together very rapidly, and then leaves the tank, again at a specified rate. Find the concentration of substance x in the tank at any time t.

Problems of this type fall under the general heading of "mixing problems," or compartment analysis. The following example illustrates how to solve these problems.

Example 1. A tank contains S_0 lb of salt dissolved in 200 gallons of water. Starting at time $t=0$, water containing $\frac{1}{2}$ lb of salt per gallon enters the tank at the rate of 4 gal/min, and the well stirred solution leaves the tank at the same rate. Find the concentration of salt in the tank at any time $t>0$.

Solution. Let $S(t)$ denote the amount of salt in the tank at time t. Then, $S'(t)$, which is the rate of change of salt in the tank at time t, must equal the rate at which salt enters the tank minus the rate at which it leaves the tank. Obviously, the rate at which salt enters the tank is

$$\tfrac{1}{2}\text{ lb/gal times } 4\text{ gal/min} = 2\text{ lb/min}.$$

After a moment's reflection, it is also obvious that the rate at which salt leaves the tank is

$$4\text{ gal/min times } \frac{S(t)}{200}.$$

Thus

$$S'(t) = 2 - \frac{S(t)}{50}, \qquad S(0) = S_0,$$

and this implies that

$$S(t) = S_0 e^{-0.02t} + 100(1 - e^{-0.02t}). \tag{5}$$

Hence, the concentration $c(t)$ of salt in the tank is given by

$$c(t) = \frac{S(t)}{200} = \frac{S_0}{200} e^{-0.02t} + \tfrac{1}{2}(1 - e^{-0.02t}). \tag{6}$$

Remark. The first term on the right-hand side of (5) represents the portion of the original amount of salt remaining in the tank at time t. This term becomes smaller and smaller with the passage of time as the original solution is drained from the tank. The second term on the right-hand side of (5) represents the amount of salt in the tank at time t due to the action of the flow process. Clearly, the amount of salt in the tank must ultimately approach the limiting value of 100 lb, and this is easily verified by letting t approach ∞ in (5).

(c) *Orthogonal trajectories*

In many physical applications, it is often necessary to find the orthogonal trajectories of a given family of curves. (A curve which intersects each member of a family of curves at right angles is called an orthogonal trajectory of the given family.) For example, a charged particle moving under the influence of a magnetic field always travels on a curve which is perpendicular to each of the magnetic field lines. The problem of computing orthogonal trajectories of a family of curves can be solved in the following manner. Let the given family of curves be described by the relation

$$F(x,y,c)=0. \tag{7}$$

Differentiating this equation yields

$$F_x+F_y y'=0, \quad \text{or} \quad y'=-\frac{F_x}{F_y}. \tag{8}$$

Next, we solve for $c=c(x,y)$ from (7) and replace every c in (8) by this value $c(x,y)$. Finally, since the slopes of curves which intersect orthogonally are negative reciprocals of each other, we see that the orthogonal trajectories of (7) are the solution curves of the equation

$$y'=\frac{F_y}{F_x}. \tag{9}$$

Example 2. Find the orthogonal trajectories of the family of parabolas

$$x=cy^2.$$

Solution. Differentiating the equation $x=cy^2$ gives $1=2cyy'$. Since $c=x/y^2$, we see that $y'=y/2x$. Thus, the orthogonal trajectories of the family

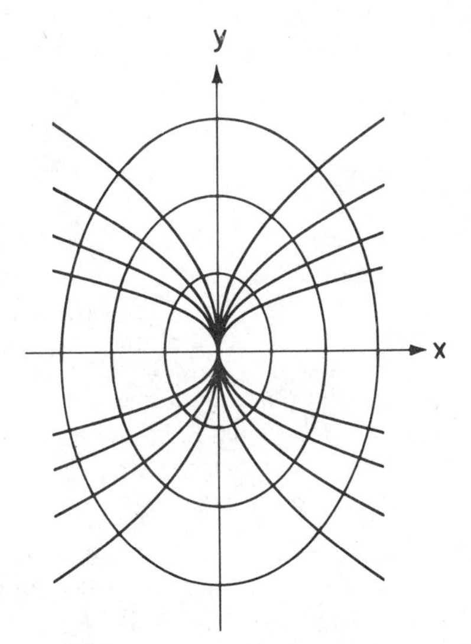

Figure 1. The parabolas $x=cy^2$ and their orthogonal trajectories

of parabolas $x = cy^2$ are the solution curves of the equation

$$y' = -\frac{2x}{y}. \tag{10}$$

This equation is separable, and its solution is

$$y^2 + 2x^2 = k^2. \tag{11}$$

Thus, the family of ellipses (11) (see Figure 1) are the orthogonal trajectories of the family of parabolas $x = cy^2$.

Reference

Burton, Alan C., Rate of growth of solid tumors as a problem of diffusion, *Growth*, 1966, vol. 30, pp. 157–176.

Exercises

1. A given substance satisfies the exponential growth law (1). Show that the graph of $\ln V$ versus t is a straight line.

2. A substance x multiplies exponentially, and a given quantity of the substance doubles every 20 years. If we have 3 lb of substance x at the present time, how many lb will we have 7 years from now?

3. A substance x decays exponentially, and only half of the given quantity of x remains after 2 years. How long does it take for 5 lb of x to decay to 1 lb?

4. The equation $p' = ap^{\alpha}$, $\alpha > 1$, is proposed as a model of the population growth of a certain species. Show that $p(t) \to \infty$ in finite time. Conclude, therefore, that this model is not accurate over a reasonable length of time.

5. A cancerous tumor satisfies the Gompertzian relation (3). Originally, when it contained 10^4 cells, the tumor was increasing at the rate of 20% per unit time. The numerical value of the retarding constant α is 0.02. What is the limiting number of cells in this tumor?

6. A *tracer dose* of radioactive iodine ^{131}I is injected into the blood stream at time $t = 0$. Assume that the original amount Q_0 of iodine is distributed evenly in the entire blood stream before any loss occurs. Let $Q(t)$ denote the amount of iodine in the blood at time $t > 0$. Part of the iodine leaves the blood and enters the urine at the rate $k_1 Q$. Another part of the iodine enters the thyroid gland at the rate $k_2 Q$. Find $Q(t)$.

7. Industrial waste is pumped into a tank containing 1000 gallons of water at the rate of 1 gal/min, and the well-stirred mixture leaves the tank at the same rate. (a) Find the concentration of waste in the tank at time t. (b) How long does it take for the concentration to reach 20%?

8. A tank contains 300 gallons of water and 100 gallons of pollutants. Fresh water is pumped into the tank at the rate of 2 gal/min, and the well-stirred mixture leaves at the same rate. How long does it take for the concentration of pollutants in the tank to decrease to 1/10 of its original value?

9. Consider a tank containing, at time $t=0$, Q_0 lb of salt dissolved in 150 gallons of water. Assume that water containing $\frac{1}{2}$ lb of salt per gallon is entering the tank at a rate of 3 gal/min, and that the well-stirred solution is leaving the tank at the same rate. Find an expression for the concentration of salt in the tank at time t.

10. A room containing 1000 cubic feet of air is originally free of carbon monoxide. Beginning at time $t=0$, cigarette smoke containing 4 percent carbon monoxide is blown into the room at the rate of 0.1 ft^3/min, and the well-circulated mixture leaves the room at the same rate. Find the time when the concentration of carbon monoxide in the room reaches 0.012 percent. (Extended exposure to this concentration of carbon monoxide is dangerous.)

11. A 500 gallon tank originally contains 100 gallons of fresh water. Beginning at time $t=0$, water containing 50 percent pollutants flows into the tank at the rate of 2 gal/min, and the well-stirred mixture leaves at the rate of 1 gal/min. Find the concentration of pollutants in the tank at the moment it overflows.

In Exercises 12–17, find the orthogonal trajectories of the given family of curves.

12. $y=cx^2$

13. $y^2-x^2=c$

14. $y=c\sin x$

15. $x^2+y^2=cx$ (see Exercise 13 of Section 1.4)

16. $y=ce^x$

17. $y=e^{cx}$

18. The presence of toxins in a certain medium destroys a strain of bacteria at a rate jointly proportional to the number of bacteria present and to the amount of toxin. Call the constant of proportionality a. If there were no toxins present, the bacteria would grow at a rate proportional to the amount present. Call this constant of proportionality b. Assume that the amount T of toxin is increasing at a constant rate c, that is, $dT/dt=c$, and that the production of toxins begins at time $t=0$. Let $y(t)$ denote the number of living bacteria present at time t.
(a) Find a first-order differential equation satisfied by $y(t)$.
(b) Solve this differential equation to obtain $y(t)$. What happens to $y(t)$ as t approaches ∞?

19. Many savings banks now advertise continuous compounding of interest. This means that the amount of money $P(t)$ on deposit at time t, satisfies the differential equation $dP(t)/dt=rP(t)$ where r is the annual interest rate and t is measured in years. Let P_0 denote the original principal.
(a) Show that $P(1)=P_0e^r$.
(b) Let $r=0.0575$, 0.065, 0.0675, and 0.075. Show that $e^r=1.05919$, 1.06716, 1.06983, and 1.07788, respectively. Thus, the effective annual yield on interest rates of $5\frac{3}{4}$, $6\frac{1}{2}$, $6\frac{3}{4}$, and $7\frac{1}{2}$% should be 5.919, 6.716, 6.983, and 7.788%, respectively. Most banks, however, advertise effective annual yields of 6, 6.81, 7.08, and 7.9%, respectively. The reason for this discrepancy is that banks calculate a daily rate of interest based on 360 days, and they pay interest for each day money is on deposit. For a year, one gets five extra

days. Thus, we must multiply the annual yields of 5.919, 6.716, 6.983, and 7.788% by 365/360, and then we obtain the advertised values.

(c) It is interesting to note that the Old Colony Cooperative Bank in Rhode Island advertises an effective annual yield of 6.72% on an annual interest rate of $6\frac{1}{2}$% (the lower value), and an effective annual yield of 7.9% on an annual interest rate of $7\frac{1}{2}$%. Thus they are inconsistent.

1.9 Exact equations, and why we cannot solve very many differential equations

When we began our study of differential equations, the only equation we could solve was $dy/dt = g(t)$. We then enlarged our inventory to include all linear and separable equations. More generally, we can solve all differential equations which are, or can be put, in the form

$$\frac{d}{dt}\phi(t,y) = 0 \tag{1}$$

for some function $\phi(t,y)$. To wit, we can integrate both sides of (1) to obtain that

$$\phi(t,y) = \text{constant} \tag{2}$$

and then solve for y as a function of t from (2).

Example 1. The equation $1 + \cos(t+y) + \cos(t+y)(dy/dt) = 0$ can be written in the form $(d/dt)[t + \sin(t+y)] = 0$. Hence,

$$\phi(t,y) = t + \sin(t+y) = c, \quad \text{and} \quad y = -t + \arcsin(c-t).$$

Example 2. The equation $\cos(t+y) + [1 + \cos(t+y)]dy/dt = 0$ can be written in the form $(d/dt)[y + \sin(t+y)] = 0$. Hence,

$$\phi(t,y) = y + \sin(t+y) = c.$$

We must leave the solution in this form though, since we cannot solve for y explicitly as a function of time.

Equation (1) is clearly the most general first-order differential equation that we can solve. Thus, it is important for us to be able to recognize when a differential equation can be put in this form. This is not as simple as one might expect. For example, it is certainly not obvious that the differential equation

$$2t + y - \sin t + (3y^2 + \cos y + t)\frac{dy}{dt} = 0$$

can be written in the form $(d/dt)(y^3 + t^2 + ty + \sin y + \cos t) = 0$. To find all those differential equations which can be written in the form (1), observe, from the chain rule of partial differentiation, that

$$\frac{d}{dt}\phi(t,y(t)) = \frac{\partial\phi}{\partial t} + \frac{\partial\phi}{\partial y}\frac{dy}{dt}.$$

Hence, the differential equation $M(t,y) + N(t,y)(dy/dt) = 0$ can be written in the form $(d/dt)\phi(t,y) = 0$ if and only if there exists a function $\phi(t,y)$ such that $M(t,y) = \partial\phi/\partial t$ and $N(t,y) = \partial\phi/\partial y$.

This now leads us to the following question. Given two functions $M(t,y)$ and $N(t,y)$, does there exist a function $\phi(t,y)$ such that $M(t,y) = \partial\phi/\partial t$ and $N(t,y) = \partial\phi/\partial y$? Unfortunately, the answer to this question is almost always no as the following theorem shows.

Theorem 1. *Let $M(t,y)$ and $N(t,y)$ be continuous and have continuous partial derivatives with respect to t and y in the rectangle R consisting of those points (t,y) with $a<t<b$ and $c<y<d$. There exists a function $\phi(t,y)$ such that $M(t,y) = \partial\phi/\partial t$ and $N(t,y) = \partial\phi/\partial y$ if, and only if,*

$$\partial M/\partial y = \partial N/\partial t$$

in R.

PROOF. Observe that $M(t,y) = \partial\phi/\partial t$ for some function $\phi(t,y)$ if, and only if,

$$\phi(t,y) = \int M(t,y)\,dt + h(y) \tag{3}$$

where $h(y)$ is an arbitrary function of y. Taking partial derivatives of both sides of (3) with respect to y, we obtain that

$$\frac{\partial\phi}{\partial y} = \int \frac{\partial M(t,y)}{\partial y}\,dt + h'(y).$$

Hence, $\partial\phi/\partial y$ will be equal to $N(t,y)$ if, and only if,

$$N(t,y) = \int \frac{\partial M(t,y)}{\partial y}\,dt + h'(y)$$

or

$$h'(y) = N(t,y) - \int \frac{\partial M(t,y)}{\partial y}\,dt. \tag{4}$$

Now $h'(y)$ is a function of y alone, while the right-hand side of (4) appears to be a function of both t and y. But a function of y alone cannot be equal to a function of both t and y. Thus Equation (4) makes sense only if the right-hand side is a function of y alone, and this is the case if, and only if,

$$\frac{\partial}{\partial t}\left[N(t,y) - \int \frac{\partial M(t,y)}{\partial y}\,dt\right] = \frac{\partial N}{\partial t} - \frac{\partial M}{\partial y} = 0.$$

Hence, if $\partial N/\partial t \neq \partial M/\partial y$, then there is no function $\phi(t,y)$ such that $M = \partial\phi/\partial t$, $N = \partial\phi/\partial y$. On the other hand, if $\partial N/\partial t = \partial M/\partial y$ then we can solve for

$$h(y) = \int\left[N(t,y) - \int\frac{\partial M(t,y)}{\partial y}dt\right]dy.$$

Consequently, $M = \partial\phi/\partial t$, and $N = \partial\phi/\partial y$ with

$$\phi(t,y) = \int M(t,y)\,dt + \int\left[N(t,y) - \int\frac{\partial M(t,y)}{\partial y}dt\right]dy. \quad \square \tag{5}$$

Definition. The differential equation

$$M(t,y) + N(t,y)\frac{dy}{dt} = 0 \tag{6}$$

is said to be *exact* if $\partial M/\partial y = \partial N/\partial t$.

The reason for this definition, of course, is that the left-hand side of (6) is the exact derivative of a known function of t and y if $\partial M/\partial y = \partial N/\partial t$.

Remark 1. It is not essential, in the statement of Theorem 1, that $\partial M/\partial y = \partial N/\partial t$ in a rectangle. It is sufficient if $\partial M/\partial y = \partial N/\partial t$ in any region R which contains no "holes". That is to say, if C is any closed curve lying entirely in R, then its interior also lies entirely in R.

Remark 2. The differential equation $dy/dt = f(t,y)$ can always be written in the form $M(t,y) + N(t,y)(dy/dt) = 0$ by setting $M(t,y) = -f(t,y)$ and $N(t,y) = 1$.

Remark 3. It is customary to say that the solution of an exact differential equation is given by $\phi(t,y) = \text{constant}$. What we really mean is that the equation $\phi(t,y) = c$ is to be solved for y as a function of t and c. Unfortunately, most exact differential equations cannot be solved explicitly for y as a function of t. While this may appear to be very disappointing, we wish to point out that it is quite simple, with the aid of a computer, to compute $y(t)$ to any desired accuracy (see Section 1.11).

In practice, we do not recommend memorizing Equation (5). Rather, we will follow one of three different methods to obtain $\phi(t,y)$.
First Method: The equation $M(t,y) = \partial\phi/\partial t$ determines $\phi(t,y)$ up to an arbitrary function of y alone, that is,

$$\phi(t,y) = \int M(t,y)\,dt + h(y).$$

The function $h(y)$ is then determined from the equation

$$h'(y)=N(t,y)-\int\frac{\partial M(t,y)}{\partial y}dt.$$

Second Method: If $N(t,y)=\partial\phi/\partial y$, then, of necessity,

$$\phi(t,y)=\int N(t,y)\,dy+k(t)$$

where $k(t)$ is an arbitrary function of t alone. Since

$$M(t,y)=\frac{\partial\phi}{\partial t}=\int\frac{\partial N(t,y)}{\partial t}dy+k'(t)$$

we see that $k(t)$ is determined from the equation

$$k'(t)=M(t,y)-\int\frac{\partial N(t,y)}{\partial t}dy.$$

Note that the right-hand side of this equation (see Exercise 2) is a function of t alone if $\partial M/\partial y=\partial N/\partial t$.
Third Method: The equations $\partial\phi/\partial t=M(t,y)$ and $\partial\phi/\partial y=N(t,y)$ imply that

$$\phi(t,y)=\int M(t,y)\,dt+h(y)\quad\text{and}\quad\phi(t,y)=\int N(t,y)\,dy+k(t).$$

Usually, we can determine $h(y)$ and $k(t)$ just by inspection.

Example 3. Find the general solution of the differential equation

$$3y+e^t+(3t+\cos y)\frac{dy}{dt}=0.$$

Solution. Here $M(t,y)=3y+e^t$ and $N(t,y)=3t+\cos y$. This equation is exact since $\partial M/\partial y=3$ and $\partial N/\partial t=3$. Hence, there exists a function $\phi(t,y)$ such that

$$\text{(i)}\ \ 3y+e^t=\frac{\partial\phi}{\partial t}\quad\text{and}\quad\text{(ii)}\ \ 3t+\cos y=\frac{\partial\phi}{\partial y}.$$

We will find $\phi(t,y)$ by each of the three methods outlined above.

First Method: From (i), $\phi(t,y)=e^t+3ty+h(y)$. Differentiating this equation with respect to y and using (ii) we obtain that

$$h'(y)+3t=3t+\cos y.$$

Thus, $h(y)=\sin y$ and $\phi(t,y)=e^t+3ty+\sin y$. (Strictly speaking, $h(y)=\sin y+\text{constant}$. However, we already incorporate this constant of integration into the solution when we write $\phi(t,y)=c$.) The general solution of the differential equation must be left in the form $e^t+3ty+\sin y=c$ since we cannot find y explicitly as a function of t from this equation.
Second Method: From (ii), $\phi(t,y)=3ty+\sin y+k(t)$. Differentiating this

expression with respect to t, and using (i) we obtain that

$$3y + k'(t) = 3y + e^t.$$

Thus, $k(t) = e^t$ and $\phi(t,y) = 3ty + \sin y + e^t$.
Third Method: From (i) and (ii)

$$\phi(t,y) = e^t + 3ty + h(y) \quad \text{and} \quad \phi(t,y) = 3ty + \sin y + k(t).$$

Comparing these two expressions for the *same* function $\phi(t,y)$ it is obvious that $h(y) = \sin y$ and $k(t) = e^t$. Hence

$$\phi(t,y) = e^t + 3ty + \sin y.$$

Example 4. Find the solution of the initial-value problem

$$3t^2y + 8ty^2 + (t^3 + 8t^2y + 12y^2)\frac{dy}{dt} = 0, \qquad y(2) = 1.$$

Solution. Here $M(t,y) = 3t^2y + 8ty^2$ and $N(t,y) = t^3 + 8t^2y + 12y^2$. This equation is exact since

$$\frac{\partial M}{\partial y} = 3t^2 + 16ty \quad \text{and} \quad \frac{\partial N}{\partial t} = 3t^2 + 16ty.$$

Hence, there exists a function $\phi(t,y)$ such that

$$\text{(i) } 3t^2y + 8ty^2 = \frac{\partial \phi}{\partial t} \quad \text{and} \quad \text{(ii) } t^3 + 8t^2y + 12y^2 = \frac{\partial \phi}{\partial y}.$$

Again, we will find $\phi(t,y)$ by each of three methods.

First Method: From (i), $\phi(t,y) = t^3y + 4t^2y^2 + h(y)$. Differentiating this equation with respect to y and using (ii) we obtain that

$$t^3 + 8t^2y + h'(y) = t^3 + 8t^2y + 12y^2.$$

Hence, $h(y) = 4y^3$ and the general solution of the differential equation is $\phi(t,y) = t^3y + 4t^2y^2 + 4y^3 = c$. Setting $t = 2$ and $y = 1$ in this equation, we see that $c = 28$. Thus, the solution of our initial-value problem is defined implicitly by the equation $t^3y + 4t^2y^2 + 4y^3 = 28$.
Second Method: From (ii), $\phi(t,y) = t^3y + 4t^2y^2 + 4y^3 + k(t)$. Differentiating this expression with respect to t and using (i) we obtain that

$$3t^2y + 8ty^2 + k'(t) = 3t^2y + 8ty^2.$$

Thus $k(t) = 0$ and $\phi(t,y) = t^3y + 4t^2y^2 + 4y^3$.
Third Method: From (i) and (ii)

$$\phi(t,y) = t^3y + 4t^2y^2 + h(y) \quad \text{and} \quad \phi(t,y) = t^3y + 4t^2y^2 + 4y^3 + k(t).$$

Comparing these two expressions for the same function $\phi(t,y)$ we see that $h(y) = 4y^3$ and $k(t) = 0$. Hence, $\phi(t,y) = t^3y + 4t^2y^2 + 4y^3$.

In most instances, as Examples 3 and 4 illustrate, the third method is the simplest to use. However, if it is much easier to integrate N with respect to y than it is to integrate M with respect to t, we should use the second method, and vice-versa.

Example 5. Find the solution of the initial-value problem

$$4t^3e^{t+y}+t^4e^{t+y}+2t+(t^4e^{t+y}+2y)\frac{dy}{dt}=0, \qquad y(0)=1.$$

Solution. This equation is exact since

$$\frac{\partial}{\partial y}(4t^3e^{t+y}+t^4e^{t+y}+2t)=(t^4+4t^3)e^{t+y}=\frac{\partial}{\partial t}(t^4e^{t+y}+2y).$$

Hence, there exists a function $\phi(t,y)$ such that

$$\text{(i)}\ 4t^3e^{t+y}+t^4e^{t+y}+2t=\frac{\partial\phi}{\partial t}$$

and

$$\text{(ii)}\ t^4e^{t+y}+2y=\frac{\partial\phi}{\partial y}.$$

Since it is much simpler to integrate $t^4e^{t+y}+2y$ with respect to y than it is to integrate $4t^3e^{t+y}+t^4e^{t+y}+2t$ with respect to t, we use the second method. From (ii), $\phi(t,y)=t^4e^{t+y}+y^2+k(t)$. Differentiating this expression with respect to t and using (i) we obtain

$$(t^4+4t^3)e^{t+y}+k'(t)=4t^3e^{t+y}+t^4e^{t+y}+2t.$$

Thus, $k(t)=t^2$ and the general solution of the differential equation is $\phi(t,y)=t^4e^{t+y}+y^2+t^2=c$. Setting $t=0$ and $y=1$ in this equation yields $c=1$. Thus, the solution of our initial-value problem is defined implicitly by the equation $t^4e^{t+y}+t^2+y^2=1$.

Suppose now that we are given a differential equation

$$M(t,y)+N(t,y)\frac{dy}{dt}=0 \tag{7}$$

which is not exact. Can we make it exact? More precisely, can we find a function $\mu(t,y)$ such that the equivalent differential equation

$$\mu(t,y)M(t,y)+\mu(t,y)N(t,y)\frac{dy}{dt}=0 \tag{8}$$

is exact? This question is simple, in principle, to answer. The condition that (8) be exact is that

$$\frac{\partial}{\partial y}(\mu(t,y)M(t,y))=\frac{\partial}{\partial t}(\mu(t,y)N(t,y))$$

or

$$M\frac{\partial\mu}{\partial y}+\mu\frac{\partial M}{\partial y}=N\frac{\partial\mu}{\partial t}+\mu\frac{\partial N}{\partial t}. \tag{9}$$

(For simplicity of writing, we have suppressed the dependence of μ, M and N on t and y in (9).) Thus, Equation (8) is exact if and only if $\mu(t,y)$ satisfies Equation (9).

Definition. A function $\mu(t,y)$ satisfying Equation (9) is called an *integrating factor* for the differential equation (7).

The reason for this definition, of course, is that if μ satisfies (9) then we can write (8) in the form $(d/dt)\phi(t,y)=0$ and this equation can be integrated immediately to yield the solution $\phi(t,y)=c$. Unfortunately, though, there are only two special cases where we can find an explicit solution of (9). These occur when the differential equation (7) has an integrating factor which is either a function of t alone, or a function of y alone. Observe that if μ is a function of t alone, then Equation (9) reduces to

$$N\frac{d\mu}{dt}=\mu\left(\frac{\partial M}{\partial y}-\frac{\partial N}{\partial t}\right) \quad \text{or} \quad \frac{d\mu}{dt}=\frac{\left(\frac{\partial M}{\partial y}-\frac{\partial N}{\partial t}\right)}{N}\mu.$$

But this equation is meaningless unless the expression

$$\frac{\frac{\partial M}{\partial y}-\frac{\partial N}{\partial t}}{N}$$

is a function of t alone, that is,

$$\frac{\frac{\partial M}{\partial y}-\frac{\partial N}{\partial t}}{N}=R(t).$$

If this is the case then $\mu(t)=\exp\left(\int R(t)\,dt\right)$ is an integrating factor for the differential equation (7).

Remark. It should be noted that the expression

$$\frac{\frac{\partial M}{\partial y}-\frac{\partial N}{\partial t}}{N}$$

is almost always a function of both t and y. Only for very special pairs of functions $M(t,y)$ and $N(t,y)$ is it a function of t alone. A similar situation occurs if μ is a function of y alone (see Exercise 17). It is for this reason that we cannot solve very many differential equations.

Example 6. Find the general solution of the differential equation

$$\frac{y^2}{2}+2ye^t+(y+e^t)\frac{dy}{dt}=0.$$

Solution. Here $M(t,y)=(y^2/2)+2ye^t$ and $N(t,y)=y+e^t$. This equation is not exact since $\partial M/\partial y = y+2e^t$ and $\partial N/\partial t = e^t$. However,

$$\frac{1}{N}\left(\frac{\partial M}{\partial y}-\frac{\partial N}{\partial t}\right)=\frac{y+e^t}{y+e^t}=1.$$

Hence, this equation has $\mu(t)=\exp\left(\int 1\,dt\right)=e^t$ as an integrating factor. This means, of course, that the equivalent differential equation

$$\frac{y^2}{2}e^t+2ye^{2t}+(ye^t+e^{2t})\frac{dy}{dt}=0$$

is exact. Therefore, there exists a function $\phi(t,y)$ such that

$$\text{(i)}\ \frac{y^2}{2}e^t+2ye^{2t}=\frac{\partial\phi}{\partial t}$$

and

$$\text{(ii)}\ ye^t+e^{2t}=\frac{\partial\phi}{\partial y}.$$

From Equations (i) and (ii),

$$\phi(t,y)=\frac{y^2}{2}e^t+ye^{2t}+h(y)$$

and

$$\phi(t,y)=\frac{y^2}{2}e^t+ye^{2t}+k(t).$$

Thus, $h(y)=0$, $k(t)=0$ and the general solution of the differential equation is

$$\phi(t,y)=\frac{y^2}{2}e^t+ye^{2t}=c.$$

Solving this equation for y as a function of t we see that

$$y(t)=-e^t\pm\left[e^{2t}+2ce^{-t}\right]^{1/2}.$$

Example 7. Use the methods of this section to find the general solution of the linear equation $(dy/dt)+a(t)y=b(t)$.

Solution. We write this equation in the form $M(t,y)+N(t,y)(dy/dt)=0$ with $M(t,y)=a(t)y-b(t)$ and $N(t,y)=1$. This equation is not exact since $\partial M/\partial y=a(t)$ and $\partial N/\partial t=0$. However, $((\partial M/\partial y)-(\partial N/\partial t))/N=a(t)$. Hence, $\mu(t)=\exp\left(\int a(t)\,dt\right)$ is an integrating factor for the first-order linear equation. Therefore, there exists a function $\phi(t,y)$ such that

$$\text{(i)}\ \mu(t)\left[a(t)y-b(t)\right]=\frac{\partial\phi}{\partial t}$$

and

$$\text{(ii)}\ \mu(t)=\frac{\partial\phi}{\partial y}.$$

Now, observe from (ii) that $\phi(t,y)=\mu(t)y+k(t)$. Differentiating this equation with respect to t and using (i) we see that

$$\mu'(t)y+k'(t)=\mu(t)a(t)y-\mu(t)b(t).$$

But, $\mu'(t)=a(t)\mu(t)$. Consequently, $k'(t)=-\mu(t)b(t)$ and

$$\phi(t,y)=\mu(t)y-\int\mu(t)b(t)\,dt.$$

Hence, the general solution of the first-order linear equation is

$$\mu(t)y-\int\mu(t)b(t)\,dt=c,$$

and this is the result we obtained in Section 1.2.

Exercises

1. Use the theorem of equality of mixed partial derivatives to show that $\partial M/\partial y=\partial N/\partial t$ if the equation $M(t,y)+N(t,y)(dy/dt)=0$ is exact.

2. Show that the expression $M(t,y)-\int(\partial N(t,y)/\partial t)\,dy$ is a function of t alone if $\partial M/\partial y=\partial N/\partial t$.

In each of Problems 3–6 find the general solution of the given differential equation.

3. $2t\sin y+y^3e^t+(t^2\cos y+3y^2e^t)\dfrac{dy}{dt}=0$

4. $1+(1+ty)e^{ty}+(1+t^2e^{ty})\dfrac{dy}{dt}=0$

5. $y\sec^2t+\sec t\tan t+(2y+\tan t)\dfrac{dy}{dt}=0$

6. $\dfrac{y^2}{2}-2ye^t+(y-e^t)\dfrac{dy}{dt}=0$

In each of Problems 7–11, solve the given initial-value problem.

7. $2ty^3+3t^2y^2\dfrac{dy}{dt}=0,\qquad y(1)=1$

8. $2t\cos y+3t^2y+(t^3-t^2\sin y-y)\dfrac{dy}{dt}=0,\qquad y(0)=2$

9. $3t^2+4ty+(2y+2t^2)\dfrac{dy}{dt}=0,\qquad y(0)=1$

10. $y(\cos 2t)e^{ty}-2(\sin 2t)e^{ty}+2t+(t(\cos 2t)e^{ty}-3)\dfrac{dy}{dt}=0,\qquad y(0)=0$

11. $3ty+y^2+(t^2+ty)\dfrac{dy}{dt}=0, \qquad y(2)=1$

In each of Problems 12–14, determine the constant a so that the equation is exact, and then solve the resulting equation.

12. $t+ye^{2ty}+ate^{2ty}\dfrac{dy}{dt}=0$

13. $\dfrac{1}{t^2}+\dfrac{1}{y^2}+\dfrac{(at+1)}{y^3}\dfrac{dy}{dt}=0$

14. $e^{at+y}+3t^2y^2+(2yt^3+e^{at+y})\dfrac{dy}{dt}=0$

15. Show that every separable equation of the form $M(t)+N(y)dy/dt=0$ is exact.

16. Find all functions $f(t)$ such that the differential equation

$$y^2\sin t+yf(t)(dy/dt)=0$$

is exact. Solve the differential equation for these $f(t)$.

17. Show that if $((\partial N/\partial t)-(\partial M/\partial y))/M=Q(y)$, then the differential equation $M(t,y)+N(t,y)dy/dt=0$ has an integrating factor $\mu(y)=\exp\left(\int Q(y)\,dy\right)$.

18. The differential equation $f(t)(dy/dt)+t^2+y=0$ is known to have an integrating factor $\mu(t)=t$. Find all possible functions $f(t)$.

19. The differential equation $e^t\sec y-\tan y+(dy/dt)=0$ has an integrating factor of the form $e^{-at}\cos y$ for some constant a. Find a, and then solve the differential equation.

20. The Bernoulli differential equation is $(dy/dt)+a(t)y=b(t)y^n$. Multiplying through by $\mu(t)=\exp\left(\int a(t)\,dt\right)$, we can rewrite this equation in the form $d/dt(\mu(t)y)=b(t)\mu(t)y^n$. Find the general solution of this equation by finding an appropriate integrating factor. *Hint*: Divide both sides of the equation by an appropriate function of y.

1.10 The existence–uniqueness theorem; Picard iteration

Consider the initial-value problem

$$\frac{dy}{dt}=f(t,y), \qquad y(t_0)=y_0 \tag{1}$$

where f is a given function of t and y. Chances are, as the remarks in Section 1.9 indicate, that we will be unable to solve (1) explicitly. This leads us to ask the following questions.

1. How are we to know that the initial-value problem (1) actually has a solution if we can't exhibit it?

2. How do we know that there is only one solution $y(t)$ of (1)? Perhaps there are two, three, or even infinitely many solutions.
3. Why bother asking the first two questions? After all, what's the use of determining whether (1) has a unique solution if we won't be able to explicitly exhibit it?

The answer to the third question lies in the observation that it is never necessary, in applications, to find the solution $y(t)$ of (1) to more than a finite number of decimal places. Usually, it is more than sufficient to find $y(t)$ to four decimal places. As we shall see in Sections 1.13–17, this can be done quite easily with the aid of a digital computer. In fact, we will be able to compute $y(t)$ to eight, and even sixteen, decimal places. Thus, the knowledge that (1) has a unique solution $y(t)$ is our hunting license to go looking for it.

To resolve the first question, we must establish the existence of a function $y(t)$ whose value at $t=t_0$ is y_0, and whose derivative at any time t equals $f(t,y(t))$. In order to accomplish this, we must find a theorem which enables us to establish the existence of a function having certain properties, without our having to exhibit this function explicitly. If we search through the Calculus, we find that we encounter such a situation exactly once, and this is in connection with the theory of limits. As we show in Appendix B, it is often possible to prove that a sequence of functions $y_n(t)$ has a limit $y(t)$, without our having to exhibit $y(t)$. For example, we can prove that the sequence of functions

$$y_n(t)=\frac{\sin \pi t}{1^2}+\frac{\sin 2\pi t}{2^2}+\dots+\frac{\sin n\pi t}{n^2}$$

has a limit $y(t)$ even though we cannot exhibit $y(t)$ explicitly. This suggests the following algorithm for proving the existence of a solution $y(t)$ of (1).

(a) Construct a sequence of functions $y_n(t)$ which come closer and closer to solving (1).

(b) Show that the sequence of functions $y_n(t)$ has a limit $y(t)$ on a suitable interval $t_0 \leqslant t \leqslant t_0+\alpha$.

(c) Prove that $y(t)$ is a solution of (1) on this interval.

We now show how to implement this algorithm.

(a) *Construction of the approximating sequence $y_n(t)$*

The problem of finding a sequence of functions that come closer and closer to satisfying a certain equation is one that arises quite often in mathematics. Experience has shown that it is often easiest to resolve this problem when our equation can be written in the special form

$$y(t)=L(t,y(t)), \tag{2}$$

where L may depend explicitly on y, and on integrals of functions of y.

For example, we may wish to find a function $y(t)$ satisfying

$$y(t)=1+\sin[t+y(t)],$$

or

$$y(t)=1+y^2(t)+\int_0^t y^3(s)\,ds.$$

In these two cases, $L(t,y(t))$ is an abbreviation for

$$1+\sin[t+y(t)]$$

and

$$1+y^2(t)+\int_0^t y^3(s)\,ds,$$

respectively.

The key to understanding what is special about Equation (2) is to view $L(t,y(t))$ as a "machine" that takes in one function and gives back another one. For example, let

$$L(t,y(t))=1+y^2(t)+\int_0^t y^3(s)\,ds.$$

If we plug the function $y(t)=t$ into this machine, (that is, if we compute $1+t^2+\int_0^t s^3\,ds$) then the machine returns to us the function $1+t^2+t^4/4$. If we plug the function $y(t)=\cos t$ into this machine, then it returns to us the function

$$1+\cos^2 t+\int_0^t \cos^3 s\,ds=1+\cos^2 t+\sin t-\frac{\sin^3 t}{3}.$$

According to this viewpoint, we can characterize all solutions $y(t)$ of (2) as those functions $y(t)$ which the machine L leaves unchanged. In other words, if we plug a function $y(t)$ into the machine L, and the machine returns to us this same function, then $y(t)$ is a solution of (2).

We can put the initial-value problem (1) into the special form (2) by integrating both sides of the differential equation $y'=f(t,y)$ with respect to t. Specifically, if $y(t)$ satisfies (1), then

$$\int_{t_0}^t \frac{dy(s)}{ds}\,ds=\int_{t_0}^t f(s,y(s))\,ds$$

so that

$$y(t)=y_0+\int_{t_0}^t f(s,y(s))\,ds. \tag{3}$$

Conversely, if $y(t)$ is continuous and satisfies (3), then $dy/dt=f(t,y(t))$. Moreover, $y(t_0)$ is obviously y_0. Therefore, $y(t)$ is a solution of (1) if, and only if, it is a continuous solution of (3).

Equation (3) is called an integral equation, and it is in the special form (2) if we set

$$L(t,y(t)) = y_0 + \int_{t_0}^{t} f(s,y(s))\,ds.$$

This suggests the following scheme for constructing a sequence of "approximate solutions" $y_n(t)$ of (3). Let us start by guessing a solution $y_0(t)$ of (3). The simplest possible guess is $y_0(t) = y_0$. To check whether $y_0(t)$ is a solution of (3), we compute

$$y_1(t) = y_0 + \int_{t_0}^{t} f(s,y_0(s))\,ds.$$

If $y_1(t) = y_0$, then $y(t) = y_0$ is indeed a solution of (3). If not, then we try $y_1(t)$ as our next guess. To check whether $y_1(t)$ is a solution of (3), we compute

$$y_2(t) = y_0 + \int_{t_0}^{t} f(s,y_1(s))\,ds,$$

and so on. In this manner, we define a sequence of functions $y_1(t)$, $y_2(t), \ldots$, where

$$y_{n+1}(t) = y_0 + \int_{t_0}^{t} f(s,y_n(s))\,ds. \tag{4}$$

These functions $y_n(t)$ are called successive approximations, or Picard iterates, after the French mathematician Picard who first discovered them. Remarkably, these Picard iterates always converge, on a suitable interval, to a solution $y(t)$ of (3).

Example 1. Compute the Picard iterates for the initial-value problem

$$y' = y, \qquad y(0) = 1,$$

and show that they converge to the solution $y(t) = e^t$.

Solution. The integral equation corresponding to this initial-value problem is

$$y(t) = 1 + \int_0^t y(s)\,ds.$$

Hence, $y_0(t) = 1$

$$y_1(t) = 1 + \int_0^t 1\,ds = 1 + t$$

$$y_2(t) = 1 + \int_0^t y_1(s)\,ds = 1 + \int_0^t (1+s)\,ds = 1 + t + \frac{t^2}{2!}$$

and, in general,

$$y_n(t)=1+\int_0^t y_{n-1}(s)\,ds=1+\int_0^t\left[1+s+\ldots+\frac{s^{n-1}}{(n-1)!}\right]ds$$

$$=1+t+\frac{t^2}{2!}+\ldots+\frac{t^n}{n!}.$$

Since $e^t=1+t+t^2/2!+\ldots$, we see that the Picard iterates $y_n(t)$ converge to the solution $y(t)$ of this initial-value problem.

Example 2. Compute the Picard iterates $y_1(t), y_2(t)$ for the initial-value problem $y'=1+y^3$, $y(1)=1$.

Solution. The integral equation corresponding to this initial-value problem is

$$y(t)=1+\int_1^t\left[1+y^3(s)\right]ds.$$

Hence, $y_0(t)=1$

$$y_1(t)=1+\int_1^t(1+1)\,ds=1+2(t-1)$$

and

$$y_2(t)=1+\int_1^t\left\{1+\left[1+2(s-1)\right]^3\right\}ds$$

$$=1+2(t-1)+3(t-1)^2+4(t-1)^3+2(t-1)^4.$$

Notice that it is already quite cumbersome to compute $y_3(t)$.

(b) *Convergence of the Picard iterates*

As was mentioned in Section 1.4, the solutions of nonlinear differential equations may not exist for all time t. Therefore, we cannot expect the Picard iterates $y_n(t)$ of (3) to converge for all t. To provide us with a clue, or estimate, of where the Picard iterates converge, we try to find an interval in which all the $y_n(t)$ are uniformly bounded (that is, $|y_n(t)|\leqslant K$ for some fixed constant K). Equivalently, we seek a rectangle R which contains the graphs of all the Picard iterates $y_n(t)$. Lemma 1 shows us how to find such a rectangle.

Lemma 1. *Choose any two positive numbers a and b, and let R be the rectangle: $t_0\leqslant t\leqslant t_0+a, |y-y_0|\leqslant b$. Compute*

$$M=\max_{(t,y)\text{ in }R}|f(t,y)|,\quad \textit{and set}\quad \alpha=\min\left(a,\frac{b}{M}\right).$$

Then,

$$|y_n(t)-y_0|\leqslant M(t-t_0) \tag{5}$$

for $t_0\leqslant t\leqslant t_0+\alpha$.

Lemma 1 states that the graph of $y_n(t)$ is sandwiched between the lines $y=y_0+M(t-t_0)$ and $y=y_0-M(t-t_0)$, for $t_0 \leqslant t \leqslant t_0+\alpha$. These lines leave the rectangle R at $t=t_0+a$ if $a \leqslant b/M$, and at $t=t_0+b/M$ if $b/M < a$ (see Figures 1a and 1b). In either case, therefore, the graph of $y_n(t)$ is contained in R for $t_0 \leqslant t \leqslant t_0+\alpha$.

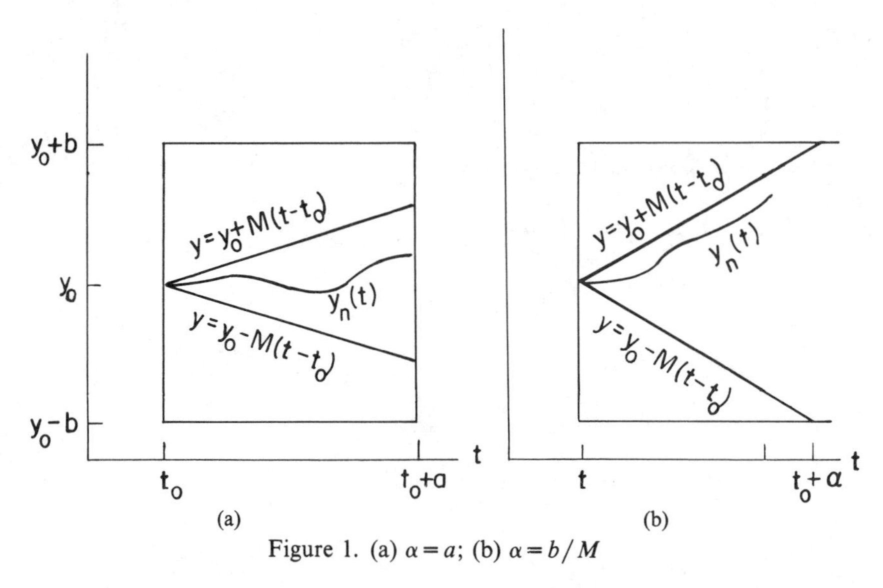

Figure 1. (a) $\alpha=a$; (b) $\alpha=b/M$

PROOF OF LEMMA 1. We establish (5) by induction on n. Observe first that (5) is obviously true for $n=0$, since $y_0(t)=y_0$. Next, we must show that (5) is true for $n=j+1$ if it is true for $n=j$. But this follows immediately, for if $|y_j(t)-y_0| \leqslant M(t-t_0)$, then

$$
\begin{aligned}
|y_{j+1}(t)-y_0| &= \left|\int_{t_0}^{t} f(s,y_j(s))\,ds\right| \\
&\leqslant \int_{t_0}^{t} |f(s,y_j(s))|\,ds \leqslant M(t-t_0)
\end{aligned}
$$

for $t_0 \leqslant t \leqslant t_0+\alpha$. Consequently, (5) is true for all n, by induction. □

We now show that the Picard iterates $y_n(t)$ of (3) converge for each t in the interval $t_0 \leqslant t \leqslant t_0+\alpha$, if $\partial f/\partial y$ exists and is continuous. Our first step is to reduce the problem of showing that the sequence of functions $y_n(t)$ converges to the much simpler problem of proving that an infinite series converges. This is accomplished by writing $y_n(t)$ in the form

$$y_n(t)=y_0(t)+\left[y_1(t)-y_0(t)\right]+\ldots+\left[y_n(t)-y_{n-1}(t)\right].$$

Clearly, the sequence $y_n(t)$ converges if, and only if, the infinite series

$$\left[y_1(t)-y_0(t)\right]+\left[y_2(t)-y_1(t)\right]+\ldots+\left[y_n(t)-y_{n-1}(t)\right]+\ldots \tag{6}$$

converges. To prove that the infinite series (6) converges, it suffices to show that

$$\sum_{n=1}^{\infty} |y_n(t)-y_{n-1}(t)|<\infty. \tag{7}$$

This is accomplished in the following manner. Observe that

$$\begin{aligned}
|y_n(t)-y_{n-1}(t)| &= \left|\int_{t_0}^{t}\left[f(s,y_{n-1}(s))-f(s,y_{n-2}(s))\right]ds\right| \\
&\leqslant \int_{t_0}^{t}|f(s,y_{n-1}(s))-f(s,y_{n-2}(s))|\,ds \\
&= \int_{t_0}^{t}\left|\frac{\partial f(s,\xi(s))}{\partial y}\right||y_{n-1}(s)-y_{n-2}(s)|\,ds,
\end{aligned}$$

where $\xi(s)$ lies between $y_{n-1}(s)$ and $y_{n-2}(s)$. (Recall that $f(x_1)-f(x_2)=f'(\xi)(x_1-x_2)$, where ξ is some number between x_1 and x_2.) It follows immediately from Lemma 1 that the points $(s,\xi(s))$ all lie in the rectangle R for $s<t_0+\alpha$. Consequently,

$$|y_n(t)-y_{n-1}(t)|\leqslant L\int_{t_0}^{t}|y_{n-1}(s)-y_{n-2}(s)|\,ds, \qquad t_0\leqslant t\leqslant t_0+\alpha, \tag{8}$$

where

$$L=\max_{(t,y)\text{ in }R}\left|\frac{\partial f(t,y)}{\partial y}\right|. \tag{9}$$

Equation (9) defines the constant L. Setting $n=2$ in (8) gives

$$\begin{aligned}
|y_2(t)-y_1(t)| &\leqslant L\int_{t_0}^{t}|y_1(s)-y_0|\,ds\leqslant L\int_{t_0}^{t}M(s-t_0)\,ds \\
&= \frac{LM(t-t_0)^2}{2}.
\end{aligned}$$

This, in turn, implies that

$$\begin{aligned}
|y_3(t)-y_2(t)| &\leqslant L\int_{t_0}^{t}|y_2(s)-y_1(s)|\,ds\leqslant ML^2\int_{t_0}^{t}\frac{(s-t_0)^2}{2}\,ds \\
&= \frac{ML^2(t-t_0)^3}{3!}.
\end{aligned}$$

Proceeding inductively, we see that

$$|y_n(t)-y_{n-1}(t)|\leqslant \frac{ML^{n-1}(t-t_0)^n}{n!}, \quad \text{for } t_0\leqslant t\leqslant t_0+\alpha. \tag{10}$$

Therefore, for $t_0 \leqslant t \leqslant t_0 + \alpha$,

$$|y_1(t) - y_0(t)| + |y_2(t) - y_1(t)| + \dots$$

$$\leqslant M(t - t_0) + \frac{ML(t - t_0)^2}{2!} + \frac{ML^2(t - t_0)^3}{3!} + \dots$$

$$\leqslant M\alpha + \frac{ML\alpha^2}{2!} + \frac{ML^2\alpha^3}{3!} + \dots$$

$$= \frac{M}{L}\left[\alpha L + \frac{(\alpha L)^2}{2!} + \frac{(\alpha L)^3}{3!} + \dots\right]$$

$$= \frac{M}{L}(e^{\alpha L} - 1).$$

This quantity, obviously, is less than infinity. Consequently, the Picard iterates $y_n(t)$ converge for each t in the interval $t_0 \leqslant t \leqslant t_0 + \alpha$. (A similar argument shows that $y_n(t)$ converges for each t in the interval $t_0 - \beta \leqslant t \leqslant t_0$, where $\beta = \min(a, b/N)$, and N is the maximum value of $|f(t,y)|$ for (t,y) in the rectangle $t_0 - a \leqslant t \leqslant t_0, |y - y_0| \leqslant b$.) We will denote the limit of the sequence $y_n(t)$ by $y(t)$. □

(c) *Proof that $y(t)$ satisfies the initial-value problem* (1)

We will show that $y(t)$ satisfies the integral equation

$$y(t) = y_0 + \int_{t_0}^{t} f(s, y(s))\, ds \tag{11}$$

and that $y(t)$ is continuous. To this end, recall that the Picard iterates $y_n(t)$ are defined recursively through the equation

$$y_{n+1}(t) = y_0 + \int_{t_0}^{t} f(s, y_n(s))\, ds. \tag{12}$$

Taking limits of both sides of (12) gives

$$y(t) = y_0 + \lim_{n \to \infty} \int_{t_0}^{t} f(s, y_n(s))\, ds. \tag{13}$$

To show that the right-hand side of (13) equals

$$y_0 + \int_{t_0}^{t} f(s, y(s))\, ds,$$

(that is, to justify passing the limit through the integral sign) we must show that

$$\left| \int_{t_0}^{t} f(s, y(s))\, ds - \int_{t_0}^{t} f(s, y_n(s))\, ds \right|$$

approaches zero as n approaches infinity. This is accomplished in the following manner. Observe first that the graph of $y(t)$ lies in the rectangle R for $t \leqslant t_0 + \alpha$, since it is the limit of functions $y_n(t)$ whose graphs lie in R.

Hence

$$\left|\int_{t_0}^{t} f(s,y(s))\,ds - \int_{t_0}^{t} f(s,y_n(s))\,ds\right|$$

$$\leqslant \int_{t_0}^{t} |f(s,y(s)) - f(s,y_n(s))|\,ds \leqslant L\int_{t_0}^{t} |y(s)-y_n(s)|\,ds$$

where L is defined by Equation (9). Next, observe that

$$y(s)-y_n(s) = \sum_{j=n+1}^{\infty} \left[y_j(s)-y_{j-1}(s)\right]$$

since

$$y(s) = y_0 + \sum_{j=1}^{\infty} \left[y_j(s)-y_{j-1}(s)\right]$$

and

$$y_n(s) = y_0 + \sum_{j=1}^{n} \left[y_j(s)-y_{j-1}(s)\right].$$

Consequently, from (10),

$$|y(s)-y_n(s)| \leqslant M \sum_{j=n+1}^{\infty} L^{j-1}\frac{(s-t_0)^j}{j!}$$

$$\leqslant M \sum_{j=n+1}^{\infty} \frac{L^{j-1}\alpha^j}{j!} = \frac{M}{L} \sum_{j=n+1}^{\infty} \frac{(\alpha L)^j}{j!}, \tag{14}$$

and

$$\left|\int_{t_0}^{t} f(s,y(s))\,ds - \int_{t_0}^{t} f(s,y_n(s))\,ds\right| \leqslant M \sum_{j=n+1}^{\infty} \frac{(\alpha L)^j}{j!} \int_{t_0}^{t} ds$$

$$\leqslant M\alpha \sum_{j=n+1}^{\infty} \frac{(\alpha L)^j}{j!}.$$

This summation approaches zero as n approaches infinity, since it is the tail end of the convergent Taylor series expansion of $e^{\alpha L}$. Hence,

$$\lim_{n\to\infty} \int_{t_0}^{t} f(s,y_n(s))\,ds = \int_{t_0}^{t} f(s,y(s))\,ds,$$

and $y(t)$ satisfies (11).

To show that $y(t)$ is continuous, we must show that for every $\varepsilon > 0$ we can find $\delta > 0$ such that

$$|y(t+h)-y(t)| < \varepsilon \quad \text{if } |h| < \delta.$$

Now, we cannot compare $y(t+h)$ with $y(t)$ directly, since we do not know $y(t)$ explicitly. To overcome this difficulty, we choose a large integer N and observe that

$$y(t+h)-y(t)=\left[y(t+h)-y_N(t+h)\right] + \left[y_N(t+h)-y_N(t)\right]+\left[y_N(t)-y(t)\right].$$

Specifically, we choose N so large that

$$\frac{M}{L}\sum_{j=N+1}^{\infty}\frac{(\alpha L)^j}{j!}<\frac{\varepsilon}{3}.$$

Then, from (14),

$$|y(t+h)-y_N(t+h)|<\frac{\varepsilon}{3}\quad\text{and}\quad|y_N(t)-y(t)|<\frac{\varepsilon}{3},$$

for $t<t_0+\alpha$, and h sufficiently small (so that $t+h<t_0+\alpha$.) Next, observe that $y_N(t)$ is continuous, since it is obtained from N repeated integrations of continuous functions. Therefore, we can choose $\delta>0$ so small that

$$|y_N(t+h)-y_N(t)|<\frac{\varepsilon}{3}\quad\text{for } |h|<\delta.$$

Consequently,

$$|y(t+h)-y(t)|\leqslant|y(t+h)-y_N(t+h)|+|y_N(t+h)-y_N(t)| + |y_N(t)-y(t)|<\frac{\varepsilon}{3}+\frac{\varepsilon}{3}+\frac{\varepsilon}{3}=\varepsilon$$

for $|h|<\delta$. Therefore, $y(t)$ is a continuous solution of the integral equation (11), and this completes our proof that $y(t)$ satisfies (1). □

In summary, we have proven the following theorem.

Theorem 2. *Let f and $\partial f/\partial y$ be continuous in the rectangle $R: t_0\leqslant t\leqslant t_0+a$, $|y-y_0|\leqslant b$. Compute*

$$M=\max_{(t,y)\text{ in }R}|f(t,y)|,\quad\textit{and set}\quad \alpha=\min\left(a,\frac{b}{M}\right).$$

Then, the initial-value problem $y'=f(t,y)$, $y(t_0)=y_0$ has at least one solution $y(t)$ on the interval $t_0\leqslant t\leqslant t_0+\alpha$. A similar result is true for $t<t_0$.

Remark. The number α in Theorem 2 depends specifically on our choice of a and b. Different choices of a and b lead to different values of α. Moreover, α doesn't necessarily increase when a and b increase, since an increase in a or b will generally result in an increase in M.

Finally, we turn our attention to the problem of uniqueness of solutions of (1). Consider the initial-value problem

$$\frac{dy}{dt}=(\sin 2t)y^{1/3},\qquad y(0)=0. \tag{15}$$

One solution of (15) is $y(t)=0$. Additional solutions can be obtained if we ignore the fact that $y(0)=0$ and rewrite the differential equation in the form

$$\frac{1}{y^{1/3}}\frac{dy}{dt}=\sin 2t,$$

or

$$\frac{d}{dt}\frac{3y^{2/3}}{2}=\sin 2t.$$

Then,

$$\frac{3y^{2/3}}{2}=\frac{1-\cos 2t}{2}=\sin^2 t$$

and $y=\pm\sqrt{8/27}\,\sin^3 t$ are two additional solutions of (15).

Now, initial-value problems that have more than one solution are clearly unacceptable in applications. Therefore, it is important for us to find out exactly what is "wrong" with the initial-value problem (15) that it has more than one solution. If we look carefully at the right-hand side of this differential equation, we see that it does not have a partial derivative with respect to y at $y=0$. This is indeed the problem, as the following theorem shows.

Theorem 2′. *Let f and $\partial f/\partial y$ be continuous in the rectangle $R: t_0 \leqslant t \leqslant t_0+a$, $|y-y_0|\leqslant b$. Compute*

$$M=\max_{(t,y)\text{ in }R}|f(t,y)|, \quad \textit{and set} \quad \alpha=\min\left(a,\frac{b}{M}\right).$$

Then, the initial-value problem

$$y'=f(t,y), \qquad y(t_0)=y_0 \tag{16}$$

has a unique solution $y(t)$ on the interval $t_0\leqslant t\leqslant t_0+\alpha$. In other words, if $y(t)$ and $z(t)$ are two solutions of (16), then $y(t)$ must equal $z(t)$ for $t_0\leqslant t \leqslant t_0+\alpha$.

Proof. Theorem 2 guarantees the existence of at least one solution $y(t)$ of (16). Suppose that $z(t)$ is a second solution of (16). Then,

$$y(t)=y_0+\int_{t_0}^{t} f(s,y(s))\,ds \quad\text{and}\quad z(t)=y_0+\int_{t_0}^{t} f(s,z(s))\,ds.$$

Subtracting these two equations gives

$$\begin{aligned}|y(t)-z(t)|&=\left|\int_{t_0}^{t}\left[f(s,y(s))-f(s,z(s))\right]ds\right|\\ &\leqslant\int_{t_0}^{t}|f(s,y(s))-f(s,z(s))|\,ds\\ &\leqslant L\int_{t_0}^{t}|y(s)-z(s)|\,ds\end{aligned}$$

where L is the maximum value of $|\partial f/\partial y|$ for (t,y) in R. As Lemma 2 below shows, this inequality implies that $y(t)=z(t)$. Hence, the initial-value problem (16) has a unique solution $y(t)$. □

Lemma 2. *Let $w(t)$ be a nonnegative function, with*

$$w(t) \leqslant L\int_{t_0}^{t} w(s)\,ds. \tag{17}$$

Then, $w(t)$ is identically zero.

FAKE PROOF. Differentiating both sides of (17) gives

$$\frac{dw}{dt} \leqslant Lw(t), \quad \text{or} \quad \frac{dw}{dt} - Lw(t) \leqslant 0.$$

Multiplying both sides of this inequality by the integrating factor $e^{-L(t-t_0)}$ gives

$$\frac{d}{dt}e^{-L(t-t_0)}w(t) \leqslant 0, \quad \text{so that} \quad e^{-L(t-t_0)}w(t) \leqslant w(t_0)$$

for $t \geqslant t_0$. But $w(t_0)$ must be zero if $w(t)$ is nonnegative and satisfies (17). Consequently, $e^{-L(t-t_0)}w(t) \leqslant 0$, and this implies that $w(t)$ is identically zero.

The error in this proof, of course, is that we cannot differentiate both sides of an inequality, and still expect to preserve the inequality. For example, the function $f_1(t)=2t-2$ is less than $f_2(t)=t$ on the interval $[0,1]$, but $f_1'(t)$ is greater than $f_2'(t)$ on this interval. We make this proof "kosher" by the clever trick of setting

$$U(t) = \int_{t_0}^{t} w(s)\,ds.$$

Then,

$$\frac{dU}{dt} = w(t) \leqslant L\int_{t_0}^{t} w(s)\,ds = LU(t).$$

Consequently, $e^{-L(t-t_0)}U(t) \leqslant U(t_0)=0$, for $t \geqslant t_0$, and thus $U(t)=0$. This, in turn, implies that $w(t)=0$ since

$$0 \leqslant w(t) \leqslant L\int_{t_0}^{t} w(s)\,ds = LU(t) = 0.$$ □

Example 3. Show that the solution $y(t)$ of the initial-value problem

$$\frac{dy}{dt} = t^2 + e^{-y^2}, \qquad y(0)=0$$

exists for $0 \leqslant t \leqslant \frac{1}{2}$, and in this interval, $|y(t)| \leqslant 1$.

Solution. Let R be the rectangle $0 \leqslant t \leqslant \frac{1}{2}$, $|y| \leqslant 1$. Computing

$$M = \max_{(t,y) \text{ in } R} t^2 + e^{-y^2} = 1 + \left(\tfrac{1}{2}\right)^2 = \tfrac{5}{4},$$

we see that $y(t)$ exists for

$$0 \leqslant t \leqslant \min\left(\frac{1}{2}, \frac{1}{5/4}\right) = \frac{1}{2},$$

and in this interval, $|y(t)| \leqslant 1$.

Example 4. Show that the solution $y(t)$ of the initial-value problem

$$\frac{dy}{dt} = e^{-t^2} + y^3, \qquad y(0) = 1$$

exists for $0 \leqslant t \leqslant 1/9$, and in this interval, $0 \leqslant y \leqslant 2$.
Solution. Let R be the rectangle $0 \leqslant t \leqslant \frac{1}{9}$, $0 \leqslant y \leqslant 2$. Computing

$$M = \max_{(t,y) \text{ in } R} e^{-t^2} + y^3 = 1 + 2^3 = 9,$$

we see that $y(t)$ exists for

$$0 \leqslant t \leqslant \min\left(\tfrac{1}{9}, \tfrac{1}{9}\right)$$

and in this interval, $0 \leqslant y \leqslant 2$.

Example 5. What is the largest interval of existence that Theorem 2 predicts for the solution $y(t)$ of the initial-value problem $y' = 1 + y^2$, $y(0) = 0$?
Solution. Let R be the rectangle $0 \leqslant t \leqslant a$, $|y| \leqslant b$. Computing

$$M = \max_{(t,y) \text{ in } R} 1 + y^2 = 1 + b^2,$$

we see that $y(t)$ exists for

$$0 \leqslant t \leqslant \alpha = \min\left(a, \frac{b}{1+b^2}\right).$$

Clearly, the largest α that we can achieve is the maximum value of the function $b/(1+b^2)$. This maximum value is $\frac{1}{2}$. Hence, Theorem 2 predicts that $y(t)$ exists for $0 \leqslant t \leqslant \frac{1}{2}$. The fact that $y(t) = \tan t$ exists for $0 \leqslant t \leqslant \pi/2$ points out the limitation of Theorem 2.

Example 6. Suppose that $|f(t,y)| \leqslant K$ in the strip $t_0 \leqslant t < \infty$, $-\infty < y < \infty$. Show that the solution $y(t)$ of the initial-value problem $y' = f(t,y)$, $y(t_0) = y_0$ exists for all $t \geqslant t_0$.
Solution. Let R be the rectangle $t_0 \leqslant t \leqslant t_0 + a$, $|y - y_0| \leqslant b$. The quantity

$$M = \max_{(t,y) \text{ in } R} |f(t,y)|$$

is at most K. Hence, $y(t)$ exists for

$$t_0 \leqslant t \leqslant t_0 + \min(a, b/K).$$

Now, we can make the quantity $\min(a,b/K)$ as large as desired by choosing a and b sufficiently large. Therefore $y(t)$ exists for $t \geqslant t_0$.

Exercises

1. Construct the Picard iterates for the initial-value problem $y' = 2t(y+1)$, $y(0) = 0$ and show that they converge to the solution $y(t) = e^{t^2} - 1$.

2. Compute the first two Picard iterates for the initial-value problem $y' = t^2 + y^2$, $y(0) = 1$.

3. Compute the first three Picard iterates for the initial-value problem $y' = e^t + y^2$, $y(0) = 0$.

In each of Problems 4–15, show that the solution $y(t)$ of the given initial-value problem exists on the specified interval.

4. $y' = y^2 + \cos t^2, \quad y(0) = 0; \quad 0 \leqslant t \leqslant \frac{1}{2}$

5. $y' = 1 + y + y^2 \cos t, \quad y(0) = 0; \quad 0 \leqslant t \leqslant \frac{1}{3}$

6. $y' = t + y^2, y(0) = 0; 0 \leqslant t \leqslant (\frac{1}{2})^{2/3}$

7. $y' = e^{-t^2} + y^2, \quad y(0) = 0; \quad 0 \leqslant t \leqslant \frac{1}{2}$

8. $y' = e^{-t^2} + y^2, y(1) = 0; 1 \leqslant t \leqslant 1 + \sqrt{e}/2$

9. $y' = e^{-t^2} + y^2, \quad y(0) = 1; \quad 0 \leqslant t \leqslant \dfrac{\sqrt{2}}{1 + (1 + \sqrt{2})^2}$

10. $y' = y + e^{-y} + e^{-t}, \quad y(0) = 0; \quad 0 \leqslant t \leqslant 1$

11. $y' = y^3 + e^{-5t}, \quad y(0) = 0.4; \quad 0 \leqslant t \leqslant \frac{3}{10}$

12. $y' = e^{(y-t)^2}, \quad y(0) = 1; \quad 0 \leqslant t \leqslant \dfrac{\sqrt{3} - 1}{2} e^{-((1+\sqrt{3})/2)^2}$

13. $y' = (4y + e^{-t^2})e^{2y}, \quad y(0) = 0; \quad 0 \leqslant t \leqslant \dfrac{1}{8\sqrt{e}}$

14. $y' = e^{-t} + \ln(1 + y^2), \quad y(0) = 0; \quad 0 \leqslant t < \infty$

15. $y' = \frac{1}{4}(1 + \cos 4t)y - \frac{1}{800}(1 - \cos 4t)y^2, \quad y(0) = 100; \quad 0 \leqslant t \leqslant 1$

16. Consider the initial-value problem

$$y' = t^2 + y^2, \qquad y(0) = 0, \qquad (*)$$

and let R be the rectangle $0 \leqslant t \leqslant a, \ -b \leqslant y \leqslant b$.
(a) Show that the solution $y(t)$ of (*) exists for

$$0 \leqslant t \leqslant \min\left(a, \frac{b}{a^2 + b^2}\right).$$

(b) Show that the maximum value of $b/(a^2+b^2)$, for a fixed, is $1/2a$.
(c) Show that $\alpha=\min(a,\frac{1}{2}a)$ is largest when $a=1/\sqrt{2}$.
(d) Conclude that the solution $y(t)$ of (*) exists for $0\leqslant t\leqslant 1/\sqrt{2}$.

17. Prove that $y(t)=-1$ is the only solution of the initial-value problem

$$y'=t(1+y), \qquad y(0)=-1.$$

18. Find a nontrivial solution of the initial-value problem $y'=ty^a$, $y(0)=0$, $a>1$. Does this violate Theorem 2′? Explain.

19. Find a solution of the initial-value problem $y'=t\sqrt{1-y^2}$, $y(0)=1$, other than $y(t)=1$. Does this violate Theorem 2′? Explain.

20. Here is an alternate proof of Lemma 2. Let $w(t)$ be a nonnegative function with

$$w(t)\leqslant L\int_{t_0}^{t} w(s)\,ds \tag{*}$$

on the interval $t_0\leqslant t\leqslant t_0+\alpha$. Since $w(t)$ is continuous, we can find a constant A such that $0\leqslant w(t)\leqslant A$ for $t_0\leqslant t\leqslant t_0+\alpha$.
(a) Show that $w(t)\leqslant LA(t-t_0)$.
(b) Use this estimate of $w(t)$ in (*) to obtain

$$w(t)\leqslant\frac{AL^2(t-t_0)^2}{2}.$$

(c) Proceeding inductively, show that $w(t)\leqslant AL^n(t-t_0)^n/n!$, for every integer n.
(d) Conclude that $w(t)=0$ for $t_0\leqslant t\leqslant t_0+\alpha$.

1.11 Finding roots of equations by iteration

Suppose that we are interested in finding the roots of an equation having the special form

$$x=f(x). \tag{1}$$

For example, we might want to find the roots of the equation

$$x=\sin x+\tfrac{1}{4}.$$

The methods introduced in the previous section suggest the following algorithm for solving this problem.

1. Try an initial guess x_0, and use this number to construct a sequence of guesses $x_1,x_2,x_3,\ldots$, where $x_1=f(x_0)$, $x_2=f(x_1)$, $x_3=f(x_2)$, and so on.
2. Show that this sequence of iterates x_n has a limit η as n approaches infinity.
3. Show that η is a root of (1); i.e., $\eta=f(\eta)$.

The following theorem tells us when this algorithm will work.

Theorem 3. *Let $f(x)$ and $f'(x)$ be continuous in the interval $a\leqslant x\leqslant b$, with $|f'(x)|\leqslant\lambda<1$ in this interval. Suppose, moreover, that the iterates x_n, de-*

fined recursively by the equation

$$x_{n+1}=f(x_n) \tag{2}$$

all lie in the interval $[a,b]$. *Then, the iterates* x_n *converge to a unique number* η *satisfying* (1).

PROOF. We convert the problem of proving that the sequence x_n converges to the simpler problem of proving that an infinite series converges by writing x_n in the form

$$x_n=x_0+(x_1-x_0)+(x_2-x_1)+\dots+(x_n-x_{n-1}).$$

Clearly, the sequence x_n converges if, and only if, the infinite series

$$(x_1-x_0)+(x_2-x_1)+\dots+(x_n-x_{n-1})+\dots=\sum_{n=1}^{\infty}(x_n-x_{n-1})$$

converges. To prove that this infinite series converges, it suffices to show that

$$|x_1-x_0|+|x_2-x_1|+\dots=\sum_{n=1}^{\infty}|x_n-x_{n-1}|<\infty.$$

This is accomplished in the following manner. By definition, $x_n=f(x_{n-1})$ and $x_{n-1}=f(x_{n-2})$. Subtracting these two equations gives

$$x_n-x_{n-1}=f(x_{n-1})-f(x_{n-2})=f'(\xi)(x_{n-1}-x_{n-2}),$$

where ξ is some number between x_{n-1} and x_{n-2}. In particular, ξ is in the interval $[a,b]$. Therefore, $|f'(\xi)|\leqslant\lambda$, and

$$|x_n-x_{n-1}|\leqslant\lambda|x_{n-1}-x_{n-2}|. \tag{3}$$

Iterating this inequality $n-1$ times gives

$$\begin{aligned}|x_n-x_{n-1}|&\leqslant\lambda|x_{n-1}-x_{n-2}|\\&\leqslant\lambda^2|x_{n-2}-x_{n-3}|\\&\;\;\vdots\\&\leqslant\lambda^{n-1}|x_1-x_0|.\end{aligned}$$

Consequently,

$$\begin{aligned}\sum_{n=1}^{\infty}|x_n-x_{n-1}|&\leqslant\sum_{n=1}^{\infty}\lambda^{n-1}|x_1-x_0|\\&=|x_1-x_0|\left[1+\lambda+\lambda^2+\dots\right]=\frac{|x_1-x_0|}{1-\lambda}.\end{aligned}$$

This quantity, obviously, is less than infinity. Therefore, the sequence of iterates x_n has a limit η as n approaches infinity. Taking limits of both sides of (2) gives

$$\eta = \lim_{n\to\infty} x_{n+1} = \lim_{n\to\infty} f(x_n) = f(\eta).$$

Hence, η is a root of (1).

Finally, suppose that η is not unique; that is, there exist two solutions η_1 and η_2 of (1) in the interval $[a,b]$. Then,

$$\eta_1 - \eta_2 = f(\eta_1) - f(\eta_2) = f'(\xi)(\eta_1 - \eta_2),$$

where ξ is some number between η_1 and η_2. This implies that $\eta_1 = \eta_2$ or $f'(\xi) = 1$. But $f'(\xi)$ cannot be one, since ξ is in the interval $[a,b]$. Therefore, $\eta_1 = \eta_2$. □

Example 1. Show that the sequence of iterates

$$x_0,\quad x_1 = 1 + \tfrac{1}{2}\arctan x_0,\quad x_2 = 1 + \tfrac{1}{2}\arctan x_1, \ldots$$

converge to a unique number η satisfying

$$\eta = 1 + \tfrac{1}{2}\arctan \eta$$

for every initial guess x_0.

Solution. Let $f(x) = 1 + \frac{1}{2}\arctan x$. Computing $f'(x) = \frac{1}{2}1/(1+x^2)$, we see that $|f'(x)|$ is always less than or equal to $\frac{1}{2}$. Hence, by Theorem 3, the sequence of iterates $x_0, x_1, x_2, \ldots$ converges to the unique root η of the equation $x = 1 + \frac{1}{2}\arctan x$, for every choice of x_0.

There are many instances where we know, a priori, that the equation $x = f(x)$ has a unique solution η in a given interval $[a,b]$. In these instances, we can use Theorem 3 to obtain a very good approximation of η. Indeed, life is especially simple in these instances, since we don't have to check that the iterates x_n all lie in a specified interval. If x_0 is sufficiently close to η, then the iterates x_n will always converge to η, as we now show.

Theorem 4. *Assume that $f(\eta) = \eta$, and that $|f'(x)| \leqslant \lambda < 1$ in the interval $|x - \eta| \leqslant \alpha$. Choose a number x_0 in this interval. Then, the sequence of iterates x_n, defined recursively by the equation $x_{n+1} = f(x_n)$, will always converge to η.*

PROOF. Denote the interval $|x - \eta| \leqslant \alpha$ by I. By Theorem 3, it suffices to show that all the iterates x_n lie in I. To this end, observe that

$$x_{j+1} - \eta = f(x_j) - f(\eta) = f'(\xi)(x_j - \eta)$$

where ξ is some number between x_j and η. In particular, ξ is in I if x_j is in

I. Thus,

$$|x_{j+1}-\eta| \leqslant \lambda |x_j-\eta| < |x_j-\eta| \tag{4}$$

if x_j is in I. This implies that x_{j+1} is in I whenever x_j is in I. By induction, therefore, all the iterates x_n lie in I. □

Equation (4) also shows that x_{n+1} is closer to η than x_n. Specifically, the error we make in approximating η by x_n decreases by at least a factor of λ each time we increase n. Thus, if λ is very small, then the convergence of x_n to η is very rapid, while if λ is close to one, then the convergence is very slow.

Example 2.
(a) Show that the equation

$$x = \sin x + \tfrac{1}{4} \tag{5}$$

has a unique root η in the interval $[\pi/4, \pi/2]$.
(b) Show that the sequence of numbers

$$x_0, \quad x_1 = \sin x_0 + \tfrac{1}{4}, \quad x_2 = \sin x_1 + \tfrac{1}{4}, \ldots$$

will converge to η if $\pi/4 \leqslant x_0 \leqslant \pi/2$.
(c) Write a computer program to evaluate the first N iterates $x_1, x_2, \ldots, x_N$.
Solution.
(a) Let $g(x) = x - \sin x - \frac{1}{4}$, and observe that $g(\pi/4)$ is negative while $g(\pi/2)$ is positive. Moreover, $g(x)$ is a monotonic increasing function of x for $\pi/4 \leqslant x \leqslant \pi/2$, since its derivative is strictly positive in this interval. Therefore, Equation (5) has a unique root $x = \eta$ in the interval $\pi/4 < x < \pi/2$.
(b) Let I denote the interval $\eta - \pi/4 \leqslant x \leqslant \eta + \pi/4$. The left endpoint of this interval is greater than zero, while the right endpoint is less than $3\pi/4$. Hence, there exists a number λ, with $0 < \lambda < 1$, such that

$$|\cos x| = \left| \frac{d}{dx}\left(\sin x + \tfrac{1}{4}\right) \right| \leqslant \lambda$$

for x in I. Clearly, the interval $[\pi/4, \pi/2]$ is contained in I. Therefore, by Theorem 4, the sequence of numbers

$$x_0, \quad x_1 = \sin x_0 + \tfrac{1}{4}, \quad x_2 = \sin x_1 + \tfrac{1}{4}, \ldots$$

will converge to η for every x_0 in the interval $[\pi/4, \pi/2]$.

(c)

APL Program

```
     ∇ ITERATE
[1]  X←Nρ0
[2]  X[1]←0.25+1○X0
[3]  K←1
[4]  X[K+1]←0.25+1○X[K]
[5]  K←K+1
[6]  →4×ιK<N
[7]  ⍉(2,ρX)ρ(ιρX),X      ∇
```

Remark. If x is a vector with N components, then ιρX is the vector 1, $2,\ldots,N$. Thus, statement [7] causes the computer to print out the two vectors $1,2,\ldots,N$ and $x_1,x_2,\ldots,x_N$ in adjacent columns.

Fortran Program

```
          DIMENSION X(200)
          READ (5,10) X0,N
     10   FORMAT (F15.8,I5)
C         COMPUTE X(1) FIRST
          X(1)=0.25+SIN(X0)
          KA=0
          KB=1
          WRITE (6,20) KA,X0,KB,X(1)
     20   FORMAT (1H1,4X,'N',10X,'X'/(1H,3X,I3,4X,F15.9))
C         COMPUTE X(2) THRU X(N)
          DO 40 K=2,N
          X(K)=0.25+SIN(X(K-1))
          WRITE (6,30) K,X(K)
     30   FORMAT (1H,3X,I3,4X,F15.9)
     40   CONTINUE
          CALL EXIT
          END
```

We ran these programs for $x_0=1$ and $N=15$, and the results are given in Table 1. This data implies that $\eta=1.17122965$ to eight significant decimal places. Moreover, we require only eleven iterations to find η correct to eight decimal places.

Table 1

n	x_n	n	x_n
0	1	8	1.17110411
1	1.09147099	9	1.17122962
2	1.13730626	10	1.17122964
3	1.15750531	11	1.17122965
4	1.16580403	12	1.17122965
5	1.16910543	13	1.17122965
6	1.17040121	14	1.17122965
7	1.17090706	15	1.17122965

In many instances, we want to compute a root of the equation $x=f(x)$ to within a certain accuracy. The easiest, and most efficient way of accomplishing this is to instruct the computer to terminate the program at $k=j$ if x_{j+1} agrees with x_j within the prescribed accuracy.

Exercises

1. Let η be the unique root of Equation (5).
   (a) Let $x_0=\pi/4$. Show that 20 iterations are required to find η to 8 significant decimal places.
   (b) Let $x_0=\pi/2$. Show that 20 iterations are required to find η to 8 decimal places.
   (c) Let $x_0=3\pi/8$. Show that 16 iterations are required to find η to 8 decimal places.

2. (a) Determine suitable values of x_0 so that the iterates x_n, defined by the equation

$$x_{n+1}=x_n-\tfrac{1}{4}(x_n^2-2)$$

   will converge to $\sqrt{2}$.
   (b) Choose $x_0=1.4$. Show that 14 iterations are required to find $\sqrt{2}$ to 8 significant decimal places. ($\sqrt{2}=1.41421356$ to 8 significant decimal places.)

3. (a) Determine suitable values of x_0 so that the iterates x_n, defined by the equation

$$x_{n+1}=x_n-\tfrac{1}{10}(x_n^2-2)$$

   will converge to $\sqrt{2}$.
   (b) Choose $x_0=1.4$. Show that 30 iterations are required to find $\sqrt{2}$ to 6 significant decimal places.

4. (a) Determine a suitable value of α so that the iterates x_n, defined by the equation

$$x_{n+1}=x_n-\alpha(x_n^2-3), \qquad x_0=1.7$$

   will converge to $\sqrt{3}$.
   (b) Find $\sqrt{3}$ to 6 significant decimal places.

5. Let η be the unique root of the equation $x=1+\frac{1}{2}\arctan x$. Find η to 5 significant decimal places.

6. (a) Show that the equation $2-x=(\ln x)/4$ has a unique root $x=\eta$ in the interval $0<x<\infty$.
(b) Let

$$x_{n+1}=2-(\ln x_n)/4, \qquad n=0,1,2,\ldots$$

Show that $1\leqslant x_n\leqslant 2$ if $1\leqslant x_0\leqslant 2$.
(c) Prove that $x_n\to\eta$ as $n\to\infty$ if $1\leqslant x_0\leqslant 2$.
(d) Compute η to 5 significant decimal places.

7. (a) Show that the equation $x=\cos x$ has a unique root $x=\eta$ in the interval $0\leqslant x\leqslant 1$.
(b) Let $x_{n+1}=\cos x_n$, $n=0,1,2,\ldots$, with $0<x_0<1$. Show that $0<x_n<1$. Conclude, therefore, that $x_n\to\eta$ as $n\to\infty$.
(c) Find η to 5 significant decimal places.

1.11.1 Newton's method

The method of iteration which we used to solve the equation $x=f(x)$ can also be used to solve the equation $g(x)=0$. To wit, any solution $x=\eta$ of the equation $g(x)=0$ is also a solution of the equation

$$x=f(x)=x-g(x), \tag{1}$$

and vice-versa. Better yet, any solution $x=\eta$ of the equation $g(x)=0$ is also a solution of the equation

$$x=f(x)=x-\frac{g(x)}{h(x)} \tag{2}$$

for any function $h(x)$. Of course, $h(x)$ must be unequal to zero for x near η.

Equation (2) has an arbitrary function $h(x)$ in it. Let us try and choose $h(x)$ so that (i) the assumptions of Theorem 4, Section 1.11 are satisfied, and (ii) the iterates

$$x_0,\quad x_1=x_0-\frac{g(x_0)}{h(x_0)},\quad x_2=x_1-\frac{g(x_1)}{h(x_1)},\ldots$$

converge as "rapidly as possible" to the desired root η. To this end, we compute

$$f'(x)=\frac{d}{dx}\left[x-\frac{g(x)}{h(x)}\right]=1-\frac{g'(x)}{h(x)}+\frac{h'(x)g(x)}{h^2(x)}$$

and observe that

$$f'(\eta)=1-\frac{g'(\eta)}{h(\eta)}.$$

This suggests that we set $h(x)=g'(x)$, since then $f'(\eta)=0$. Consequently, the iterates x_n, defined recursively by the equation

$$x_{n+1}=x_n-\frac{g(x_n)}{g'(x_n)}, \qquad n=0,1,2,\ldots \tag{3}$$

will converge to η if the initial guess x_0 is sufficiently close to η. (If $f'(\eta)=0$, then $|f'(x)|\leqslant\lambda<1$ for $|x-\eta|$ sufficiently small.) Indeed, the choice of $h(x)=f'(x)$ is an *optimal* choice of $h(x)$, since the convergence of x_n to η will be extremely rapid. This follows immediately from the fact that the number λ in Equation 4, Section 1.11 can be taken arbitrarily small, as x_n approaches η.

The iteration scheme (3) is known as Newton's method for solving the equation $g(x)=0$. It can be shown that if $g(\eta)=0$, and x_0 is sufficiently close to η, then

$$|x_{n+1}-\eta|\leqslant c|x_n-\eta|^2,$$

for some positive constant c. In other words, the error we make in approximating η by x_{n+1} is proportional to the square of the error we make in approximating η by x_n. This type of convergence is called quadratic convergence, and it implies that the iterates x_n converge extremely rapidly to η. In many instances, only five or six iterations are required to find η to eight or more significant decimal places.

Example 1. Use Newton's method to compute $\sqrt{2}$.
Solution. The square root of two is a solution of the equation

$$g(x)=x^2-2=0.$$

Hence, Newton's scheme for this problem is

$$x_{n+1}=x_n-\frac{g(x_n)}{g'(x_n)}=x_n-\frac{(x_n{}^2-2)}{2x_n}$$

$$=\frac{x_n}{2}+\frac{1}{x_n}, \qquad n=0,1,2,\ldots. \tag{4}$$

Sample APL and Fortran programs to compute the first N iterates of an initial guess x_0 are given below.

APL Program

```
     ∇ NEWTON
[1]  X←Nρ0
[2]  X[1]←(X0÷2)+÷X0
[3]  K←1
[4]  X[K+1]←(X[K]÷2)+÷X[K]
[5]  K←K+1
[6]  →4×ιK<N
[7]  ⍉(2,ρX)ρ(ιρX),X      ∇
```

Table 1

n	x_n	n	x_n
0	1.4	3	1.41421356
1	1.41428571	4	1.41421356
2	1.41421356	5	1.41421356

Fortran Program

We need only replace the instructions for computing X(1) and X(K) in the Fortran program of Section 1.11 by

```
X(1)=(X0/2)+1/X0
```

and

```
X(K)=(X(K-1)/2)+1/X(K-1)
```

We ran these programs for $x_0=1.4$ and $N=5$, and the results are given in Table 1. Notice that Newton's method requires only 2 iterations to find $\sqrt{2}$ to eight significant decimal places.

Example 2. Use Newton's method to find the impact velocity of the drums in Section 1.7.

Solution. The impact velocity of the drums satisfies the equation

$$g(v)=v+\frac{300cg}{W}+\frac{W-B}{c}\ln\left[\frac{W-B-cv}{W-B}\right]=0 \tag{5}$$

where

$$c=0.08, \quad g=32.2, \quad W=527.436, \quad \text{and} \quad B=470.327.$$

Setting $a=(W-B)/c$ and $d=300cg/W$ puts (5) in the simpler form

$$g(v)=v+d+a\ln(1-v/a)=0. \tag{6}$$

Newton's iteration scheme for this problem is

$$v_{n+1}=v_n-\frac{g(v_n)}{g'(v_n)}=v_n+\frac{(1-v_n/a)\left[v_n+d+a\ln(1-v_n/a)\right]}{v_n/a}$$

$$=v_n+\frac{a-v_n}{v_n}\left[v_n+d+a\ln(1-v_n/a)\right], \qquad n=0,1,2,\ldots.$$

Sample APL and Fortran programs to compute the first N iterates of v_0 are given below.

APL Program

```
     ∇ NEWTON
[1]  V←Nρ0
[2]  V1←V0+D+A× ⍟1−V0÷A
[3]  V[1]←V0+(A−V0)×V1÷V0
[4]  K←1
[5]  VK←V[K]+D+A× ⍟ 1−V[K]÷A
[6]  V[K+1]←V[K]+(A−V[K])×VK÷V[K]
[7]  K←K+1
[8]  →5×ιK<N
[9]  ⍉(2,ρV)ρ(ιρV),V      ∇
```

Fortran Program

Change every X to V, and replace the instructions for X(1) and X(K) in the Fortran program of Section 1.11 by

```
V(1)=V0+((A−V0)/V0)*(V0+D+A*A LOG(1 − (V0 / A)))
```

and

```
V(K)=V(K−1)+((A−V(K−1))/V(K−1))* (V(K−1)+D
     +A*A LOG (1−(V(K−1) / A)))
```

(Before running these programs, of course, we must instruct the computer to evaluate the constants $a=(W-B)/c$ and $d=300\, cg/W$.)

As was shown in Section 1.7, $v_0=45.7$ is a very good approximation of v. We set $v_0=45.7$ in the above programs, and the iterates v_n converged very rapidly to $v=45.1$ ft/s. Thus, the drums can indeed break upon impact.

In general, it is not possible to determine, a priori, how many iterations will be required to achieve a certain accuracy. In practice, we usually take N very large, and instruct the computer to terminate the program if one of the iterates agrees with its predecessor to the desired accuracy.

Exercises

1. Show that the iterates x_n defined by (4) will converge to $\sqrt{2}$ if

$$\sqrt{2/3} < x_0 < \sqrt{2} + (\sqrt{2} - \sqrt{2/3}\,).$$

2. Use Newton's method to find the following numbers to 8 significant decimal places. (a) $\sqrt{3}$, (b) $\sqrt{5}$, (c) $\sqrt{7}$.

3. The number π is a root of the equation

$$\tan\frac{x}{4}-\cot\frac{x}{4}=0.$$

Use Newton's method to find π to 8 significant decimal places.

Show that each of the following equations has a unique solution in the given interval, and use Newton's method to find it to 5 significant decimal places.

4. $2x-\tan x=0;\quad \pi\leqslant x\leqslant 3\pi/2$

5. $\frac{1}{2}-x+\frac{1}{5}\sin x=0;\quad \frac{1}{2}\leqslant x\leqslant 1$

6. $\ln x+(x+1)^3=0;\quad 0<x<1$

7. $2\sqrt{x}=\cos\frac{\pi x}{2};\quad 0\leqslant x\leqslant 1$

8. $(x-1)^2-\frac{1}{2}e^x=0;\quad 0<x<1$

9. $x-e^{-x^2}=1;\quad 0\leqslant x\leqslant 2.$

1.12 Difference equations, and how to compute the interest due on your student loans

In Sections 1.13–1.16 we will construct various approximations of the solution of the initial-value problem $dy/dt=f(t,y)$, $y(t_0)=y_0$. In determining how good these approximations are, we will be confronted with the following problem: How large can the numbers $E_1,\ldots,E_N$ be if

$$E_{n+1}\leqslant AE_n+B,\qquad n=0,1,\ldots,N-1 \tag{1}$$

for some positive constants A and B, and $E_0=0$? This is a very difficult problem since it deals with *inequalities*, rather than *equalities*. Fortunately, though, we can convert the problem of solving the inequalities (1) into the simpler problem of solving a system of equalities. This is the content of the following lemma.

Lemma 1. *Let $E_1,\ldots,E_N$ satisfy the inequalities*

$$E_{n+1}\leqslant AE_n+B,\qquad E_0=0$$

for some positive constants A and B. Then, E_n is less than or equal to y_n, where

$$y_{n+1}=Ay_n+B,\qquad y_0=0. \tag{2}$$

PROOF. We prove Lemma 1 by induction on n. To this end, observe that Lemma 1 is obviously true for $n=0$. Next, we assume that Lemma 1 is true for $n=j$. We must show that Lemma 1 is also true for $n=j+1$. That is to say, we must prove that $E_j\leqslant y_j$ implies $E_{j+1}\leqslant y_{j+1}$. But this follows immediately, for if $E_j\leqslant y_j$ then

$$E_{j+1}\leqslant AE_j+B\leqslant Ay_j+B=y_{j+1}.$$

By induction, therefore, $E_n\leqslant y_n$, $n=0,1,\ldots,N$. □

Our next task is to solve Equation (2), which is often referred to as a difference equation. We will accomplish this in two steps. First we will solve the "simple" difference equation

$$y_{n+1}=y_n+B_n, \qquad y_0=y_0. \tag{3}$$

Then we will reduce the difference equation (2) to the difference equation (3) by a clever change of variables.

Equation (3) is trivial to solve. Observe that

$$\begin{aligned} y_1-y_0 &= B_0 \\ y_2-y_1 &= B_1 \\ &\vdots \\ y_{n-1}-y_{n-2} &= B_{n-2} \\ y_n-y_{n-1} &= B_{n-1}. \end{aligned}$$

Adding these equations gives

$$(y_n-y_{n-1})+(y_{n-1}-y_{n-2})+\ldots+(y_1-y_0)=B_0+B_1+\ldots+B_{n-1}.$$

Hence,

$$y_n=y_0+B_0+\ldots+B_{n-1}=y_0+\sum_{j=0}^{n-1}B_j.$$

Next, we reduce the difference equation (2) to the simpler equation (3) in the following clever manner. Let

$$z_n=\frac{y_n}{A^n}, \qquad n=0,1,\ldots,N.$$

Then, $z_{n+1}=y_{n+1}/A^{n+1}$. But $y_{n+1}=Ay_n+B$. Consequently,

$$z_{n+1}=\frac{y_n}{A^n}+\frac{B}{A^{n+1}}=z_n+\frac{B}{A^{n+1}}.$$

Therefore,

$$z_n=z_0+\sum_{j=0}^{n-1}\frac{B}{A^{j+1}}=y_0+\frac{B}{A}\left[\frac{1-\left(\frac{1}{A}\right)^n}{1-\frac{1}{A}}\right]$$

$$=y_0+\frac{B}{A-1}\left[1-\left(\frac{1}{A}\right)^n\right]$$

and

$$y_n=A^nz_n=A^ny_0+\frac{B}{A-1}(A^n-1). \tag{4}$$

Finally, returning to the inequalities (1), we see that

$$E_n\leqslant\frac{B}{A-1}(A^n-1), \qquad n=1,2,\ldots,N. \tag{5}$$

While collecting material for this book, this author was approached by a colleague with the following problem. He had just received a bill from the bank for the first payment on his wife's student loan. This loan was to be repaid in 10 years in 120 equal monthly installments. According to his rough estimate, the bank was overcharging him by at least 20%. Before confronting the bank's officers, though, he wanted to compute exactly the monthly payments due on this loan.

This problem can be put in the following more general framework. Suppose that P dollars are borrowed from a bank at an annual interest rate of $R\%$. This loan is to be repaid in n years in equal monthly installments of x dollars. Find x.

Our first step in solving this problem is to compute the interest due on the loan. To this end observe that the interest I_1 owed when the first payment is due is $I_1=(r/12)P$, where $r=R/100$. The principal outstanding during the second month of the loan is $(x-I_1)$ less than the principal outstanding during the first month. Hence, the interest I_2 owed during the second month of the loan is

$$I_2=I_1-\frac{r}{12}(x-I_1).$$

Similarly, the interest I_{j+1} owed during the $(j+1)$st month is

$$I_{j+1}=I_j-\frac{r}{12}(x-I_j)=\left(1+\frac{r}{12}\right)I_j-\frac{r}{12}x, \tag{6}$$

where I_j is the interest owed during the jth month.

Equation (6) is a difference equation for the numbers

$$I_1=\frac{r}{12}P, I_2,\ldots,I_{12n}.$$

Its solution (see Exercise 4) is

$$I_j=\frac{r}{12}P\left(1+\frac{r}{12}\right)^{j-1}+x\left[1-\left(1+\frac{r}{12}\right)^{j-1}\right].$$

Hence, the total amount of interest paid on the loan is

$$\begin{aligned}I=I_1+I_2+\ldots+I_{12n}&=\sum_{j=1}^{12n}I_j\\&=\frac{r}{12}P\sum_{j=1}^{12n}\left(1+\frac{r}{12}\right)^{j-1}+12nx-x\sum_{j=1}^{12n}\left(1+\frac{r}{12}\right)^{j-1}.\end{aligned}$$

Now,

$$\sum_{j=1}^{12n}\left(1+\frac{r}{12}\right)^{j-1}=\frac{12}{r}\left[\left(1+\frac{r}{12}\right)^{12n}-1\right].$$

Therefore,

$$I=P\left[\left(1+\frac{r}{12}\right)^{12n}-1\right]+12nx-\frac{12x}{r}\left[\left(1+\frac{r}{12}\right)^{12n}-1\right]$$

$$=12nx-P+P\left(1+\frac{r}{12}\right)^{12n}-\frac{12x}{r}\left[\left(1+\frac{r}{12}\right)^{12n}-1\right].$$

But, $12nx-P$ must equal I, since $12nx$ is the amount of money paid the bank and P was the principal loaned. Consequently,

$$P\left(1+\frac{r}{12}\right)^{12n}-\frac{12x}{r}\left[\left(1+\frac{r}{12}\right)^{12n}-1\right]=0$$

and this equation implies that

$$x=\frac{\frac{r}{12}P\left(1+\frac{r}{12}\right)^{12n}}{\left(1+\frac{r}{12}\right)^{12n}-1}. \tag{7}$$

Epilog. Using Equation (7), this author computed x for his wife's and his colleague's wife's student loans. In both cases the bank was right—to the penny.

Exercises

1. Solve the difference equation $y_{n+1}=-7y_n+2, y_0=1$.

2. Find y_{37} if $y_{n+1}=3y_n+1$, $y_0=0$, $n=0,1,\ldots,36$.

3. Estimate the numbers $E_0,E_1,\ldots,E_N$ if $E_0=0$ and
(a) $E_{n+1}\leqslant 3E_n+1$, $n=0,1,\ldots,N-1$;
(b) $E_{n+1}\leqslant 2E_n+2$, $n=0,1,\ldots,N-1$.

4. (a) Show that the transformation $y_j=I_{j+1}$ transforms the difference equation

$$I_{j+1}=\left(1+\frac{r}{12}\right)I_j-\frac{r}{12}x, \qquad I_1=\frac{r}{12}P$$

into the difference equation

$$y_{j+1}=\left(1+\frac{r}{12}\right)y_j-\frac{r}{12}x, \qquad y_0=\frac{r}{12}P.$$

(b) Use Equation (4) to find $y_{j-1}=I_j$.

5. Solve the difference equation $y_{n+1}=a_ny_n+b_n$, $y_1=\alpha$. *Hint*: Set $z_1=y_1$ and $z_n=y_n/a_1\ldots a_{n-1}$ for $n\geqslant 2$. Observe that

$$z_{n+1}=\frac{y_{n+1}}{a_1\ldots a_n}=\frac{a_ny_n}{a_1\ldots a_n}+\frac{b_n}{a_1\ldots a_n}$$
$$=z_n+\frac{b_n}{a_1\ldots a_n}.$$

Hence, conclude that $z_n=z_1+\sum_{j=1}^{n-1}b_j/a_1\ldots a_j$.

6. Solve the difference equation $y_{n+1}-ny_n=1-n$, $y_1=2$.

7. Find y_{25} if $y_1=1$ and $(n+1)y_{n+1}-ny_n=2^n$, $n=1,\ldots,24$.

8. A student borrows P dollars at an annual interest rate of $R\%$. This loan is to be repayed in n years in equal monthly installments of x dollars. Find x if
(a) $P=4250$, $R=3$, and $n=5$;
(b) $P=5000$, $R=7$, and $n=10$.

9. A home buyer takes out a $30,000 mortgage at an annual interest rate of 9%. This loan is to be repaid over 20 years in 240 equal monthly installments of x dollars.
(a) Compute x.
(b) Find x if the annual interest rate is 10%.

10. The quantity supplied of some commodity in a given week is obviously an increasing function of its price the previous week, while the quantity demanded in a given week is a function of its current price. Let S_j and D_j denote, respectively, the quantities supplied and demanded in the jth week, and let P_j denote the price of the commodity in the jth week. We assume that there exist positive constants a, b, and c such that

$$S_j=aP_{j-1} \quad \text{and} \quad D_j=b-cP_j.$$

(a) Show that $P_j=b/(a+c)+(-a/c)^j(P_0-b/(a+c))$, if supply always equals demand.
(b) Show that P_j approaches $b/(a+c)$ as j approaches infinity if $a/c<1$.
(c) Show that $P=b/(a+c)$ represents an equilibrium situation. That is to say, if supply always equals demand, and if the price ever reaches the level $b/(a+c)$, then it will always remain at that level.

1.13 Numerical approximations; Euler's method

In Section 1.9 we showed that it is not possible, in general, to solve the initial-value problem

$$\frac{dy}{dt}=f(t,y), \qquad y(t_0)=y_0. \tag{1}$$

Therefore, in order that differential equations have any practical value for us, we must devise ways of obtaining accurate approximations of the solution $y(t)$ of (1). In Sections 1.13–1.16 we will derive algorithms, which can be implemented on a digital computer, for obtaining accurate approximations of $y(t)$.

Now, a computer obviously cannot approximate a function on an entire interval $t_0 \leqslant t \leqslant t_0+a$ since this would require an infinite amount of information. At best it can compute approximate values $y_1,\ldots,y_N$ of $y(t)$ at a finite number of points $t_1,t_2,\ldots,t_N$. However, this is sufficient for our purpose since we can use the numbers $y_1,\ldots,y_N$ to obtain an accurate approximation of $y(t)$ on the entire interval $t_0 \leqslant t \leqslant t_0+a$. To wit, let $\hat{y}(t)$ be

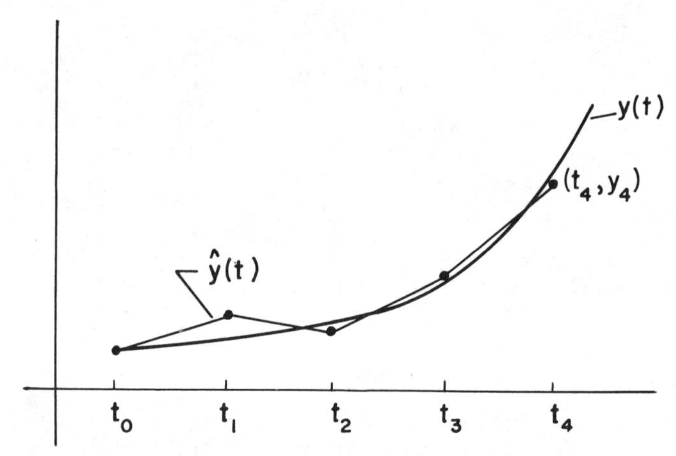

Figure 1. Comparison of $\hat{y}(t)$ and $y(t)$

the function whose graph on each interval $[t_j, t_{j+1}]$ is the straight line connecting the points (t_j, y_j) and (t_{j+1}, y_{j+1}) (see Figure 1). We can express $\hat{y}(t)$ analytically by the equation

$$\hat{y}(t) = y_j + \frac{1}{h}(t - t_j)(y_{j+1} - y_j), \qquad t_j \leq t \leq t_{j+1}.$$

If $\hat{y}(t)$ is close to $y(t)$ at $t = t_j$; that is, if y_j is close to $y(t_j)$, and if t_{j+1} is close to t_j, then $\hat{y}(t)$ is close to $y(t)$ on the entire interval $t_j \leq t \leq t_{j+1}$. This follows immediately from the continuity of both $y(t)$ and $\hat{y}(t)$. Thus, we need only devise schemes for obtaining accurate approximations of $y(t)$ at a discrete number of points $t_1, \ldots, t_N$ in the interval $t_0 \leq t \leq t_0 + a$. For simplicity, we will require that the points $t_1, \ldots, t_N$ be equally spaced. This is achieved by choosing a large integer N and setting $t_k = t_0 + k(a/N)$, $k = 1, \ldots, N$. Alternately, we may write $t_{k+1} = t_k + h$ where $h = a/N$.

Now, the only thing we know about $y(t)$ is that it satisfies a certain differential equation, and that its value at $t = t_0$ is y_0. We will use this information to compute an approximate value y_1 of y at $t = t_1 = t_0 + h$. Then, we will use this approximate value y_1 to compute an approximate value y_2 of y at $t = t_2 = t_1 + h$, and so on. In order to accomplish this we must find a theorem which enables us to compute the value of y at $t = t_k + h$ from the knowledge of y at $t = t_k$. This theorem, of course, is Taylor's Theorem, which states that

$$y(t_k + h) = y(t_k) + h\frac{dy(t_k)}{dt} + \frac{h^2}{2!}\frac{d^2y(t_k)}{dt^2} + \ldots. \tag{2}$$

Thus, if we know the value of y and its derivatives at $t = t_k$, then we can compute the value of y at $t = t_k + h$. Now, $y(t)$ satisfies the initial-value problem (1). Hence, its derivative, when evaluated at $t = t_k$, must equal

$f(t_k, y(t_k))$. Moreover, by repeated use of the chain rule of partial differentiation (see Appendix A), we can evaluate

$$\frac{d^2y(t_k)}{dt^2} = \left[\frac{\partial f}{\partial t} + f\frac{\partial f}{\partial y}\right](t_k, y(t_k))$$

and all other higher-order derivatives of $y(t)$ at $t = t_k$. Hence, we can rewrite (2) in the form

$$y(t_{k+1}) = y(t_k) + hf(t_k, y(t_k)) + \frac{h^2}{2!}\left[\frac{\partial f}{\partial t} + f\frac{\partial f}{\partial y}\right](t_k, y(t_k)) + \dots. \tag{3}$$

The simplest approximation of $y(t_{k+1})$ is obtained by truncating the Taylor series (3) after the second term. This gives rise to the numerical scheme

$$y_1 = y_0 + hf(t_0, y_0), \qquad y_2 = y_1 + hf(t_1, y_1),$$

and, in general,

$$y_{k+1} = y_k + hf(t_k, y_k), \qquad y_0 = y(t_0). \tag{4}$$

Notice how we use the initial-value y_0 and the fact that $y(t)$ satisfies the differential equation $dy/dt = f(t,y)$ to compute an approximate value y_1 of $y(t)$ at $t = t_1$. Then, we use this approximate value y_1 to compute an approximate value y_2 of $y(t)$ at $t = t_2$, and so on.

Equation (4) is known as *Euler's scheme*. It is the simplest numerical scheme for obtaining approximate values $y_1, \dots, y_N$ of the solution $y(t)$ at times $t_1, \dots, t_N$. Of course, it is also the least accurate scheme, since we have only retained two terms in the Taylor series expansion for $y(t)$. As we shall see shortly, Euler's scheme is not accurate enough to use in many problems. However, it is an excellent introduction to the more complicated schemes that will follow.

Example 1. Let $y(t)$ be the solution of the initial-value problem

$$dy/dt = 1 + (y-t)^2, \qquad y(0) = \tfrac{1}{2}.$$

Use Euler's scheme to compute approximate values $y_1, \dots, y_N$ of $y(t)$ at the points $t_1 = 1/N, t_2 = 2/N, \dots, t_N = 1$.

Solution. Euler's scheme for this problem is

$$y_{k+1} = y_k + h\left[1 + (y_k - t_k)^2\right], \qquad k = 0, 1, \dots, N-1, \quad h = 1/N$$

with $y_0 = \frac{1}{2}$. Sample APL and Fortran programs to compute $y_1, \dots, y_N$ are given below. These programs, as well as all subsequent programs, have variable values for t_0, y_0, a, and N, so that they may also be used to solve the more general initial-value problem $dy/dt = 1 + (y-t)^2$, $y(t_0) = y_0$ on any desired interval. Moreover, these same programs work even if we change the differential equation; if we change the function $f(t,y)$ then we need only change lines 5 and 8 in the APL program and the expressions for Y(1) and Y(K) in Section B of the Fortran program.

APL Program

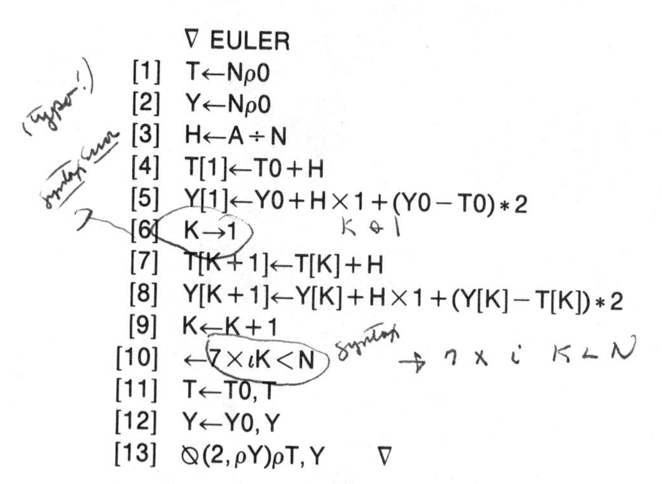

```
     ∇ EULER
[1]  T←Nρ0
[2]  Y←Nρ0
[3]  H←A÷N
[4]  T[1]←T0+H
[5]  Y[1]←Y0+H×1+(Y0−T0)*2
[6]  K→1
[7]  T[K+1]←T[K]+H
[8]  Y[K+1]←Y[K]+H×1+(Y[K]−T[K])*2
[9]  K←K+1
[10] ←7×ιK<N
[11] T←T0,T
[12] Y←Y0,Y
[13] ⍉(2,ρY)ρT,Y     ∇
```

Fortran Program

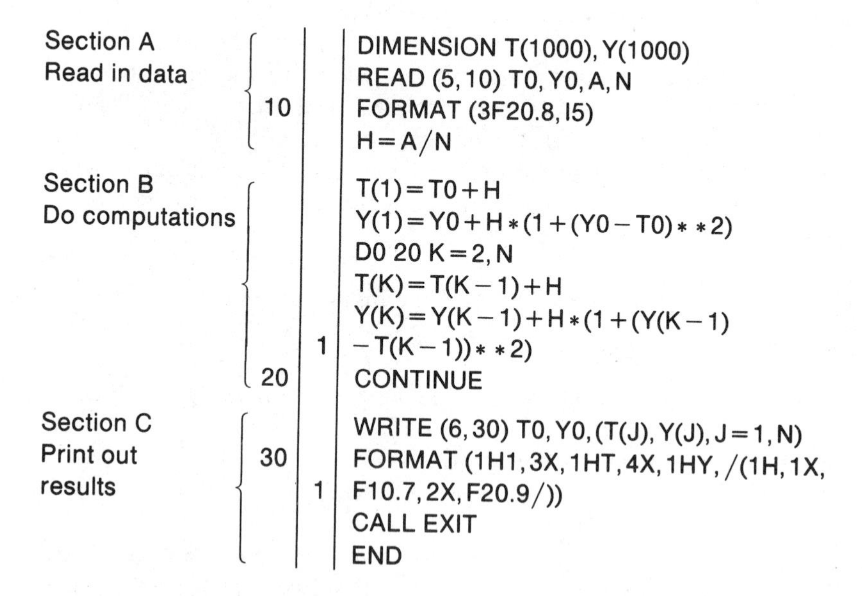

```
Section A              DIMENSION T(1000), Y(1000)
Read in data           READ (5,10) T0, Y0, A, N
                 10    FORMAT (3F20.8, I5)
                       H=A/N
Section B              T(1)=T0+H
Do computations        Y(1)=Y0+H*(1+(Y0−T0)**2)
                       DO 20 K=2, N
                       T(K)=T(K−1)+H
                       Y(K)=Y(K−1)+H*(1+(Y(K−1)
                     1 −T(K−1))**2)
                 20    CONTINUE
Section C              WRITE (6,30) T0, Y0, (T(J), Y(J), J=1, N)
Print out        30    FORMAT (1H1,3X,1HT,4X,1HY,/(1H,1X,
results              1 F10.7,2X,F20.9/))
                       CALL EXIT
                       END
```

Table 1 below gives the results of these computations for $a=1$, $N=10$, $t_0=0$, and $y_0=\frac{1}{2}$. All of these computations, and all subsequent computations, were carried out on an IBM 360 computer using 16 decimal places accuracy. The results have been rounded to 8 significant decimal places.

Table 1

t	y	t	y
0	0.5	0.6	1.29810115
0.1	0.625	0.7	1.44683567
0.2	0.7525625	0.8	1.60261202
0.3	0.88309503	0.9	1.76703063
0.4	1.01709501	1	1.94220484
0.5	1.15517564		

The exact solution of this initial-value problem (see Exercise 7) is

$$y(t)=t+1/(2-t).$$

Thus, the error we make in approximating the value of the solution at $t=1$ by y_{10} is approximately 0.06, since $y(1)=2$. If we run this program for $N=20$ and $N=40$, we obtain that $y_{20}=1.96852339$ and $y_{40}=1.9835109$. Hence, the error we make in approximating $y(1)$ by y_{40} is already less than 0.02.

Exercises

Using Euler's method with step size $h=0.1$, determine an approximate value of the solution at $t=1$ for each of the initial-value problems 1–5. Repeat these computations with $h=0.025$ and compare the results with the given value of the solution.

1. $\dfrac{dy}{dt}=1+t-y,\quad y(0)=0;\ (y(t)=t)$

2. $\dfrac{dy}{dt}=2ty,\quad y(0)=2;\ (y(t)=2e^{t^2})$

3. $\dfrac{dy}{dt}=1+y^2-t^2,\quad y(0)=0;\ (y(t)=t)$

4. $\dfrac{dy}{dt}=te^{-y}+\dfrac{t}{1+t^2},\quad y(0)=0;\ (y(t)=\ln(1+t^2))$

5. $\dfrac{dy}{dt}=-1+2t+\dfrac{y^2}{(1+t^2)^2},\quad y(0)=1;\ (y(t)=1+t^2)$

6. Using Euler's method with $h=\pi/40$, determine an approximate value of the solution of the initial-value problem

$$\frac{dy}{dt}=2\sec^2 t-(1+y^2),\qquad y(0)=0$$

at $t=\pi/4$. Repeat these computations with $h=\pi/160$ and compare the results with the number one which is the value of the solution $y(t)=\tan t$ at $t=\pi/4$.

7. (a) Show that the substitution $y=t+z$ reduces the initial-value problem $y'=1+(y-t)^2$, $y(0)=0.5$ to the simpler initial-value problem $z'=z^2$, $z(0)=0.5$.
(b) Show that $z(t)=1/(2-t)$. Hence, $y(t)=t+1/(2-t)$.

1.13.1 Error analysis for Euler's method

One of the nice features of Euler's method is that it is relatively simple to estimate the error we make in approximating $y(t_k)$ by y_k. Unfortunately, though, we must make the severe restriction that $t_1,\ldots,t_N$ do not exceed $t_0+\alpha$, where α is the number defined in the existence–uniqueness theorem of Section 1.10. More precisely, let a and b be two positive numbers and assume that the functions f, $\partial f/\partial t$, and $\partial f/\partial y$ are defined and continuous in the rectangle $t_0\leqslant t\leqslant t_0+a$, $y_0-b\leqslant y\leqslant y_0+b$. We will denote this rectangle by R. Let M be the maximum value of $|f(t,y)|$ for (t,y) in R, and set $\alpha=\min(a,b/M)$. We will determine the error committed in approximating $y(t_k)$ by y_k, for $t_k\leqslant t_0+\alpha$.

To this end observe that the numbers $y_0,y_1,\ldots,y_N$ satisfy the difference equation

$$y_{k+1}=y_k+hf(t_k,y_k), \qquad k=0,1,\ldots,N-1 \tag{1}$$

while the numbers $y(t_0),y(t_1),\ldots,y(t_N)$ satisfy the difference equation

$$y(t_{k+1})=y(t_k)+hf(t_k,y(t_k))+\frac{h^2}{2}\left[\frac{\partial f}{\partial t}+f\frac{\partial f}{\partial y}\right](\xi_k,y(\xi_k)) \tag{2}$$

where ξ_k is some number between t_k and t_{k+1}. Equation (2) follows from the identity

$$\frac{d^2y}{dt^2}=\frac{\partial f}{\partial t}+f\frac{\partial f}{\partial y}$$

and the fact that

$$y(t+h)=y(t)+h\frac{dy(t)}{dt}+\frac{h^2}{2}\frac{d^2y(\tau)}{dt^2},$$

for some number τ between t and $t+h$. Subtracting Equation (1) from Equation (2) gives

$$\begin{aligned}y(t_{k+1})-y_{k+1}&=y(t_k)-y_k+h\big[f(t_k,y(t_k))-f(t_k,y_k)\big]\\&\quad+\frac{h^2}{2}\left[\frac{\partial f}{\partial t}+f\frac{\partial f}{\partial y}\right](\xi_k,y(\xi_k)).\end{aligned}$$

Next, observe that

$$f(t_k,y(t_k))-f(t_k,y_k)=\frac{\partial f(t_k,\eta_k)}{\partial y}\big[y(t_k)-y_k\big]$$

where η_k is some number between $y(t_k)$ and y_k. Consequently,

$$\begin{aligned}|y(t_{k+1})-y_{k+1}|&\leqslant|y(t_k)-y_k|+h\left|\frac{\partial f(t_k,\eta_k)}{\partial y}\right||y(t_k)-y_k|\\&\quad+\frac{h^2}{2}\left|\left[\frac{\partial f}{\partial t}+f\frac{\partial f}{\partial y}\right](\xi_k,y(\xi_k))\right|.\end{aligned}$$

In order to proceed further, we must obtain estimates of the quantities $(\partial f(t_k,\eta_k))/\partial y$ and $[(\partial f/\partial t)+f(\partial f/\partial y)](\xi_k,y(\xi_k))$. To this end observe that the points $(\xi_k,y(\xi_k))$ and (t_k,y_k) all lie in the rectangle R. (It was shown in Section 1.10 that the points $(\xi_k,y(\xi_k))$ lie in R. In addition, a simple induction argument (see Exercise 9) shows that the points (t_k,y_k) all lie in R.) Consequently, the points (t_k,η_k) must also lie in R. Let L and D be two positive numbers such that

$$\max_{(t,y)\text{ in }R}\left|\frac{\partial f}{\partial y}\right|\leqslant L$$

and

$$\max_{(t,y)\text{ in }R}\left|\frac{\partial f}{\partial t}+f\frac{\partial f}{\partial y}\right|\leqslant D.$$

Such numbers always exist if f, $\partial f/\partial t$, and $\partial f/\partial y$ are continuous in R. Then,

$$|y(t_{k+1})-y_{k+1}|\leqslant|y(t_k)-y_k|+hL|y(t_k)-y_k|+\frac{Dh^2}{2}. \tag{3}$$

Now, set $E_k=|y(t_k)-y_k|$, $k=0,1,\ldots,N$. The number E_k is the error we make at the kth step in approximating $y(t_k)$ by y_k. From (3)

$$E_{k+1}\leqslant(1+hL)E_k+\frac{Dh^2}{2},\qquad k=0,1,\ldots,N-1. \tag{4}$$

Moreover, $E_0=0$ since $y(t_0)=y_0$. Thus, the numbers $E_0,E_1,\ldots,E_N$ satisfy the set of inequalities

$$E_{k+1}\leqslant AE_k+B,\qquad E_0=0$$

with $A=1+hL$ and $B=Dh^2/2$. Consequently, (see Section 1.12)

$$E_k\leqslant\frac{B}{A-1}(A^k-1)=\frac{Dh}{2L}\left[(1+hL)^k-1\right]. \tag{5}$$

We can also obtain an estimate for E_k that is independent of k. Observe that $1+hL\leqslant e^{hL}$. This follows from the fact that

$$\begin{aligned}e^{hL}&=1+hL+\frac{(hL)^2}{2!}+\frac{(hL)^3}{3!}+\ldots\\&=(1+hL)+\text{“something positive”}.\end{aligned}$$

Therefore,

$$E_k\leqslant\frac{Dh}{2L}\left[(e^{hL})^k-1\right]=\frac{Dh}{2L}\left[e^{khL}-1\right].$$

Finally, since $kh\leqslant\alpha$, we see that

$$E_k\leqslant\frac{Dh}{2L}\left[e^{\alpha L}-1\right],\qquad k=1,\ldots,N. \tag{6}$$

Equation (6) says that the error we make in approximating the solution $y(t)$ at time $t=t_k$ by y_k is at most a fixed constant times h. This suggests, as a rule of thumb, that our error should decrease by approximately $\frac{1}{2}$ if we decrease h by $\frac{1}{2}$. We can verify this directly in Example 1 of the previous section where our error at $t=1$ for $h=0.1$, 0.05, and 0.025 is 0.058, 0.032, and 0.017 respectively.

Example 1. Let $y(t)$ be the solution of the initial-value problem

$$\frac{dy}{dt}=\frac{t^2+y^2}{2}, \qquad y(0)=0.$$

(a) Show that $y(t)$ exists at least for $0\leqslant t\leqslant 1$, and that in this interval, $-1\leqslant y(t)\leqslant 1$.
(b) Let N be a large positive integer. Set up Euler's scheme to find approximate values of y at the points $t_k=k/N$, $k=0,1,\ldots,N$.
(c) Determine a step size $h=1/N$ so that the error we make in approximating $y(t_k)$ by y_k does not exceed 0.0001.
Solution. (a) Let R be the rectangle $0\leqslant t\leqslant 1$, $-1\leqslant y\leqslant 1$. The maximum value that $(t^2+y^2)/2$ achieves for (t,y) in R is 1. Hence, by the existence–uniqueness theorem of Section 1.10, $y(t)$ exists at least for

$$0\leqslant t\leqslant \alpha=min\left(1,\tfrac{1}{1}\right)=1,$$

and in this interval, $-1\leqslant y\leqslant 1$.

(b) $$y_{k+1}=y_k+h\left(\frac{t_k^2+y_k^2}{2}\right)=y_k+\frac{1}{2N}\left[\left(\frac{k}{N}\right)^2+y_k^2\right]$$

with $y_0=0$. The integer k runs from 0 to $N-1$.
(c) Let $f(t,y)=(t^2+y^2)/2$, and compute

$$\frac{\partial f}{\partial y}=y \quad\text{and}\quad \frac{\partial f}{\partial t}+f\frac{\partial f}{\partial y}=t+\frac{y}{2}(t^2+y^2).$$

From (6), $|y(t_k)-y_k|\leqslant(Dh/2L)(e^L-1)$ where L and D are two positive numbers such that

$$\max_{(t,y)\text{ in }R}|y|\leqslant L$$

and

$$\max_{(t,y)\text{ in }R}\left|t+\frac{y}{2}(t^2+y^2)\right|\leqslant D.$$

Now, the maximum values of the functions $|y|$ and $|t+(y/2)(t^2+y^2)|$ for (t,y) in R are clearly 1 and 2 respectively. Hence,

$$|y(t_k)-y_k|\leqslant\frac{2h}{2}(e-1)=h(e-1).$$

This implies that the step size h should be smaller than $0.0001/(e-1)$.

Equivalently, N should be larger than $(e-1)10^4 = 17{,}183$. Thus, we must iterate the equation

$$y_{k+1} = y_k + \frac{1}{2(17{,}183)}\left[\left(\frac{k}{17{,}183}\right)^2 + y_k^2\right]$$

17,183 times to be sure that $y(1)$ is correct to four decimal places.

Example 2. Let $y(t)$ be the solution of the initial-value problem

$$\frac{dy}{dt} = t^2 + e^{-y^2}, \qquad y(0) = 1.$$

(a) Show that $y(t)$ exists at least for $0 \leqslant t \leqslant 1$, and that in this interval, $-1 \leqslant y \leqslant 3$.
(b) Let N be a large positive integer. Set up Euler's scheme to find approximate values of $y(t)$ at the points $t_k = k/N$, $k = 0, 1, \ldots, N$.
(c) Determine a step size h so that the error we make in approximating $y(t_k)$ by y_k does not exceed 0.0001.
Solution.
(a) Let R be the rectangle $0 \leqslant t \leqslant 1$, $|y-1| \leqslant 2$. The maximum value that $t^2 + e^{-y^2}$ achieves for (t,y) in R is 2. Hence, $y(t)$ exists at least for $0 \leqslant t \leqslant \min(1, 2/2) = 1$, and in this interval, $-1 \leqslant y \leqslant 3$.
(b) $y_{k+1} = y_k + h(t_k{}^2 + e^{-y_k^2}) = y_k + (1/N)[(k/N)^2 + e^{-y_k^2}]$ with $y_0 = 1$. The integer k runs from 0 to $N-1$.
(c) Let $f(t,y) = t^2 + e^{-y^2}$ and compute

$$\frac{\partial f}{\partial y} = -2ye^{-y^2}, \quad \text{and} \quad \frac{\partial f}{\partial t} + f\frac{\partial f}{\partial y} = 2t - 2y(t^2 + e^{-y^2})e^{-y^2}.$$

From (6), $|y(t_k) - y_k| \leqslant (Dh/2L)(e^L - 1)$ where L and D are two positive numbers such that

$$\max_{(t,y)\text{ in } R} |-2ye^{-y^2}| \leqslant L$$

and

$$\max_{(t,y)\text{ in } R} |2t - 2y(t^2 + e^{-y^2})e^{-y^2}| \leqslant D.$$

Now, it is easily seen that the maximum value of $|2ye^{-y^2}|$ for $-1 \leqslant y \leqslant 3$ is $\sqrt{2/e}$. Thus, we take $L = \sqrt{2/e}$. Unfortunately, though, it is extremely difficult to compute the maximum value of the function

$$|2t - 2y(t^2 + e^{-y^2})e^{-y^2}|$$

for (t,y) in R. However, we can still find an acceptable value D by observing that for (t,y) in R,

$$\begin{aligned} \max|2t - 2v(t^2 + e^{-y^2})e^{-y^2}| &\leqslant \max|2t| + \max|2y(t^2 + e^{-y^2})e^{-y^2}| \\ &\leqslant \max|2t| + \max|2ye^{-y^2}| \times \max(t^2 + e^{-y^2}) \\ &= 2 + 2\sqrt{2/e} = 2\left(1 + \sqrt{2/e}\,\right). \end{aligned}$$

Hence, we may choose $D=2(1+\sqrt{2/e}\,)$. Consequently,

$$|y(t_k)-y_k| \leqslant \frac{2\left(1+\sqrt{2/e}\,\right)h\left[e^{\sqrt{2/e}}-1\right]}{2\sqrt{2/e}}.$$

This implies that the step size h must be smaller than

$$\frac{\sqrt{2/e}}{1+\sqrt{2/e}} \times \frac{0.0001}{e^{\sqrt{2/e}}-1}.$$

Examples 1 and 2 show that Euler's method is not very accurate since approximately 20,000 iterations are required to achieve an accuracy of four decimal places. One obvious disadvantage of a scheme which requires so many iterations is the cost. The going rate for computer usage at present is about \$1200.00 per hour. A second, and much more serious disadvantage, is that y_k may be very far away from $y(t_k)$ if N is exceptionally large. To wit, a digital computer can never perform a computation exactly since it only retains a finite number of decimal places. Consequently, every time we perform an arithmetic operation on the computer, we must introduce a "round off" error. This error, of course, is small. However, if we perform too many operations then the accumulated round off error may become so large as to make our results meaningless. Exercise 8 gives an illustration of this for Euler's method.

Exercises

1. Determine an upper bound on the error we make in using Euler's method with step size h to find an approximate value of the solution of the initial-value problem

$$\frac{dy}{dt}=\frac{t^2+y^2}{2}, \qquad y(0)=1$$

at any point t in the interval $[0,\frac{2}{5}]$. *Hint*: Let R be the rectangle $0 \leqslant t \leqslant 1$, $0 \leqslant y \leqslant 2$.

2. Determine an upper bound on the error we make in using Euler's method with step size h to find an approximate value of the solution of the initial-value problem

$$\frac{dy}{dt}=t-y^4, \qquad y(0)=0$$

at any point t in the interval $[0,1]$. *Hint*: Let R be the rectangle $0 \leqslant t \leqslant 1$, $-1 \leqslant y \leqslant 1$.

3. Determine an upper bound on the error we make in using Euler's method with step size h to find an approximate value of the solution of the initial-value problem

$$\frac{dy}{dt}=t+e^y, \qquad y(0)=0$$

at any point t in the interval $[0,1/(e+1)]$. *Hint*: Let R be the rectangle $0\leqslant t\leqslant 1$, $-1\leqslant y\leqslant 1$.

4. Determine a suitable value of h so that the error we make in using Euler's method with step size h to find an approximate value of the solution of the initial-value problem

$$\frac{dy}{dt}=e^t-y^2, \qquad y(0)=0$$

at any point t in the interval $[0,1/e]$ is at most 0.0001. *Hint*: Let R be the rectangle $0\leqslant t\leqslant 1$, $-1\leqslant y\leqslant 1$.

5. Determine a suitable value of h so that the error we make in using Euler's method with step size h to find an approximate value of the solution of the initial-value problem

$$\frac{dy}{dt}=t^2+\tan^2 y, \qquad y(0)=0$$

at any point t in the interval $[0,\frac{1}{2}]$ is at most 0.00001. *Hint*: Let R be the rectangle $0\leqslant t\leqslant \frac{1}{2}$, $-\pi/4\leqslant y\leqslant \pi/4$.

6. Determine a suitable value of h so that the error we make in using Euler's method with step size h to find an approximate value of the solution of the initial-value problem

$$\frac{dy}{dt}=\frac{1}{1+t^2+y^2}, \qquad y(0)=0$$

at any point t in the interval $[0,1]$ is at most 0.0001. *Hint*: Let R be the rectangle $0\leqslant t\leqslant 1$, $-1\leqslant y\leqslant 1$.

7. Let $y(t)$ be the solution of the initial-value problem

$$\frac{dy}{dt}=f(t,y), \qquad y(0)=0.$$

Suppose that $|f(t,y)|\leqslant 1$, $|\partial f/\partial y|\leqslant 1$, and $|(\partial f/\partial t)+f(\partial f/\partial y)|\leqslant 2$ in the rectangle $0\leqslant t\leqslant 1$, $-1\leqslant y\leqslant 1$. When the Euler scheme

$$y_{k+1}=y_k+hf(t_k,y_k), \qquad h=\frac{1}{N}$$

is used with $N=10$, the value of y_5 is $-0.15[(\frac{11}{10})^5-1]$, and the value of y_6 is $0.12[(\frac{11}{10})^6-1]$. Prove that $y(t)$ is zero at least once in the interval $(\frac{1}{2},\frac{3}{5})$.

8. Let $y(t)$ be the solution of the initial-value problem

$$y'=f(t,y), \qquad y(t_0)=y_0.$$

Euler's method for finding approximate values of $y(t)$ is $y_{k+1}=y_k+hf(t_k,y_k)$. However, the quantity $y_k+hf(t_k,y_k)$ is never computed exactly: we always introduce an error ε_k with $|\varepsilon_k|<\varepsilon$. That is to say, the computer computes numbers $\tilde{y}_1,\tilde{y}_2,\ldots$, such that

$$\tilde{y}_{k+1}=\tilde{y}_k+hf(t_k,\tilde{y}_k)+\varepsilon_k$$

with $\tilde{y}_0=y_0$. Suppose that $|\partial f/\partial y|\leqslant L$ and $|(\partial f/\partial t)+f(\partial f/\partial y)|\leqslant D$ for all t and y.

(a) Show that

$$E_{k+1}\equiv|y(t_{k+1})-\tilde{y}_{k+1}|\leqslant(1+hL)E_k+\frac{D}{2}h^2+\varepsilon$$

(b) Conclude from (a) that

$$E_k \leqslant \left[\frac{Dh}{2} + \frac{\varepsilon}{h} \right] \frac{e^{\alpha L} - 1}{L}$$

for $kh \leqslant \alpha$.

(c) Choose h so that the error E_k is minimized. Notice that the error E_k may be very large if h is very small.

9. Let $y_1, y_2, \ldots$ satisfy the recursion relation

$$y_{k+1} = y_k + hf(t_k, y_k).$$

Let R be the rectangle $t_0 \leqslant t \leqslant t_0 + a, y_0 - b \leqslant y \leqslant y_0 + b$, and assume that $|f(t,y)| \leqslant M$ for (t,y) in R. Finally, let $\alpha = \min(a, b/M)$.

(a) Prove that $|y_j - y_0| \leqslant jhM$, as long as $jh \leqslant \alpha$. *Hint*: Use induction.

(b) Conclude from (a) that the points (t_j, y_j) all lie in R as long as $j \leqslant \alpha/h$.

1.14 The three term Taylor series method

Euler's method was derived by truncating the Taylor series

$$y(t_{k+1}) = y(t_k) + hf(t_k, y(t_k)) + \frac{h^2}{2}\left[\frac{\partial f}{\partial t} + f\frac{\partial f}{\partial y} \right](t_k, y(t_k)) + \ldots \tag{1}$$

after the second term. The most obvious way of obtaining better numerical schemes is to retain more terms in Equation (1). If we truncate this Taylor series after three terms then we obtain the numerical scheme

$$y_{k+1} = y_k + hf(t_k, y_k) + \frac{h^2}{2}\left[\frac{\partial f}{\partial t} + f\frac{\partial f}{\partial y} \right](t_k, y_k), \qquad k = 0, \ldots, N-1 \tag{2}$$

with $y_0 = y(t_0)$.

Equation (2) is called the *three term Taylor series method.* It is obviously more accurate than Euler's method. Hence, for fixed h, we would expect that the numbers y_k generated by Equation (2) are better approximations of $y(t_k)$ than the numbers y_k generated by Euler's scheme. This is indeed the case, for it can be shown that $|y(t_k) - y_k|$ is proportional to h^2 whereas the error we make using Euler's method is only proportional to h. The quantity h^2 is much less than h if h is very small. Thus, the three term Taylor series method is a significant improvement over Euler's method.

Example 1. Let $y(t)$ be the solution of the initial-value problem

$$\frac{dy}{dt} = 1 + (y - t)^2, \qquad y(0) = \frac{1}{2}.$$

Use the three term Taylor series method to compute approximate values of $y(t)$ at the points $t_k = k/N$, $k = 1, \ldots, N$.

Solution. Let $f(t,y)=1+(y-t)^2$. Then,

$$\frac{\partial f}{\partial t}+f\frac{\partial f}{\partial y}=-2(y-t)+2(y-t)\left[1+(y-t)^2\right]=2(y-t)^3.$$

Hence, the three term Taylor series scheme is

$$y_{k+1}=y_k+h\left[1+(y_k-t_k)^2\right]+h^2(y_k-t_k)^3$$

with $h=1/N$ and $y_0=\frac{1}{2}$. The integer k runs from 0 to $N-1$. Sample APL and Fortran programs to compute $y_1,\ldots,y_N$ are given below. Again, these programs have variable values for t_0, y_0, a, and N.

APL Program

```
      ∇ TAYLOR
 [1]  T←Nρ0
 [2]  Y←Nρ0
 [3]  H←A÷N
 [4]  T[1]←T0+H
 [5]  D2Y←H×(Y0−T0)∗3
 [6]  Y[1]←Y0+H×D2Y+1+(Y0−T0)∗2
 [7]  K←1
 [8]  T[K+1]←T[K]+H
 [9]  D2Y←H×(Y[K]−T[K])∗3
[10]  Y[K+1]←Y[K]+H×D2Y+1+(Y[K]−T[K])∗2
[11]  K←K+1
[12]  →8×ιK<N
[13]  T←T0,T
[14]  Y←Y0,Y
[15]  ⍉(2,ρY)ρT,Y      ∇
```

Fortran Program

Replace Section B of the Fortran program in Section 1.13 by the following:

```
      T(1)=T0+H
      D2Y=H*(Y0-T0)**3
      Y(1)=Y0+H*(D2Y+1+(Y0-T0)**2)
      DO 20 K=2,N
      T(K)=T(K-1)+H
      D2Y=H*(Y(K-1)-T(K-1))**3
      Y(K)=Y(K-1)+H*(D2Y+1+(Y(K-1)-T(K-1))**2)
20    CONTINUE
```

Table 1 below shows the results of these computations for $a=1$, $N=10$, $t_0=0$, and $y_0=0.5$.

Table 1

t	y	t	y
0	0.5	0.6	1.31331931
0.1	0.62625	0.7	1.4678313
0.2	0.7554013	0.8	1.63131465
0.3	0.88796161	0.9	1.80616814
0.4	1.02456407	1	1.99572313
0.5	1.1660084		

Now Euler's method with $N=10$ predicted a value of 1.9422 for $y(1)$. Notice how much closer the number 1.9957 is to the correct value 2. If we run this program for $N=20$ and $N=40$, we obtain that $y_{20}=1.99884247$ and $y_{40}=1.99969915$. These numbers are also much more accurate than the values 1.96852339 and 1.9835109 predicted by Euler's method.

EXERCISES

Using the three term Taylor series method with $h=0.1$, determine an approximate value of the solution at $t=1$ for each of the initial-value problems 1–5. Repeat these computations with $h=0.025$ and compare the results with the given value of the solution.

1. $dy/dt=1+t-y, \quad y(0)=0; \quad (y(t)=t)$

2. $dy/dt=2ty, \quad y(0)=2; \quad (y(t)=2e^{t^2})$

3. $dy/dt=1+y^2-t^2, \quad y(0)=0; \quad (y(t)=t)$

4. $dy/dt=te^{-y}+t/(1+t^2), \quad y(0)=0; \quad (y(t)=\ln(1+t^2))$

5. $dy/dt=-1+2t+y^2/(1+t^2)^2, \quad y(0)=1; \quad (y(t)=1+t^2)$

6. Using the three term Taylor series method with $h=\pi/40$, determine an approximate value of the solution of the initial-value problem

$$\frac{dy}{dt}=2\sec^2 t-(1+y^2), \qquad y(0)=0$$

at $t=\pi/4$. Repeat these computations with $h=\pi/160$ and compare the results with the number one which is the value of the solution $y(t)=\tan t$ at $t=\pi/4$.

1.15 An improved Euler method

The three term Taylor series method is a significant improvement over Euler's method. However, it has the serious disadvantage of requiring us to compute partial derivatives of $f(t,y)$, and this can be quite difficult if the function $f(t,y)$ is fairly complicated. For this reason we would like to derive numerical schemes which do not require us to compute partial derivatives of $f(t,y)$. One approach to this problem is to integrate both sides of the differential equation $y'=f(t,y)$ between t_k and t_k+h to obtain that

$$y(t_{k+1})=y(t_k)+\int_{t_k}^{t_k+h} f(t,y(t))\,dt. \tag{1}$$

This reduces the problem of finding an approximate value of $y(t_{k+1})$ to the much simpler problem of approximating the area under the curve $f(t,y(t))$ between t_k and t_k+h. A crude approximation of this area is $hf(t_k,y(t_k))$, which is the area of the rectangle R in Figure 1a. This gives rise to the numerical scheme

$$y_{k+1}=y_k+hf(t_k,y_k)$$

which, of course, is Euler's method.

A much better approximation of this area is

$$\frac{h}{2}\left[f(t_k,y(t_k))+f(t_{k+1},y(t_{k+1}))\right]$$

which is the area of the trapezoid T in Figure 1b. This gives rise to the numerical scheme

$$y_{k+1}=y_k+\frac{h}{2}\left[f(t_k,y_k)+f(t_{k+1},y_{k+1})\right]. \tag{2}$$

However, we cannot use this scheme to determine y_{k+1} from y_k since y_{k+1} also appears on the right-hand side of (2). A very clever way of overcoming this difficulty is to replace y_{k+1} in the right-hand side of (2) by the value $y_k+hf(t_k,y_k)$ predicted for it by Euler's method. This gives rise to the numerical scheme

$$y_{k+1}=y_k+\frac{h}{2}\left[f(t_k,y_k)+f(t_k+h,y_k+hf(t_k,y_k))\right], \qquad y_0=y(t_0). \tag{3}$$

Equation (3) is known as the *improved Euler method*. It can be shown that $|y(t_k)-y_k|$ is at most a fixed constant times h^2. Hence, the improved Euler method gives us the same accuracy as the three term Taylor series method without requiring us to compute partial derivatives.

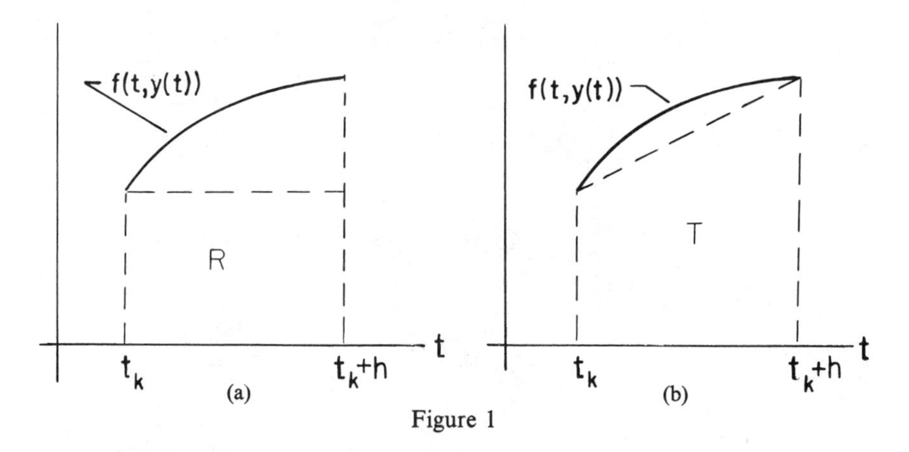

Figure 1

Example 1. Let $y(t)$ be the solution of the initial-value problem

$$\frac{dy}{dt}=1+(y-t)^2, \qquad y(0)=\frac{1}{2}.$$

Use the improved Euler method to compute approximate values of $y(t)$ at the points $t_k=k/N$, $k=1,\dots,N$.

Solution. The improved Euler scheme for this problem is

$$y_{k+1}=y_k+\frac{h}{2}\left\{1+(y_k-t_k)^2+1+\left[y_k+h\left(1+(y_k-t_k)^2\right)-t_{k+1}\right]^2\right\}$$

with $h=1/N$ and $y_0=0.5$. The integer k runs from 0 to $N-1$. Sample APL and Fortran programs to compute $y_1,\dots,y_N$ are given below. Again, these programs have variable values for t_0, y_0, a, and N.

APL Program

```
     ∇ IMPROVED
 [1]  T←Nρ0
 [2]  Y←Nρ0
 [3]  H←A÷N
 [4]  T[1]←T0+H
 [5]  R←1+(Y0−T0)*2
 [6]  Y[1]←Y0+(H÷2)×R+1+(Y0+(H×R)−T[1])*2
 [7]  K←1
 [8]  T[K+1]←T[K]+H
 [9]  R←1+(Y[K]−T[K])*2
[10]  Y[K+1]←Y[K]+(H÷2)×R+1+(Y[K]+(H×R)−T[K+1])*2
[11]  K←K+1
[12]  →8×ιK<N
[13]  T←T0,T
[14]  Y←Y0,Y
[15]  ⍉(2,ρY)ρT,Y      ∇
```

Fortran Program

Replace Section B of the Fortran program in Example 1 of Section 1.13 by the following:

```
      T(1)=T0+H
      R=1+(Y0−T0)**2
      Y(1)=Y0+(H/2)*(R+1+(Y0+(H*R)−T(1))**2)
      DO 20 K=2,N
      T(K)=T(K−1)+H
      R=1+(Y(K−1)−T(K−1))**2
      Y(K)=Y(K−1)+(H/2)*(R+1+(Y(K−1)+(H*R)−T(K))**2)
20    CONTINUE
```

Table 1 below shows the results of these computations for $a=1$, $N=10$, $t_0=0$, and $y_0=0.5$. If we run this program for $N=20$ and $N=40$ we obtain that $y_{20}=1.99939944$ and $y_{40}=1.99984675$. Hence the values y_{10}, y_{20}, and y_{40} computed by the improved Euler method are even closer to the correct value 2 than the corresponding values 1.99572313, 1.99884246, and 1.99969915 computed by the three term Taylor series method.

Table 1

t	y	t	y
0	0.5	0.6	1.31377361
0.1	0.62628125	0.7	1.46848715
0.2	0.75547445	0.8	1.63225727
0.3	0.88809117	0.9	1.80752701
0.4	1.02477002	1	1.99770114
0.5	1.16631867		

EXERCISES

Using the improved Euler method with $h=0.1$, determine an approximate value of the solution at $t=1$ for each of the initial-value problems 1–5. Repeat these computations with $h=0.025$ and compare the results with the given value of the solution.

1. $dy/dt=1+t-y,\quad y(0)=0;\quad (y(t)=t)$

2. $dy/dt=2ty,\quad y(0)=2;\quad (y(t)=2e^{t^2})$

3. $dy/dt=1+y^2-t^2,\quad y(0)=0;\quad (y(t)=t)$

4. $dy/dt=te^{-y}+t/(1+t^2),\quad y(0)=0;\quad (y(t)=\ln(1+t^2))$

5. $dy/dt=-1+2t+y^2/(1+t^2)^2,\quad y(0)=1;\quad (y(t)=1+t^2)$

6. Using the improved Euler method with $h=\pi/40$, determine an approximate value of the solution of the initial-value problem

$$\frac{dy}{dt}=2\sec^2 t-(1+y^2),\qquad y(0)=0$$

at $t=\pi/4$. Repeat these computations with $h=\pi/160$ and compare the results with the number one which is the value of the solution $y(t)=\tan t$ at $t=\pi/4$.

1.16 The Runge–Kutta method

We now present, without proof, a very powerful scheme which was developed around 1900 by the mathematicians Runge and Kutta. Because of its simplicity and great accuracy, the Runge–Kutta method is still one of the most widely used numerical schemes for solving differential equations. It is defined by the equation

$$y_{k+1}=y_k+\frac{h}{6}\left[L_{k,1}+2L_{k,2}+2L_{k,3}+L_{k,4}\right], \qquad k=0,1,\ldots,N-1$$

where $y_0=y(t_0)$ and

$$L_{k,1}=f(t_k,y_k), \qquad L_{k,2}=f\left(t_k+\tfrac{1}{2}h,y_k+\tfrac{1}{2}hL_{k,1}\right)$$
$$L_{k,3}=f\left(t_k+\tfrac{1}{2}h,y_k+\tfrac{1}{2}hL_{k,2}\right), \qquad L_{k,4}=f(t_k+h,y_k+hL_{k,3}).$$

This formula involves a weighted average of values of $f(t,y)$ taken at different points. Hence the sum $\frac{1}{6}[L_{k,1}+2L_{k,2}+2L_{k,3}+L_{k,4}]$ can be interpreted as an average slope It can be shown that the error $|y(t_k)-y_k|$ is at most a fixed constant times h^4. Thus, the Runge–Kutta method is much more accurate than Euler's method, the three term Taylor series method and the improved Euler method.

Example 1. Let $y(t)$ be the solution of the initial-value problem

$$\frac{dy}{dt}=1+(y-t)^2, \qquad y(0)=\tfrac{1}{2}.$$

Use the Runge–Kutta method to find approximate values $y_1,\ldots,y_N$ of y at the points $t_k=k/N$, $k=1,\ldots,N$.

Solution. Sample APL and Fortran programs to compute $y_1,\ldots,y_N$ by the Runge–Kutta method are given below. These programs differ from our previous programs in that they do not compute y_1 separately. Rather, they compute y_1 in the same "loop" as they compute $y_2,\ldots,y_N$. This is accomplished by relabeling the numbers t_0 and y_0 as t_1 and y_1 respectively.

APL Program

```
      ∇ RUNKUT
 [1]  T←,T0
 [2]  Y←,Y0
 [3]  H←A÷N
 [4]  K←1
 [5]  T←T,T[K]+H
 [6]  LK1←1+(Y[K]−T[K])*2
 [7]  LK2←1+(Y[K]+(H×LK1÷2)−T[K]+H÷2)*2
 [8]  LK3←1+(Y[K]+(H×LK2÷2)−T[K]+H÷2)*2
 [9]  LK4←1+(Y[K]+(H×LK3)−T[K]+H)*2
[10]  Y←Y,Y[K]+(H÷6)×LK1+LK4+2×LK2+LK3
[11]  K←K+1
[12]  →5×⍳K≤N
[13]  ⍉(2,⍴Y)⍴T,Y  ∇
```

Fortran Program

```
      DIMENSION T(1000), Y(1000)
      READ (5,10)  T(1), Y(1), A, N
 10   FORMAT (3F20.8, I5)
      H=A/N
      DO 20 K=1,N
      T(K+1)=T(K)+H
      REAL  LK1,  LK2,  LK3,  LK4
      LK1 =1 +(Y(K)-T(K))* *2
      LK2=1 +((Y(K)+(H/2)*LK1)-(T(K)+H/2))* *2
      LK3=1 +((Y(K)+(H/2)*LK2)-(T(K)+H/2))* *2
      LK4=1 +((Y(K)+H*LK3)-(T(K)+H))* *2
      Y(K+1)=Y(K)+(H/6)*(LK1 +LK4+2*(LK2+LK3))
 20   CONTINUE
      NA=N+1
      WRITE (6,30)  (T(J), Y(J), J=1, NA)
 30   FORMAT (1H1, 3X, 1HT, 4X, 1HY, /(1H, 1X, F10.7, 2X, F20.9/))
      CALL EXIT
      END
```

Table 1 below shows the results of these computations for $a=1$, $N=10$, $t_0=0$, and $y_0=0.5$.

Table 1

t	y	t	y
0	0.5	0.6	1.31428555
0.1	0.62631578	0.7	1.4692305
0.2	0.75555536	0.8	1.6333329
0.3	0.88823526	0.9	1.8090902
0.4	1.02499993	1	1.9999988
0.5	1.16666656		

Notice how much closer the number $y_{10}=1.9999988$ computed by the Runge–Kutta method is to the correct value 2 than the numbers $y_{10}=1.94220484$, $y_{10}=1.99572312$, and $y_{10}=1.99770114$ computed by the Euler, three term Taylor series and improved Euler methods, respectively. If we run this program for $N=20$ and $N=40$, we obtain that $y_{20}=1.99999992$ and $y_{40}=2$. Thus, our approximation of $y(1)$ is already correct to eight decimal places when $h=0.025$. Equivalently, we need only choose $N \geqslant 40$ to achieve eight decimal places accuracy.

To put the accuracy of the various schemes into proper perspective, let us say that we have three different schemes for numerically solving the initial-value problem $dy/dt=f(t,y)$, $y(0)=0$ on the interval $0 \leqslant t \leqslant 1$, and that the error we make in using these schemes is $3h$, $11h^2$, and $42h^4$, respectively. If our problem is such that we require eight decimal places accuracy, then the step sizes h_1, h_2, and h_3 of these three schemes must satisfy the inequalities $3h_1 \leqslant 10^{-8}$, $11h_2^2 \leqslant 10^{-8}$, and $42h_3^4 \leqslant 10^{-8}$. Hence, the number of iterations N_1, N_2, and N_3 of these three schemes must satisfy the inequalities

$$N_1 \geqslant 3\times 10^8 = 300{,}000{,}000, \qquad N_2 \geqslant \sqrt{11}\times 10^4 \approx 34{,}000$$

and

$$N_3 \geqslant (42)^{1/4}\times 10^2 \approx 260.$$

This is a striking example of the difference between the Runge–Kutta method and the Euler, improved Euler and three term Taylor series methods.

Remark. It should be noted that we perform four functional evaluations at each step in the Runge–Kutta method, whereas we only perform one functional evaluation at each step in Euler's method. Nevertheless, the Runge–Kutta method still beats the heck out of Euler's method, the three term Taylor series method, and the improved Euler method.

Exercises

Using the Runge–Kutta method with $h=0.1$, determine an approximate value of the solution at $t=1$ for each of the initial-value problems 1–5. Repeat these computations with $h=0.025$ and compare the results with the given value of the solution.

1. $dy/dt = 1+t-y, \quad y(0)=0; \quad (y(t)=t)$

2. $dy/dt = 2ty, \quad y(0)=2; \quad (y(t)=2e^{t^2})$

3. $dy/dt = 1+y^2-t^2, \quad y(0)=0; \quad (y(t)=t)$

4. $dy/dt = te^{-y}+t/(1+t^2), \quad y(0)=0; \quad (y(t)=\ln(1+t^2))$

5. $dy/dt = -1+2t+y^2/((1+t^2)^2), \quad y(0)=1; \quad (y(t)=1+t^2)$

6. Using the Runge–Kutta method with $h=\pi/40$, determine an approximate value of the solution of the initial-value problem

$$\frac{dy}{dt} = 2\sec^2 t-(1+y^2), \qquad y(0)=0$$

at $t=\pi/4$. Repeat these computations with $h=\pi/160$ and compare the results with the number one which is the value of the solution $y(t)=\tan t$ at $t=\pi/4$.

1.17 What to do in practice

In this section we discuss some of the practical problems which arise when we attempt to solve differential equations on the computer. First, and foremost, is the problem of estimating the error that we make. It is not too difficult to show that the error we make using Euler's method, the three term Taylor series method, the improved Euler method and the Runge–Kutta method with step size h is at most c_1h, c_2h^2, c_3h^2, and c_4h^4, respectively. With one exception, though, it is practically impossible to find the constants c_1, c_2, c_3, and c_4. The one exception is Euler's method where we can explicitly estimate (see Section 1.13.1) the error we make in approximating $y(t_k)$ by y_k. However, this estimate is not very useful, since it is only valid for t_k sufficiently close to t_0, and we are usually interested in the values of y at times t much larger than t_0. Thus, we usually do not know, a priori, how small to choose the step size h so as to achieve a desired accuracy. We only know that the approximate values y_k that we compute get closer and closer to $y(t_k)$ as h gets smaller and smaller.

One way of resolving this difficulty is as follows. Using one of the schemes presented in the previous section, we choose a step size h and compute numbers $y_1,\ldots,y_N$. We then repeat the computations with a step size $h/2$ and compare the results. If the changes are greater than we are willing to accept, then it is necessary to use a smaller step size. We keep repeating this process until we achieve a desired accuracy. For example, suppose that we require the solution of the initial-value problem $y'=f(t,y)$, $y(0)=y_0$ at $t=1$ to four decimal places accuracy. We choose a step size $h=1/100$, say, and compute $y_1,\ldots,y_{100}$. We then repeat these computations with $h=1/200$ and obtain new approximations $z_1,\ldots,z_{200}$. If y_{100} and z_{200} agree in their first four decimal places then we take z_{200} as our approximation of $y(1)$.* If y_{100} and z_{200} do not agree in their first four decimal places, then we repeat our computations with step size $h=1/400$.

Example 1. Find the solution of the initial-value problem

$$\frac{dy}{dt}=y(1+e^{-y})+e^{t}, \qquad y(0)=0$$

at $t=1$ to four decimal places accuracy.

Solution. We illustrate how to try and solve this problem using Euler's method, the three term Taylor series method, the improved Euler method, and the Runge–Kutta method.

*This does not guarantee that z_{200} agrees with $y(1)$ to four decimal places. As an added precaution, we might halve the step size again. If the first four decimal places still remain unchanged, then we can be reasonably certain that z_{200} agrees with $y(1)$ to four decimal places.

However, it is certainly possible for "machine convergence" to produce a wrong answer!

(i) *Euler's method*:

APL Program

```
    ∇ EULER
[1]  T←,T0
[2]  Y←,Y0
[3]  H←A÷N
[4]  K←1
[5]  T←T,T[K]+H
[6]  Y←Y,Y[K]+H×(Y[K]×1+*−Y[K])+*T[K]
[7]  K←K+1
[8]  →5×ιK≤N
[9]  N,H,Y[N+1]  ∇
```

Fortran Program

```
Section A             |    |      DIMENSION T(1000), Y(1000)
Read in data          |    |      READ (5,10) T(1),Y(1),A,N,
                      | 10 |      FORMAT (3F20.8,I5)
                      |    |      H=A/N

Section B             |    |      DO20 K=1,N
Do computations       |    |      T(K+1)=T(K)+H
                      |    |      Y(K+1)=Y(K)+H*(Y(K)*(1+EXP(-Y(K)))
                      |    |    1          +EXP(T(K)))
                      | 20 |      CONTINUE

Section C             |    |      WRITE (6,30) N,H,Y(N+1)
Print out             | 30 |      FORMAT (1H,1X,I5,2X,F10.7,4X,F20.9)
results               |    |      CALL EXIT
                      |    |      END
```

We set $A=1, T0=0, Y0=0$ ($T(1)=Y(1)=0$ in the Fortran program) and ran these programs for $N=10$, 20, 40, 80, 160, 320, and 640. The results of these computations are given in Table 1. Notice that even with a step size h

Table 1

N	h	y_N
10	0.1	2.76183168
20	0.05	2.93832741
40	0.025	3.03202759
80	0.0125	3.08034440
160	0.00625	3.10488352
320	0.003125	3.11725009
640	0.0015625	3.12345786

as small as $1/640$, we can only guarantee an accuracy of one decimal place. This points out the limitation of Euler's method. Since N is so large already, it is wiser to use a more accurate scheme than to keep choosing smaller and smaller step sizes h for Euler's method.

(ii) *The three term Taylor series method*

APL Program

```
      ∇ TAYLOR
[1]   T←T0
[2]   Y←,Y0
[3]   H←A÷N
[4]   K←1
[5]   T←T,T[K]+H
[6]   DY1←1+(1−Y[K])×*−Y[K]
[7]   DY2←(Y[K]×1+*−Y[K])+*T[K]
[8]   Y←Y,Y[K]+(H×DY2)+(H×H÷2)×(*T[K])+DY1×DY2
[9]   K←K+1
[10]  →5×⍳K≤N
[11]  N,H,Y[N+1]  ∇
```

Fortran Program

Replace Section B of the previous Fortran program by

```
      DO20 K=1,N
      T(K+1)=T(K)+H
      DY1=1+(1−Y(K))*EXP(−Y(K))
      DY2=Y(K)*(1+EXP(−Y(K)))+EXP(T(K))
      Y(K+1)=Y(K)+H*DY2+(H*H/2)*(EXP(T(K))+DY1*DY2)
   20 CONTINUE
```

We set $A=1$, $T0=0$, and $Y0=0$ ($T(1)=0$, and $Y(1)=0$ in the Fortran program) and ran these programs for $N=10$, 20, 40, 60, 80, 160, and 320. The results of these computations are given in Table 2. Observe that y_{160}

Table 2

N	h	y_N
10	0.1	3.11727674
20	0.05	3.12645293
40	0.025	3.12885845
80	0.0125	3.12947408
160	0.00625	3.12962979
320	0.003125	3.12966689

and y_{320} agree in their first four decimal places. Hence the approximation $y(1)=3.12966689$ is correct to four decimal places.

(iii) *The improved Euler method*

APL Program

```
      ∇ IMPROVED
 [1]  T←,T0
 [2]  Y←,Y0
 [3]  H←A÷N
 [4]  K←1
 [5]  T←T,T[K]+H
 [6]  R←H×(Y[K]×1+*−Y[K])+*T[K]
 [7]  Y←Y,Y[K]+(÷2)×R+H×((Y[K]+R)×1+*−Y[K]+R)+*T[K+1]
 [8]  K←K+1
 [9]  →5×ιK≤N
[10]  N,H,Y[N+1]  ∇
```

Fortran Program

Replace Section B of the first Fortran program in this section by

```
      DO 20 K=1,N
      T(K+1)=T(K)+H
      R1=Y(K)*(1+EXP(−Y(K)))+EXP(T(K))
      R2=(Y(K)+H*R1)*(1+EXP(−(Y(K)+H*R1)))+EXP(T(K+1))
      Y(K+1)=Y(K)+(H/2)*(R1+R2)
 20   CONTINUE
```

We set $A=1$, $T0=0$ and $Y0=0$ ($T(1)=0$ and $Y(1)=0$ in the Fortran program) and ran these programs for $N=10$, 20, 40, 80, 160, and 320. The results of these computations are given in Table 3. Observe that y_{160} and y_{320}

Table 3

N	h	y_N
10	0.1	3.11450908
20	0.05	3.12560685
40	0.025	3.1286243
80	0.0125	3.12941247
160	0.00625	3.12961399
320	0.003125	3.12964943

agree in their first four decimal places. Hence the approximation $y(1) = 3.12964943$ is correct to four decimal places.

(iv) *The method of Runge–Kutta*

APL Program

```
      ∇ RUNKUT
 [1]  T←,T0
 [2]  Y←,Y0
 [3]  H←A÷N
 [4]  K←1
 [5]  T←T,T[K]+H
 [6]  LK1←(Y[K]×1+*−Y[K])+*T[K]
 [7]  LK2←((Y[K]+(H÷2)×LK1)×1+*−Y[K]+(H÷2)×LK1)+*T[K]+
      H÷2
 [8]  LK3←((Y[K]+(H÷2)×LK2)×1+*−Y[K]+(H÷2)×LK2)+*T[K]+
      H÷2
 [9]  LK4←((Y[K]+H×LK3)×1+*−Y[K]+H×LK3)+*T[K]+H
[10]  Y←Y,Y[K]+(H÷6)×LK1+LK4+2×LK2+LK3
[11]  K←K+1
[12]  →5×ιK≤N
[13]  N,H,Y[N+1]  ∇
```

Fortran Program

Replace Section B of the first Fortran program in this section by

```
      DO20 K=1,N
      T(K+1)=T(K)+H
      LK1=Y(K)*(1+EXP(-Y(K))+EXP(T(K))
      LK2=(Y(K)+(H/2)*LK1)*(1+EXP(-(Y(K)+(H/2)*LK1)))
     1 +EXP(T(K)+(H/2))
      LK3=(Y(K)+(H/2)*LK2)*(1+EXP(-(Y(K)+(H/2)*LK2)))
     1 +EXP(T(K)+(H/2))
      LK4=(Y(K)+H*LK3)*(1+EXP(-(Y(K)+H*LK3)))
     1 +EXP(T(K+1))
   20 Y(K+1)=Y(K)+(H/6)*(LK1+2*LK2+2*LK3+LK4)
      CONTINUE
```

We set $A = 1$, $T0 = 0$ and $Y0 = 0$ ($T(1) = 0$ and $Y(1) = 0$ in the Fortran program) and ran these programs for $N = 10$, 20, 40, 80, 160, and 320. The results of these computations are given in Table 4. Notice that our approximation of $y(1)$ is already correct to four decimal places with $h = 0.1$, and that it is already correct to eight decimal places with $h=0.00625 (N=160)$. This example again illustrates the power of the Runge–Kutta method.

Table 4

N	h	y_N
10	0.1	3.1296517
20	0.05	3.12967998
40	0.025	3.1296819
80	0.0125	3.12968203
160	0.00625	3.12968204
320	0.003125	3.12968204

We conclude this section with two examples which point out some additional difficulties which may arise when we solve initial-value problems on a digital computer.

Example 2. Use the Runge–Kutta method to find approximate values of the solution of the initial-value problem

$$\frac{dy}{dt}=t^2+y^2, \qquad y(0)=1$$

at the points $t_k=k/N$, $k=1,\ldots,N$.
Solution.

APL Program

```
      ∇ RUNKUT
 [1]  T←,T0
 [2]  Y←,Y0
 [3]  H←A÷N
 [4]  K←1
 [5]  T←T,T[K]+H
 [6]  LK1←(T[K]*2)+Y[K]*2
 [7]  LK2←((T[K]+H÷2)*2)+(Y[K]+H×LK1÷2)*2
 [8]  LK3←((T[K]+H÷2)*2)+(Y[K]+H×LK2÷2)*2
 [9]  LK4←((T[K]+H)*2)+(Y[K]+H×LK3)*2
[10]  Y←Y,Y[K]+(H÷6)×LK1+LK4+2×LK2+LK3
[11]  K←K+1
[12]  →5×⍳K≤N
[13]  ⍉(2,⍴Y)⍴T,Y  ∇
```

Fortran Program

Replace Sections B and C of the first Fortran program in this section by

```
Section B              DO20 K=1,N
Do computa-            T(K+1)=T(K)+H
  tions                LK1=T(K)**2+Y(K)**2
                       LK2=(T(K)+(H/2))**2+(Y(K)+(H/2)*LK1)**2
                       LK3=(T(K)+(H/2))**2+(Y(K)+(H/2)*LK2)**2
                       LK4=(T(K)+H)**2+(Y(K)+H*LK3)**2
                       Y(K+1)=Y(K)+(H/6)*(LK1+2*LK2+2*LK3+LK4)
                20     CONTINUE

Section C              NA=N+1
Print out              WRITE (6,30) (T(J),Y(J),J=1,NA)
results         30     FORMAT (1H1,3X,1HT,4X,1HY/(1H,1X,F9.7,
                     1 2X,F20.9/))
                       CALL EXIT
                       END
```

We attempted to run these programs with $A=1$, $T0=0$, $Y0=1$ ($T(1)=0$, and $Y(1)=1$ in the Fortran program) and $N=10$, but we received an error message that the numbers being computed exceeded the domain of the computer. That is to say, they were larger than 10^{38}. This indicates that the solution $y(t)$ goes to infinity somewhere in the interval $[0,1]$. We can prove this analytically, and even obtain an estimate of where $y(t)$ goes to infinity, by the following clever argument. Observe that for $0 \leqslant t \leqslant 1$, $y(t)$ is never less than the solution $\phi_1(t)=1/(1-t)$ of the initial-value problem

$$\frac{dy}{dt}=y^2, \qquad y(0)=1.$$

In addition, $y(t)$ never exceeds the solution $\phi_2(t)=\tan(t+\pi/4)$ of the initial-value problem $dy/dt=1+y^2$, $y(0)=1$. Hence, for $0 \leqslant t \leqslant 1$,

$$\frac{1}{1-t} \leqslant y(t) \leqslant \tan(t+\pi/4).$$

This situation is described graphically in Figure 1. Since $\phi_1(t)$ and $\phi_2(t)$ become infinite at $t=1$ and $t=\pi/4$ respectively, we conclude that $y(t)$ becomes infinite somewhere between $\pi/4$ and 1.

The solutions of most initial-value problems which arise in physical and biological applications exist for all future time. Thus, we need not be overly concerned with the problem of solutions going to infinity in finite time or the problem of solutions becoming exceedingly large. On the other hand, though, there are several instances in economics where this problem is of paramount importance. In these instances, we are often interested in

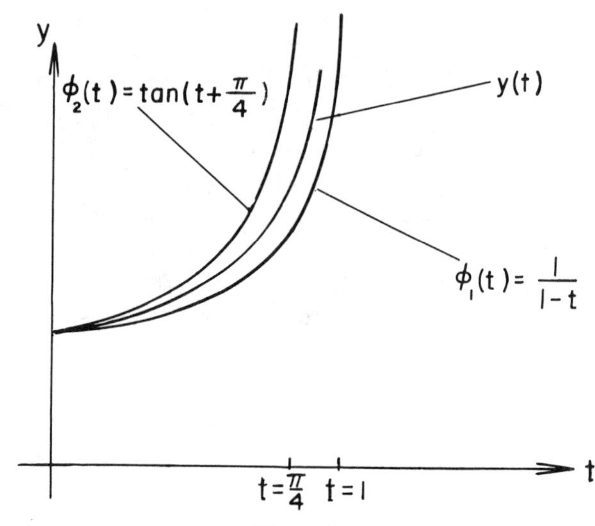

Figure 1

determining whether certain differential equations can accurately model a given economic phenomenon. It is often possible to eliminate several of these equations by showing that they allow solutions which are unrealistically large.

Example 3. Use Euler's method to determine approximate values of the solution of the initial-value problem

$$\frac{dy}{dt}=y|y|^{-3/4}+t\sin\frac{\pi}{t}, \qquad y(0)=0 \tag{1}$$

at the points $1/N, 2/N, \ldots, 2$.

Solution. The programming for this problem is simplified immensely by observing that

$$y|y|^{-3/4}=(\operatorname{sgn} y)|y|^{1/4}, \quad \text{where} \quad \operatorname{sgn} y=\begin{cases} 1, & y>0 \\ 0, & y=0 \\ -1, & y<0 \end{cases}$$

APL Program

```
    ∇ EULER
[1]  T←Nρ0
[2]  Y←Nρ0
[3]  H←2÷N
[4]  T[1]←H
[5]  K←1
[6]  T[K+1]←T[K]+H
```

```
 [7]  Y[K+1]←Y[K]+H×((×Y[K])×(|Y[K])*0.25)+T[K]×1○○÷T[K]
 [8]  K←K+1
 [9]  →6×ιK<N
[10]  T←0,T
[11]  Y←0,Y
[12]  ⍉(2,ρY)ρT,Y  ∇
```

Fortran Program

```
         DIMENSION T(1000),Y(1000)
         READ (5,10) N
10       FORMAT (I5)
         H=2/N
         T(1)=H
         Y(1)=0
         D0 20 K=2,N
         T(K)=T(K-1)+H
         Y(K)=Y(K-1)+H*(SIGN(Y(K-1)**0.25,Y(K-1))+T(K-1)*
       1 SIN(3.141592654/T(K-1))
20       CONTINUE
         WRITE(6,30) 0,0,(T(J),Y(J),J=1,N)
30       FORMAT (1H1,3X,1HT,4X,1HY/(1H,1X,F10.7,2X,F20.9/))
         CALL EXIT
         END
```

When we set $N=25$ we obtained the value 2.4844172 for $y(2)$, but when we set $N=27$, we obtained the value -0.50244575 for $y(2)$. Moreover, all the y_k were positive for $N=25$ and negative for $N=27$. We repeated these computations with $N=89$ and $N=91$ and obtained the values 2.64286349 and -0.6318074 respectively. In addition, all the y_k were again positive for $N=89$ and negative for $N=91$. Indeed, it is possible, but rather difficult, to prove that all the y_k will be positive if $N=1,5,9,13,17,\ldots$ and negative if $N=3,7,11,15,\ldots$. This suggests that the solution of the initial-value problem (1) is not unique. We cannot prove this analytically, since we cannot solve the differential equation explicitly. It should be noted though, that the existence–uniqueness theorem of Section 1.10 does not apply here, since the partial derivative with respect to y of the function $|y|^{-3/4}y+t\sin\pi/t$ does not exist at $y=0$.

Most of the initial-value problems that arise in applications have unique solutions. Thus, we need not be overly concerned with the problem of non-uniqueness of solutions. However, we should always bear in mind that initial-value problems which do not obey the hypotheses of the existence–uniqueness theorem of Section 1.10 might possess more than one solution, for the consequences of picking the wrong solution in these rare instances can often be catastrophic.

EXERCISES

In each of Problems 1–5, find the solution of the given initial-value problem at $t=1$ to four decimal places accuracy.

1. $\dfrac{dy}{dt}=y+e^{-y}+2t,\quad y(0)=0$

2. $\dfrac{dy}{dt}=1-t+y^2,\quad y(0)=0$

3. $\dfrac{dy}{dt}=\dfrac{t^2+y^2}{1+t+y^2},\quad y(0)=0$

4. $\dfrac{dy}{dt}=e^t y^2-2y,\quad y(0)=1$

5. $\dfrac{dy}{dt}=ty^3-y,\quad y(0)=1$

Second-order linear differential equations 2

2.1 Algebraic properties of solutions

A second-order differential equation is an equation of the form

$$\frac{d^2y}{dt^2}=f\left(t,y,\frac{dy}{dt}\right). \tag{1}$$

For example, the equation

$$\frac{d^2y}{dt^2}=\sin t+3y+\left(\frac{dy}{dt}\right)^2$$

is a second-order differential equation. A function $y=y(t)$ is a solution of (1) if $y(t)$ satisfies the differential equation; that is

$$\frac{d^2y(t)}{dt^2}=f\left(t,y(t),\frac{dy(t)}{dt}\right).$$

Thus, the function $y(t)=\cos t$ is a solution of the second-order equation $d^2y/dt^2=-y$ since $d^2(\cos t)/dt^2=-\cos t$.

Second-order differential equations arise quite often in applications. The most famous second-order differential equation is Newton's second law of motion (see Section 1.7)

$$m\frac{d^2y}{dt^2}=F\left(t,y,\frac{dy}{dt}\right)$$

which governs the motion of a particle of mass m moving under the influence of a force F. In this equation, m is the mass of the particle, $y=y(t)$ is its position at time t, dy/dt is its velocity, and F is the total force acting on the particle. As the notation suggests, the force F may depend on the position and velocity of the particle, as well as on time.

In addition to the differential equation (1), we will often impose initial conditions on $y(t)$ of the form

$$y(t_0)=y_0, \qquad y'(t_0)=y'_0. \tag{1'}$$

The differential equation (1) together with the initial conditions (1′) is referred to as an initial-value problem. For example, let $y(t)$* denote the position at time t of a particle moving under the influence of gravity. Then, $y(t)$ satisfies the initial-value problem

$$\frac{d^2y}{dt^2}=-g; \qquad y(t_0)=y_0, \quad y'(t_0)=y'_0,$$

where y_0 is the initial position of the particle and y'_0 is the initial velocity of the particle.

Second-order differential equations are extremely difficult to solve. This should not come as a great surprise to us after our experience with first-order equations. We will only succeed in solving the special differential equation

$$\frac{d^2y}{dt^2}+p(t)\frac{dy}{dt}+q(t)y=g(t). \tag{2}$$

Fortunately, though, many of the second-order equations that arise in applications are of this form.

The differential equation (2) is called a second-order linear differential equation. We single out this equation and call it linear because both y and dy/dt appear by themselves. For example, the differential equations

$$\frac{d^2y}{dt^2}+3t\frac{dy}{dt}+(\sin t)y=e^t$$

and

$$\frac{d^2y}{dt^2}+e^t\frac{dy}{dt}+2y=1$$

are linear, while the differential equations

$$\frac{d^2y}{dt^2}+3\frac{dy}{dt}+\sin y=t^3$$

and

$$\frac{d^2y}{dt^2}+\left(\frac{dy}{dt}\right)^2=1$$

are both nonlinear, due to the presence of the $\sin y$ and $(dy/dt)^2$ terms, respectively.

We consider first the second-order linear homogeneous equation

$$\frac{d^2y}{dt^2}+p(t)\frac{dy}{dt}+q(t)y=0 \tag{3}$$

*The positive direction of y is taken upwards.

which is obtained from (2) by setting $g(t)=0$. It is certainly not obvious at this point how to find all the solutions of (3), or how to solve the initial-value problem

$$\frac{d^2y}{dt^2}+p(t)\frac{dy}{dt}+q(t)y=0; \qquad y(t_0)=y_0,\ \ y'(t_0)=y_0'. \tag{4}$$

Therefore, before trying to develop any elaborate procedures for solving (4), we should first determine whether it actually has a solution. This information is contained in the following theorem, whose proof will be indicated in Chapter 4.

Theorem 1. (Existence–uniqueness Theorem). *Let the functions $p(t)$ and $q(t)$ be continuous in the open interval $\alpha<t<\beta$. Then, there exists one, and only one function $y(t)$ satisfying the differential equation* (3) *on the entire interval $\alpha<t<\beta$, and the prescribed initial conditions $y(t_0)=y_0$, $y'(t_0)=y_0'$. In particular, any solution $y=y(t)$ of* (3) *which satisfies $y(t_0)=0$ and $y'(t_0)=0$ at some time $t=t_0$ must be identically zero.*

Theorem 1 is an extremely important theorem for us. On the one hand, it is our hunting license to find the unique solution $y(t)$ of (4). And, on the other hand, we will actually use Theorem 1 to help us find all the solutions of (3).

We begin our analysis of Equation (3) with the important observation that the left-hand side

$$y''+p(t)y'+q(t)y \qquad \left(y'=\frac{dy}{dt},y''=\frac{d^2y}{dt^2}\right)$$

of the differential equation can be viewed as defining a "function of a function": with each function y having two derivatives, we associate another function, which we'll call $L[y]$, by the relation

$$L[y](t)=y''(t)+p(t)y'(t)+q(t)y(t).$$

In mathematical terminology, L is an operator which operates on functions; that is, there is a prescribed recipe for associating with each function y a new function $L[y]$.

Example 1. Let $p(t)=0$ and $q(t)=t$. Then,

$$L[y](t)=y''(t)+ty(t).$$

If $y(t)=\cos t$, then

$$L[y](t)=(\cos t)''+t\cos t=(t-1)\cos t,$$

and if $y(t)=t^3$, then

$$L[y](t)=(t^3)''+t(t^3)=t^4+6t.$$

Thus, the operator L assigns the function $(t-1)\cos t$ to the function $\cos t$, and the function $6t+t^4$ to the function t^3.

The concept of an operator acting on functions, or a "function of a function" is analogous to that of a function of a single variable t. Recall the definition of a function f on an interval I: with each number t in I we associate a new number called $f(t)$. In an exactly analogous manner, we associate with each function y having two derivatives a new function called $L[y]$. This is an extremely sophisticated mathematical concept, because in a certain sense, we are treating a function exactly as we do a point. Admittedly, this is quite difficult to grasp. It's not surprising, therefore, that the concept of a "function of a function" was not developed till the beginning of this century, and that many of the "high powered" theorems of mathematical analysis were proved only after this concept was mastered.

We now derive several important properties of the operator L, which we will use to great advantage shortly.

Property 1. $L[cy]=cL[y]$, *for any constant* c.

PROOF.
$$\begin{aligned} L[cy](t) &= (cy)''(t)+p(t)(cy)'(t)+q(t)(cy)(t) \\ &= cy''(t)+cp(t)y'(t)+cq(t)y(t) \\ &= c\left[y''(t)+p(t)y'(t)+q(t)y(t)\right] \\ &= cL[y](t). \end{aligned}$$ □

The meaning of Property 1 is that the operator L assigns to the function (cy) c times the function it assigns to y. For example, let

$$L[y](t)=y''(t)+6y'(t)-2y(t).$$

This operator L assigns the function

$$(t^2)''+6(t^2)'-2(t^2)=2+12t-2t^2$$

to the function t^2. Hence, L must assign the function $5(2+12t-2t^2)$ to the function $5t^2$.

Property 2. $L[y_1+y_2]=L[y_1]+L[y_2]$.

PROOF.

$$\begin{aligned} L[y_1+y_2](t) &= (y_1+y_2)''(t)+p(t)(y_1+y_2)'(t)+q(t)(y_1+y_2)(t) \\ &= y_1''(t)+y_2''(t)+p(t)y_1'(t)+p(t)y_2'(t)+q(t)y_1(t)+q(t)y_2(t) \\ &= \left[y_1''(t)+p(t)y_1'(t)+q(t)y_1(t)\right]+\left[y_2''(t)+p(t)y_2'(t)+q(t)y_2(t)\right] \\ &= L[y_1](t)+L[y_2](t). \end{aligned}$$ □

The meaning of Property 2 is that the operator L assigns to the function y_1+y_2 the sum of the functions it assigns to y_1 and y_2. For example, let

$$L[y](t)=y''(t)+2y'(t)-y(t).$$

This operator L assigns the function

$$(\cos t)'' + 2(\cos t)' - \cos t = -2\cos t - 2\sin t$$

to the function $\cos t$, and the function

$$(\sin t)'' + 2(\sin t)' - \sin t = 2\cos t - 2\sin t$$

to the function $\sin t$. Hence, L assigns the function

$$(-2\cos t - 2\sin t) + 2\cos t - 2\sin t = -4\sin t$$

to the function $\sin t + \cos t$.

Definition. An operator L which assigns functions to functions and which satisfies Properties 1 and 2 is called a linear operator. All other operators are nonlinear. An example of a nonlinear operator is

$$L[y](t) = y''(t) - 2t[y(t)]^4.$$

This operator assigns the function

$$\left(\frac{1}{t}\right)'' - 2t\left(\frac{1}{t}\right)^4 = \frac{2}{t^3} - \frac{2}{t^3} = 0$$

to the function $1/t$, and the function

$$\left(\frac{c}{t}\right)'' - 2t\left(\frac{c}{t}\right)^4 = \frac{2c}{t^3} - \frac{2c^4}{t^3} = \frac{2c(1-c^3)}{t^3}$$

to the function c/t. Hence, for $c \neq 0, 1$, and $y(t) = 1/t$, we see that $L[cy] \neq cL[y]$.

The usefulness of Properties 1 and 2 lies in the observation that the solutions $y(t)$ of the differential equation (3) are exactly those functions y for which

$$L[y](t) = y''(t) + p(t)y'(t) + q(t)y(t) = 0.$$

In other words, the solutions $y(t)$ of (3) are exactly those functions y to which the operator L assigns the zero function.* Hence, if $y(t)$ is a solution of (3) then so is $cy(t)$, since

$$L[cy](t) = cL[y](t) = 0.$$

If $y_1(t)$ and $y_2(t)$ are solutions of (3), then $y_1(t) + y_2(t)$ is also a solution of (3), since

$$L[y_1 + y_2](t) = L[y_1](t) + L[y_2](t) = 0 + 0 = 0.$$

Combining Properties 1 and 2, we see that all linear combinations

$$c_1 y_1(t) + c_2 y_2(t)$$

of solutions of (3) are again solutions of (3).

*The zero function is the function whose value at any time t is zero.

The preceding argument shows that we can use our knowledge of two solutions $y_1(t)$ and $y_2(t)$ of (3) to generate infinitely many other solutions. This statement has some very interesting implications. Consider, for example, the differential equation

$$\frac{d^2y}{dt^2}+y=0. \tag{5}$$

Two solutions of (5) are $y_1(t)=\cos t$ and $y_2(t)=\sin t$. Hence,

$$y(t)=c_1\cos t+c_2\sin t \tag{6}$$

is also a solution of (5), for every choice of constants c_1 and c_2. Now, Equation (6) contains two arbitrary constants. It is natural to suspect, therefore, that this expression represents the general solution of (5); that is, every solution $y(t)$ of (5) must be of the form (6). This is indeed the case, as we now show. Let $y(t)$ be any solution of (5). By the existence–uniqueness theorem, $y(t)$ exists for all t. Let $y(0)=y_0$, $y'(0)=y_0'$, and consider the function

$$\phi(t)=y_0\cos t+y_0'\sin t.$$

This function is a solution of (5) since it is a linear combination of solutions of (5). Moreover, $\phi(0)=y_0$ and $\phi'(0)=y_0'$. Thus, $y(t)$ and $\phi(t)$ satisfy the same second-order linear homogeneous equation and the same initial conditions. Therefore, by the uniqueness part of Theorem 1, $y(t)$ must be identically equal to $\phi(t)$, so that

$$y(t)=y_0\cos t+y_0'\sin t.$$

Thus, Equation (6) is indeed the general solution of (5).

Let us return now to the general linear equation (3). Suppose, in some manner, that we manage to find two solutions $y_1(t)$ and $y_2(t)$ of (3). Then, every function

$$y(t)=c_1y_1(t)+c_2y_2(t) \tag{7}$$

is again a solution of (3). Does the expression (7) represent the general solution of (3)? That is to say, does every solution $y(t)$ of (3) have the form (7)? The following theorem answers this question.

Theorem 2. *Let $y_1(t)$ and $y_2(t)$ be two solutions of* (3) *on the interval $\alpha<t<\beta$, with*

$$y_1(t)y_2'(t)-y_1'(t)y_2(t)$$

unequal to zero in this interval. Then,

$$y(t)=c_1y_1(t)+c_2y_2(t)$$

is the general solution of (3).

PROOF. Let $y(t)$ be any solution of (3). We must find constants c_1 and c_2 such that $y(t)=c_1y_1(t)+c_2y_2(t)$. To this end, pick a time t_0 in the interval

(α,β) and let y_0 and y_0' denote the values of y and y' at $t=t_0$. The constants c_1 and c_2, if they exist, must satisfy the two equations

$$c_1y_1(t_0)+c_2y_2(t_0)=y_0$$
$$c_1y_1'(t_0)+c_2y_2'(t_0)=y_0'.$$

Multiplying the first equation by $y_2'(t_0)$, the second equation by $y_2(t_0)$ and subtracting gives

$$c_1\left[y_1(t_0)y_2'(t_0)-y_1'(t_0)y_2(t_0)\right]=y_0y_2'(t_0)-y_0'y_2(t_0).$$

Similarly, multiplying the first equation by $y_1'(t_0)$, the second equation by $y_1(t_0)$ and subtracting gives

$$c_2\left[y_1'(t_0)y_2(t_0)-y_1(t_0)y_2'(t_0)\right]=y_0y_1'(t_0)-y_0'y_1(t_0).$$

Hence,

$$c_1=\frac{y_0y_2'(t_0)-y_0'y_2(t_0)}{y_1(t_0)y_2'(t_0)-y_1'(t_0)y_2(t_0)}$$

and

$$c_2=\frac{y_0'y_1(t_0)-y_0y_1'(t_0)}{y_1(t_0)y_2'(t_0)-y_1'(t_0)y_2(t_0)}$$

if $y_1(t_0)y_2'(t_0)-y_1'(t_0)y_2(t_0)\neq 0$. Now, let

$$\phi(t)=c_1y_1(t)+c_2y_2(t)$$

for this choice of constants c_1,c_2. We know that $\phi(t)$ satisfies (3), since it is a linear combination of solutions of (3). Moreover, by construction, $\phi(t_0)=y_0$ and $\phi'(t_0)=y_0'$. Thus, $y(t)$ and $\phi(t)$ satisfy the same second-order linear homogeneous equation and the same initial conditions. Therefore, by the uniqueness part of Theorem 1, $y(t)$ must be identically equal to $\phi(t)$; that is,

$$y(t)=c_1y_1(t)+c_2y_2(t),\qquad \alpha<t<\beta. \qquad \square$$

Theorem 2 is an extremely useful theorem since it reduces the problem of finding all solutions of (3), of which there are infinitely many, to the much simpler problem of finding just two solutions $y_1(t),y_2(t)$. The only condition imposed on the solutions $y_1(t)$ and $y_2(t)$ is that the quantity $y_1(t)y_2'(t)-y_1'(t)y_2(t)$ be unequal to zero for $\alpha<t<\beta$. When this is the case, we say that $y_1(t)$ and $y_2(t)$ are a *fundamental* set of solutions of (3), since all other solutions of (3) can be obtained by taking linear combinations of $y_1(t)$ and $y_2(t)$.

Definition. The quantity $y_1(t)y_2'(t)-y_1'(t)y_2(t)$ is called the *Wronskian* of y_1 and y_2, and is denoted by $W(t)=W[y_1,y_2](t)$.

Theorem 2 requires that $W[y_1,y_2](t)$ be unequal to zero at all points in the interval (α,β). In actual fact, the Wronskian of any two solutions

$y_1(t), y_2(t)$ of (3) is either identically zero, or is never zero, as we now show.

Theorem 3. *Let $p(t)$ and $q(t)$ be continuous in the interval $\alpha < t < \beta$, and let $y_1(t)$ and $y_2(t)$ be two solutions of (3). Then, $W[y_1, y_2](t)$ is either identically zero, or is never zero, on the interval $\alpha < t < \beta$.*

We prove Theorem 3 with the aid of the following lemma.

Lemma 1. *Let $y_1(t)$ and $y_2(t)$ be two solutions of the linear differential equation $y'' + p(t)y' + q(t)y = 0$. Then, their Wronskian*

$$W(t) = W[y_1, y_2](t) = y_1(t)y_2'(t) - y_1'(t)y_2(t)$$

satisfies the first-order differential equation

$$W' + p(t)W = 0.$$

PROOF. Observe that

$$\begin{aligned} W'(t) &= \frac{d}{dt}(y_1 y_2' - y_1' y_2) \\ &= y_1 y_2'' + y_1' y_2' - y_1' y_2' - y_1'' y_2 \\ &= y_1 y_2'' - y_1'' y_2. \end{aligned}$$

Since y_1 and y_2 are solutions of $y'' + p(t)y' + q(t)y = 0$, we know that

$$y_2'' = -p(t)y_2' - q(t)y_2$$

and

$$y_1'' = -p(t)y_1' - q(t)y_1.$$

Hence,

$$\begin{aligned} W'(t) &= y_1[-p(t)y_2' - q(t)y_2] - y_2[-p(t)y_1' - q(t)y_1] \\ &= -p(t)[y_1 y_2' - y_1' y_2] \\ &= -p(t)W(t). \end{aligned}$$

□

We can now give a very simple proof of Theorem 3.

PROOF OF THEOREM 3. Pick any t_0 in the interval (α, β). From Lemma 1,

$$W[y_1, y_2](t) = W[y_1, y_2](t_0)\exp\left(-\int_{t_0}^{t} p(s)\,ds\right).$$

Now, $\exp\left(-\int_{t_0}^{t} p(s)\,ds\right)$ is unequal to zero for $\alpha < t < \beta$. Therefore, $W[y_1, y_2](t)$ is either identically zero, or is never zero. □

The simplest situation where the Wronskian of two functions $y_1(t),y_2(t)$ vanishes identically is when one of the functions is identically zero. More generally, the Wronskian of two functions $y_1(t),y_2(t)$ vanishes identically if one of the functions is a constant multiple of the other. If $y_2=cy_1$, say, then

$$W[y_1,y_2](t)=y_1(cy_1)'-y_1'(cy_1)=0.$$

Conversely, suppose that the Wronskian of two *solutions* $y_1(t),y_2(t)$ of (3) vanishes identically. Then, one of these solutions must be a constant multiple of the other, as we now show.

Theorem 4. *Let $y_1(t)$ and $y_2(t)$ be two solutions of* (3) *on the interval $\alpha<t<\beta$, and suppose that $W[y_1,y_2](t_0)=0$ for some t_0 in this interval. Then, one of these solutions is a constant multiple of the other.*

PROOF #1. Suppose that $W[y_1,y_2](t_0)=0$. Then, the equations

$$c_1y_1(t_0)+c_2y_2(t_0)=0$$
$$c_1y_1'(t_0)+c_2y_2'(t_0)=0$$

have a nontrivial solution c_1,c_2; that is, a solution c_1,c_2 with $|c_1|+|c_2|\neq 0$. Let $y(t)=c_1y_1(t)+c_2y_2(t)$, for this choice of constants c_1,c_2. We know that $y(t)$ is a solution of (3), since it is a linear combination of $y_1(t)$ and $y_2(t)$. Moreover, by construction, $y(t_0)=0$ and $y'(t_0)=0$. Therefore, by Theorem 1, $y(t)$ is identically zero, so that

$$c_1y_1(t)+c_2y_2(t)=0, \qquad \alpha<t<\beta,$$

If $c_1\neq 0$, then $y_1(t)=-(c_2/c_1)y_2(t)$, and if $c_2\neq 0$, then $y_2(t)=-(c_1/c_2)y_1(t)$. In either case, one of these solutions is a constant multiple of the other. □

PROOF #2. Suppose that $W[y_1,y_2](t_0)=0$. Then, by Theorem 3, $W[y_1,y_2](t)$ is identically zero. Assume that $y_1(t)y_2(t)\neq 0$ for $\alpha<t<\beta$. Then, dividing both sides of the equation

$$y_1(t)y_2'(t)-y_1'(t)y_2(t)=0$$

by $y_1(t)y_2(t)$ gives

$$\frac{y_2'(t)}{y_2(t)}-\frac{y_1'(t)}{y_1(t)}=0.$$

This equation implies that $y_1(t)=cy_2(t)$ for some constant c.

Next, suppose that $y_1(t)y_2(t)$ is zero at some point $t=t^*$ in the interval $\alpha<t<\beta$. Without loss of generality, we may assume that $y_1(t^*)=0$, since otherwise we can relabel y_1 and y_2. In this case it is simple to show (see Exercise 19) that either $y_1(t)\equiv 0$, or $y_2(t)=[y_2'(t^*)/y_1'(t^*)]y_1(t)$. This completes the proof of Theorem 4. □

Definition. The functions $y_1(t)$ and $y_2(t)$ are said to be *linearly dependent* on an interval I if one of these functions is a constant multiple of the other on I. The functions $y_1(t)$ and $y_2(t)$ are said to be *linearly independent* on an interval I if they are not linearly dependent on I.

Corollary to Theorem 4. *Two solutions $y_1(t)$ and $y_2(t)$ of* (3) *are linearly independent on the interval $\alpha<t<\beta$ if, and only if, their Wronskian is unequal to zero on this interval. Thus, two solutions $y_1(t)$ and $y_2(t)$ form a fundamental set of solutions of* (3) *on the interval $\alpha<t<\beta$ if, and only if, they are linearly independent on this interval.*

EXERCISES

1. Let $L[y](t)=y''(t)-3ty'(t)+3y(t)$. Compute
(a) $L[e^t]$, (b) $L[\cos\sqrt{3}\,t]$, (c) $L[2e^t+4\cos\sqrt{3}\,t]$,
(d) $L[t^2]$, (e) $L[5t^2]$, (f) $L[t]$, (g) $L[t^2+3t]$.

2. Let $L[y](t)=y''(t)-6y'(t)+5y(t)$. Compute
(a) $L[e^t]$, (b) $L[e^{2t}]$, (c) $L[e^{3t}]$, (d) $L[e^{rt}]$,
(e) $L[t]$, (f) $L[t^2]$, (g) $L[t^2+2t]$.

3. Show that the operator L defined by

$$L[y](t)=\int_a^t s^2 y(s)\,ds$$

is linear; that is, $L[cy]=cL[y]$ and $L[y_1+y_2]=L[y_1]+L[y_2]$.

4. Let $L[y](t)=y''(t)+p(t)y'(t)+q(t)y(t)$, and suppose that $L[t^2]=t+1$ and $L[t]=2t+2$. Show that $y(t)=t-2t^2$ is a solution of $y''+p(t)y'+q(t)y=0$.

5. (a) Show that $y_1(t)=\sqrt{t}$ and $y_2(t)=1/t$ are solutions of the differential equation

$$2t^2y''+3ty'-y=0 \qquad (*)$$

on the interval $0<t<\infty$.
(b) Compute $W[y_1,y_2](t)$. What happens as t approaches zero?
(c) Show that $y_1(t)$ and $y_2(t)$ form a fundamental set of solutions of (*) on the interval $0<t<\infty$.
(d) Solve the initial-value problem $2t^2y''+3ty'-y=0$; $y(1)=2$, $y'(1)=1$.

6. (a) Show that $y_1(t)=e^{-t^2/2}$ and $y_2(t)=e^{-t^2/2}\int_0^t e^{s^2/2}\,ds$ are solutions of

$$y''+ty'+y=0 \qquad (*)$$

on the interval $-\infty<t<\infty$.
(b) Compute $W[y_1,y_2](t)$.
(c) Show that y_1 and y_2 form a fundamental set of solutions of (*) on the interval $-\infty<t<\infty$.
(d) Solve the initial-value problem $y''+ty'+y=0$; $y(0)=0$, $y'(0)=1$.

7. Compute the Wronskian of the following pairs of functions.
(a) $\sin at, \cos bt$ (b) $\sin^2 t, 1-\cos 2t$
(c) e^{at}, e^{bt} (d) e^{at}, te^{at}
(e) $t, t\ln t$ (f) $e^{at}\sin bt, e^{at}\cos bt$

8. Let $y_1(t)$ and $y_2(t)$ be solutions of (3) on the interval $-\infty < t < \infty$ with $y_1(0) = 3$, $y_1'(0) = 1$, $y_2(0) = 1$, and $y_2'(0) = \frac{1}{3}$. Show that $y_1(t)$ and $y_2(t)$ are linearly dependent on the interval $-\infty < t < \infty$.

9. (a) Let $y_1(t)$ and $y_2(t)$ be solutions of (3) on the interval $\alpha < t < \beta$, with $y_1(t_0) = 1$, $y_1'(t_0) = 0$, $y_2(t_0) = 0$, and $y_2'(t_0) = 1$. Show that $y_1(t)$ and $y_2(t)$ form a fundamental set of solutions of (3) on the interval $\alpha < t < \beta$.
(b) Show that $y(t) = y_0 y_1(t) + y_0' y_2(t)$ is the solution of (3) satisfying $y(t_0) = y_0$ and $y'(t_0) = y_0'$.

10. Show that $y(t) = t^2$ can never be a solution of (3) if the functions $p(t)$ and $q(t)$ are continuous at $t = 0$.

11. Let $y_1(t) = t^2$ and $y_2(t) = t|t|$.
(a) Show that y_1 and y_2 are linearly dependent on the interval $0 \leqslant t \leqslant 1$.
(b) Show that y_1 and y_2 are linearly independent on the interval $-1 \leqslant t \leqslant 1$.
(c) Show that $W[y_1, y_2](t)$ is identically zero.
(d) Show that y_1 and y_2 can never be two solutions of (3) on the interval $-1 < t < 1$ if both p and q are continuous in this interval.

12. Suppose that y_1 and y_2 are linearly independent on an interval I. Prove that $z_1 = y_1 + y_2$ and $z_2 = y_1 - y_2$ are also linearly independent on I.

13. Let y_1 and y_2 be solutions of Bessel's equation

$$t^2 y'' + ty' + (t^2 - n^2)y = 0$$

on the interval $0 < t < \infty$, with $y_1(1) = 1$, $y_1'(1) = 0$, $y_2(1) = 0$, and $y_2'(1) = 1$. Compute $W[y_1, y_2](t)$.

14. Suppose that the Wronskian of any two solutions of (3) is constant in time. Prove that $p(t) = 0$.

In Problems 15–18, assume that p and q are continuous, and that the functions y_1 and y_2 are solutions of the differential equation

$$y'' + p(t)y' + q(t)y = 0$$

on the interval $\alpha < t < \beta$.

15. Prove that if y_1 and y_2 vanish at the same point in the interval $\alpha < t < \beta$, then they cannot form a fundamental set of solutions on this interval.

16. Prove that if y_1 and y_2 achieve a maximum or minimum at the same point in the interval $\alpha < t < \beta$, then they cannot form a fundamental set of solutions on this interval.

17. Prove that if y_1 and y_2 are a fundamental set of solutions, then they cannot have a common point of inflection in $\alpha < t < \beta$ unless p and q vanish simultaneously there.

18. Suppose that y_1 and y_2 are a fundamental set of solutions on the interval $-\infty < t < \infty$. Show that there is one and only one zero of y_1 between consecutive zeros of y_2. *Hint*: Differentiate the quantity y_2/y_1 and use Rolle's Theorem.

19. Suppose that $W[y_1,y_2](t^*)=0$, and, in addition, $y_1(t^*)=0$. Prove that either $y_1(t)\equiv 0$ or $y_2(t)=[y_2'(t^*)/y_1'(t^*)]y_1(t)$. *Hint*: If $W[y_1,y_2](t^*)=0$ and $y_1(t^*)=0$, then $y_2(t^*)y_1'(t^*)=0$.

2.2 Linear equations with constant coefficients

We consider now the homogeneous linear second-order equation with constant coefficients

$$L[y]=a\frac{d^2y}{dt^2}+b\frac{dy}{dt}+cy=0 \tag{1}$$

where a, b, and c are constants, with $a\neq 0$. Theorem 2 of Section 2.1 tells us that we need only find two independent solutions y_1 and y_2 of (1); all other solutions of (1) are then obtained by taking linear combinations of y_1 and y_2. Unfortunately, Theorem 2 doesn't tell us how to find two solutions of (1). Therefore, we will try an educated guess. To this end, observe that a function $y(t)$ is a solution of (1) if a constant times its second derivative, plus another constant times its first derivative, plus a third constant times itself is identically zero. In other words, the three terms ay'', by', and cy must cancel each other. In general, this can only occur if the three functions $y(t)$, $y'(t)$, and $y''(t)$ are of the "same type". For example, the function $y(t)=t^5$ can never be a solution of (1) since the three terms $20at^3$, $5bt^4$, and ct^5 are polynomials in t of different degree, and therefore cannot cancel each other. On the other hand, the function $y(t)=e^{rt}$, r constant, has the property that both $y'(t)$ and $y''(t)$ are multiples of $y(t)$. This suggests that we try $y(t)=e^{rt}$ as a solution of (1). Computing

$$\begin{aligned}L[e^{rt}]&=a(e^{rt})''+b(e^{rt})'+c(e^{rt})\\&=(ar^2+br+c)e^{rt},\end{aligned}$$

we see that $y(t)=e^{rt}$ is a solution of (1) if, and only if

$$ar^2+br+c=0. \tag{2}$$

Equation (2) is called the *characteristic equation* of (1). It has two roots r_1,r_2 given by the quadratic formula

$$r_1=\frac{-b+\sqrt{b^2-4ac}}{2a},\qquad r_2=\frac{-b-\sqrt{b^2-4ac}}{2a}.$$

If b^2-4ac is positive, then r_1 and r_2 are real and distinct. In this case, $y_1(t)=e^{r_1t}$ and $y_2(t)=e^{r_2t}$ are two distinct solutions of (1). These solutions are clearly linearly independent (on any interval I), since e^{r_2t} is obviously not a constant multiple of e^{r_1t} for $r_2\neq r_1$. (If the reader is unconvinced of this he can compute

$$W[e^{r_1t},e^{r_2t}]=(r_2-r_1)e^{(r_1+r_2)t},$$

and observe that W is never zero. Hence, e^{r_1t} and e^{r_2t} are linearly independent on any interval I.)

Example 1. Find the general solution of the equation

$$\frac{d^2y}{dt^2}+5\frac{dy}{dt}+4y=0. \tag{3}$$

Solution. The characteristic equation $r^2+5r+4=(r+4)(r+1)=0$ has two distinct roots $r_1=-4$ and $r_2=-1$. Thus, $y_1(t)=e^{-4t}$ and $y_2(t)=e^{-t}$ form a fundamental set of solutions of (3), and every solution $y(t)$ of (3) is of the form

$$y(t)=c_1e^{-4t}+c_2e^{-t}$$

for some choice of constants c_1,c_2.

Example 2. Find the solution $y(t)$ of the initial-value problem

$$\frac{d^2y}{dt^2}+4\frac{dy}{dt}-2y=0; \qquad y(0)=1, \quad y'(0)=2.$$

Solution. The characteristic equation $r^2+4r-2=0$ has 2 roots

$$r_1=\frac{-4+\sqrt{16+8}}{2}=-2+\sqrt{6}$$

and

$$r_2=\frac{-4-\sqrt{16+8}}{2}=-2-\sqrt{6}\,.$$

Hence, $y_1(t)=e^{r_1t}$ and $y_2(t)=e^{r_2t}$ are a fundamental set of solutions of $y''+4y'-2y=0$, so that

$$y(t)=c_1e^{(-2+\sqrt{6}\,)t}+c_2e^{(-2-\sqrt{6}\,)t}$$

for some choice of constants c_1,c_2. The constants c_1 and c_2 are determined from the initial conditions

$$c_1+c_2=1 \quad\text{and}\quad (-2+\sqrt{6}\,)c_1+(-2-\sqrt{6}\,)c_2=2.$$

From the first equation, $c_2=1-c_1$. Substituting this value of c_2 into the second equation gives

$$(-2+\sqrt{6}\,)c_1-(2+\sqrt{6}\,)(1-c_1)=2, \quad\text{or}\quad 2\sqrt{6}\,c_1=4+\sqrt{6}\,.$$

Therefore, $c_1=2/\sqrt{6}+\frac{1}{2}$, $c_2=1-c_1=\frac{1}{2}-2/\sqrt{6}$, and

$$y(t)=\left(\frac{1}{2}+\frac{2}{\sqrt{6}}\right)e^{(-2+\sqrt{6}\,)t}+\left(\frac{1}{2}-\frac{2}{\sqrt{6}}\right)e^{-(2+\sqrt{6}\,)t}.$$

EXERCISES

Find the general solution of each of the following equations.

1. $\dfrac{d^2y}{dt^2}-y=0$

2. $6\dfrac{d^2y}{dt^2}-7\dfrac{dy}{dt}+y=0$

3. $\dfrac{d^2y}{dt^2}-3\dfrac{dy}{dt}+y=0$

4. $3\dfrac{d^2y}{dt^2}+6\dfrac{dy}{dt}+2y=0$

Solve each of the following initial-value problems.

5. $\dfrac{d^2y}{dt^2}-3\dfrac{dy}{dt}-4y=0;\quad y(0)=1,\ y'(0)=0$

6. $2\dfrac{d^2y}{dt^2}+\dfrac{dy}{dt}-10y=0;\quad y(1)=5,\ y'(1)=2$

7. $5\dfrac{d^2y}{dt^2}+5\dfrac{dy}{dt}-y=0;\quad y(0)=0,\ y'(0)=1$

8. $\dfrac{d^2y}{dt^2}-6\dfrac{dy}{dt}+y=0;\quad y(2)=1,\ y'(2)=1$

Remark. In doing Problems 6 and 8, observe that $e^{r(t-t_0)}$ is also a solution of the differential equation $ay''+by'+cy=0$ if $ar^2+br+c=0$. Thus, to find the solution $y(t)$ of the initial-value problem $ay''+by'+cy=0$; $y(t_0)=y_0, y'(t_0)=y_0'$, we would write $y(t)=c_1e^{r_1(t-t_0)}+c_2e^{r_2(t-t_0)}$ and solve for c_1 and c_2 from the initial conditions.

9. Let $y(t)$ be the solution of the initial-value problem

$$\frac{d^2y}{dt^2}+5\frac{dy}{dt}+6y=0;\qquad y(0)=1,\quad y'(0)=V.$$

For what values of V does $y(t)$ remain nonnegative for all $t\geqslant 0$?

10. The differential equation

$$L[y]=t^2y''+\alpha ty'+\beta y=0 \tag{*}$$

is known as Euler's equation. Observe that t^2y'', ty', and y are all multiples of t^r if $y=t^r$. This suggests that we try $y=t^r$ as a solution of (*). Show that $y=t^r$ is a solution of (*) if $r^2+(\alpha-1)r+\beta=0$.

11. Find the general solution of the equation

$$t^2y''+5ty'-5y=0,\qquad t>0$$

12. Solve the initial-value problem

$$t^2y''-ty'-2y=0;\qquad y(1)=0,\quad y'(1)=1$$

on the interval $0<t<\infty$.

2.2.1 Complex roots

If b^2-4ac is negative, then the characteristic equation $ar^2+br+c=0$ has complex roots

$$r_1=\frac{-b+i\sqrt{4ac-b^2}}{2a} \quad \text{and} \quad r_2=\frac{-b-i\sqrt{4ac-b^2}}{2a}.$$

We would like to say that e^{r_1t} and e^{r_2t} are solutions of the differential equation

$$a\frac{d^2y}{dt^2}+b\frac{dy}{dt}+cy=0. \tag{1}$$

However, this presents us with two serious difficulties. On the one hand, the function e^{rt} is not defined, as yet, for r complex. And on the other hand, even if we succeed in defining e^{r_1t} and e^{r_2t} as complex-valued solutions of (1), we are still faced with the problem of finding two *real-valued* solutions of (1).

We begin by resolving the second difficulty, since otherwise there's no sense tackling the first problem. Assume that $y(t)=u(t)+iv(t)$ is a complex-valued solution of (1). This means, of course, that

$$a\left[u''(t)+iv''(t)\right]+b\left[u'(t)+iv'(t)\right]+c\left[u(t)+iv(t)\right]=0. \tag{2}$$

This complex-valued solution of (1) gives rise to *two* real-valued solutions, as we now show.

Lemma 1. *Let $y(t)=u(t)+iv(t)$ be a complex-valued solution of* (1), *with a, b, and c real. Then, $y_1(t)=u(t)$ and $y_2(t)=v(t)$ are two real-valued solutions of* (1). *In other words, both the real and imaginary parts of a complex-valued solution of* (1) *are solutions of* (1). (*The imaginary part of the complex number $\alpha+i\beta$ is β. Similarly, the imaginary part of the function $u(t)+iv(t)$ is $v(t)$.*)

PROOF. From Equation (2),

$$\left[au''(t)+bu'(t)+cu(t)\right]+i\left[av''(t)+bv'(t)+cv(t)\right]=0. \tag{3}$$

Now, if a complex number is zero, then both its real and imaginary parts must be zero. Consequently,

$$au''(t)+bu'(t)+cu(t)=0 \quad \text{and} \quad av''(t)+bv'(t)+cv(t)=0,$$

and this proves Lemma 1. □

The problem of defining e^{rt} for r complex can also be resolved quite easily. Let $r=\alpha+i\beta$. By the law of exponents,

$$e^{rt}=e^{\alpha t}e^{i\beta t}. \tag{4}$$

Thus, we need only define the quantity $e^{i\beta t}$, for β real. To this end, recall that

$$e^x = 1 + x + \frac{x^2}{2!} + \frac{x^3}{3!} + \dots. \tag{5}$$

Equation (5) makes sense, formally, even for x complex. This suggests that we set

$$e^{i\beta t} = 1 + i\beta t + \frac{(i\beta t)^2}{2!} + \frac{(i\beta t)^3}{3!} + \dots.$$

Next, observe that

$$\begin{aligned}
1 + i\beta t + \frac{(i\beta t)^2}{2!} + \dots &= 1 + i\beta t - \frac{\beta^2 t^2}{2!} - \frac{i\beta^3 t^3}{3!} + \frac{\beta^4 t^4}{4!} + \frac{i\beta^5 t^5}{5!} + \dots \\
&= \left[1 - \frac{\beta^2 t^2}{2!} + \frac{\beta^4 t^4}{4!} + \dots\right] + i\left[\beta t - \frac{\beta^3 t^3}{3!} + \frac{\beta^5 t^5}{5!} + \dots\right] \\
&= \cos\beta t + i\sin\beta t.
\end{aligned}$$

Hence,

$$e^{(\alpha + i\beta)t} = e^{\alpha t} e^{i\beta t} = e^{\alpha t}(\cos\beta t + i\sin\beta t). \tag{6}$$

Returning to the differential equation (1), we see that

$$\begin{aligned}
y(t) &= e^{[-b + i\sqrt{4ac - b^2}\,]t/2a} \\
&= e^{-bt/2a}\left[\cos\sqrt{4ac - b^2}\; t/2a + i\sin\sqrt{4ac - b^2}\; t/2a\right]
\end{aligned}$$

is a complex-valued solution of (1) if $b^2 - 4ac$ is negative. Therefore, by Lemma 1,

$$y_1(t) = e^{-bt/2a}\cos\beta t \quad \text{and} \quad y_2(t) = e^{-bt/2a}\sin\beta t, \qquad \beta = \frac{\sqrt{4ac - b^2}}{2a}$$

are two real-valued solutions of (1). These two functions are linearly independent on any interval I, since their Wronskian (see Exercise 10) is never zero. Consequently, the general solution of (1) for $b^2 - 4ac < 0$ is

$$y(t) = e^{-bt/2a}\left[c_1\cos\beta t + c_2\sin\beta t\right], \qquad \beta = \frac{\sqrt{4ac - b^2}}{2a}.$$

Remark 1. Strictly speaking, we must verify that the formula

$$\frac{d}{dt}e^{rt} = re^{rt}$$

is true even for r complex, before we can assert that $e^{r_1 t}$ and $e^{r_2 t}$ are complex-valued solutions of (1). To this end, we compute

$$\begin{aligned}
\frac{d}{dt}e^{(\alpha + i\beta)t} &= \frac{d}{dt}e^{\alpha t}\left[\cos\beta t + i\sin\beta t\right] \\
&= e^{\alpha t}\left[(\alpha\cos\beta t - \beta\sin\beta t) + i(\alpha\sin\beta t + \beta\cos\beta t)\right]
\end{aligned}$$

and this equals $(\alpha+i\beta)e^{(\alpha+i\beta)t}$, since

$$\begin{aligned}(\alpha+i\beta)e^{(\alpha+i\beta)t} &= (\alpha+i\beta)e^{\alpha t}[\cos\beta t+i\sin\beta t]\\ &= e^{\alpha t}[(\alpha\cos\beta t-\beta\sin\beta t)+i(\alpha\sin\beta t+\beta\cos\beta t)].\end{aligned}$$

Thus, $(d/dt)e^{rt}=re^{rt}$, even for r complex.

Remark 2. At first glance, one might think that e^{r_2t} would give rise to two additional solutions of (1). This is not the case, though, since

$$\begin{aligned}e^{r_2t} &= e^{-(b/2a)t}e^{-i\beta t}, \qquad \beta=\sqrt{4ac-b^2}\,/2a\\ &= e^{-bt/2a}[\cos(-\beta t)+i\sin(-\beta t)]=e^{-bt/2a}[\cos\beta t-i\sin\beta t].\end{aligned}$$

Hence,

$$\mathrm{Re}\{e^{r_2t}\}=e^{-bt/2a}\cos\beta t=y_1(t)$$

and

$$\mathrm{Im}\{e^{r_2t}\}=-e^{-bt/2a}\sin\beta t=-y_2(t).$$

Example 1. Find two linearly independent real-valued solutions of the differential equation

$$4\frac{d^2y}{dt^2}+4\frac{dy}{dt}+5y=0. \tag{7}$$

Solution. The characteristic equation $4r^2+4r+5=0$ has complex roots $r_1=-\frac{1}{2}+i$ and $r_2=-\frac{1}{2}-i$. Consequently,

$$e^{r_1t}=e^{(-1/2+i)t}=e^{-t/2}\cos t+ie^{-t/2}\sin t$$

is a complex-valued solution of (7). Therefore, by Lemma 1,

$$\mathrm{Re}\{e^{r_1t}\}=e^{-t/2}\cos t \quad\text{and}\quad \mathrm{Im}\{e^{r_1t}\}=e^{-t/2}\sin t$$

are two linearly independent real-valued solutions of (7).

Example 2. Find the solution $y(t)$ of the initial-value problem

$$\frac{d^2y}{dt^2}+2\frac{dy}{dt}+4y=0; \qquad y(0)=1, \quad y'(0)=1.$$

Solution. The characteristic equation $r^2+2r+4=0$ has complex roots $r_1=-1+\sqrt{3}\,i$ and $r_2=-1-\sqrt{3}\,i$. Hence,

$$e^{r_1t}=e^{(-1+\sqrt{3}\,i)t}=e^{-t}\cos\sqrt{3}\,t+ie^{-t}\sin\sqrt{3}\,t$$

is a complex-valued solution of $y''+2y'+4y=0$. Therefore, by Lemma 1, both

$$\mathrm{Re}\{e^{r_1t}\}=e^{-t}\cos\sqrt{3}\,t \quad\text{and}\quad \mathrm{Im}\{e^{r_1t}\}=e^{-t}\sin\sqrt{3}\,t$$

are real-valued solutions. Consequently,

$$y(t)=e^{-t}[c_1\cos\sqrt{3}\,t+c_2\sin\sqrt{3}\,t]$$

for some choice of constants c_1, c_2. The constants c_1 and c_2 are determined from the initial conditions

$$1 = y(0) = c_1$$

and

$$1 = y'(0) = -c_1 + \sqrt{3}\, c_2.$$

This implies that

$$c_1 = 1, c_2 = \frac{2}{\sqrt{3}} \quad \text{and} \quad y(t) = e^{-t}\left[\cos\sqrt{3}\, t + \frac{2}{\sqrt{3}}\sin\sqrt{3}\, t\right].$$

Exercises

Find the general solution of each of the following equations.

1. $\dfrac{d^2y}{dt^2} + \dfrac{dy}{dt} + y = 0$

2. $2\dfrac{d^2y}{dt^2} + 3\dfrac{dy}{dt} + 4y = 0$

3. $\dfrac{d^2y}{dt^2} + 2\dfrac{dy}{dt} + 3y = 0$

4. $4\dfrac{d^2y}{dt^2} - \dfrac{dy}{dt} + y = 0$

Solve each of the following initial-value problems.

5. $\dfrac{d^2y}{dt^2} + \dfrac{dy}{dt} + 2y = 0; \quad y(0) = 1,\ y'(0) = -2$

6. $\dfrac{d^2y}{dt^2} + 2\dfrac{dy}{dt} + 5y = 0; \quad y(0) = 0,\ y'(0) = 2$

7. Assume that $b^2 - 4ac < 0$. Show that

$$y_1(t) = e^{(-b/2a)(t-t_0)}\cos\beta(t - t_0)$$

and

$$y_2(t) = e^{(-b/2a)(t-t_0)}\sin\beta(t - t_0), \qquad \beta = \frac{\sqrt{4ac - b^2}}{2a}$$

are solutions of (1), for any number t_0.

Solve each of the following initial-value problems.

8. $2\dfrac{d^2y}{dt^2} - \dfrac{dy}{dt} + 3y = 0; \quad y(1) = 1,\ y'(1) = 1$

9. $3\dfrac{d^2y}{dt^2} - 2\dfrac{dy}{dt} + 4y = 0; \quad y(2) = 1,\ y'(2) = -1$

10. Verify that $W[e^{\alpha t}\cos\beta t, e^{\alpha t}\sin\beta t] = \beta e^{2\alpha t}$.

11. Show that $e^{i\omega t}$ is a complex-valued solution of the differential equation $y'' + \omega^2 y = 0$. Find two real-valued solutions.

12. Show that $(\cos t + i\sin t)^r = \cos rt + i\sin rt$. Use this result to obtain the double angle formulas $\sin 2t = 2\sin t\cos t$ and $\cos 2t = \cos^2 t - \sin^2 t$.

13. Show that

$$(\cos t_1 + i\sin t_1)(\cos t_2 + i\sin t_2) = \cos(t_1+t_2) + i\sin(t_1+t_2).$$

Use this result to obtain the trigonometric identities

$$\cos(t_1+t_2) = \cos t_1 \cos t_2 - \sin t_1 \sin t_2,$$
$$\sin(t_1+t_2) = \sin t_1 \cos t_2 + \cos t_1 \sin t_2.$$

14. Show that any complex number $a+ib$ can be written in the form $Ae^{i\theta}$, where $A=\sqrt{a^2+b^2}$ and $\tan\theta = b/a$.

15. Defining the two possible square roots of a complex number $Ae^{i\theta}$ as $\pm\sqrt{A}\, e^{i\theta/2}$, compute the square roots of i, $1+i$, $-i$, $\sqrt{i}$.

16. Use Problem 14 to find the three cube roots of i.

17. (a) Let $r_1 = \lambda + i\mu$ be a complex root of $r^2 + (\alpha-1)r + \beta = 0$. Show that

$$t^{\lambda+i\mu} = t^{\lambda}t^{i\mu} = t^{\lambda}e^{(\ln t)i\mu} = t^{\lambda}[\cos\mu\ln t + i\sin\mu\ln t]$$

is a complex-valued solution of Euler's equation

$$t^2\frac{d^2y}{dt^2} + \alpha t\frac{dy}{dt} + \beta y = 0. \tag{*}$$

(b) Show that $t^{\lambda}\cos\mu\ln t$ and $t^{\lambda}\sin\mu\ln t$ are real-valued solutions of (*).

Find the general solution of each of the following equations.

18. $t^2\dfrac{d^2y}{dt^2} + t\dfrac{dy}{dt} + y = 0, \quad t>0$

19. $t^2\dfrac{d^2y}{dt^2} + 2t\dfrac{dy}{dt} + 2y = 0, \quad t>0$

2.2.2 *Equal roots; reduction of order*

If $b^2 = 4ac$, then the characteristic equation $ar^2 + br + c = 0$ has real equal roots $r_1 = r_2 = -b/2a$. In this case, we obtain only one solution

$$y_1(t) = e^{-bt/2a}$$

of the differential equation

$$a\frac{d^2y}{dt^2} + b\frac{dy}{dt} + cy = 0. \tag{1}$$

Our problem is to find a second solution which is independent of y_1. One approach to this problem is to try some additional guesses. A second, and much more clever approach is to try and use our knowledge of $y_1(t)$ to help us find a second independent solution. More generally, suppose that we know one solution $y = y_1(t)$ of the second-order linear equation

$$L[y] = \frac{d^2y}{dt^2} + p(t)\frac{dy}{dt} + q(t)y = 0. \tag{2}$$

Can we use this solution to help us find a second independent solution?

The answer to this question is yes. Once we find one solution $y = y_1(t)$ of (2), we can reduce the problem of finding all solutions of (2) to that of solving a first-order linear homogeneous equation. This is accomplished by defining a new dependent variable v through the substitution

$$y(t) = y_1(t)v(t).$$

Then

$$\frac{dy}{dt} = v\frac{dy_1}{dt} + y_1\frac{dv}{dt}$$

and

$$\frac{d^2y}{dt^2} = v\frac{d^2y_1}{dt^2} + 2\frac{dv}{dt}\frac{dy_1}{dt} + y_1\frac{d^2v}{dt^2}.$$

Consequently.

$$\begin{aligned} L[y] &= v\frac{d^2y_1}{dt^2} + 2\frac{dv}{dt}\frac{dy_1}{dt} + y_1\frac{d^2v}{dt^2} + p(t)\left[v\frac{dy_1}{dt} + y_1\frac{dv}{dt}\right] + q(t)vy_1 \\ &= y_1\frac{d^2v}{dt^2} + \left[2\frac{dy_1}{dt} + p(t)y_1\right]\frac{dv}{dt} + \left[\frac{d^2y_1}{dt^2} + p(t)\frac{dy_1}{dt} + q(t)y_1\right]v \\ &= y_1\frac{d^2v}{dt^2} + \left[2\frac{dy_1}{dt} + p(t)y_1\right]\frac{dv}{dt}, \end{aligned}$$

since $y_1(t)$ is a solution of $L[y] = 0$. Hence, $y(t) = y_1(t)v(t)$ is a solution of (2) if v satisfies the differential equation

$$y_1\frac{d^2v}{dt^2} + \left[2\frac{dy_1}{dt} + p(t)y_1\right]\frac{dv}{dt} = 0. \tag{3}$$

Now, observe that Equation (3) is really a first-order linear equation for dv/dt. Its solution is

$$\begin{aligned} \frac{dv}{dt} &= c\exp\left[-\int\left[2\frac{y_1'(t)}{y_1(t)} + p(t)\right]dt\right] \\ &= c\exp\left(-\int p(t)\,dt\right)\exp\left[-2\int\frac{y_1'(t)}{y_1(t)}dt\right] \\ &= \frac{c\exp\left(-\int p(t)\,dt\right)}{y_1^2(t)}. \end{aligned} \tag{4}$$

Since we only need one solution $v(t)$ of (3), we set $c = 1$ in (4). Integrating this equation with respect to t, and setting the constant of integration equal to zero, we obtain that $v(t)=\int u(t)\,dt$, where

$$u(t)=\frac{\exp\left(-\int p(t)\,dt\right)}{y_1^2(t)}. \tag{5}$$

Hence,

$$y_2(t)=v(t)y_1(t)=y_1(t)\int u(t)\,dt \tag{6}$$

is a second solution of (2). This solution is independent of y_1, for if $y_2(t)$ were a constant multiple of $y_1(t)$ then $v(t)$ would be constant, and consequently, its derivative would vanish identically. However, from (4)

$$\frac{dv}{dt}=\frac{\exp\left(-\int p(t)\,dt\right)}{y_1^2(t)},$$

and this quantity is never zero.

Remark 1. In writing $v(t)=\int u(t)\,dt$, we set the constant of integration equal to zero. Choosing a nonzero constant of integration would only add a constant multiple of $y_1(t)$ to $y_2(t)$. Similarly, the effect of choosing a constant c other than one in Equation (4) would be to multiply $y_2(t)$ by c.

Remark 2. The method we have just presented for solving Equation (2) is known as the method of *reduction of order*, since the substitution $y(t)=y_1(t)v(t)$ reduces the problem of solving the second-order equation (2) to that of solving a first-order equation.

Application to the case of equal roots: In the case of equal roots, we found $y_1(t)=e^{-bt/2a}$ as one solution of the equation

$$a\frac{d^2y}{dt^2}+b\frac{dy}{dt}+cy=0. \tag{7}$$

We can find a second solution from Equations (5) and (6). It is important to realize though, that Equations (5) and (6) were derived under the assumption that our differential equation was written in the form

$$\frac{d^2y}{dt^2}+p(t)\frac{dy}{dt}+q(t)y=0;$$

that is, the coefficient of y'' was one. In our equation, the coefficient of y'' is a. Hence, we must divide Equation (7) by a to obtain the equivalent equation

$$\frac{d^2y}{dt^2}+\frac{b}{a}\frac{dy}{dt}+\frac{c}{a}y=0.$$

Now, we can insert $p(t)=b/a$ into (5) to obtain that

$$u(t)=\frac{\exp\left(-\int\frac{b}{a}dt\right)}{\left[e^{-bt/2a}\right]^2}=\frac{e^{-bt/a}}{e^{-bt/a}}=1.$$

Hence,

$$y_2(t)=y_1(t)\int dt=ty_1(t)$$

is a second solution of (7). The functions $y_1(t)$ and $y_2(t)$ are clearly linearly independent on the interval $-\infty<t<\infty$. Therefore, the general solution of (7) in the case of equal roots is

$$y(t)=c_1e^{-bt/2a}+c_2te^{-bt/2a}=\left[c_1+c_2t\right]e^{-bt/2a}$$

Example 1. Find the solution $y(t)$ of the initial-value problem

$$\frac{d^2y}{dt^2}+4\frac{dy}{dt}+4y=0;\qquad y(0)=1,\quad y'(0)=3.$$

Solution. The characteristic equation $r^2+4r+4=(r+2)^2=0$ has two equal roots $r_1=r_2=-2$. Hence,

$$y(t)=c_1e^{-2t}+c_2te^{-2t}$$

for some choice of constants c_1, c_2. The constants c_1 and c_2 are determined from the initial conditions

$$1=y(0)=c_1$$

and

$$3=y'(0)=-2c_1+c_2.$$

This implies that $c_1=1$ and $c_2=5$, so that $y(t)=(1+5t)e^{-2t}$.

Example 2. Find the solution $y(t)$ of the initial-value problem

$$(1-t^2)\frac{d^2y}{dt^2}+2t\frac{dy}{dt}-2y=0;\qquad y(0)=3,\quad y'(0)=-4$$

on the interval $-1<t<1$.

Solution. Clearly, $y_1(t)=t$ is one solution of the differential equation

$$(1-t^2)\frac{d^2y}{dt^2}+2t\frac{dy}{dt}-2y=0.\tag{8}$$

We will use the method of reduction of order to find a second solution $y_2(t)$ of (8). To this end, divide both sides of (8) by $1-t^2$ to obtain the equivalent equation

$$\frac{d^2y}{dt^2}+\frac{2t}{1-t^2}\frac{dy}{dt}-\frac{2}{1-t^2}y=0.$$

Then, from (5)

$$u(t)=\frac{\exp\left(-\int\frac{2t}{1-t^2}\,dt\right)}{y_1^2(t)}=\frac{e^{\ln(1-t^2)}}{t^2}=\frac{1-t^2}{t^2},$$

and

$$y_2(t)=t\int\frac{1-t^2}{t^2}\,dt=-t\left(\frac{1}{t}+t\right)=-(1+t^2)$$

is a second solution of (8). Therefore,

$$y(t)=c_1t-c_2(1+t^2)$$

for some choice of constants c_1, c_2. (Notice that all solutions of (9) are continuous at $t=\pm 1$ even though the differential equation is not defined at these points. Thus, it does not necessarily follow that the solutions of a differential equation are discontinuous at a point where the differential equation is not defined—but this is often the case.) The constants c_1 and c_2 are determined from the initial conditions

$$3=y(0)=-c_2 \quad\text{and}\quad -4=y'(0)=c_1.$$

Hence, $y(t)=-4t+3(1+t^2)$.

Exercises

Find the general solution of each of the following equations

1. $\dfrac{d^2y}{dt^2}-6\dfrac{dy}{dt}+9y=0$

2. $4\dfrac{d^2y}{dt^2}-12\dfrac{dy}{dt}+9y=0$

Solve each of the following initial-value problems.

3. $9\dfrac{d^2y}{dt^2}+6\dfrac{dy}{dt}+y=0;\quad y(0)=1,\ y'(0)=0$

4. $4\dfrac{d^2y}{dt^2}-4\dfrac{dy}{dt}+y=0;\quad y(0)=0,\ y'(0)=3$

5. Suppose $b^2=4ac$. Show that

$$y_1(t)=e^{-b(t-t_0)/2a} \quad\text{and}\quad y_2(t)=(t-t_0)e^{-b(t-t_0)/2a}$$

are solutions of (1) for every choice of t_0.

Solve the following initial-value problems.

6. $\dfrac{d^2y}{dt^2}+2\dfrac{dy}{dt}+y=0;\quad y(2)=1,\ y'(2)=-1$

7. $9\dfrac{d^2y}{dt^2}-12\dfrac{dy}{dt}+4y=0;\quad y(\pi)=0,\ y'(\pi)=2$

8. Let a, b and c be positive numbers. Prove that every solution of the differential equation $ay''+by'+cy=0$ approaches zero as t approaches infinity.

9. Here is an alternate and very elegant way of finding a second solution $y_2(t)$ of (1).

(a) Assume that $b^2=4ac$. Show that

$$L[e^{rt}]=a(e^{rt})''+b(e^{rt})'+ce^{rt}=a(r-r_1)^2e^{rt}$$

for $r_1=-b/2a$.

(b) Show that

$$(\partial/\partial r)L[e^{rt}]=L[(\partial/\partial r)e^{rt}]=L[te^{rt}]=2a(r-r_1)e^{rt}+at(r-r_1)^2e^{rt}.$$

(c) Conclude from (a) and (b) that $L[te^{r_1t}]=0$. Hence, $y_2(t)=te^{r_1t}$ is a second solution of (1).

Use the method of reduction of order to find the general solution of the following differential equations.

10. $\dfrac{d^2y}{dt^2}-\dfrac{2(t+1)}{(t^2+2t-1)}\dfrac{dy}{dt}+\dfrac{2}{(t^2+2t-1)}y=0 \quad (y_1(t)=t+1)$

11. $\dfrac{d^2y}{dt^2}-4t\dfrac{dy}{dt}+(4t^2-2)y=0 \quad (y_1(t)=e^{t^2})$

12. $(1-t^2)\dfrac{d^2y}{dt^2}-2t\dfrac{dy}{dt}+2y=0 \quad (y_1(t)=t)$

13. $(1+t^2)\dfrac{d^2y}{dt^2}-2t\dfrac{dy}{dt}+2y=0 \quad (y_1(t)=t)$

14. $(1-t^2)\dfrac{d^2y}{dt^2}-2t\dfrac{dy}{dt}+6y=0 \quad (y_1(t)=3t^2-1)$

15. $(2t+1)\dfrac{d^2y}{dt^2}-4(t+1)\dfrac{dy}{dt}+4y=0 \quad (y_1(t)=t+1)$

16. $t^2\dfrac{d^2y}{dt^2}+t\dfrac{dy}{dt}+\left(t^2-\dfrac{1}{4}\right)y=0 \quad \left(y_1(t)=\dfrac{\sin t}{\sqrt{t}}\right)$

17. Given that the equation

$$t\frac{d^2y}{dt^2}-(1+3t)\frac{dy}{dt}+3y=0$$

has a solution of the form e^{ct}, for some constant c, find the general solution.

18. (a) Show that t^r is a solution of Euler's equation

$$t^2y''+\alpha ty'+\beta y=0, \quad t>0$$

if $r^2+(\alpha-1)r+\beta=0$.

(b) Suppose that $(\alpha-1)^2=4\beta$. Using the method of reduction of order, show that $(\ln t)t^{(1-\alpha)/2}$ is a second solution of Euler's equation.

Find the general solution of each of the following equations.

19. $t^2\dfrac{d^2y}{dt^2}+3t\dfrac{dy}{dt}+y=0$

20. $t^2\dfrac{d^2y}{dt^2}-t\dfrac{dy}{dt}+y=0$

2.3 The nonhomogeneous equation

We turn our attention now to the nonhomogeneous equation

$$L[y] = \frac{d^2y}{dt^2} + p(t)\frac{dy}{dt} + q(t)y = g(t) \tag{1}$$

where the functions $p(t)$, $q(t)$ and $g(t)$ are continuous on an open interval $\alpha < t < \beta$. An important clue as to the nature of all solutions of (1) is provided by the first-order linear equation

$$\frac{dy}{dt} - 2ty = -t. \tag{2}$$

The general solution of this equation is

$$y(t) = ce^{t^2} + \tfrac{1}{2}.$$

Now, observe that this solution is the sum of two terms: the first term, ce^{t^2}, is the general solution of the homogeneous equation

$$\frac{dy}{dt} - 2ty = 0 \tag{3}$$

while the second term, $\frac{1}{2}$, is a solution of the nonhomogeneous equation (2). In other words, every solution $y(t)$ of (2) is the sum of a particular solution, $\psi(t) = \frac{1}{2}$, with a solution ce^{t^2} of the homogeneous equation. A similar situation prevails in the case of second-order equations, as we now show.

Theorem 5. *Let $y_1(t)$ and $y_2(t)$ be two linearly independent solutions of the homogeneous equation*

$$L[y] = \frac{d^2y}{dt^2} + p(t)\frac{dy}{dt} + q(t)y = 0 \tag{4}$$

and let $\psi(t)$ be any particular solution of the nonhomogeneous equation (1). *Then, every solution $y(t)$ of* (1) *must be of the form*

$$y(t) = c_1y_1(t) + c_2y_2(t) + \psi(t)$$

for some choice of constants c_1, c_2.

The proof of Theorem 5 relies heavily on the following lemma.

Lemma 1. *The difference of any two solutions of the nonhomogeneous equation* (1) *is a solution of the homogeneous equation* (4).

PROOF. Let $\psi_1(t)$ and $\psi_2(t)$ be two solutions of (1). By the linearity of L,

$$L[\psi_1 - \psi_2](t) = L[\psi_1](t) - L[\psi_2](t) = g(t) - g(t) = 0.$$

Hence, $\psi_1(t) - \psi_2(t)$ is a solution of the homogeneous equation (4). □

We can now give a very simple proof of Theorem 5.

PROOF OF THEOREM 5. Let $y(t)$ be any solution of (1). By Lemma 1, the function $\phi(t)=y(t)-\psi(t)$ is a solution of the homogeneous equation (4). But every solution $\phi(t)$ of the homogeneous equation (4) is of the form $\phi(t)=c_1y_1(t)+c_2y_2(t)$, for some choice of constants c_1, c_2. Therefore,

$$y(t)=\phi(t)+\psi(t)=c_1y_1(t)+c_2y_2(t)+\psi(t). \qquad \square$$

Remark. Theorem 5 is an extremely useful theorem since it reduces the problem of finding all solutions of (1) to the much simpler problem of finding just two solutions of the homogeneous equation (4), and one solution of the nonhomogeneous equation (1).

Example 1. Find the general solution of the equation

$$\frac{d^2y}{dt^2}+y=t. \tag{5}$$

Solution. The functions $y_1(t)=\cos t$ and $y_2(t)=\sin t$ are two linearly independent solutions of the homogeneous equation $y''+y=0$. Moreover, $\psi(t)=t$ is obviously a particular solution of (5). Therefore, by Theorem 5, every solution $y(t)$ of (5) must be of the form

$$y(t)=c_1\cos t+c_2\sin t+t.$$

Example 2. Three solutions of a certain second-order nonhomogeneous linear equation are

$$\psi_1(t)=t, \qquad \psi_2(t)=t+e^t, \quad \text{and} \quad \psi_3(t)=1+t+e^t.$$

Find the general solution of this equation.
Solution. By Lemma 1, the functions

$$\psi_2(t)-\psi_1(t)=e^t \quad \text{and} \quad \psi_3(t)-\psi_2(t)=1$$

are solutions of the corresponding homogeneous equation. Moreover, these functions are obviously linearly independent. Therefore, by Theorem 5, every solution $y(t)$ of this equation must be of the form

$$y(t)=c_1e^t+c_2+t.$$

EXERCISES

1. Three solutions of a certain second-order nonhomogeneous linear equation are

$$\psi_1(t)=t^2, \psi_2(t)=t^2+e^{2t}$$

and

$$\psi_3(t)=1+t^2+2e^{2t}.$$

Find the general solution of this equation.

2. Three solutions of a certain second-order linear nonhomogeneous equation are

$$\psi_1(t)=1+e^{t^2},\ \psi_2(t)=1+te^{t^2}$$

and

$$\psi_3(t)=(t+1)e^{t^2}+1$$

Find the general solution of this equation.

3. Three solutions of a second-order linear equation $L[y]=g(t)$ are

$$\psi_1(t)=3e^t+e^{t^2},\ \psi_2(t)=7e^t+e^{t^2}$$

and

$$\psi_3(t)=5e^t+e^{-t^3}+e^{t^2}.$$

Find the solution of the initial-value problem

$$L[y]=g; \qquad y(0)=1, \quad y'(0)=2.$$

4. Let a, b and c be positive constants. Show that the difference of any two solutions of the equation

$$ay''+by'+cy=g(t)$$

approaches zero as t approaches infinity.

5. Let $\psi(t)$ be a solution of the nonhomogeneous equation (1), and let $\phi(t)$ be a solution of the homogeneous equation (4). Show that $\psi(t)+\phi(t)$ is again a solution of (1).

2.4 The method of variation of parameters

In this section we describe a very general method for finding a particular solution $\psi(t)$ of the nonhomogeneous equation

$$L[y]=\frac{d^2y}{dt^2}+p(t)\frac{dy}{dt}+q(t)y=g(t), \tag{1}$$

once the solutions of the homogeneous equation

$$L[y]=\frac{d^2y}{dt^2}+p(t)\frac{dy}{dt}+q(t)y=0 \tag{2}$$

are known. The basic principle of this method is to use our knowledge of the solutions of the homogeneous equation to help us find a solution of the nonhomogeneous equation.

Let $y_1(t)$ and $y_2(t)$ be two linearly independent solutions of the homogeneous equation (2). We will try to find a particular solution $\psi(t)$ of the nonhomogeneous equation (1) of the form

$$\psi(t) = u_1(t)y_1(t) + u_2(t)y_2(t); \tag{3}$$

that is, we will try to find functions $u_1(t)$ and $u_2(t)$ so that the linear combination $u_1(t)y_1(t) + u_2(t)y_2(t)$ is a solution of (1). At first glance, this would appear to be a dumb thing to do, since we are replacing the problem of finding one unknown function $\psi(t)$ by the seemingly harder problem of finding two unknown functions $u_1(t)$ and $u_2(t)$. However, by playing our cards right, we will be able to find $u_1(t)$ and $u_2(t)$ as the solutions of two very simple first-order equations. We accomplish this in the following manner. Observe that the differential equation (1) imposes only one condition on the two unknown functions $u_1(t)$ and $u_2(t)$. Therefore, we have a certain "freedom" in choosing $u_1(t)$ and $u_2(t)$. Our goal is to impose an additional condition on $u_1(t)$ and $u_2(t)$ which will make the expression $L[u_1y_1+u_2y_2]$ as simple as possible. Computing

$$\begin{aligned}\frac{d}{dt}\psi(t)&=\frac{d}{dt}[u_1y_1+u_2y_2]\\&=[u_1y_1'+u_2y_2']+[u_1'y_1+u_2'y_2]\end{aligned}$$

we see that $d^2\psi/dt^2$, and consequently $L[\psi]$, will contain no second-order derivatives of u_1 and u_2 if

$$y_1(t)u_1'(t)+y_2(t)u_2'(t)=0. \tag{4}$$

This suggests that we impose the condition (4) on the functions $u_1(t)$ and $u_2(t)$. In this case, then,

$$\begin{aligned}L[\psi]&=[u_1y_1'+u_2y_2']'+p(t)[u_1y_1'+u_2y_2']+q(t)[u_1y_1+u_2y_2]\\&=u_1'y_1'+u_2'y_2'+u_1[y_1''+p(t)y_1'+q(t)y_1]+u_2[y_2''+p(t)y_2'+q(t)y_2]\\&=u_1'y_1'+u_2'y_2'\end{aligned}$$

since both $y_1(t)$ and $y_2(t)$ are solutions of the homogeneous equation $L[y]=0$. Consequently, $\psi(t)=u_1y_1+u_2y_2$ is a solution of the nonhomogeneous equation (1) if $u_1(t)$ and $u_2(t)$ satisfy the two equations

$$\begin{aligned}y_1(t)u_1'(t)+y_2(t)u_2'(t)&=0\\y_1'(t)u_1'(t)+y_2'(t)u_2'(t)&=g(t).\end{aligned}$$

Multiplying the first equation by $y_2'(t)$, the second equation by $y_2(t)$, and subtracting gives

$$[y_1(t)y_2'(t)-y_1'(t)y_2(t)]u_1'(t)=-g(t)y_2(t),$$

while multiplying the first equation by $y_1'(t)$, the second equation by $y_1(t)$, and subtracting gives

$$[y_1(t)y_2'(t)-y_1'(t)y_2(t)]u_2'(t)=g(t)y_1(t).$$

Hence,

$$u_1'(t)=-\frac{g(t)y_2(t)}{W[y_1,y_2](t)}\quad\text{and}\quad u_2'(t)=\frac{g(t)y_1(t)}{W[y_1,y_2](t)}. \tag{5}$$

Finally, we obtain $u_1(t)$ and $u_2(t)$ by integrating the right-hand sides of (5).

Remark. The general solution of the homogeneous equation (2) is

$$y(t)=c_1y_1(t)+c_2y_2(t).$$

By letting c_1 and c_2 vary with time, we obtain a solution of the nonhomogeneous equation. Hence, this method is known as the method of variation of parameters.

Example 1.
(a) Find a particular solution $\psi(t)$ of the equation

$$\frac{d^2y}{dt^2}+y=\tan t \tag{6}$$

on the interval $-\pi/2<t<\pi/2$.
(b) Find the solution $y(t)$ of (6) which satisfies the initial conditions $y(0)=1$, $y'(0)=1$.
Solution.
(a) The functions $y_1(t)=\cos t$ and $y_2(t)=\sin t$ are two linearly independent solutions of the homogeneous equation $y''+y=0$ with

$$W[y_1,y_2](t)=y_1y_2'-y_1'y_2=(\cos t)\cos t-(-\sin t)\sin t=1.$$

Thus, from (5),

$$u_1'(t)=-\tan t\sin t \quad\text{and}\quad u_2'(t)=\tan t\cos t. \tag{7}$$

Integrating the first equation of (7) gives

$$\begin{aligned}u_1(t)&=-\int\tan t\sin t\,dt=-\int\frac{\sin^2 t}{\cos t}dt\\&=\int\frac{\cos^2 t-1}{\cos t}dt=\sin t-\ln|\sec t+\tan t|.\\&=\sin t-\ln(\sec t+\tan t),\ -\frac{\pi}{2}<t<\frac{\pi}{2}\end{aligned}$$

while integrating the second equation of (7) gives

$$u_2(t)=\int\tan t\cos t\,dt=\int\sin t\,dt=-\cos t.$$

Consequently,

$$\begin{aligned}\psi(t)&=\cos t\left[\sin t-\ln(\sec t+\tan t)\right]+\sin t(-\cos t)\\&=-\cos t\ln(\sec t+\tan t)\end{aligned}$$

is a particular solution of (6) on the interval $-\pi/2<t<\pi/2$.
(b) By Theorem 5 of Section 2.3,

$$y(t)=c_1\cos t+c_2\sin t-\cos t\ln(\sec t+\tan t)$$

for some choice of constants c_1, c_2. The constants c_1 and c_2 are determined from the initial conditions

$$1=y(0)=c_1 \quad\text{and}\quad 1=y'(0)=c_2-1.$$

Hence, $c_1=1$, $c_2=2$ and

$$y(t)=\cos t+2\sin t-\cos t\ln(\sec t+\tan t).$$

Remark. Equation (5) determines $u_1(t)$ and $u_2(t)$ up to two constants of integration. We usually take these constants to be zero, since the effect of choosing nonzero constants is to add a solution of the homogeneous equation to $\psi(t)$.

Exercises

Find the general solution of each of the following equations.

1. $\dfrac{d^2y}{dt^2}+y=\sec t, \quad -\dfrac{\pi}{2}<t<\dfrac{\pi}{2}$

2. $\dfrac{d^2y}{dt^2}-4\dfrac{dy}{dt}+4y=te^{2t}$

3. $2\dfrac{d^2y}{dt^2}-3\dfrac{dy}{dt}+y=(t^2+1)e^t$

4. $\dfrac{d^2y}{dt^2}-3\dfrac{dy}{dt}+2y=te^{3t}+1$

Solve each of the following initial-value problems.

5. $3y''+4y'+y=(\sin t)e^{-t};\quad y(0)=1,\ y'(0)=0$

6. $y''+4y'+4y=t^{5/2}e^{-2t};\quad y(0)=y'(0)=0$

7. $y''-3y'+2y=\sqrt{t+1}\ ;\quad y(0)=y'(0)=0$

8. $y''-y=f(t);\quad y(0)=y'(0)=0$

Warning. It must be remembered, while doing Problems 3 and 5, that Equation (5) was derived under the assumption that the coefficient of y'' was one.

9. Find two linearly independent solutions of $t^2y''-2y=0$ of the form $y(t)=t^r$. Using these solutions, find the general solution of $t^2y''-2y=t^2$.

10. One solution of the equation

$$y''+p(t)y'+q(t)y=0 \tag{*}$$

is $(1+t)^2$, and the Wronskian of any two solutions of (*) is constant. Find the general solution of

$$y''+p(t)y'+q(t)y=1+t.$$

11. Find the general solution of $y''+(1/4t^2)y=f\cos t$, $t>0$, given that $y_1(t)=\sqrt{t}$ is a solution of the homogeneous equation.

12. Find the general solution of the equation

$$\frac{d^2y}{dt^2} - \frac{2t}{1+t^2}\frac{dy}{dt} + \frac{2}{1+t^2}y = 1+t^2.$$

13. Show that $\sec t + \tan t$ is positive for $-\pi/2 < t < \pi/2$.

2.5 The method of judicious guessing

A serious disadvantage of the method of variation of parameters is that the integrations required are often quite difficult. In certain cases, it is usually much simpler to guess a particular solution. In this section we will establish a systematic method for guessing solutions of the equation

$$a\frac{d^2y}{dt^2} + b\frac{dy}{dt} + cy = g(t) \tag{1}$$

where a, b and c are constants, and $g(t)$ has one of several special forms.

Consider first the differential equation

$$L[y] = a\frac{d^2y}{dt^2} + b\frac{dy}{dt} + cy = a_0 + a_1t + \ldots + a_nt^n. \tag{2}$$

We seek a function $\psi(t)$ such that the three functions $a\psi''$, $b\psi'$ and $c\psi$ add up to a given polynomial of degree n. The obvious choice for $\psi(t)$ is a polynomial of degree n. Thus, we set

$$\psi(t) = A_0 + A_1t + \ldots + A_nt^n \tag{3}$$

and compute

$$\begin{aligned}L[\psi](t) &= a\psi''(t) + b\psi'(t) + c\psi(t)\\ &= a\left[2A_2 + \ldots + n(n-1)A_nt^{n-2}\right] + b\left[A_1 + \ldots + nA_nt^{n-1}\right]\\ &\quad + c\left[A_0 + A_1t + \ldots + A_nt^n\right]\\ &= cA_nt^n + (cA_{n-1} + nbA_n)t^{n-1} + \ldots + (cA_0 + bA_1 + 2aA_2).\end{aligned}$$

Equating coefficients of like powers of t in the equation

$$L[\psi](t) = a_0 + a_1t + \ldots + a_nt^n$$

gives

$$cA_n = a_n,\; cA_{n-1} + nbA_n = a_{n-1}, \ldots, cA_0 + bA_1 + 2aA_2 = a_0. \tag{4}$$

The first equation determines $A_n = a_n/c$, for $c \neq 0$, and the remaining equations then determine $A_{n-1}, \ldots, A_0$ successively. Thus, Equation (1) has a particular solution $\psi(t)$ of the form (3), for $c \neq 0$.

We run into trouble when $c = 0$, since then the first equation of (4) has no solution A_n. This difficulty is to be expected though, for if $c = 0$, then $L[\psi] = a\psi'' + b\psi'$ is a polynomial of degree $n-1$, while the right hand side of (2) is a polynomial of degree n. To guarantee that $a\psi'' + b\psi'$ is a polynomial of degree n, we must take ψ as a polynomial of degree $n+1$. Thus, we set

$$\psi(t) = t[A_0 + A_1 t + \ldots + A_n t^n]. \tag{5}$$

We have omitted the constant term in (5) since $y=$ constant is a solution of the homogeneous equation $ay'' + by' = 0$, and thus can be subtracted from $\psi(t)$. The coefficients $A_0, A_1, \ldots, A_n$ are determined uniquely (see Exercise 19) from the equation

$$a\psi'' + b\psi' = a_0 + a_1 t + \ldots + a_n t^n$$

if $b \neq 0$.

Finally, the case $b = c = 0$ is trivial to handle since the differential equation (2) can then be integrated immediately to yield a particular solution $\psi(t)$ of the form

$$\psi(t) = \frac{1}{a}\left[\frac{a_0 t^2}{1\cdot 2} + \frac{a_1 t^3}{2\cdot 3} + \ldots + \frac{a_n t^{n+2}}{(n+1)(n+2)}\right].$$

Summary. The differential equation (2) has a solution $\psi(t)$ of the form

$$\psi(t) = \begin{cases} A_0 + A_1 t + \ldots + A_n t^n, & c \neq 0 \\ t(A_0 + A_1 t + \ldots + A_n t^n), & c = 0, b \neq 0. \\ t^2(A_0 + A_1 t + \ldots + A_n t^n), & c = b = 0 \end{cases}$$

Example 1. Find a particular solution $\psi(t)$ of the equation

$$L[y] = \frac{d^2 y}{dt^2} + \frac{dy}{dt} + y = t^2. \tag{6}$$

Solution. We set $\psi(t) = A_0 + A_1 t + A_2 t^2$ and compute

$$\begin{aligned} L[\psi](t) &= \psi''(t) + \psi'(t) + \psi(t) \\ &= 2A_2 + (A_1 + 2A_2 t) + A_0 + A_1 t + A_2 t^2 \\ &= (A_0 + A_1 + 2A_2) + (A_1 + 2A_2)t + A_2 t^2. \end{aligned}$$

Equating coefficients of like powers of t in the equation $L[\psi](t) = t^2$ gives

$$A_2 = 1, \qquad A_1 + 2A_2 = 0$$

and

$$A_0 + A_1 + 2A_2 = 0.$$

The first equation tells us that $A_2 = 1$, the second equation then tells us that $A_1 = -2$, and the third equation then tells us that $A_0 = 0$. Hence,

$$\psi(t) = -2t + t^2$$

is a particular solution of (6).

Let us now re-do this problem using the method of variation of parameters. It is easily verified that

$$y_1(t)=e^{-t/2}\cos\sqrt{3}\,t/2 \quad\text{and}\quad y_2(t)=e^{-t/2}\sin\sqrt{3}\,t/2$$

are two solutions of the homogeneous equation $L[y]=0$. Hence,

$$\psi(t)=u_1(t)e^{-t/2}\cos\sqrt{3}\,t/2+u_2(t)e^{-t/2}\sin\sqrt{3}\,t/2$$

is a particular solution of (6), where

$$u_1(t)=\int\frac{-t^2e^{-t/2}\sin\sqrt{3}\,t/2}{W[y_1,y_2](t)}dt=\frac{-2}{\sqrt{3}}\int t^2e^{t/2}\sin\sqrt{3}\,t/2\,dt$$

and

$$u_2(t)=\int\frac{t^2e^{-t/2}\cos\sqrt{3}\,t/2}{W[y_1,y_2](t)}dt=\frac{2}{\sqrt{3}}\int t^2e^{t/2}\cos\sqrt{3}\,t/2\,dt.$$

These integrations are extremely difficult to perform. Thus, the method of guessing is certainly preferrable, in this problem at least, to the method of variation of parameters.

Consider now the differential equation

$$L[y]=a\frac{d^2y}{dt^2}+b\frac{dy}{dt}+cy=(a_0+a_1t+\ldots+a_nt^n)e^{\alpha t}. \tag{7}$$

We would like to remove the factor $e^{\alpha t}$ from the right-hand side of (7), so as to reduce this equation to Equation (2). This is accomplished by setting $y(t)=e^{\alpha t}v(t)$. Then,

$$y'=e^{\alpha t}(v'+\alpha v) \quad\text{and}\quad y''=e^{\alpha t}(v''+2\alpha v'+\alpha^2v)$$

so that

$$L[y]=e^{\alpha t}\left[av''+(2a\alpha+b)v'+(a\alpha^2+b\alpha+c)v\right].$$

Consequently, $y(t)=e^{\alpha t}v(t)$ is a solution of (7) if, and only if,

$$a\frac{d^2v}{dt^2}+(2a\alpha+b)\frac{dv}{dt}+(a\alpha^2+b\alpha+c)v=a_0+a_1t+\ldots+a_nt^n. \tag{8}$$

In finding a particular solution $v(t)$ of (8), we must distinguish as to whether (i) $a\alpha^2+b\alpha+c\neq0$; (ii) $a\alpha^2+b\alpha+c=0$, but $2a\alpha+b\neq0$; and (iii) both $a\alpha^2+b\alpha+c$ and $2a\alpha+b=0$. The first case means that α is not a root of the characteristic equation

$$ar^2+br+c=0. \tag{9}$$

In other words, $e^{\alpha t}$ is not a solution of the homogeneous equation $L[y]=0$. The second condition means that α is a single root of the characteristic equation (9). This implies that $e^{\alpha t}$ is a solution of the homogeneous equation, but $te^{\alpha t}$ is not. Finally, the third condition means that α is a double root of the characteristic equation (9), so that both $e^{\alpha t}$ and $te^{\alpha t}$ are solutions of the homogeneous equation. Hence, Equation (7) has a particular solution $\psi(t)$ of the form (i) $\psi(t)=(A_0+\dots+A_nt^n)e^{\alpha t}$, if $e^{\alpha t}$ is not a solution of the homogeneous equation; (ii) $\psi(t)=t(A_0+\dots+A_nt^n)e^{\alpha t}$, if $e^{\alpha t}$ is a solution of the homogeneous equation but $te^{\alpha t}$ is not; and (iii) $\psi(t)=t^2(A_0+\dots+A_nt^n)e^{\alpha t}$ if both $e^{\alpha t}$ and $te^{\alpha t}$ are solutions of the homogeneous equation.

Remark. There are two ways of computing a particular solution $\psi(t)$ of (7). Either we make the substitution $y=e^{\alpha t}v$ and find $v(t)$ from (8), or we guess a solution $\psi(t)$ of the form $e^{\alpha t}$ times a suitable polynomial in t. If α is a double root of the characteristic equation (9), or if $n\geqslant 2$, then it is advisable to set $y=e^{\alpha t}v$ and then find $v(t)$ from (8). Otherwise, we guess $\psi(t)$ directly.

Example 2. Find the general solution of the equation

$$\frac{d^2y}{dt^2}-4\frac{dy}{dt}+4y=(1+t+\dots+t^{27})e^{2t}. \tag{10}$$

Solution. The characteristic equation $r^2-4r+4=0$ has equal roots $r_1=r_2=2$. Hence, $y_1(t)=e^{2t}$ and $y_2(t)=te^{2t}$ are solutions of the homogeneous equation $y''-4y'+4y=0$. To find a particular solution $\psi(t)$ of (10), we set $y=e^{2t}v$. Then, of necessity,

$$\frac{d^2v}{dt^2}=1+t+t^2+\dots+t^{27}.$$

Integrating this equation twice, and setting the constants of integration equal to zero gives

$$v(t)=\frac{t^2}{1\cdot 2}+\frac{t^3}{2\cdot 3}+\dots+\frac{t^{29}}{28\cdot 29}.$$

Hence, the general solution of (10) is

$$\begin{aligned} y(t)&=c_1e^{2t}+c_2te^{2t}+e^{2t}\left[\frac{t^2}{1\cdot 2}+\dots+\frac{t^{29}}{28\cdot 29}\right]\\ &=e^{2t}\left[c_1+c_2t+\frac{t^2}{1\cdot 2}+\dots+\frac{t^{29}}{28\cdot 29}\right]. \end{aligned}$$

It would be sheer madness (and a terrible waste of paper) to plug the expression

$$\psi(t)=t^2(A_0+A_1t+\dots+A_{27}t^{27})e^{2t}$$

into (10) and then solve for the coefficients $A_0, A_1,\dots,A_{27}$.

Example 3. Find a particular solution $\psi(t)$ of the equation

$$L[y]=\frac{d^2y}{dt^2}-3\frac{dy}{dt}+2y=(1+t)e^{3t}.$$

Solution. In this case, e^{3t} is not a solution of the homogeneous equation $y''-3y'+2y=0$. Thus, we set $\psi(t)=(A_0+A_1t)e^{3t}$. Computing

$$\begin{aligned}L[\psi](t)&=\psi''-3\psi'+2\psi\\&=e^{3t}\left[(9A_0+6A_1+9A_1t)-3(3A_0+A_1+3A_1t)+2(A_0+A_1t)\right]\\&=e^{3t}\left[(2A_0+3A_1)+2A_1t\right]\end{aligned}$$

and cancelling off the factor e^{3t} from both sides of the equation

$$L[\psi](t)=(1+t)e^{3t},$$

gives

$$2A_1t+(2A_0+3A_1)=1+t.$$

This implies that $2A_1=1$ and $2A_0+3A_1=1$. Hence, $A_1=\frac{1}{2}$, $A_0=-\frac{1}{4}$ and $\psi(t)=(-\frac{1}{4}+t/2)e^{3t}$.

Finally, we consider the differential equation

$$L[y]=a\frac{d^2y}{dt^2}+b\frac{dy}{dt}+cy=(a_0+a_1t+\ldots+a_nt^n)\times\begin{cases}\cos\omega t\\ \sin\omega t\end{cases}. \tag{11}$$

We can reduce the problem of finding a particular solution $\psi(t)$ of (11) to the simpler problem of finding a particular solution of (7) with the aid of the following simple but extremely useful lemma.

Lemma 1. *Let $y(t)=u(t)+iv(t)$ be a complex-valued solution of the equation*

$$L[y]=a\frac{d^2y}{dt^2}+b\frac{dy}{dt}+cy=g(t)=g_1(t)+ig_2(t) \tag{12}$$

where a, b and c are real. This means, of course, that

$$a\left[u''(t)+iv''(t)\right]+b\left[u'(t)+iv'(t)\right]+c\left[u(t)+iv(t)\right]=g_1(t)+ig_2(t). \tag{13}$$

Then, $L[u](t)=g_1(t)$ and $L[v](t)=g_2(t)$.

PROOF. Equating real and imaginary parts in (13) gives

$$au''(t)+bu'(t)+cu(t)=g_1(t)$$

and

$$av''(t)+bv'(t)+cv(t)=g_2(t). \qquad\square$$

Now, let $\phi(t)=u(t)+iv(t)$ be a particular solution of the equation

$$a\frac{d^2y}{dt^2}+b\frac{dy}{dt}+cy=(a_0+\ldots+a_nt^n)e^{i\omega t}. \tag{14}$$

The real part of the right-hand side of (14) is $(a_0+\ldots+a_nt^n)\cos\omega t$, while the imaginary part is $(a_0+\dots+a_nt^n)\sin\omega t$. Hence, by Lemma 1

$$u(t)=\mathrm{Re}\{\phi(t)\}$$

is a solution of

$$ay''+by'+cy=(a_0+\dots+a_nt^n)\cos\omega t$$

while

$$v(t)=\mathrm{Im}\{\phi(t)\}$$

is a solution of

$$ay''+by'+cy=(a_0+\dots+a_nt^n)\sin\omega t.$$

Example 4. Find a particular solution $\psi(t)$ of the equation

$$L[y]=\frac{d^2y}{dt^2}+4y=\sin 2t. \tag{15}$$

Solution. We will find $\psi(t)$ as the imaginary part of a complex-valued solution $\phi(t)$ of the equation

$$L[y]=\frac{d^2y}{dt^2}+4y=e^{2it}. \tag{16}$$

To this end, observe that the characteristic equation $r^2+4=0$ has complex roots $r=\pm 2i$. Therefore, Equation (16) has a particular solution $\phi(t)$ of the form $\phi(t)=A_0te^{2it}$. Computing

$$\phi'(t)=A_0(1+2it)e^{2it} \quad \text{and} \quad \phi''(t)=A_0(4i-4t)e^{2it}$$

we see that

$$L[\phi](t)=\phi''(t)+4\phi(t)=4iA_0e^{2it}.$$

Hence, $A_0=1/4i=-i/4$ and

$$\phi(t)=-\frac{it}{4}e^{2it}=-\frac{it}{4}(\cos 2t+i\sin 2t)=\frac{t}{4}\sin 2t-i\frac{t}{4}\cos 2t.$$

Therefore, $\psi(t)=\mathrm{Im}\{\phi(t)\}=-(t/4)\cos 2t$ is a particular solution of (15).

Example 5. Find a particular solution $\psi(t)$ of the equation

$$\frac{d^2y}{dt^2}+4y=\cos 2t. \tag{17}$$

Solution. From Example 4, $\phi(t)=(t/4)\sin 2t-i(t/4)\cos 2t$ is a complex-valued solution of (16). Therefore,

$$\psi(t)=\mathrm{Re}\{\phi(t)\}=\frac{t}{4}\sin 2t$$

is a particular solution of (17).

Example 6. Find a particular solution $\psi(t)$ of the equation

$$L[y]=\frac{d^2y}{dt^2}+2\frac{dy}{dt}+y=te^t\cos t. \tag{18}$$

Solution. Observe that $te^t\cos t$ is the real part of $te^{(1+i)t}$. Therefore, we can find $\psi(t)$ as the real part of a complex-valued solution $\phi(t)$ of the equation

$$L[y]=\frac{d^2y}{dt^2}+2\frac{dy}{dt}+y=te^{(1+i)t}. \tag{19}$$

To this end, observe that $1+i$ is not a root of the characteristic equation $r^2+2r+1=0$. Therefore, Equation (19) has a particular solution $\phi(t)$ of the form $\phi(t)=(A_0+A_1t)e^{(1+i)t}$. Computing $L[\phi]=\phi''+2\phi'+\phi$, and using the identity

$$(1+i)^2+2(1+i)+1=\left[(1+i)+1\right]^2=(2+i)^2$$

we see that

$$\left[(2+i)^2A_1t+(2+i)^2A_0+2(2+i)A_1\right]=t.$$

Equating coefficients of like powers of t in this equation gives

$$(2+i)^2A_1=1$$

and

$$(2+i)A_0+2A_1=0.$$

This implies that $A_1=1/(2+i)^2$ and $A_0=-2/(2+i)^3$, so that

$$\phi(t)=\left[\frac{-2}{(2+i)^3}+\frac{t}{(2+i)^2}\right]e^{(1+i)t}.$$

After a little algebra, we find that

$$\phi(t)=\frac{e^t}{125}\{\left[(15t-4)\cos t+(20t-22)\sin t\right]$$
$$+i\left[(22-20t)\cos t+(15t-4)\sin t\right]\}.$$

Hence,

$$\psi(t)=\mathrm{Re}\{\phi(t)\}=\frac{e^t}{125}\left[(15t-4)\cos t+(20t-22)\sin t\right].$$

Remark. The method of judicious guessing also applies to the equation

$$L[y]=a\frac{d^2y}{dt^2}+b\frac{dy}{dt}+cy=\sum_{j=1}^{n}p_j(t)e^{\alpha_jt} \tag{20}$$

where the $p_j(t), j=1,\ldots,n$ are polynomials in t. Let $\psi_j(t)$ be a particular solution of the equation

$$L[y]=p_j(t)e^{\alpha_j t}, \qquad j=1,\dots,n.$$

Then, $\psi(t)=\sum_{j=1}^n\psi_j(t)$ is a solution of (20) since

$$L[\psi]=L\left[\sum_{j=1}^n \psi_j\right]=\sum_{j=1}^n L[\psi_j]=\sum_{j=1}^n p_j(t)e^{\alpha_j t}.$$

Thus, to find a particular solution of the equation

$$y''+y'+y=e^t+t\sin t$$

we find particular solutions $\psi_1(t)$ and $\psi_2(t)$ of the equations

$$y''+y'+y=e^t \quad \text{and} \quad y''+y'+y=t\sin t$$

respectively, and then add these two solutions together.

Exercises

Find a particular solution of each of the following equations.

1. $y''+3y=t^3-1$

2. $y''+4y'+4y=te^{\alpha t}$

3. $y''-y=t^2e^t$

4. $y''+y'+y=1+t+t^2$

5. $y''+2y'+y=e^{-t}$

6. $y''+5y'+4y=t^2e^{7t}$

7. $y''+4y=t\sin 2t$

8. $y''-6y'+9y=(3t^7-5t^4)e^{3t}$

9. $y''-2y'+5y=2\cos^2 t$

10. $y''-2y'+5y=2(\cos^2 t)e^t$

11. $y''+y'-6y=\sin t+te^{2t}$

12. $y''+y'+4y=t^2+(2t+3)(1+\cos t)$

13. $y''-3y'+2y=e^t+e^{2t}$

14. $y''+2y'=1+t^2+e^{-2t}$

15. $y''+y=\cos t\cos 2t$

16. $y''+y=\cos t\cos 2t\cos 3t.$

17. (a) Show that $\cos^3\omega t=\frac{1}{4}\mathrm{Re}\{e^{3i\omega t}+3e^{i\omega t}\}$.
Hint: $\cos\omega t=(e^{i\omega t}+e^{-i\omega t})/2$.
(b) Find a particular solution of the equation

$$10y''+0.2y'+1000y=5+20\cos^3 10t$$

18. (a) Let $L[y]=y''-2r_1y'+r_1^2y$. Show that

$$L\left[e^{r_1t}v(t)\right]=e^{r_1t}v''(t).$$

(b) Find the general solution of the equation

$$y''-6y'+9y=t^{3/2}e^{3t}.$$

19. Let $\psi(t)=t(A_0+\dots+A_nt^n)$, and assume that $b\neq 0$. Show that the equation $a\psi''+b\psi'=a_0+\dots+a_nt^n$ determines $A_0,\dots,A_n$ uniquely.

2.6 Mechanical vibrations

Consider the case where a small object of mass m is attached to an elastic spring of length l, which is suspended from a rigid horizontal support (see Figure 1). (An elastic spring has the property that if it is stretched or compressed a distance Δl which is small compared to its natural length l, then it will exert a restoring force of magnitude $k\Delta l$. The constant k is called the spring-constant, and is a measure of the stiffness of the spring.) In addition, the mass and spring may be immersed in a medium, such as oil, which impedes the motion of an object through it. Engineers usually refer to such systems as spring-mass-dashpot systems, or as seismic instruments, since they are similar, in principle, to a seismograph which is used to detect motions of the earth's surface.

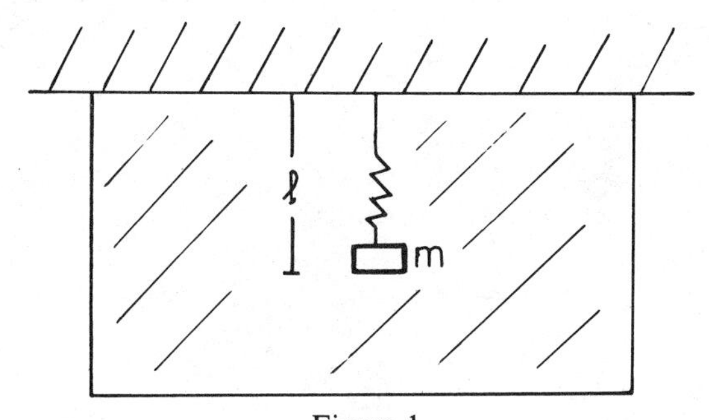

Figure 1

Spring-mass-dashpot systems have many diverse applications. For example, the shock absorbers in our automobiles are simple spring-mass-dashpot systems. Also, most heavy gun emplacements are attached to such systems so as to minimize the "recoil" effect of the gun. The usefulness of these devices will become apparent after we set up and solve the differential equation of motion of the mass m.

In calculating the motion of the mass m, it will be convenient for us to measure distances from the equilibrium position of the mass, rather than the horizontal support. The equilibrium position of the mass is that point where the mass will hang at rest if no external forces act upon it. In equilibrium, the weight mg of the mass is exactly balanced by the restoring force of the spring. Thus, in its equilibrium position, the spring has been stretched a distance Δl, where $k\Delta l = mg$. We let $y=0$ denote this equilibrium position, and we take the downward direction as positive. Let $y(t)$ denote the position of the mass at time t. To find $y(t)$, we must compute the total force acting on the mass m. This force is the sum of four separate forces W, R, D and F.

(i) The force $W = mg$ is the weight of the mass pulling it downward. This force is positive, since the downward direction is the positive y direction.

(ii) The force R is the restoring force of the spring, and it is proportional to the elongation, or compression, $\Delta l + y$ of the spring. It always acts to restore the spring to its natural length. If $\Delta l + y > 0$, then R is negative, so that $R = -k(\Delta l + y)$, and if $\Delta l + y < 0$, then R is positive, so that $R = -k(\Delta l + y)$. In either case,

$$R = -k(\Delta l + y).$$

(iii) The force D is the damping, or drag force, which the medium exerts on the mass m. (Most media, such as oil and air, tend to resist the motion of an object through it.) This force always acts in the direction opposite the direction of motion, and is usually directly proportional to the magnitude of the velocity dy/dt. If the velocity is positive; that is, the mass is moving in the downward direction, then $D = -c\,dy/dt$, and if the velocity is negative, then $D = -c\,dy/dt$. In either case,

$$D = -c\,dy/dt.$$

(iv) The force F is the external force applied to the mass. This force is directed upward or downward, depending as to whether F is positive or negative. In general, this external force will depend explicitly on time.

From Newton's second law of motion (see Section 1.7)

$$\begin{aligned} m\frac{d^2y}{dt^2} &= W + R + D + F \\ &= mg - k(\Delta l + y) - c\frac{dy}{dt} + F(t) \\ &= -ky - c\frac{dy}{dt} + F(t), \end{aligned}$$

since $mg = k\,\Delta l$. Hence, the position $y(t)$ of the mass satisfies the second-order linear differential equation

$$m\frac{d^2y}{dt^2} + c\frac{dy}{dt} + ky = F(t) \tag{1}$$

where m, c and k are nonnegative constants. We adopt here the mks system of units so that F is measured in newtons, y is measured in meters, and t is measured in seconds. In this case, the units of k are N/m, the units of c are N·s/m, and the units of m are kilograms (N·s^2/m)

(a) *Free vibrations*:

We consider first the simplest case of free undamped motion. In this case, Equation (1) reduces to

$$m\frac{d^2y}{dt^2} + ky = 0 \quad \text{or} \quad \frac{d^2y}{dt^2} + \omega_0^2 y = 0 \tag{2}$$

where $\omega_0^2 = k/m$. The general solution of (2) is

$$y(t) = a\cos\omega_0 t + b\sin\omega_0 t. \tag{3}$$

In order to analyze the solution (3), it is convenient to rewrite it as a single cosine function. This is accomplished by means of the following lemma.

Lemma 1. *Any function $y(t)$ of the form* (3) *can be written in the simpler form*

$$y(t) = R\cos(\omega_0 t - \delta) \tag{4}$$

where $R = \sqrt{a^2+b^2}$ and $\delta = \tan^{-1} b/a$.

PROOF. We will verify that the two expressions (3) and (4) are equal. To this end, compute

$$R\cos(\omega_0 t - \delta) = R\cos\omega_0 t\cos\delta + R\sin\omega_0 t\sin\delta$$

and observe from Figure 2 that $R\cos\delta = a$ and $R\sin\delta = b$. Hence,

$$R\cos(\omega_0 t - \delta) = a\cos\omega_0 t + b\sin\omega_0 t. \qquad \square$$

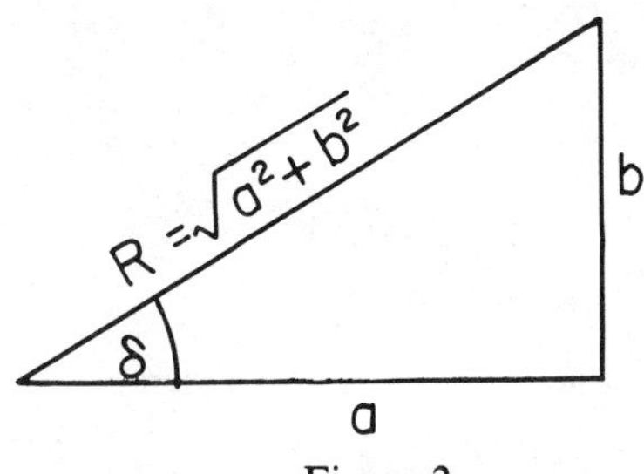

Figure 2

In Figure 3 we have graphed the function $y = R\cos(\omega_0 t - \delta)$. Notice that $y(t)$ always lies between $-R$ and $+R$, and that the motion of the mass is periodic—it repeats itself over every time interval of length $2\pi/\omega_0$. This type of motion is called simple harmonic motion; R is called the amplitude of the motion, δ the phase angle of the motion, $T_0 = 2\pi/\omega_0$ the natural period of the motion, and $\omega_0 = \sqrt{k/m}$ the natural frequency of the system.

(b) *Damped free vibrations*:

If we now include the effect of damping, then the differential equation governing the motion of the mass is

$$m\frac{d^2y}{dt^2} + c\frac{dy}{dt} + ky = 0. \tag{5}$$

The roots of the characteristic equation $mr^2 + cr + k = 0$ are

$$r_1 = \frac{-c + \sqrt{c^2 - 4km}}{2m} \quad \text{and} \quad r_2 = \frac{-c - \sqrt{c^2 - 4km}}{2m}.$$

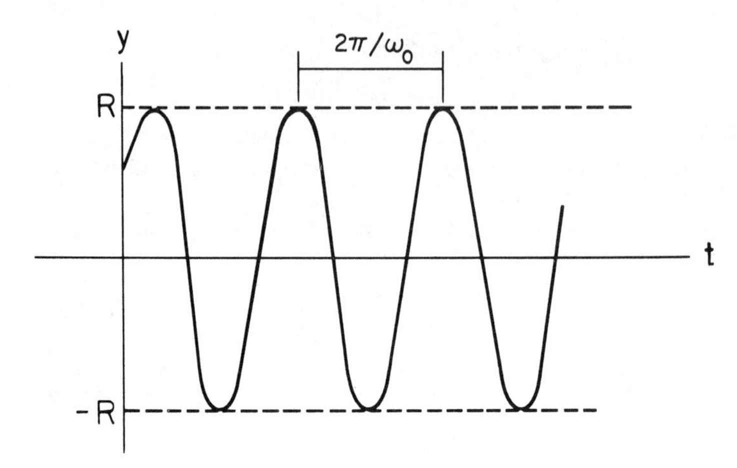

Figure 3. Graph of $y(t)=R\cos(\omega_0 t-\delta)$

Thus, there are three cases to consider, depending as to whether c^2-4km is positive, negative or zero.

(i) $c^2-4km>0$. In this case both r_1 and r_2 are negative, and every solution $y(t)$ of (5) has the form

$$y(t)=ae^{r_1t}+be^{r_2t}.$$

(ii) $c^2-4km=0$. In this case, every solution $y(t)$ of (5) is of the form

$$y(t)=(a+bt)e^{-ct/2m}.$$

(iii) $c^2-4km<0$. In this case, every solution $y(t)$ of (5) is of the form

$$y(t)=e^{-ct/2m}\left[a\cos\mu t+b\sin\mu t\right], \qquad \mu=\frac{\sqrt{4km-c^2}}{2m}.$$

The first two cases are referred to as overdamped and critically damped, respectively. They represent motions in which the originally displaced mass creeps back to its equilibrium position. Depending on the initial conditions, it may be possible to overshoot the equilibrium position once, but no more than once (see Exercises 2-3). The third case, which is referred to as an underdamped motion, occurs quite often in mechanical systems and represents a damped vibration. To see this, we use Lemma 1 to rewrite the function

$$y(t)=e^{-ct/2m}\left[a\cos\mu t+b\sin\mu t\right]$$

in the form

$$y(t)=Re^{-ct/2m}\cos(\mu t-\delta).$$

The displacement y oscillates between the curves $y=\pm Re^{-ct/2m}$, and thus represents a cosine curve with decreasing amplitude, as shown in Figure 4.

Now, observe that the motion of the mass always dies out eventually if there is damping in the system. In other words, any initial disturbance of

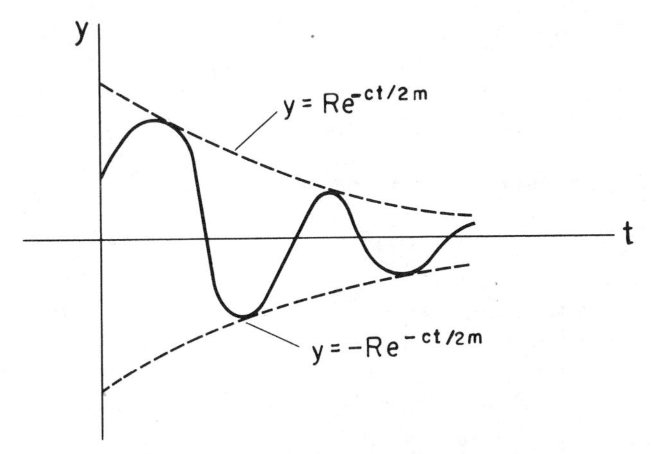

Figure 4. Graph of $\mathrm{Re}^{-ct/2m}\cos(\mu t-\delta)$

the system is dissipated by the damping present in the system. This is one reason why spring-mass-dashpot systems are so useful in mechanical systems: they can be used to damp out any undesirable disturbances. For example, the shock transmitted to an automobile by a bump in the road is dissipated by the shock absorbers in the car, and the momentum from the recoil of a gun barrel is dissipated by a spring-mass-dashpot system attached to the gun.

(c) *Damped forced vibrations*:

If we now introduce an external force $F(t)=F_0\cos\omega t$, then the differential equation governing the motion of the mass is

$$m\frac{d^2y}{dt^2}+c\frac{dy}{dt}+ky=F_0\cos\omega t. \tag{6}$$

Using the method of judicious guessing, we can find a particular solution $\psi(t)$ of (6) of the form

$$\begin{aligned}\psi(t)&=\frac{F_0}{(k-m\omega^2)^2+c^2\omega^2}\left[(k-m\omega^2)\cos\omega t+c\omega\sin\omega t\right]\\&=\frac{F_0}{(k-m\omega^2)^2+c^2\omega^2}\left[(k-m\omega^2)^2+c^2\omega^2\right]^{1/2}\cos(\omega t-\delta)\\&=\frac{F_0\cos(\omega t-\delta)}{\left[(k-m\omega^2)^2+c^2\omega^2\right]^{1/2}}\end{aligned} \tag{7}$$

where $\tan\delta=c\omega/(k-m\omega^2)$. Hence, every solution $y(t)$ of (6) must be of the form

$$y(t)=\phi(t)+\psi(t)=\phi(t)+\frac{F_0\cos(\omega t-\delta)}{\left[(k-m\omega^2)^2+c^2\omega^2\right]^{1/2}} \tag{8}$$

where $\phi(t)$ is a solution of the homogeneous equation

$$m\frac{d^2y}{dt^2}+c\frac{dy}{dt}+ky=0. \tag{9}$$

We have already seen though, that every solution $y=\phi(t)$ of (9) approaches zero as t approaches infinity. Thus, for large t, the equation $y(t)=\psi(t)$ describes very accurately the position of the mass m, regardless of its initial position and velocity. For this reason, $\psi(t)$ is called the steady state part of the solution (8), while $\phi(t)$ is called the transient part of the solution.

(d) *Forced free vibrations*:

We now remove the damping from our system and consider the case of forced free vibrations where the forcing term is periodic and has the form $F(t)=F_0\cos\omega t$. In this case, the differential equation governing the motion of the mass m is

$$\frac{d^2y}{dt^2}+\omega_0^2y=\frac{F_0}{m}\cos\omega t, \qquad \omega_0^2=k/m. \tag{10}$$

The case $\omega\neq\omega_0$ is uninteresting; every solution $y(t)$ of (10) has the form

$$y(t)=c_1\cos\omega_0t+c_2\sin\omega_0t+\frac{F_0}{m\left(\omega_0^2-\omega^2\right)}\cos\omega t,$$

and thus is the sum of two periodic functions of different periods. The interesting case is when $\omega=\omega_0$; that is, when the frequency ω of the external force equals the natural frequency of the system. This case is called the *resonance* case, and the differential equation of motion for the mass m is

$$\frac{d^2y}{dt^2}+\omega_0^2y=\frac{F_0}{m}\cos\omega_0t. \tag{11}$$

We will find a particular solution $\psi(t)$ of (11) as the real part of a complex-valued solution $\phi(t)$ of the equation

$$\frac{d^2y}{dt^2}+\omega_0^2y=\frac{F_0}{m}e^{i\omega_0t}. \tag{12}$$

Since $e^{i\omega_0t}$ is a solution of the homogeneous equation $y''+\omega_0^2y=0$, we know that (12) has a particular solution $\phi(t)=Ate^{i\omega_0t}$, for some constant A. Computing

$$\phi''+\omega_0^2\phi=2i\omega_0Ae^{i\omega_0t}$$

we see that

$$A = \frac{1}{2i\omega_0}\frac{F_0}{m} = \frac{-iF_0}{2m\omega_0}.$$

Hence,

$$\begin{aligned}\phi(t) &= \frac{-iF_0t}{2m\omega_0}(\cos\omega_0 t + i\sin\omega_0 t)\\ &= \frac{F_0t}{2m\omega_0}\sin\omega_0 t - i\frac{F_0t}{2m\omega_0}\cos\omega_0 t\end{aligned}$$

is a particular solution of (12), and

$$\psi(t) = \mathrm{Re}\{\phi(t)\} = \frac{F_0t}{2m\omega_0}\sin\omega_0 t$$

is a particular solution of (11). Consequently, every solution $y(t)$ of (11) is of the form

$$y(t) = c_1\cos\omega_0 t + c_2\sin\omega_0 t + \frac{F_0t}{2m\omega_0}\sin\omega_0 t \tag{13}$$

for some choice of constants c_1, c_2.

Now, the sum of the first two terms in (13) is a periodic function of time. The third term, though, represents an oscillation with increasing amplitude, as shown in Figure 5. Thus, the forcing term $F_0\cos\omega t$, if it is in resonance with the natural frequency of the system, will always cause unbounded oscillations. Such a phenomenon was responsible for the collapse of the Tacoma Bridge, (see Section 2.6.1) and many other mechanical catastrophes.

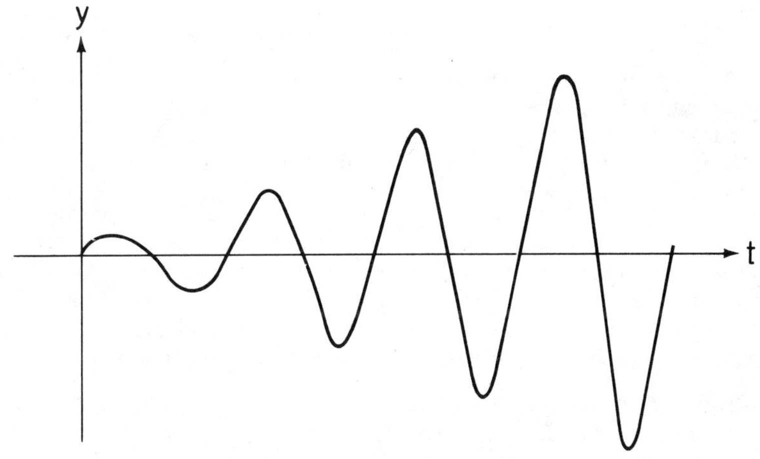

Figure 5. Graph of $f(t) = At\sin\omega_0 t$

EXERCISES

1. It is found experimentally that a 1 kg mass stretches a spring 49/320 m. If the mass is pulled down an additional 1/4 m and released, find the amplitude, period and frequency of the resulting motion, neglecting air resistance (use $g = 9.8\text{ m/s}^2$).

2. Let $y(t) = Ae^{r_1 t} + Be^{r_2 t}$, with $|A| + |B| \neq 0$.
(a) Show that $y(t)$ is zero at most once.
(b) Show that $y'(t)$ is zero at most once.

3. Let $y(t) = (A + Bt)e^{rt}$, with $|A| + |B| \neq 0$.
(a) Show that $y(t)$ is zero at most once.
(b) Show that $y'(t)$ is zero at most once.

4. A small object of mass 1 kg is attached to a spring with spring constant 2N/m. This spring–mass system is immersed in a viscous medium with damping constant 3 N·s/m. At time $t = 0$, the mass is lowered 1/2 m below its equilibrium position, and released. Show that the mass will creep back to its equilibrium position as t approaches infinity.

5. A small object of mass 1 kg is attached to a spring with spring-constant 1 N/m and is immersed in a viscous medium with damping constant 2 N·s/m. At time $t = 0$, the mass is lowered 1/4 m and given an initial velocity of 1 m/s in the upward direction. Show that the mass will overshoot its equilibrium position once, and then creep back to equilibrium.

6. A small object of mass 4 kg is attached to an elastic spring with spring-constant 64 N/m, and is acted upon by an external force $F(t) = A\cos^3 \omega t$. Find all values of ω at which resonance occurs.

7. The gun of a U.S. M60 tank is attached to a spring–mass–dashpot system with spring-constant $100\alpha^2$ and damping constant 200α, in their appropriate units. The mass of the gun is 100 kg. Assume that the displacement $y(t)$ of the gun from its rest position after being fired at time $t = 0$ satisfies the initial-value problem

$$100y'' + 200\alpha y' + 100\alpha^2 y = 0; \; y(0) = 0, \, y'(0) = 100\text{ m/s}.$$

It is desired that one second later, the quantity $y^2 + (y')^2$ be less than .01. How large must α be to guarantee that this is so? (The spring–mass–dashpot mechanism in the M60 tanks supplied by the U.S. to Israel are critically damped, for this situation is preferable in desert warfare where one has to fire again as quickly as possible).

8. A spring–mass–dashpot system has the property that the spring constant k is 9 times its mass m, and the damping constant c is 6 times its mass. At time $t = 0$, the mass, which is hanging at rest, is acted upon by an external force $F(t) = (3\sin 3t)$ N. The spring will break if it is stretched an additional 5 m from its equilibrium position. Show that the spring will not break if $m \geqslant 1/5$ kg.

9. A spring–mass–dashpot system with $m = 1$, $k = 2$ and $c = 2$ (in their respective units) hangs in equilibrium. At time $t = 0$, an external force $F(t) = \pi - t$ N acts for a time interval π. Find the position of the mass at anytime $t > \pi$.

10. A 1 kg mass is attached to a spring with spring constant $k=64$ N/m. With the mass on the spring at rest in the equilibrium position at time $t=0$, an external force $F(t)=(\frac{1}{2}t)$ N is applied until time $t_1=7\pi/16$ seconds, at which time it is removed. Assuming no damping, find the frequency and amplitude of the resulting oscillation.

11. A 1 kg mass is attached to a spring with spring constant $k=4$ N/m, and hangs in equilibrium. An external force $F(t)=(1+t+\sin 2t)$ N is applied to the mass beginning at time $t=0$. If the spring is stretched a length $(1/2+\pi/4)$ m or more from its equilibrium position, then it will break. Assuming no damping present, find the time at which the spring breaks.

12. A small object of mass 1 kg is attached to a spring with spring constant $k=1$ N/m. This spring-mass system is then immersed in a viscous medium with damping constant c. An external force $F(t)=(3-\cos t)$ N is applied to the system. Determine the minimum positive value of c so that the magnitude of the steady state solution does not exceed 5 m.

13. Determine a particular solution $\psi(t)$ of $my''+cy'+ky=F_0\cos\omega t$, of the form $\psi(t)=A\cos(\omega t-\phi)$. Show that the amplitude A is a maximum when $\omega^2=\omega_0^2-\frac{1}{2}(c/m)^2$. This value of ω is called the *resonant frequency* of the system. What happens when $\omega_0^2<\frac{1}{2}(c/m)^2$?

2.6.1 The Tacoma Bridge disaster

On July 1, 1940, the Tacoma Narrows Bridge at Puget Sound in the state of Washington was completed and opened to traffic. From the day of its opening the bridge began undergoing vertical oscillations, and it soon was nicknamed "Galloping Gertie." Strange as it may seem, traffic on the bridge increased tremendously as a result of its novel behavior. People came from hundreds of miles in their cars to enjoy the curious thrill of riding over a galloping, rolling bridge. For four months, the bridge did a thriving business. As each day passed, the authorities in charge became more and more confident of the safety of the bridge—so much so, in fact, that they were planning to cancel the insurance policy on the bridge.

Starting at about 7:00 on the morning of November 7, 1940, the bridge began undulating persistently for three hours. Segments of the span were heaving periodically up and down as much as three feet. At about 10:00 a.m., something seemed to snap and the bridge began oscillating wildly. At one moment, one edge of the roadway was twenty-eight feet higher than the other; the next moment it was twenty-eight feet lower than the other edge. At 10:30 a.m. the bridge began cracking, and finally, at 11:10 a.m. the entire bridge came crashing down. Fortunately, only one car was on the bridge at the time of its failure. It belonged to a newspaper reporter who had to abandon the car and its sole remaining occupant, a pet dog, when the bridge began its violent twisting motion. The reporter reached safety, torn and bleeding, by crawling on hands and knees, desperately clutching the curb of the bridge. His dog went down with the car and the span—the only life lost in the disaster.

There were many humorous and ironic incidents associated with the collapse of the Tacoma Bridge. When the bridge began heaving violently, the authorities notified Professor F. B. Farquharson of the University of Washington. Professor Farquharson had conducted numerous tests on a simulated model of the bridge and had assured everyone of its stability. The professor was the last man on the bridge. Even when the span was tilting more than twenty-eight feet up and down, he was making scientific observations with little or no anticipation of the imminent collapse of the bridge. When the motion increased in violence, he made his way to safety by scientifically following the yellow line in the middle of the roadway. The professor was one of the most surprised men when the span crashed into the water.

One of the insurance policies covering the bridge had been written by a local travel agent who had pocketed the premium and had neglected to report the policy, in the amount of \$800,000, to his company. When he later received his prison sentence, he ironically pointed out that his embezzlement would never have been discovered if the bridge had only remained up for another week, at which time the bridge officials had planned to cancel all of the policies.

A large sign near the bridge approach advertised a local bank with the slogan "as safe as the Tacoma Bridge." Immediately following the collapse of the bridge, several representatives of the bank rushed out to remove the billboard.

After the collapse of the Tacoma Bridge, the governor of the state of Washington made an emotional speech, in which he declared "We are going to build the exact same bridge, exactly as before." Upon hearing this, the noted engineer Von Karman sent a telegram to the governor stating "If you build the exact same bridge exactly as before, it will fall into the exact same river exactly as before."

The collapse of the Tacoma Bridge was due to an aerodynamical phenomenon known as *stall flutter*. This can be explained very briefly in the following manner. If there is an obstacle in a stream of air, or liquid, then a "vortex street" is formed behind the obstacle, with the vortices flowing off at a definite periodicity, which depends on the shape and dimension of the structure as well as on the velocity of the stream (see Figure 1). As a result of the vortices separating alternately from either side of the obstacle, it is acted upon by a periodic force perpendicular to the direction of the stream, and of magnitude $F_0 \cos \omega t$. The coefficient F_0 depends on the shape of the structure. The poorer the streamlining of the structure; the larger the coefficient F_0, and hence the amplitude of the force. For example, flow around an airplane wing at small angles of attack is very smooth, so that the vortex street is not well defined and the coefficient F_0 is very small. The poorly streamlined structure of a suspension bridge is another

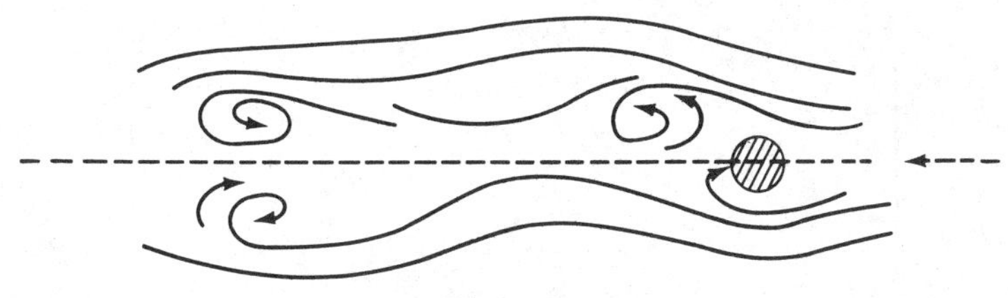

Figure 1

matter, and it is natural to expect that a force of large amplitude will be set up. Thus, a structure suspended in an air stream experiences the effect of this force and hence goes into a state of forced vibrations. The amount of danger from this type of motion depends on how close the natural frequency of the structure (remember that bridges are made of steel, a highly elastic material) is to the frequency of the driving force. If the two frequencies are the same, resonance occurs, and the oscillations will be destructive if the system does not have a sufficient amount of damping. It has now been established that oscillations of this type were responsible for the collapse of the Tacoma Bridge. In addition, resonances produced by the separation of vortices have been observed in steel factory chimneys, and in the periscopes of submarines.

The phenomenon of resonance was also responsible for the collapse of the Broughton suspension bridge near Manchester, England in 1831. This occurred when a column of soldiers marched in cadence over the bridge, thereby setting up a periodic force of rather large amplitude. The frequency of this force was equal to the natural frequency of the bridge. Thus, very large oscillations were induced, and the bridge collapsed. It is for this reason that soldiers are ordered to break cadence when crossing a bridge.

Epilog. The father of one of my students is an engineer who worked on the construction of the Bronx Whitestone Bridge in New York City. He informed me that the original plans for this bridge were very similar to those of the Tacoma Bridge. These plans were hastily redrawn following the collapse of the Tacoma Bridge.

2.6.2 Electrical networks

We now briefly study a simple series circuit, as shown in Figure 1 below. The symbol E represents a source of electromotive force. This may be a battery or a generator which produces a potential difference (or voltage), that causes a current I to flow through the circuit when the switch S is closed. The symbol R represents a resistance to the flow of current such as that produced by a lightbulb or toaster. When current flows through a coil of wire L, a magnetic field is produced which opposes any change in the current through the coil. The change in voltage produced by the coil is proportional to the rate of change of the current, and the constant of propor-

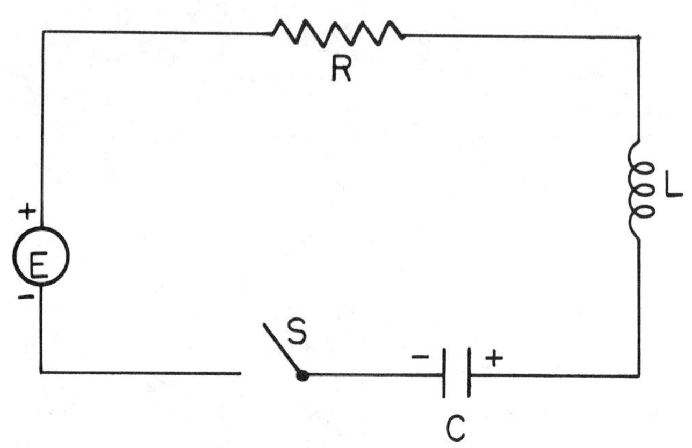

Figure 1. A simple series circuit

tionality is called the inductance L of the coil. A capacitor, or condenser, indicated by C, usually consists of two metal plates separated by a material through which very little current can flow. A capacitor has the effect of reversing the flow of current as one plate or the other becomes charged.

Let $Q(t)$ be the charge on the capacitor at time t. To derive a differential equation which is satisfied by $Q(t)$ we use the following.

Kirchoff's second law: In a closed circuit, the impressed voltage equals the sum of the voltage drops in the rest of the circuit.

Now,

(i) The voltage drop across a resistance of R ohms equals RI (Ohm's law).
(ii) The voltage drop across an inductance of L henrys equals $L(dI/dt)$.
(iii) The voltage drop across a capacitance of C farads equals Q/C.

Hence,

$$E(t)=L\frac{dI}{dt}+RI+\frac{Q}{C},$$

and since $I(t)=dQ(t)/dt$, we see that

$$L\frac{d^2Q}{dt^2}+R\frac{dQ}{dt}+\frac{Q}{C}=E(t). \tag{1}$$

Notice the resemblance of Equation (1) to the equation of a vibrating mass. Among the similarities with mechanical vibrations, electrical circuits also have the property of resonance. Unlike mechanical systems, though, resonance is put to good use in electrical systems. For example, the tuning knob of a radio is used to vary the capacitance in the tuning circuit. In this manner, the resonant frequency (see Exercise 13, Section 2.6) is changed until it agrees with the frequency of one of the incoming radio signals. The amplitude of the current produced by this signal will be much greater than that of all other signals. In this way, the tuning circuit picks out the desired station.

Exercises

1. Suppose that a simple series circuit has no resistance and no impressed voltage. Show that the charge Q on the capacitor is periodic in time, with frequency $\omega_0 = \sqrt{1/LC}$. The quantity $\sqrt{1/LC}$ is called the natural frequency of the circuit.

2. Suppose that a simple series circuit consisting of an inductor, a resistor and a capacitor is open, and that there is an initial charge $Q_0 = 10^{-8}$ coulombs on the capacitor. Find the charge on the capacitor and the current flowing in the circuit after the switch is closed for each of the following cases.
(a) $L = 0.5$ henrys, $C = 10^{-5}$ farads, $R = 1000$ ohms
(b) $L = 1$ henry, $C = 10^{-4}$ farads, $R = 200$ ohms
(c) $L = 2$ henrys, $C = 10^{-6}$ farads, $R = 2000$ ohms

3. A simple series circuit has an inductor of 1 henry, a capacitor of 10^{-6} farads, and a resistor of 1000 ohms. The initial charge on the capacitor is zero. If a 12 volt battery is connected to the circuit, and the circuit is closed at $t = 0$, find the charge on the capacitor 1 second later, and the steady state charge.

4. A capacitor of 10^{-3} farads is in series with an electromotive force of 12 volts and an inductor of 1 henry. At $t = 0$, both Q and I are zero.
(a) Find the natural frequency and period of the electrical oscillations.
(b) Find the maximum charge on the capacitor, and the maximum current flowing in the circuit.

5. Show that if there is no resistance in a circuit, and the impressed voltage is of the form $E_0 \sin \omega t$, then the charge on the capacitor will become unbounded as $t \to \infty$ if $\omega = \sqrt{1/LC}$. This is the phenomenon of resonance.

6. Consider the differential equation

$$L\ddot{Q} + R\dot{Q} + \frac{Q}{C} = E_0 \cos \omega t. \qquad \text{(i)}$$

We find a particular solution $\psi(t)$ of (i) as the real part of a particular solution $\phi(t)$ of

$$L\ddot{Q} + R\dot{Q} + \frac{Q}{C} = E_0 e^{i\omega t}. \qquad \text{(ii)}$$

(a) Show that

$$i\omega\phi(t) = \frac{E_0}{R + i\left(\omega L - \dfrac{1}{\omega C}\right)} e^{i\omega t}.$$

(b) The quantity $Z = R + i(\omega L - 1/\omega C)$ is known as the complex impedance of the circuit. The reciprocal of Z is called the admittance, and the real and imaginary parts of $1/Z$ are called the conductance and susceptance. Determine the admittance, conductance and susceptance.

7. Consider a simple series circuit with given values of L, R and C, and an impressed voltage $E_0 \sin \omega t$. For which value of ω will the steady state current be a maximum?

2.7 A model for the detection of diabetes

Diabetes mellitus is a disease of metabolism which is characterized by too much sugar in the blood and urine. In diabetes, the body is unable to burn off all its sugars, starches, and carbohydrates because of an insufficient supply of insulin. Diabetes is usually diagnosed by means of a glucose tolerance test (GTT). In this test the patient comes to the hospital after an overnight fast and is given a large dose of glucose (sugar in the form in which it usually appears in the bloodstream). During the next three to five hours several measurements are made of the concentration of glucose in the patient's blood, and these measurements are used in the diagnosis of diabetes. A very serious difficulty associated with this method of diagnosis is that there is no universally accepted criterion for interpreting the results of a glucose tolerance test. Three physicians interpreting the results of a GTT may come up with three different diagnoses. In one case recently, in Rhode Island, one physician, after reviewing the results of a GTT, came up with a diagnosis of diabetes. A second physician declared the patient to be normal. To settle the question, the results of the GTT were sent to a specialist in Boston. After examining these results, the specialist concluded that the patient was suffering from a pituitary tumor.

In the mid 1960's Drs. Rosevear and Molnar of the Mayo Clinic and Ackerman and Gatewood of the University of Minnesota discovered a fairly reliable criterion for interpreting the results of a glucose tolerance test. Their discovery arose from a very simple model they developed for the blood glucose regulatory system. Their model is based on the following simple and fairly well known facts of elementary biology.

1. Glucose plays an important role in the metabolism of any vertebrate since it is a source of energy for all tissues and organs. For each individual there is an optimal blood glucose concentration, and any excessive deviation from this optimal concentration leads to severe pathological conditions and potentially death.

2. While blood glucose levels tend to be autoregulatory, they are also influenced and controlled by a wide variety of hormones and other metabolites. Among these are the following.

(i) *Insulin*, a hormone secreted by the β cells of the pancreas. After we eat any carbohydrates, our G.I. tract sends a signal to the pancreas to secrete more insulin. In addition, the glucose in our blood directly stimulates the β cells of the pancreas to secrete insulin. It is generally believed that insulin facilitates tissue uptake of glucose by attaching itself to the impermeable membrane walls, thus allowing glucose to pass through the membranes to the center of the cells, where most of the biological and chemical activity takes place. Without sufficient insulin, the body cannot avail itself of all the energy it needs.

(ii) *Glucagon*, a hormone secreted by the α cells of the pancreas. Any excess glucose is stored in the liver in the form of glycogen. In times of need this glycogen is converted back into glucose. The hormone glucagon increases the rate of breakdown of glycogen into glucose. Evidence collected thus far clearly indicates that hypoglycemia (low blood sugar) and fasting promote the secretion of glucagon while increased blood glucose levels suppress its secretion.

(iii) *Epinephrine* (adrenalin), a hormone secreted by the adrenal medulla. Epinephrine is part of an emergency mechanism to quickly increase the concentration of glucose in the blood in times of extreme hypoglycemia. Like glucagon, epinephrine increases the rate of breakdown of glycogen into glucose. In addition, it directly inhibits glucose uptake by muscle tissue; it acts directly on the pancreas to inhibit insulin secretion; and it aids in the conversion of lactate to glucose in the liver.

(iv) *Glucocorticoids*, hormones such as cortisol which are secreted by the adrenal cortex. Glucocorticoids play an important role in the metabolism of carbohydrates.

(v) *Thyroxin*, a hormone secreted by the thyroid gland. This hormone aids the liver in forming glucose from non-carbohydrate sources such as glycerol, lactate and amino acids.

(vi) *Growth hormone* (somatotropin), a hormone secreted by the anterior pituitary gland. Not only does growth hormone affect glucose levels in a direct manner, but it also tends to "block" insulin. It is believed that growth hormone decreases the sensitivity of muscle and adipose membrane to insulin, thereby reducing the effectiveness of insulin in promoting glucose uptake.

The aim of Ackerman et al was to construct a model which would accurately describe the blood glucose regulatory system during a glucose tolerance test, and in which one or two parameters would yield criteria for distinguishing normal individuals from mild diabetics and pre-diabetics. Their model is a very simplified one, requiring only a limited number of blood samples during a GTT. It centers attention on two concentrations, that of glucose in the blood, labelled G, and that of the net hormonal concentration, labelled H. The latter is interpreted to represent the cumulative effect of all the pertinent hormones. Those hormones such as insulin which decrease blood glucose concentrations are considered to increase H, while those hormones such as cortisol which increase blood glucose concentrations are considered to decrease H. Now there are two reasons why such a simplified model can still provide an accurate description of the blood glucose regulatory system. First, studies have shown that under normal, or close to normal conditions, the interaction of one hormone, namely insulin, with blood glucose so predominates that a simple "lumped parameter model" is quite adequate. Second, evidence indicates that normoglycemia does not depend, necessarily, on the normalcy of each kinetic mechanism of the blood glucose regulatory system. Rather, it depends on the overall performance of the blood glucose regulatory system, and this system is dominated by insulin-glucose interactions.

The basic model is described analytically by the equations

$$\frac{dG}{dt}=F_1(G,H)+J(t) \tag{1}$$

$$\frac{dH}{dt}=F_2(G,H). \tag{2}$$

The dependence of F_1 and F_2 on G and H signify that changes in G and H are determined by the values of both G and H. The function $J(t)$ is the external rate at which the blood glucose concentration is being increased. Now, we assume that G and H have achieved optimal values G_0 and H_0 by the time the fasting patient has arrived at the hospital. This implies that $F_1(G_0,H_0)=0$ and $F_2(G_0,H_0)=0$. Since we are interested here in the deviations of G and H from their optimal values, we make the substitution

$$g=G-G_0, \qquad h=H-H_0.$$

Then,

$$\frac{dg}{dt}=F_1(G_0+g,H_0+h)+J(t),$$

$$\frac{dh}{dt}=F_2(G_0+g,H_0+h).$$

Now, observe that

$$F_1(G_0+g,H_0+h)=F_1(G_0,H_0)+\frac{\partial F_1(G_0,H_0)}{\partial G}g+\frac{\partial F_1(G_0,H_0)}{\partial H}h+e_1$$

and

$$F_2(G_0+g,H_0+h)=F_2(G_0,H_0)+\frac{\partial F_2(G_0,H_0)}{\partial G}g+\frac{\partial F_2(G_0,H_0)}{\partial H}h+e_2$$

where e_1 and e_2 are very small compared to g and h. Hence, assuming that G and H deviate only slightly from G_0 and H_0, and therefore neglecting the terms e_1 and e_2, we see that

$$\frac{dg}{dt}=\frac{\partial F_1(G_0,H_0)}{\partial G}g+\frac{\partial F_1(G_0,H_0)}{\partial H}h+J(t) \tag{3}$$

$$\frac{dh}{dt}=\frac{\partial F_2(G_0,H_0)}{\partial G}g+\frac{\partial F_2(G_0,H_0)}{\partial H}h. \tag{4}$$

Now, there are no means, a priori, of determining the numbers

$$\frac{\partial F_1(G_0,H_0)}{\partial G},\ \frac{\partial F_1(G_0,H_0)}{\partial H},\ \frac{\partial F_2(G_0,H_0)}{\partial G}\ \text{ and }\ \frac{\partial F_2(G_0,H_0)}{\partial H}.$$

However, we can determine their signs. Referring to Figure 1, we see that dg/dt is negative for $g>0$ and $h=0$, since the blood glucose concentration

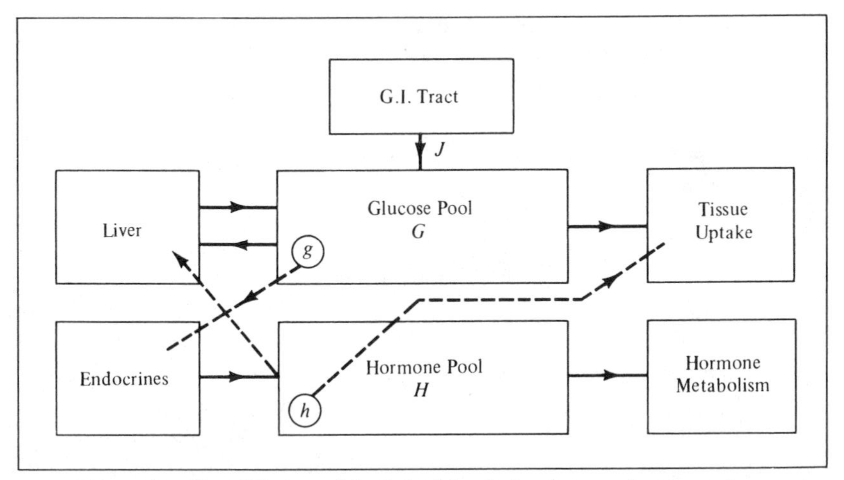

Figure 1. Simplified model of the blood glucose regulatory system

will be decreasing through tissue uptake of glucose and the storing of excess glucose in the liver in the form of glycogen. Consequently $\partial F_1(G_0,H_0)/\partial G$ must be negative. Similarly, $\partial F_1(G_0,H_0)/\partial H$ is negative since a positive value of h tends to decrease blood glucose levels by facilitating tissue uptake of glucose and by increasing the rate at which glucose is converted to glycogen. The number $\partial F_2(G_0,H_0)/\partial G$ must be positive since a positive value of g causes the endocrine glands to secrete those hormones which tend to increase H. Finally, $\partial F_2(G_0,H_0)/\partial H$ must be negative, since the concentration of hormones in the blood decreases through hormone metabolism.

Thus, we can write Equations (3) and (4) in the form

$$\frac{dg}{dt} = -m_1 g - m_2 h + J(t) \tag{5}$$

$$\frac{dh}{dt} = -m_3 h + m_4 g \tag{6}$$

where m_1, m_2, m_3, and m_4 are positive constants. Equations (5) and (6) are two first-order equations for g and h. However, since we only measure the concentration of glucose in the blood, we would like to remove the variable h. This can be accomplished as follows: Differentiating (5) with respect to t gives

$$\frac{d^2g}{dt^2} = -m_1\frac{dg}{dt} - m_2\frac{dh}{dt} + \frac{dJ}{dt}.$$

Substituting for dh/dt from (6) we obtain that

$$\frac{d^2g}{dt^2} = -m_1\frac{dg}{dt} + m_2m_3h - m_2m_4 g + \frac{dJ}{dt}. \tag{7}$$

Next, observe from (5) that $m_2h = (-dg/dt) - m_1 g + J(t)$. Consequently,

$g(t)$ satisfies the second-order linear differential equation

$$\frac{d^2g}{dt^2}+(m_1+m_3)\frac{dg}{dt}+(m_1m_3+m_2m_4)g=m_3J+\frac{dJ}{dt}.$$

We rewrite this equation in the form

$$\frac{d^2g}{dt^2}+2\alpha\frac{dg}{dt}+\omega_0^2g=S(t) \tag{8}$$

where $\alpha=(m_1+m_3)/2$, $\omega_0^2=m_1m_3+m_2m_4$, and $S(t)=m_3J+dJ/dt$.

Notice that the right-hand side of (8) is identically zero except for the very short time interval in which the glucose load is being ingested. We will learn to deal with such functions in Section 2.12. For our purposes here, let $t=0$ be the time at which the glucose load has been completely ingested. Then, for $t\geqslant 0$, $g(t)$ satisfies the second-order linear homogeneous equation

$$\frac{d^2g}{dt^2}+2\alpha\frac{dg}{dt}+\omega_0^2g=0. \tag{9}$$

This equation has positive coefficients. Hence, by the analysis in Section 2.6, (see also Exercise 8, Section 2.2.2) $g(t)$ approaches zero as t approaches infinity. Thus our model certainly conforms to reality in predicting that the blood glucose concentration tends to return eventually to its optimal concentration.

The solutions $g(t)$ of (9) are of three different types, depending as to whether $\alpha^2-\omega_0^2$ is positive, negative, or zero. These three types, of course, correspond to the overdamped, critically damped and underdamped cases discussed in Section 2.6. We will assume that $\alpha^2-\omega_0^2$ is negative; the other two cases are treated in a similar manner. If $\alpha^2-\omega_0^2<0$, then the characteristic equation of Equation (9) has complex roots. It is easily verified in this case (see Exercise 1) that every solution $g(t)$ of (9) is of the form

$$g(t)=Ae^{-\alpha t}\cos(\omega t-\delta),\qquad \omega^2=\omega_0^2-\alpha^2. \tag{10}$$

Consequently,

$$G(t)=G_0+Ae^{-\alpha t}\cos(\omega t-\delta). \tag{11}$$

Now there are five unknowns G_0, A, α, ω_0, and δ in (11). One way of determining them is as follows. The patient's blood glucose concentration before the glucose load is ingested is G_0. Hence, we can determine G_0 by measuring the patient's blood glucose concentration immediately upon his arrival at the hospital. Next, if we take four additional measurements G_1, G_2, G_3, and G_4 of the patient's blood glucose concentration at times t_1, t_2, t_3, and t_4, then we can determine A, α, ω_0, and δ from the four equations

$$G_j=G_0+Ae^{-\alpha t_j}\cos(\omega t_j-\delta),\qquad j=1,2,3,4.$$

A second, and better method of determining G_0, A, α, ω_0, and δ is to take n measurements $G_1,G_2,\dots,G_n$ of the patient's blood glucose concentration at times $t_1, t_2, \ldots, t_n$. Typically n is 6 or 7. We then find optimal values for G_0, A, α, ω_0, and δ such that the least square error

$$E = \sum_{j=1}^{n} \left[G_j - G_0 - Ae^{-\alpha t_j} \cos(\omega t_j - \delta) \right]^2$$

is minimized. The problem of minimizing E can be solved on a digital computer, and Ackerman et al (see reference at end of section) provide a complete Fortran program for determining optimal values for G_0, A, α, ω_0, and δ. This method is preferrable to the first method since Equation (11) is only an approximate formula for $G(t)$. Consequently, it is possible to find values G_0, A, α, ω_0, and δ so that Equation (11) is satisfied exactly at four points t_1, t_2, t_3, and t_4 but yields a poor fit to the data at other times. The second method usually offers a better fit to the data on the entire time interval since it involves more measurements.

In numerous experiments, Ackerman et al observed that a slight error in measuring G could produce a very large error in the value of α. Hence, any criterion for diagnosing diabetes that involves the parameter α is unreliable. However, the parameter ω_0, the natural frequency of the system, was relatively insensitive to experimental errors in measuring G. Thus, we may regard a value of ω_0 as the basic descriptor of the response to a glucose tolerance test. For discussion purposes, it is more convenient to use the corresponding natural period $T_0 = 2\pi/\omega_0$. The remarkable fact is that data from a variety of sources indicated that *a value of less than four hours for T_0 indicated normalcy, while appreciably more than four hours implied mild diabetes.*

Remark 1. The usual period between meals in our culture is about 4 hours. This suggests the interesting possibility that sociological factors may also play a role in the blood glucose regulatory system.

Remark 2. We wish to emphasize that the model described above can only be used to diagnose mild diabetes or pre-diabetes, since we have assumed throughout that the deviation g of G from its optimal value G_0 is small. Very large deviations of G from G_0 usually indicate severe diabetes or diabetes insipidus, which is a disorder of the posterior lobe of the pituitary gland.

A serious shortcoming of this simplified model is that it sometimes yields a poor fit to the data in the time period three to five hours after ingestion of the glucose load. This indicates, of course, that variables such as epinephrine and glucagon play an important role in this time period. Thus these variables should be included as separate variables in our model, rather than being lumped together with insulin. In fact, evidence indicates that levels of epinephrine may rise dramatically during the recovery phase of the GTT response, when glucose levels have been lowered below fasting

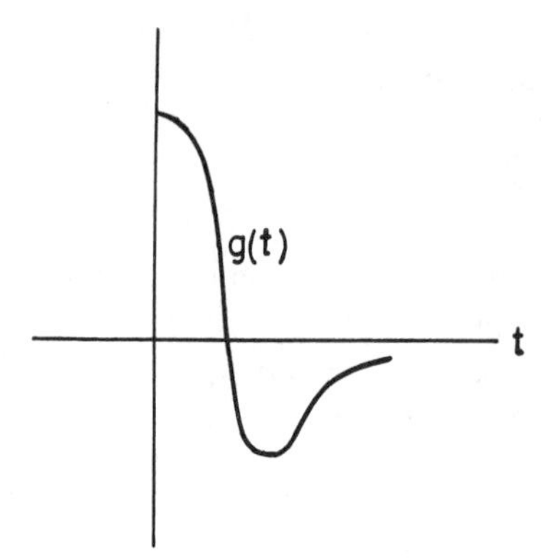

Figure 2. Graph of $g(t)$ if $\alpha^2-\omega_0^2>0$

levels. This can also be seen directly from Equation (9). If $\alpha^2-\omega_0^2>0$, then $g(t)$ may have the form described in Figure 2. Note that $g(t)$ drops very rapidly from a fairly high value to a negative one. It is quite conceivable, therefore, that the body will interpret this as an extreme emergency and thereby secrete a large amount of epinephrine.

Medical researchers have long recognized the need of including epinephrine as a separate variable in any model of the blood glucose regulatory system. However, they were stymied by the fact that there was no reliable method of measuring the concentration of epinephrine in the blood. Thus, they had to assume, for all practical purposes, the the level of epinephrine remained constant during the course of a glucose tolerance test. This author has just been informed that researchers at Rhode Island Hospital have devised an accurate method of measuring the concentration of epinephrine in the blood. Thus we will be able to develop and test more accurate models of the blood glucose regulatory system. Hopefully, this will lead to more reliable criteria for the diagnosis of diabetes.

Reference

E. Ackerman, L. Gatewood, J. Rosevear, and G. Molnar, Blood glucose regulation and diabetes, Chapter 4 in *Concepts and Models of Biomathematics*, F. Heinmets, ed., Marcel Dekker, 1969, 131–156.

Exercises

1. Derive Equation (10).

2. A patient arrives at the hospital after an overnight fast with a blood glucose concentration of 70 mg glucose/100 ml blood (mg glucose/100 ml blood = milligrams of glucose per 100 milliliters of blood). His blood glucose concentration 1 hour, 2 hours, and 3 hours after fully absorbing a large amount of glucose is 95, 65, and 75 mg glucose/100 ml blood, respectively. Show that this patient is normal. *Hint*: In the underdamped case, the time interval between two successive zeros of $G-G_0$ exceeds one half the natural period.

According to a famous diabetologist, the blood glucose concentration of a nondiabetic who has just absorbed a large amount of glucose will be at or below the fasting level in 2 hours or less. Exercises 3 and 4 compare the diagnoses of this diabetologist with those of Ackerman et al.

3. The deviation $g(t)$ of a patient's blood glucose concentration from its optimal concentration satisfies the differential equation $(d^2g/dt^2)+2\alpha(dg/dt)+\alpha^2 g=0$ immediately after he fully absorbs a large amount of glucose. The time t is measured in minutes, so that the units of α are reciprocal minutes. Show that this patient is normal according to Ackerman et al, if $\alpha>\pi/120$ (min), and that this patient is normal according to the famous diabetologist if

$$g'(0)<-\left(\tfrac{1}{120}+\alpha\right)g(0).$$

4. A patient's blood glucose concentration $G(t)$ satisfies the initial-value problem

$$\frac{d^2G}{dt^2}+\frac{1}{20\text{ (min)}}\frac{dG}{dt}+\frac{1}{2500\text{ (min)}^2}G$$

$$=\frac{1}{2500\text{ (min)}^2}75\text{ mg glucose}/100\text{ ml blood};$$

$$G(0)=150\text{ mg glucose}/100\text{ ml blood},$$

$$G'(0)=-\alpha G(0)/(\text{min});\qquad \alpha\geqslant\frac{1}{200}\,\frac{1-4e^{18/5}}{1-e^{18/5}}$$

immediately after he fully absorbs a large amount of glucose. This patient's optimal blood glucose concentration is 75 mg glucose/100 ml blood. Show that this patient is a diabetic according to Ackerman et al, but is normal according to the famous diabetologist.

2.8 Series solutions

We return now to the general homogeneous linear second-order equation

$$L[y]=P(t)\frac{d^2y}{dt^2}+Q(t)\frac{dy}{dt}+R(t)y=0 \tag{1}$$

with $P(t)$ unequal to zero in the interval $\alpha<t<\beta$. It was shown in Section 2.1 that every solution $y(t)$ of (1) can be written in the form $y(t)=c_1y_1(t)+c_2y_2(t)$, where $y_1(t)$ and $y_2(t)$ are any two linearly independent solutions of (1). Thus, the problem of finding all solutions of (1) is reduced to the simpler problem of finding just two solutions. In Section 2.2 we handled the special case where P, Q, and R are constants. The next simplest case is when $P(t)$, $Q(t)$, and $R(t)$ are polynomials in t. In this case, the form of the differential equation suggests that we guess a polynomial solution $y(t)$ of (1). If $y(t)$ is a polynomial in t, then the three functions $P(t)y''(t)$, $Q(t)y'(t)$, and $R(t)y(t)$ are again polynomials in t. Thus, in principle, we can determine a polynomial solution $y(t)$ of (1) by setting the sums of the coefficients of like powers of t in the expression $L[y](t)$ equal to zero. We illustrate this method with the following example.

Example 1. Find two linearly independent solutions of the equation

$$L[y]=\frac{d^2y}{dt^2}-2t\frac{dy}{dt}-2y=0. \tag{2}$$

Solution. We will try to find 2 polynomial solutions of (2). Now, it is not obvious, a priori, what the degree of any polynomial solution of (2) should be. Nor is it evident that we will be able to get away with a polynomial of finite degree. Therefore, we set

$$y(t)=a_0+a_1t+a_2t^2+\ldots=\sum_{n=0}^{\infty}a_nt^n.$$

Computing

$$\frac{dy}{dt}=a_1+2a_2t+3a_3t^2+\ldots=\sum_{n=0}^{\infty}na_nt^{n-1}$$

and

$$\frac{d^2y}{dt^2}=2a_2+6a_3t+\ldots=\sum_{n=0}^{\infty}n(n-1)a_nt^{n-2},$$

we see that $y(t)$ is a solution of (2) if

$$\begin{aligned}L[y](t)&=\sum_{n=0}^{\infty}n(n-1)a_nt^{n-2}-2t\sum_{n=0}^{\infty}na_nt^{n-1}-2\sum_{n=0}^{\infty}a_nt^n\\&=\sum_{n=0}^{\infty}n(n-1)a_nt^{n-2}-2\sum_{n=0}^{\infty}na_nt^n-2\sum_{n=0}^{\infty}a_nt^n=0.\end{aligned} \tag{3}$$

Our next step is to rewrite the first summation in (3) so that the exponent of the general term is n, instead of $n-2$. This is accomplished by increasing every n underneath the summation sign by 2, and decreasing the lower limit by 2, that is,

$$\sum_{n=0}^{\infty}n(n-1)a_nt^{n-2}=\sum_{n=-2}^{\infty}(n+2)(n+1)a_{n+2}t^n.$$

(If you don't believe this, you can verify it by writing out the first few terms in both summations. If you still don't believe this and want a formal proof, set $m=n-2$. When n is zero, m is -2 and when n is infinity, m is infinity. Therefore

$$\sum_{n=0}^{\infty}n(n-1)a_nt^{n-2}=\sum_{m=-2}^{\infty}(m+2)(m+1)a_{m+2}t^m,$$

and since m is a dummy variable, we may replace it by n.) Moreover, observe that the contribution to this sum from $n=-2$ and $n=-1$ is zero since the factor $(n+2)(n+1)$ vanishes in both these instances. Hence,

$$\sum_{n=0}^{\infty} n(n-1)a_n t^{n-2} = \sum_{n=0}^{\infty} (n+2)(n+1)a_{n+2}t^n$$

and we can rewrite (3) in the form

$$\sum_{n=0}^{\infty} (n+2)(n+1)a_{n+2}t^n - 2\sum_{n=0}^{\infty} na_n t^n - 2\sum_{n=0}^{\infty} a_n t^n = 0. \tag{4}$$

Setting the sum of the coefficients of like powers of t in (4) equal to zero gives

$$(n+2)(n+1)a_{n+2} - 2na_n - 2a_n = 0$$

so that

$$a_{n+2} = \frac{2(n+1)a_n}{(n+2)(n+1)} = \frac{2a_n}{n+2}. \tag{5}$$

Equation (5) is a recurrence formula for the coefficients $a_0, a_1, a_2, a_3, \ldots$. The coefficient a_n determines the coefficient a_{n+2}. Thus, a_0 determines a_2 through the relation $a_2 = 2a_0/2 = a_0$; a_2, in turn, determines a_4 through the relation $a_4 = 2a_2/(2+2) = a_0/2$; and so on. Similarly, a_1 determines a_3 through the relation $a_3 = 2a_1/(2+1) = 2a_1/3$; a_3, in turn, determines a_5 through the relation $a_5 = 2a_3/(3+2) = 4a_1/3\cdot 5$; and so on. Consequently, all the coefficients are determined uniquely once a_0 and a_1 are prescribed. The values of a_0 and a_1 are completely arbitrary. This is to be expected, though, for if

$$y(t) = a_0 + a_1 t + a_2 t^2 + \ldots$$

then the values of y and y' at $t=0$ are a_0 and a_1 respectively. Thus, the coefficients a_0 and a_1 must be arbitrary until specific initial conditions are imposed on y.

To find two solutions of (2), we choose two different sets of values of a_0 and a_1. The simplest possible choices are (i) $a_0 = 1, a_1 = 0$; (ii) $a_0 = 0$, $a_1 = 1$.

(i) $$a_0 = 1, \qquad a_1 = 0.$$

In this case, all the odd coefficients $a_1, a_3, a_5, \ldots$ are zero since $a_3 = 2a_1/3 = 0$, $a_5 = 2a_3/5 = 0$, and so on. The even coefficients are determined from the relations

$$a_2 = a_0 = 1, \qquad a_4 = \frac{2a_2}{4} = \frac{1}{2}, \qquad a_6 = \frac{2a_4}{6} = \frac{1}{2\cdot 3},$$

and so on. Proceeding inductively, we find that

$$a_{2n} = \frac{1}{2\cdot 3 \cdots n} = \frac{1}{n!}.$$

Hence,

$$y_1(t) = 1 + t^2 + \frac{t^4}{2!} + \frac{t^6}{3!} + \ldots = e^{t^2}$$

is one solution of (2).

(ii) $$a_0=0, \qquad a_1=1.$$

In this case, all the even coefficients are zero, and the odd coefficients are determined from the relations

$$a_3=\frac{2a_1}{3}=\frac{2}{3}, \qquad a_5=\frac{2a_3}{5}=\frac{2}{5}\frac{2}{3}, \qquad a_7=\frac{2a_5}{7}=\frac{2}{7}\frac{2}{5}\frac{2}{3},$$

and so on. Proceeding inductively, we find that

$$a_{2n+1}=\frac{2^n}{3\cdot5\cdots(2n+1)}.$$

Thus,

$$y_2(t)=t+\frac{2t^3}{3}+\frac{2^2t^5}{3\cdot5}+\ldots=\sum_{n=0}^{\infty}\frac{2^n t^{2n+1}}{3\cdot5\cdots(2n+1)}$$

is a second solution of (2).

Now, observe that $y_1(t)$ and $y_2(t)$ are polynomials of infinite degree, even though the coefficients $P(t)=1$, $Q(t)=-2t$, and $R(t)=-2$ are polynomials of finite degree. Such polynomials are called power series. Before proceeding further, we will briefly review some of the important properties of power series.

1. An infinite series

$$y(t)=a_0+a_1(t-t_0)+a_2(t-t_0)^2+\ldots=\sum_{n=0}^{\infty}a_n(t-t_0)^n \tag{6}$$

is called a power series about $t=t_0$.

2. All power series have an interval of convergence. This means that there exists a nonnegative number ρ such that the infinite series (6) converges for $|t-t_0|<\rho$, and diverges for $|t-t_0|>\rho$. The number ρ is called the radius of convergence of the power series.
3. The power series (6) can be differentiated and integrated term by term, and the resultant series have the same interval of convergence.
4. The simplest method (if it works) for determining the interval of convergence of the power series (6) is the Cauchy ratio test. Suppose that the absolute value of a_{n+1}/a_n approaches a limit λ as n approaches infinity. Then, the power series (6) converges for $|t-t_0|<1/\lambda$, and diverges for $|t-t_0|>1/\lambda$.
5. The product of two power series $\sum_{n=0}^{\infty}a_n(t-t_0)^n$ and $\sum_{n=0}^{\infty}b_n(t-t_0)^n$ is again a power series of the form $\sum_{n=0}^{\infty}c_n(t-t_0)^n$, with $c_n=a_0b_n+a_1b_{n-1}+\ldots+a_nb_0$. The quotient

$$\frac{a_0+a_1t+a_2t^2+\ldots}{b_0+b_1t+b_2t^2+\ldots}$$

of two power series is again a power series, provided that $b_0\neq0$.

6. Many of the functions $f(t)$ that arise in applications can be expanded in power series; that is, we can find coefficients $a_0, a_1, a_2, \ldots$ so that

$$f(t) = a_0 + a_1(t-t_0) + a_2(t-t_0)^2 + \ldots = \sum_{n=0}^{\infty} a_n(t-t_0)^n. \tag{7}$$

   Such functions are said to be *analytic* at $t = t_0$, and the series (7) is called the Taylor series of f about $t = t_0$. It can easily be shown that if f admits such an expansion, then, of necessity, $a_n = f^{(n)}(t_0)/n!$, where $f^{(n)}(t) = d^n f(t)/dt^n$.
7. The interval of convergence of the Taylor series of a function $f(t)$, about t_0, can be determined directly through the Cauchy ratio test and other similar methods, or indirectly, through the following theorem of complex analysis.

Theorem 6. *Let the variable t assume complex values, and let z_0 be the point closest to t_0 at which f or one of its derivatives fails to exist. Compute the distance ρ, in the complex plane, between t_0 and z_0. Then, the Taylor series of f about t_0 converges for $|t-t_0| < \rho$, and diverges for $|t-t_0| > \rho$.*

As an illustration of Theorem 6, consider the function $f(t) = 1/(1+t^2)$. The Taylor series of f about $t = 0$ is

$$\frac{1}{1+t^2} = 1 - t^2 + t^4 - t^6 + \ldots,$$

and this series has radius of convergence one. Although the function $(1+t^2)^{-1}$ is perfectly well behaved for t real, it goes to infinity when $t = \pm i$, and the distance of each of these points from the origin is one.

A second application of Theorem 6 is that the radius of convergence of the Taylor series about $t = 0$ of the quotient of two polynomials $a(t)$ and $b(t)$, is the magnitude of the smallest zero of $b(t)$.

At this point we make the important observation that it really wasn't necessary to assume that the functions $P(t)$, $Q(t)$, and $R(t)$ in (1) are polynomials. The method used to solve Example 1 should also be applicable to the more general differential equation

$$L[y] = P(t)\frac{d^2y}{dt^2} + Q(t)\frac{dy}{dt} + R(t)y = 0$$

where $P(t)$, $Q(t)$, and $R(t)$ are power series about t_0. (Of course, we would expect the algebra to be much more cumbersome in this case.) If

$$P(t) = p_0 + p_1(t-t_0) + \ldots, \qquad Q(t) = q_0 + q_1(t-t_0) + \ldots,$$
$$R(t) = r_0 + r_1(t-t_0) + \ldots$$

and $y(t) = a_0 + a_1(t-t_0) + \ldots$, then $L[y](t)$ will be the sum of three power series about $t = t_0$. Consequently, we should be able to find a recurrence formula for the coefficients a_n by setting the sum of the coefficients of like powers of t in the expression $L[y](t)$ equal to zero. This is the content of the following theorem, which we quote without proof.

Theorem 7. *Let the functions $Q(t)/P(t)$ and $R(t)/P(t)$ have convergent Taylor series expansions about $t=t_0$, for $|t-t_0|<\rho$. Then, every solution $y(t)$ of the differential equation*

$$P(t)\frac{d^2y}{dt^2}+Q(t)\frac{dy}{dt}+R(t)y=0 \tag{8}$$

is analytic at $t=t_0$, and the radius of convergence of its Taylor series expansion about $t=t_0$ is at least ρ. The coefficients $a_2,a_3,\ldots$, in the Taylor series expansion

$$y(t)=a_0+a_1(t-t_0)+a_2(t-t_0)^2+\ldots \tag{9}$$

are determined by plugging the series (9) *into the differential equation* (8) *and setting the sum of the coefficients of like powers of t in this expression equal to zero.*

Remark. The interval of convergence of the Taylor series expansion of any solution $y(t)$ of (8) is determined, usually, by the interval of convergence of the power series $Q(t)/P(t)$ and $R(t)/P(t)$, rather than by the interval of convergence of the power series $P(t)$, $Q(t)$, and $R(t)$. This is because the differential equation (8) must be put in the standard form

$$\frac{d^2y}{dt^2}+p(t)\frac{dy}{dt}+q(t)y=0$$

whenever we examine questions of existence and uniqueness.

Example 2.
(a) Find two linearly independent solutions of

$$L[y]=\frac{d^2y}{dt^2}+\frac{3t}{1+t^2}\frac{dy}{dt}+\frac{1}{1+t^2}y=0. \tag{10}$$

(b) Find the solution $y(t)$ of (10) which satisfies the initial conditions $y(0)=2$, $y'(0)=3$.

Solution.
(a) The *wrong* way to do this problem is to expand the functions $3t/(1+t^2)$ and $1/(1+t^2)$ in power series about $t=0$. The right way to do this problem is to multiply both sides of (10) by $1+t^2$ to obtain the equivalent equation

$$L[y]=(1+t^2)\frac{d^2y}{dt^2}+3t\frac{dy}{dt}+y=0.$$

We do the problem this way because the algebra is much less cumbersome when the coefficients of the differential equation (8) are polynomials than when they are power series. Setting $y(t)=\sum_{n=0}^{\infty}a_nt^n$, we compute

$$\begin{aligned}L[y](t)&=(1+t^2)\sum_{n=0}^{\infty}n(n-1)a_nt^{n-2}+3t\sum_{n=0}^{\infty}na_nt^{n-1}+\sum_{n=0}^{\infty}a_nt^n\\&=\sum_{n=0}^{\infty}n(n-1)a_nt^{n-2}+\sum_{n=0}^{\infty}\left[n(n-1)+3n+1\right]a_nt^n\\&=\sum_{n=0}^{\infty}(n+2)(n+1)a_{n+2}t^n+\sum_{n=0}^{\infty}(n+1)^2a_nt^n.\end{aligned}$$

Setting the sum of the coefficients of like powers of t equal to zero gives $(n+2)(n+1)a_{n+2}+(n+1)^2a_n=0$. Hence,

$$a_{n+2}=-\frac{(n+1)^2a_n}{(n+2)(n+1)}=-\frac{(n+1)a_n}{n+2}. \tag{11}$$

Equation (11) is a recurrence formula for the coefficients $a_2,a_3,\ldots$ in terms of a_0 and a_1. To find two linearly independent solutions of (10), we choose the two simplest cases (i) $a_0=1$, $a_1=0$; and (ii) $a_0=0$, $a_1=1$.

(i) $$a_0=1,\qquad a_1=0.$$

In this case, all the odd coefficients are zero since $a_3=-2a_1/3=0$, $a_5=-4a_3/5=0$, and so on. The even coefficients are determined from the relations

$$a_2=-\frac{a_0}{2}=-\frac{1}{2},\qquad a_4=-\frac{3a_2}{4}=\frac{1\cdot 3}{2\cdot 4},\qquad a_6=-\frac{5a_4}{6}=-\frac{1\cdot 3\cdot 5}{2\cdot 4\cdot 6}$$

and so on. Proceeding inductively, we find that

$$a_{2n}=(-1)^n\frac{1\cdot 3\cdots(2n-1)}{2\cdot 4\cdots 2n}=(-1)^n\frac{1\cdot 3\cdots(2n-1)}{2^nn!}.$$

Thus,

$$y_1(t)=1-\frac{t^2}{2}+\frac{1\cdot 3}{2\cdot 4}t^4+\ldots=\sum_{n=0}^{\infty}(-1)^n\frac{1\cdot 3\cdots(2n-1)}{2^nn!}t^{2n} \tag{12}$$

is one solution of (10). The ratio of the $(n+1)$st term to the nth term of $y_1(t)$ is

$$-\frac{1\cdot 3\cdots(2n-1)(2n+1)t^{2n+2}}{2^{n+1}(n+1)!}\times\frac{2^nn!}{1\cdot 3\cdots(2n-1)t^{2n}}=\frac{-(2n+1)t^2}{2(n+1)},$$

and the absolute value of this quantity approaches t^2 as n approaches infinity. Hence, by the Cauchy ratio test, the infinite series (12) converges for $|t|<1$, and diverges for $|t|>1$.

(ii) $$a_0=0,\qquad a_1=1.$$

In this case, all the even coefficients are zero, and the odd coefficients are determined from the relations

$$a_3=-\frac{2a_1}{3}=-\frac{2}{3},\qquad a_5=-\frac{4a_3}{5}=\frac{2\cdot 4}{3\cdot 5},\qquad a_7=-\frac{6a_5}{7}=-\frac{2\cdot 4\cdot 6}{3\cdot 5\cdot 7},$$

and so on. Proceeding inductively, we find that

$$a_{2n+1}=(-1)^n\frac{2\cdot 4\cdots 2n}{3\cdot 5\cdots(2n+1)}=\frac{(-1)^n2^n n!}{3\cdot 5\cdots(2n+1)}.$$

Thus,

$$y_2(t)=t-\frac{2}{3}t^3+\frac{2\cdot 4}{3\cdot 5}t^5+\ldots=\sum_{n=0}^{\infty}\frac{(-1)^n2^n n!}{3\cdot 5\cdots(2n+1)}t^{2n+1}\tag{13}$$

is a second solution of (10), and it is easily verified that this solution, too, converges for $|t|<1$, and diverges for $|t|>1$. This, of course, is not very surprising, since the Taylor series expansions about $t=0$ of the functions $3t/(1+t^2)$ and $1/(1+t^2)$ only converge for $|t|<1$.

(b) The solution $y_1(t)$ satisfies the initial conditions $y(0)=1, y'(0)=0$, while $y_2(t)$ satisfies the initial conditions $y(0)=0, y'(0)=1$. Hence $y(t)=2y_1(t)+3y_2(t)$.

Example 3. Solve the initial-value problem

$$L[y]=\frac{d^2y}{dt^2}+t^2\frac{dy}{dt}+2ty=0;\qquad y(0)=1,\quad y'(0)=0.$$

Solution. Setting $y(t)=\sum_{n=0}^{\infty}a_nt^n$, we compute

$$\begin{aligned}L[y](t)&=\sum_{n=0}^{\infty}n(n-1)a_nt^{n-2}+t^2\sum_{n=0}^{\infty}na_nt^{n-1}+2t\sum_{n=0}^{\infty}a_nt^n\\&=\sum_{n=0}^{\infty}n(n-1)a_nt^{n-2}+\sum_{n=0}^{\infty}na_nt^{n+1}+2\sum_{n=0}^{\infty}a_nt^{n+1}\\&=\sum_{n=0}^{\infty}n(n-1)a_nt^{n-2}+\sum_{n=0}^{\infty}(n+2)a_nt^{n+1}.\end{aligned}$$

Our next step is to rewrite the first summation so that the exponent of the general term is $n+1$ instead of $n-2$. This is accomplished by increasing every n underneath the summation sign by 3, and decreasing the lower limit by 3; that is,

$$\begin{aligned}\sum_{n=0}^{\infty}n(n-1)a_nt^{n-2}&=\sum_{n=-3}^{\infty}(n+3)(n+2)a_{n+3}t^{n+1}\\&=\sum_{n=-1}^{\infty}(n+3)(n+2)a_{n+3}t^{n+1}.\end{aligned}$$

Therefore,

$$\begin{aligned}L[y](t) &= \sum_{n=-1}^{\infty}(n+3)(n+2)a_{n+3}t^{n+1} + \sum_{n=0}^{\infty}(n+2)a_n t^{n+1}\\ &= 2a_2 + \sum_{n=0}^{\infty}(n+3)(n+2)a_{n+3}t^{n+1} + \sum_{n=0}^{\infty}(n+2)a_n t^{n+1}.\end{aligned}$$

Setting the sums of the coefficients of like powers of t equal to zero gives

$$2a_2=0, \quad \text{and} \quad (n+3)(n+2)a_{n+3}+(n+2)a_n=0; \qquad n=0,1,2,\ldots$$

Consequently,

$$a_2=0, \quad \text{and} \quad a_{n+3}=-\frac{a_n}{n+3}; \qquad n\geqslant 0. \tag{14}$$

The recurrence formula (14) determines a_3 in terms of a_0, a_4 in terms of a_1, a_5 in terms of a_2, and so on. Since $a_2=0$, we see that $a_5, a_8, a_{11}, \ldots$ are all zero, regardless of the values of a_0 and a_1. To satisfy the initial conditions, we set $a_0=1$ and $a_1=0$. Then, from (14), $a_4, a_7, a_{10}, \ldots$ are all zero, while

$$a_3=-\frac{a_0}{3}=-\frac{1}{3}, \qquad a_6=-\frac{a_3}{6}=\frac{1}{3\cdot 6}, \qquad a_9=-\frac{a_6}{9}=-\frac{1}{3\cdot 6\cdot 9}$$

and so on. Proceeding inductively, we find that

$$a_{3n}=\frac{(-1)^n}{3\cdot 6\cdots 3n}=\frac{(-1)^n}{3^n 1\cdot 2\cdots n}=\frac{(-1)^n}{3^n n!}.$$

Hence,

$$y(t)=1-\frac{t^3}{3}+\frac{t^6}{3\cdot 6}-\frac{t^9}{3\cdot 6\cdot 9}+\ldots=\sum_{n=0}^{\infty}\frac{(-1)^n t^{3n}}{3^n n!}.$$

By Theorem 7, this series converges for all t, since the power series t^2 and $2t$ obviously converge for all t. (We could also verify this directly using the Cauchy ratio test.)

Example 4. Solve the initial-value problem

$$L[y]=(t^2-2t)\frac{d^2y}{dt^2}+5(t-1)\frac{dy}{dt}+3y=0; \qquad y(1)=7, \quad y'(1)=3. \tag{15}$$

Solution. Since the initial conditions are given at $t=1$, we will express the coefficients of the differential equation (15) as polynomials in $(t-1)$, and then we will find $y(t)$ as a power series centered about $t=1$. To this end, observe that

$$t^2-2t=t(t-2)=\left[(t-1)+1\right]\left[(t-1)-1\right]=(t-1)^2-1.$$

Hence, the differential equation (15) can be written in the form

$$L[y]=\left[(t-1)^2-1\right]\frac{d^2y}{dt^2}+5(t-1)\frac{dy}{dt}+3y=0.$$

Setting $y(t)=\sum_{n=0}^{\infty}a_n(t-1)^n$, we compute

$$L[y](t)=\left[(t-1)^2-1\right]\sum_{n=0}^{\infty}n(n-1)a_n(t-1)^{n-2}$$

$$+5(t-1)\sum_{n=0}^{\infty}na_n(t-1)^{n-1}+3\sum_{n=0}^{\infty}a_n(t-1)^n$$

$$=-\sum_{n=0}^{\infty}n(n-1)a_n(t-1)^{n-2}$$

$$+\sum_{n=0}^{\infty}n(n-1)a_n(t-1)^n+\sum_{n=0}^{\infty}(5n+3)a_n(t-1)^n$$

$$=-\sum_{n=0}^{\infty}(n+2)(n+1)a_{n+2}(t-1)^n+\sum_{n=0}^{\infty}(n^2+4n+3)a_n(t-1)^n.$$

Setting the sums of the coefficients of like powers of t equal to zero gives $-(n+2)(n+1)a_{n+2}+(n^2+4n+3)a_n=0$, so that

$$a_{n+2}=\frac{n^2+4n+3}{(n+2)(n+1)}a_n=\frac{n+3}{n+2}a_n,\qquad n\geq 0. \tag{16}$$

To satisfy the initial conditions, we set $a_0=7$ and $a_1=3$. Then, from (16),

$$a_2=\frac{3}{2}a_0=\frac{3}{2}\cdot 7,\qquad a_4=\frac{5}{4}a_2=\frac{5\cdot 3}{4\cdot 2}\cdot 7,\qquad a_6=\frac{7}{6}a_4=\frac{7\cdot 5\cdot 3}{6\cdot 4\cdot 2}\cdot 7,\ldots$$

$$a_3=\frac{4}{3}a_1=\frac{4}{3}\cdot 3,\qquad a_5=\frac{6}{5}a_3=\frac{6\cdot 4}{5\cdot 3}\cdot 3,\qquad a_7=\frac{8}{7}a_5=\frac{8\cdot 6\cdot 4}{7\cdot 5\cdot 3}\cdot 3,\ldots$$

and so on. Proceeding inductively, we find that

$$a_{2n}=\frac{3\cdot 5\cdots(2n+1)}{2\cdot 4\cdots(2n)}\cdot 7\quad\text{and}\quad a_{2n+1}=\frac{4\cdot 6\cdots(2n+2)}{3\cdot 5\cdots(2n+1)}\cdot 3\qquad(\text{for } n\geq 1).$$

Hence,

$$y(t)=7+3(t-1)+\frac{3}{2}\cdot 7(t-1)^2+\frac{4}{3}\cdot 3(t-1)^3+\ldots$$

$$=7+7\sum_{n=1}^{\infty}\frac{3\cdot 5\cdots(2n+1)(t-1)^{2n}}{2^n n!}+3(t-1)+3\sum_{n=1}^{\infty}\frac{2^n(n+1)!(t-1)^{2n+1}}{3\cdot 5\cdots(2n+1)}.$$

Example 5. Solve the initial-value problem

$$L[y]=(1-t)\frac{d^2y}{dt^2}+\frac{dy}{dt}+(1-t)y=0;\qquad y(0)=1,\ \ y'(0)=1.$$

Solution. Setting $y(t)=\sum_{n=0}^{\infty}a_nt^n$, we compute

$$\begin{aligned}
L[y](t)&=(1-t)\sum_{n=0}^{\infty}n(n-1)a_nt^{n-2}\\
&\quad+\sum_{n=0}^{\infty}na_nt^{n-1}+(1-t)\sum_{n=0}^{\infty}a_nt^n\\
&=\sum_{n=0}^{\infty}n(n-1)a_nt^{n-2}-\sum_{n=0}^{\infty}n(n-1)a_nt^{n-1}\\
&\quad+\sum_{n=0}^{\infty}na_nt^{n-1}+\sum_{n=0}^{\infty}a_nt^n-\sum_{n=0}^{\infty}a_nt^{n+1}\\
&=\sum_{n=0}^{\infty}(n+2)(n+1)a_{n+2}t^n-\sum_{n=0}^{\infty}n(n-2)a_nt^{n-1}\\
&\quad+\sum_{n=0}^{\infty}a_nt^n-\sum_{n=0}^{\infty}a_nt^{n+1}\\
&=\sum_{n=0}^{\infty}(n+2)(n+1)a_{n+2}t^n-\sum_{n=0}^{\infty}(n+1)(n-1)a_{n+1}t^n\\
&\quad+\sum_{n=0}^{\infty}a_nt^n-\sum_{n=1}^{\infty}a_{n-1}t^n\\
&=2a_2+a_1+a_0\\
&\quad+\sum_{n=1}^{\infty}\{(n+2)(n+1)a_{n+2}-(n+1)(n-1)a_{n+1}+a_n-a_{n-1}\}t^n.
\end{aligned}$$

Setting the coefficients of each power of t equal to zero gives

$$a_2=-\frac{a_1+a_0}{2}\quad\text{and}\quad a_{n+2}=\frac{(n+1)(n-1)a_{n+1}-a_n+a_{n-1}}{(n+2)(n+1)},\qquad n\geq 1. \tag{17}$$

To satisfy the initial conditions, we set $a_0=1$ and $a_1=1$. Then, from (17),

$$a_2=-1,\qquad a_3=\frac{-a_1+a_0}{6}=0,\qquad a_4=\frac{3a_3-a_2+a_1}{12}=\frac{1}{6},$$

$$a_5=\frac{8a_4-a_3+a_2}{20}=\frac{1}{60},\qquad a_6=\frac{15a_5-a_4+a_3}{30}=\frac{1}{360}$$

and so on. Unfortunately, though, we cannot discern a general pattern for the coefficients a_n as we did in the previous examples. (This is because the coefficient a_{n+2} depends on the values of a_{n+1}, a_n, and a_{n-1}, while in our previous examples, the coefficient a_{n+2} depended on only one of its predecessors.) This is not a serious problem, though, for we can find the coefficients a_n quite easily with the aid of a digital computer. Sample APL and Fortran programs to compute the coefficients $a_2,\dots,a_n$ in terms of a_0

and a_1, and to evaluate the "approximate" solution

$$y(t) \cong a_0 + a_1 t + \ldots + a_n t^n$$

at any point t are given below. These programs have variable values for a_0 and a_1, so they can also be used to solve the more general initial-value problem

$$(1-t)\frac{d^2y}{dt^2} + \frac{dy}{dt} + (1-t)y = 0; \qquad y(0) = a_0, \quad y'(0) = a_1.$$

APL Program

```
      ∇ SERIES
 [1]  A←Nρ0
 [2]  A[1]←A1
 [3]  A[2]←−(A1+A0)÷2
 [4]  A[3]←(A0−A1)÷6
 [5]  SUM←A0+(A1×T)+(A[2]×T*2)+A[3]×T*3
 [6]  K←2
 [7]  A[K+2]←((A[K−1]−A[K])+(K−1)×(K+1)×A[K+1])÷(K+1)×
      K+2
 [8]  SUM←SUM+A[K+2]×T*K+2
 [9]  K←K+1
[10]  →7×ιK≤N−2
[11]  SUM  ∇
```

Fortran Program

```
         DIMENSION A(200)
         READ (5, 10) A0, A(1), T, N
 10      FORMAT (3F15.8, I5)
         A(2)=−0.5*(A(1)+A0)
         A(3)=(A0−A(1))/2.*3.
         SUM=A0+A(1)*T+A(2)*T**2+A(3)*T**3
         NA=N−2
         DO20 K=2,NA
         A(K+2)=(A(K−1)−A(K)+(K+1.)*(K−1.)*
      1  A(K+1))/(K+1.)*(K+2.)
         SUM=SUM+A(K+2)*T**(K+2)
 20      CONTINUE
         WRITE (6,30) N,T,SUM
 30      FORMAT (1H1, 'FOR N=', I3, ', AND T=',F10.4/1H, 'THE
      1  SUM IS', F20.9)
         CALL EXIT
         END
```

Setting $A0=1, A1=1, (A(1)=1$ for the Fortran program), $T=0.5$, and $N=20$ in these programs gives

$$y\left(\tfrac{1}{2}\right) \cong a_0 + a_1\left(\tfrac{1}{2}\right) + \dots + a_{20}\left(\tfrac{1}{2}\right)^{20} = 1.26104174.$$

This result is correct to eight significant decimal places, since any larger value of N yields the same result.

Exercises

Find the general solution of each of the following equations.

1. $y'' + ty' + y = 0$

2. $y'' - ty = 0$

3. $(2+t^2)y'' - ty' - 3y = 0$

4. $y'' - t^3y = 0$

Solve each of the following initial-value problems.

5. $t(2-t)y'' - 6(t-1)y' - 4y = 0;\quad y(1)=1,\ y'(1)=0$

6. $y'' + t^2y = 0;\quad y(0)=2,\ y'(0)=-1$

7. $y'' - t^3y = 0;\quad y(0)=0,\ y'(0)=-2$

8. $y'' + (t^2+2t+1)y' - (4+4t)y = 0;\quad y(-1)=0,\ y'(-1)=1$

9. The equation $y'' - 2ty' + \lambda y = 0$, λ constant, is known as the Hermite differential equation, and it appears in many areas of mathematics and physics.
(a) Find 2 linearly independent solutions of the Hermite equation.
(b) Show that the Hermite equation has a polynomial solution of degree n if $\lambda = 2n$. This polynomial, when properly normalized; that is, when multiplied by a suitable constant, is known as the Hermite polynomial $H_n(t)$.

10. The equation $(1-t^2)y'' - 2ty' + \alpha(\alpha+1)y = 0$, α constant, is known as the Legendre differential equation, and it appears in many areas of mathematics and physics.
(a) Find 2 linearly independent solutions of the Legendre equation.
(b) Show that the Legendre differential equation has a polynomial solution of degree n if $\alpha = n$.
(c) The Legendre polynomial $P_n(t)$ is defined as the polynomial solution of the Legendre equation with $\alpha = n$ which satisfies the condition $P_n(1) = 1$. Find $P_0(t)$, $P_1(t)$, $P_2(t)$, and $P_3(t)$.

11. The equation $(1-t^2)y'' - ty' + \alpha^2y = 0$, α constant, is known as the Tchebycheff differential equation, and it appears in many areas of mathematics and physics.
(a) Find 2 linearly independent solutions of the Tchebycheff equation.
(b) Show that the Tchebycheff equation has a polynomial solution of degree n if $\alpha = n$. These polynomials, when properly normalized, are called the Tchebycheff polynomials.

12. (a) Find 2 linearly independent solutions of

$$y'' + t^3y' + 3t^2y = 0.$$

(b) Find the first 5 terms in the Taylor series expansion about $t=0$ of the solution $y(t)$ of the initial-value problem

$$y''+t^3y'+3t^2y=e^t; \qquad y(0)=0, \quad y'(0)=0.$$

In each of Problems 13–17, (a) Find the first 5 terms in the Taylor series expansion $\sum_{n=0}^{\infty}a_nt^n$ of the solution $y(t)$ of the given initial-value problem. (b) Write a computer program to find the first $N+1$ coefficients $a_0, a_1, \ldots, a_N$, and to evaluate the polynomial $a_0+a_1t+\ldots+a_Nt^N$. Then, obtain an approximation of $y(\frac{1}{2})$ by evaluating $\sum_{n=0}^{20}a_n(\frac{1}{2})^n$.

13. $(1-t)y''+ty'+y=0; \quad y(0)=1, \ y'(0)=0$

14. $y''+y'+ty=0; \quad y(0)=-1, \ y'(0)=2$

15. $y''+ty'+e^ty=0; \quad y(0)=1, \ y'(0)=0$

16. $y''+y'+e^ty=0; \quad y(0)=0, \ y'(0)=-1$

17. $y''+y'+e^{-t}y=0; \quad y(0)=3, \ y'(0)=5$

Remark. In APL, the instruction $!N$ causes the computer to evaluate $N!$.)

2.8.1 Singular points, Euler equations

The differential equation

$$L[y]=P(t)\frac{d^2y}{dt^2}+Q(t)\frac{dy}{dt}+R(t)y=0 \tag{1}$$

is said to be singular at $t=t_0$ if $P(t_0)=0$. Solutions of (1) frequently become very large, or oscillate very rapidly, in a neighborhood of the singular point t_0. Thus, solutions of (1) may not even be continuous, let along analytic at t_0, and the method of power series solution will fail to work, in general.

Our goal is to find a class of singular equations which we can solve for t near t_0. To this end we will first study a very simple equation, known as Euler's equation, which is singular, but easily solvable. We will then use the Euler equation to motivate a more general class of singular equations which are also solvable in the vicinity of the singular point.

Definition. The differential equation

$$L[y]=t^2\frac{d^2y}{dt^2}+\alpha t\frac{dy}{dt}+\beta y=0. \tag{2}$$

where α and β are constants is known as Euler's equation.

We will assume at first, for simplicity, that $t>0$. Observe that t^2y'' and ty' are both multiples of t^r if $y=t^r$. This suggests that we try $y=t^r$ as a solution of (2). Computing

$$\frac{d}{dt}t^r = rt^{r-1} \quad \text{and} \quad \frac{d^2}{dt^2}t^r = r(r-1)t^{r-2}$$

we see that

$$\begin{aligned} L[t^r] &= r(r-1)t^r + \alpha r t^r + \beta t^r \\ &= [r(r-1) + \alpha r + \beta]t^r \\ &= F(r)t^r \end{aligned} \tag{3}$$

where

$$\begin{aligned} F(r) &= r(r-1) + \alpha r + \beta \\ &= r^2 + (\alpha-1)r + \beta. \end{aligned} \tag{4}$$

Hence, $y=t^r$ is a solution of (2) if, and only if, r is a solution of the quadratic equation

$$r^2 + (\alpha-1)r + \beta = 0. \tag{5}$$

The solutions r_1, r_2 of (5) are

$$\begin{aligned} r_1 &= -\tfrac{1}{2}\left[(\alpha-1) + \sqrt{(\alpha-1)^2 - 4\beta}\right] \\ r_2 &= -\tfrac{1}{2}\left[(\alpha-1) - \sqrt{(\alpha-1)^2 - 4\beta}\right]. \end{aligned}$$

Just as in the case of constant coefficients, we must examine separately the cases where $(\alpha-1)^2 - 4\beta$ is positive, negative, and zero.

Case 1. $(\alpha-1)^2 - 4\beta > 0$. In this case Equation (5) has two real, unequal roots, and thus (2) has two solutions of the form $y_1(t) = t^{r_1}$, $y_2(t) = t^{r_2}$. Clearly, t^{r_1} and t^{r_2} are independent if $r_1 \neq r_2$. Thus the general solution of (2) is (for $t>0$)

$$y(t) = c_1 t^{r_1} + c_2 t^{r_2}.$$

Example 1. Find the general solution of

$$L[y] = t^2\frac{d^2y}{dt^2} + 4t\frac{dy}{dt} + 2y = 0, \qquad t>0. \tag{6}$$

Solution. Substituting $y=t^r$ in (6) gives

$$\begin{aligned} L[t^r] &= r(r-1)t^r + 4rt^r + 2t^r \\ &= [r(r-1) + 4r + 2]t^r \\ &= (r^2 + 3r + 2)t^r \\ &= (r+1)(r+2)t^r \end{aligned}$$

Hence $r_1 = -1$, $r_2 = -2$ and

$$y(t) = c_1 t^{-1} + c_2 t^{-2} = \frac{c_1}{t} + \frac{c_2}{t^2}$$

is the general solution of (6).

Case 2. $(\alpha-1)^2 - 4\beta = 0$. In this case

$$r_1 = r_2 = \frac{1-\alpha}{2}$$

and we have only one solution $y = t^{r_1}$ of (2). A second solution (see Exercise 11) can be found by the method of reduction of order. However, we would like to present here an alternate method of obtaining y_2 which will generalize very nicely in Section 2.8.3. Observe that $F(r) = (r-r_1)^2$ in the case of equal roots. Hence

$$L[t^r] = (r-r_1)^2 t^r. \tag{7}$$

Taking partial derivatives of both sides of (7) with respect to r gives

$$\frac{\partial}{\partial r} L[t^r] = L\left[\frac{\partial}{\partial r} t^r\right] = \frac{\partial}{\partial r}\left[(r-r_1)^2 t^r\right].$$

Since $\partial(t^r)/\partial r = t^r \ln t$, we see that

$$L[t^r \ln t] = (r-r_1)^2 t^r \ln t + 2(r-r_1)t^r. \tag{8}$$

The right hand side of (8) vanishes when $r = r_1$. Hence,

$$L[t^{r_1} \ln t] = 0$$

which implies that $y_2(t) = t^{r_1} \ln t$ is a second solution of (2). Since t^{r_1} and $t^{r_1} \ln t$ are obviously linearly independent, the general solution of (2) in the case of equal roots is

$$y(t) = (c_1 + c_2 \ln t)t^r, \qquad t > 0.$$

Example 2. Find the general solution of

$$L[y] = t^2 \frac{d^2 y}{dt^2} - 5t\frac{dy}{dt} + 9y = 0, \qquad t > 0. \tag{9}$$

Solution. Substituting $y = t^r$ in (9) gives

$$\begin{aligned} L[t^r] &= r(r-1)t^r - 5rt^r + 9t^r \\ &= [r(r-1) - 5r + 9]t^r \\ &= (r^2 - 6r + 9)t^r \\ &= (r-3)^2 t^r. \end{aligned}$$

The equation $(r-3)^2=0$ has $r=3$ as a double root. Hence,

$$y_1(t)=t^3,\ y_2(t)=t^3\ln t$$

and the general solution of (9) is

$$y(t)=(c_1+c_2\ln t)t^3, \qquad t>0.$$

Case 3. $(\alpha-1)^2-4\beta<0$. In this case.

$$r_1=\lambda+i\mu \quad \text{and} \quad r_2=\lambda-i\mu$$

with

$$\lambda=\frac{1-\alpha}{2},\ \mu=\frac{\left[4\beta-(\alpha-1)^2\right]^{1/2}}{2} \tag{10}$$

are complex roots. Hence,

$$\begin{aligned}\phi(t)&=t^{\lambda+i\mu}=t^{\lambda}t^{i\mu}\\ &=t^{\lambda}(e^{\ln t})^{i\mu}=t^{\lambda}e^{i\mu\ln t}\\ &=t^{\lambda}\left[\cos(\mu\ln t)+i\sin(\mu\ln t)\right]\end{aligned}$$

is a complex-valued solution of (2). But then (see Section 2.2.1)

$$y_1(t)=\mathrm{Re}\{\phi(t)\}=t^{\lambda}\cos(\mu\ln t)$$

and

$$y_2(t)=\mathrm{Im}\{\phi(t)\}=t^{\lambda}\sin(\mu\ln t)$$

are two real-valued independent solutions of (2). Hence, the general solution of (2), in the case of complex roots, is

$$y(t)=t^{\lambda}\left[c_1\cos(\mu\ln t)+c_2\sin(\mu\ln t)\right]$$

with λ and μ given by (10).

Example 3. Find the general solution of the equation

$$L[y]=t^2y''-5ty'+25y=0, \qquad t>0. \tag{11}$$

Solution. Substituting $y=t^r$ in (11) gives

$$\begin{aligned}L[t^r]&=r(r-1)t^r-5rt^r+25t^r\\ &=\left[r(r-1)-5r+25\right]t^r\\ &=\left[r^2-6r+25\right]t^r\end{aligned}$$

The roots of the equation $r^2-6r+25=0$ are

$$\frac{6\pm\sqrt{36-100}}{2}=3\pm4i$$

so that

$$\begin{aligned}\phi(t)&=t^{3+4i}=t^3t^{4i}\\&=t^3e^{(\ln t)4i}=t^3e^{i(4\ln t)}\\&=t^3[\cos(4\ln t)+i\sin(4\ln t)]\end{aligned}$$

is a complex-valued solution of (11). Consequently,

$$y_1(t)=\mathrm{Re}\{\phi(t)\}=t^3\cos(4\ln t)$$

and

$$y_2(t)=\mathrm{Im}\{\phi(t)\}=t^3\sin(4\ln t)$$

are two independent solutions of (11), and the general solution is

$$y(t)=t^3[c_1\cos(4\ln t)+c_2\sin(4\ln t)],\qquad t>0.$$

Let us now return to the case $t<0$. One difficulty is that t^r may not be defined if t is negative. For example, $(-1)^{1/2}$ equals i, which is imaginary. A second difficulty is that $\ln t$ is not defined for negative t. We overcome both of these difficulties with the following clever change of variable. Set

$$t=-x,\qquad x>0,$$

and let $y=u(x)$, $x>0$. Observe, from the chain rule, that

$$\frac{dy}{dt}=\frac{du}{dx}\frac{dx}{dt}=-\frac{du}{dx}$$

and

$$\frac{d^2y}{dt^2}=\frac{d}{dt}\left(-\frac{du}{dx}\right)=\frac{d}{dx}\left(-\frac{du}{dx}\right)\frac{dx}{dt}=\frac{d^2u}{dx^2}.$$

Thus, we can rewrite (2) in the form

$$(-x)^2\frac{d^2u}{dx^2}+\alpha(-x)\left(-\frac{du}{dx}\right)+\beta u=0$$

or

$$x^2\frac{d^2u}{dx^2}+\alpha x\frac{du}{dx}+\beta u=0,\qquad x>0 \tag{12}$$

But Equation (12) is exactly the same as (2) with t replaced by x and y replaced by u. Hence, Equation (12) has solutions of the form

$$u(x)=\begin{cases}c_1x^{r_1}+c_2x^{r_2}\\(c_1+c_2\ln x)x^{r_1}\\ [c_1\cos(\mu\ln x)+c_2\sin(\mu\ln x)]x^{\lambda}\end{cases} \tag{13}$$

depending on whether $(\alpha-1)^2-4\beta$ is positive, zero, or negative. Observe now that

$$x=-t=|t|$$

for negative t. Thus, for negative t, the solutions of (2) have one of the forms

$$\begin{cases} c_1|t|^{r_1}+c_2|t|^{r_2} \\ [c_1+c_2\ln|t|]|t|^{r_1} \\ [c_1\cos(\mu\ln|t|)+c_2\sin(\mu\ln|t|)]|t|^{\lambda} \end{cases}$$

Remark. The equation

$$(t-t_0)^2\frac{d^2y}{dt^2}+\alpha(t-t_0)\frac{dy}{dt}+\beta y=0 \tag{14}$$

is also an Euler equation, with a singularity at $t=t_0$ instead of $t=0$. In this case we look for solutions of the form $(t-t_0)^r$. Alternately, we can reduce (14) to (2) by the change of variable $x=t-t_0$.

Exercises

In Problems 1–8, find the general solution of the given equation.

1. $t^2y''+5ty'-5y=0$

2. $2t^2y''+3ty'-y=0$

3. $(t-1)^2y''-2(t-1)y'+2y=0$

4. $t^2y''+3ty'+y=0$

5. $t^2y''-ty'+y=0$

6. $(t-2)^2y''+5(t-2)y'+4y=0$

7. $t^2y''+ty'+y=0$

8. $t^2y''+3ty'+2y=0$

9. Solve the initial-value problem

$$t^2y''-ty'-2y=0;\quad y(1)=0,\quad y'(1)=1$$

on the interval $0<t<\infty$.

10. Solve the initial-value problem

$$t^2y''-3ty'+4y=0;\quad y(1)=1,\quad y'(1)=0$$

on the interval $0<t<\infty$.

11. Use the method of reduction of order to show that $y_2(t)=t^{r_1}\ln t$ in the case of equal roots.

2.8.2 Regular singular points, the method of Frobenius

Our goal now is to find a class of singular differential equations which is more general than the Euler equation

$$t^2\frac{d^2y}{dt^2}+\alpha t\frac{dy}{dt}+\beta y=0 \tag{1}$$

but which is also solvable by analytical techniques. To this end we rewrite (1) in the form

$$\frac{d^2y}{dt^2}+\frac{\alpha}{t}\frac{dy}{dt}+\frac{\beta}{t^2}y=0. \tag{2}$$

A very natural generalization of (2) is the equation

$$L[y]=\frac{d^2y}{dt^2}+p(t)\frac{dy}{dt}+q(t)y=0 \tag{3}$$

where $p(t)$ and $q(t)$ can be expanded in series of the form

$$p(t)=\frac{p_0}{t}+p_1+p_2t+p_3t^2+\cdots$$

$$q(t)=\frac{q_0}{t^2}+\frac{q_1}{t}+q_2+q_3t+q_4t^2+\cdots \tag{4}$$

Definition. The equation (3) is said to have a *regular singular point* at $t=0$ if $p(t)$ and $q(t)$ have series expansions of the form (4). Equivalently, $t=0$ is a regular singular point of (3) if the functions $tp(t)$ and $t^2q(t)$ are analytic at $t=0$. Equation (3) is said to have a regular singular point at $t=t_0$ if the functions $(t-t_0)p(t)$ and $(t-t_0)^2q(t)$ are analytic at $t=t_0$. A singular point of (3) which is not regular is called *irregular*.

Example 1. Classify the singular points of Bessel's equation of order ν

$$t^2\frac{d^2y}{dt^2}+t\frac{dy}{dt}+(t^2-\nu^2)y=0, \tag{5}$$

where ν is a constant.

Solution. Here $P(t)=t^2$ vanishes at $t=0$. Hence, $t=0$ is the only singular point of (5). Dividing both sides of (5) by t^2 gives

$$\frac{d^2y}{dt^2}+\frac{1}{t}\frac{dy}{dt}+\left(1-\frac{\nu^2}{t^2}\right)y=0.$$

Observe that

$$tp(t)=1 \quad \text{and} \quad t^2q(t)=t^2-\nu^2$$

are both analytic at $t=0$. Hence Bessel's equation of order ν has a regular singular point at $t=0$.

Example 2. Classify the singular points of the Legendre equation

$$(1-t^2)\frac{d^2y}{dt^2}-2t\frac{dy}{dt}+\alpha(\alpha+1)y=0 \tag{6}$$

where α is a constant.

Solution. Since $1-t^2$ vanishes when $t=1$ and -1, we see that (6) is singular at $t=\pm 1$. Dividing both sides of (6) by $1-t^2$ gives

$$\frac{d^2y}{dt^2}-\frac{2t}{1-t^2}\frac{dy}{dt}+\alpha\frac{(\alpha+1)}{1-t^2}y=0.$$

Observe that

$$(t-1)p(t)=-(t-1)\frac{2t}{1-t^2}=\frac{2t}{1+t}$$

and

$$(t-1)^2q(t)=\alpha(\alpha+1)\frac{(t-1)^2}{1-t^2}=\alpha(\alpha+1)\frac{1-t}{1+t}$$

are analytic at $t=1$. Similarly, both $(t+1)p(t)$ and $(t+1)^2q(t)$ are analytic at $t=-1$. Hence, $t=1$ and $t=-1$ are regular singular points of (6).

Example 3. Show that $t=0$ is an irregular singular point of the equation

$$t^2\frac{d^2y}{dt^2}+3\frac{dy}{dt}+ty=0. \tag{7}$$

Solution. Dividing through by t^2 gives

$$\frac{d^2y}{dt^2}+\frac{3}{t^2}\frac{dy}{dt}+\frac{1}{t}y=0.$$

In this case, the function

$$tp(t)=t\left(\frac{3}{t^2}\right)=\frac{3}{t}$$

is not analytic at $t=0$. Hence $t=0$ is an irregular singular point of (7).

We return now to the equation

$$L[y]=\frac{d^2y}{dt^2}+p(t)\frac{dy}{dt}+q(t)y=0 \tag{8}$$

where $t=0$ is a regular singular point. For simplicity, we will restrict ourselves to the interval $t>0$. Multiplying (8) through by t^2 gives the equivalent equation

$$L[y]=t^2\frac{d^2y}{dt^2}+t(tp(t))\frac{dy}{dt}+t^2q(t)y=0. \tag{9}$$

We can view Equation (9) as being obtained from (1) by adding higher powers of t to the coefficients α and β. This suggests that we might be able to obtain solutions of (9) by adding terms of the form $t^{r+1}, t^{r+2},\ldots$ to the solutions t^r of (1). Specifically, we will try to obtain solutions of (9) of the form

$$y(t)=\sum_{n=0}^{\infty}a_nt^{n+r}=t^r\sum_{n=0}^{\infty}a_nt^n.$$

Example 4. Find two linearly independent solutions of the equation

$$L[y]=2t\frac{d^2y}{dt^2}+\frac{dy}{dt}+ty=0, \qquad 0<t<\infty. \tag{10}$$

Solution. Let

$$y(t)=\sum_{n=0}^{\infty} a_n t^{n+r}, \qquad a_0\neq 0.$$

Computing

$$y'(t)=\sum_{n=0}^{\infty}(n+r)a_n t^{n+r-1}$$

and

$$y''(t)=\sum_{n=0}^{\infty}(n+r)(n+r-1)a_n t^{n+r-2}$$

we see that

$$\begin{aligned}
L[y]&=t^r\left[2\sum_{n=0}^{\infty}(n+r)(n+r-1)a_n t^{n-1}\right.\\
&\qquad\left.+\sum_{n=0}^{\infty}(n+r)a_n t^{n-1}+\sum_{n=0}^{\infty}a_n t^{n+1}\right]\\
&=t^r\left[2\sum_{n=0}^{\infty}(n+r)(n+r-1)a_n t^{n-1}\right.\\
&\qquad\left.+\sum_{n=0}^{\infty}(n+r)a_n t^{n-1}+\sum_{n=2}^{\infty}a_{n-2} t^{n-1}\right]\\
&=[2r(r-1)a_0+ra_0]t^{r-1}+[2(1+r)ra_1+(1+r)a_1]t^r\\
&\qquad+\sum_{n=2}^{\infty}[2(n+r)(n+r-1)a_n+(n+r)a_n+a_{n-2}]t^{n+r-1}.
\end{aligned}$$

Setting the coefficients of each power of t equal to zero gives

(i) $2r(r-1)a_0+ra_0=r(2r-1)a_0=0$,
(ii) $2(r+1)ra_1+(r+1)a_1=(r+1)(2r+1)a_1=0$,
and
(iii) $2(n+r)(n+r-1)a_n+(n+r)a_n=(n+r)[2(n+r)-1]a_n=-a_{n-2}$, $n\geqslant 2$.

The first equation determines r; it implies that $r=0$ or $r=\frac{1}{2}$. The second equation then forces a_1 to be zero, and the third equation determines a_n for $n\geqslant 2$.

(i) $r=0$. In this case, the recurrence formula (iii) is

$$a_n=\frac{-a_{n-2}}{n(2n-1)}, \qquad n\geq 2.$$

Since $a_1=0$, we see that all of the odd coefficients are zero. The even coefficients are determined from the relations

$$a_2=\frac{-a_0}{2\cdot 3}, \quad a_4=\frac{-a_2}{4\cdot 7}=\frac{a_0}{2\cdot 4\cdot 3\cdot 7}, \quad a_6=\frac{-a_4}{6\cdot 11}=\frac{-a_0}{2\cdot 4\cdot 6\cdot 3\cdot 7\cdot 11}$$

and so on. Setting $a_0=1$, we see that

$$y_1(t)=1-\frac{t^2}{2\cdot 3}+\frac{t^4}{2\cdot 4\cdot 3\cdot 7}+\cdots=1+\sum_{n=1}^{\infty}\frac{(-1)^n t^{2n}}{2^n n!3\cdot 7\cdots(4n-1)}$$

is one solution of (10). It is easily verified, using the Cauchy ratio test, that this series converges for all t.

(ii) $r=\frac{1}{2}$. In this case, the recurrence formula (iii) is

$$a_n=\frac{-a_{n-2}}{(n+\frac{1}{2})[2(n+\frac{1}{2})-1]}=\frac{-a_{n-2}}{n(2n+1)}, \qquad n\geq 2.$$

Again, all of the odd coefficients are zero. The even coefficients are determined from the relations

$$a_2=\frac{-a_0}{2\cdot 5}, \quad a_4=\frac{-a_2}{4\cdot 9}=\frac{a_0}{2\cdot 4\cdot 5\cdot 9}, \quad a_6=\frac{-a_4}{6\cdot 13}=\frac{-a_0}{2\cdot 4\cdot 6\cdot 5\cdot 9\cdot 13}$$

and so on. Setting $a_0=1$, we see that

$$y_2(t)=t^{1/2}\left[1-\frac{t^2}{2\cdot 5}+\frac{t^4}{2\cdot 4\cdot 5\cdot 9}+\cdots\right]$$
$$=t^{1/2}\left[1+\sum_{n=1}^{\infty}\frac{(-1)^n t^{2n}}{2^n n!5\cdot 9\cdots(4n+1)}\right]$$

is a second solution of (10) on the interval $0<t<\infty$.

Remark. Multiplying both sides of (10) by t gives

$$2t^2\frac{d^2y}{dt^2}+t\frac{dy}{dt}+t^2y=0.$$

This equation can be viewed as a generalization of the Euler equation

$$2t^2\frac{d^2y}{dt^2}+t\frac{dy}{dt}=0. \tag{11}$$

Equation (11) has solutions of the form t^r, where

$$2r(r-1)+r=0.$$

This equation is equivalent to Equation (i) which determined r for the solutions of (10).

Let us now see whether our technique, which is known as the method of Frobenius, works in general for Equation (9). (We will assume throughout this section that $t>0$.) By assumption, this equation can be written in the form

$$L[y]=t^2\frac{d^2y}{dt^2}+t[p_0+p_1t+p_2t^2+\cdots]\frac{dy}{dt}+[q_0+q_1t+q_2t^2+\cdots]y=0.$$

Set

$$y(t)=\sum_{n=0}^{\infty}a_nt^{n+r},\text{ with } a_0\neq 0.$$

Computing

$$y'(t)=\sum_{n=0}^{\infty}(n+r)a_nt^{n+r-1}$$

and

$$y''(t)=\sum_{n=0}^{\infty}(n+r)(n+r-1)a_nt^{n+r-2}$$

we see that

$$L[y]=t^r\left\{\sum_{n=0}^{\infty}(n+r)(n+r-1)a_nt^n+\left(\sum_{m=0}^{\infty}p_mt^m\right)\left[\sum_{n=0}^{\infty}(n+r)a_nt^n\right]\right.$$
$$\left.+\left(\sum_{m=0}^{\infty}q_mt^m\right)\left(\sum_{n=0}^{\infty}a_nt^n\right)\right\}.$$

Multiplying through and collecting terms gives

$$\begin{aligned}L[y]=&[r(r-1)+p_0r+q_0]a_0t^r\\&+\{[(1+r)r+p_0(1+r)+q_0]a_1+(rp_1+q_1)a_0\}t^{r+1}\\&\vdots\\&+\left\{[(n+r)(n+r-1)+p_0(n+r)+q_0]a_n\right.\\&\left.+\sum_{k=0}^{n-1}[(k+r)p_{n-k}+q_{n-k}]a_k\right\}t^{n+r}\\&+\cdots.\end{aligned}$$

This expression can be simplified if we set

$$F(r)=r(r-1)+p_0r+q_0. \tag{12}$$

Then,

$$L[y]=a_0F(r)t^r+[a_1F(1+r)+(rp_1+q_1)a_0]t^{1+r}+\cdots$$
$$+a_nF(n+r)t^{n+r}+\left\{\sum_{k=0}^{n-1}[(k+r)p_{n-k}+q_{n-k}]a_k\right\}t^{n+r}$$
$$+\cdots.$$

Setting the coefficient of each power of t equal to zero gives

$$F(r)=r(r-1)+p_0r+q_0=0 \tag{13}$$

and

$$F(n+r)a_n=-\sum_{k=0}^{n-1}[(k+r)p_{n-k}+q_{n-k}]a_k, \quad n\geqslant 1. \tag{14}$$

Equation (13) is called the *indicial* equation of (9). It is a quadratic equation in r, and its roots determine the two possible values r_1 and r_2 of r for which there may be solutions of (9) of the form

$$\sum_{n=0}^{\infty} a_nt^{n+r}.$$

Note that the indicial equation (13) is exactly the equation we would obtain in looking for solutions t^r of the Euler equation

$$t^2\frac{d^2y}{dt^2}+p_0t\frac{dy}{dt}+q_0y=0.$$

Equation (14) shows that, in general, a_n depends on r and all the preceding coefficients $a_0, a_1,\dots,a_{n-1}$. We can solve it recursively for a_n provided that $F(1+r)$, $F(2+r),\dots,F(n+r)$ are not zero. Observe though that if $F(n+r)=0$ for some positive integer n, then $n+r$ is a root of the indicial equation (13). Consequently, if (13) has two real roots r_1, r_2 with $r_1>r_2$ and r_1-r_2 not an integer, then Equation (9) has two solutions of the form

$$y_1(t)=t^{r_1}\sum_{n=0}^{\infty}a_n(r_1)t^n,\ y_2(t)=t^{r_2}\sum_{n=0}^{\infty}a_n(r_2)t^n,$$

and these solutions can be shown to converge wherever $tp(t)$ and $t^2q(t)$ both converge.

Remark. We have introduced the notation $a_n(r_1)$ and $a_n(r_2)$ to emphasize that a_n is determined after we choose $r=r_1$ or r_2.

Example 5. Find the general solution of the equation

$$L[y]=4t\frac{d^2y}{dt^2}+3\frac{dy}{dt}+3y=0. \tag{15}$$

Solution. Equation (15) has a regular singular point at $t=0$ since

$$tp(t)=\tfrac{3}{4} \quad \text{and} \quad t^2q(t)=\tfrac{3}{4}t$$

are both analytic at $t=0$. Set

$$y(t)=\sum_{n=0}^{\infty} a_n t^{n+r}, \quad a_0 \neq 0.$$

Computing

$$y'(t)=\sum_{n=0}^{\infty} (n+r)a_n t^{n+r-1}$$

and

$$y''(t)=\sum_{n=0}^{\infty} (n+r)(n+r-1)a_n t^{n+r-2}$$

we see that

$$\begin{aligned} L[y]=4&\sum_{n=0}^{\infty} (n+r)(n+r-1)a_n t^{n+r-1} \\ &+3\sum_{n=0}^{\infty} (n+r)a_n t^{n+r-1}+3\sum_{n=0}^{\infty} a_n t^{n+r} \\ =&\sum_{n=0}^{\infty} [4(n+r)(n+r-1)+3(n+r)]a_n t^{n+r-1}+\sum_{n=1}^{\infty} 3a_{n-1}t^{n+r-1}. \end{aligned}$$

Setting the sum of coefficients of like powers of t equal to zero gives

$$4r(r-1)+3r=4r^2-r=r(4r-1)=0 \tag{16}$$

and

$$[4(n+r)(n+r-1)+3(n+r)]a_n \equiv (n+r)[4(n+r)-1]a_n=-3a_{n-1}, \quad n \geq 1. \tag{17}$$

Equation (16) is the indicial equation, and it implies that $r=0$ or $r=\frac{1}{4}$. Since these roots do not differ by an integer, we can find two solutions of (15) of the form

$$\sum_{n=0}^{\infty} a_n t^{n+r}$$

with a_n determined from (17).

$r=0$. In this case the recurrence relation (17) reduces to

$$a_n=-3\frac{a_{n-1}}{4n(n-1)+3n}=\frac{-3a_{n-1}}{n(4n-1)}.$$

Setting $a_0=1$ gives

$$a_1=-1,\quad a_2=\frac{-3a_1}{2\cdot 7}=3\frac{1}{2\cdot 7},$$
$$a_3=\frac{-3a_2}{3\cdot 11}=-3^2\frac{1}{2\cdot 3\cdot 7\cdot 11},$$
$$a_4=\frac{-3a_3}{4\cdot 15}=3^3\frac{1}{2\cdot 3\cdot 4\cdot 7\cdot 11\cdot 15},$$

and, in general,

$$a_n=\frac{(-1)^n 3^{n-1}}{n!7\cdot 11\cdot 15\cdots(4n-1)}.$$

Hence,

$$y_1(t)=\sum_{n=0}^{\infty}\frac{(-1)^n 3^{n-1}}{n!7\cdot 11\cdot 15\cdots(4n-1)}t^n \tag{18}$$

is one solution of (15). It is easily seen, using the Cauchy ratio test, that $y_1(t)$ converges for all t. Hence $y_1(t)$ is an analytic solution of (15).

$r=\frac{1}{4}$. In this case the recurrence relation (17) reduces to

$$a_n=\frac{-3a_{n-1}}{(n+\frac{1}{4})[4(n-\frac{3}{4})+3]}=\frac{-3a_{n-1}}{n(4n+1)},\quad n\geq 1.$$

Setting $a_0=1$ gives

$$a_1=\frac{-3}{5},\quad a_2=\frac{3^2}{2\cdot 5\cdot 9},\quad a_3=\frac{-3^3}{2\cdot 3\cdot 5\cdot 9\cdot 13},$$
$$a_4=\frac{3^4}{2\cdot 3\cdot 4\cdot 5\cdot 9\cdot 13\cdot 17},\ldots.$$

Proceeding inductively, we see that

$$a_n=\frac{(-1)^n 3^n}{n!5\cdot 9\cdot 13\cdots(4n+1)}.$$

Hence,

$$y_2(t)=t^{1/4}\sum_{n=0}^{\infty}\frac{(-1)^n 3^n}{n!5\cdot 9\cdot 13\cdots(4n+1)}t^n$$

is a second solution of (15). It can easily be shown, using the Cauchy ratio test, that this solution converges for all positive t. Note, however, that $y_2(t)$ is not differentiable at $t=0$.

The method of Frobenius hits a snag in two separate instances. The first instance occurs when the indicial equation (13) has equal roots $r_1=r_2$. In this case we can only find one solution of (9) of the form

$$y_1(t)=t^{r_1}\sum_{n=0}^{\infty}a_nt^n.$$

In the next section we will prove that (9) has a second solution $y_2(t)$ of the form

$$y_2(t)=y_1(t)\ln t+t^{r_1}\sum_{n=0}^{\infty}b_nt^n$$

and show how to compute the coefficients b_n. The computation of the b_n is usually a very formidable problem. We wish to point out here, though, that in many physical applications the solution $y_2(t)$ is rejected on the grounds that it is singular. Thus, it often suffices to find $y_1(t)$ alone. It is also possible to find a second solution $y_2(t)$ by the method of reduction of order, but this too is usually very cumbersome.

The second snag in the method of Frobenius occurs when the roots r_1, r_2 of the indicial equation differ by a positive integer. Suppose that $r_1=r_2+N$, where N is a positive integer. In this case, we can find one solution of the form

$$y_1(t)=t^{r_1}\sum_{n=0}^{\infty}a_nt^n.$$

However, it may not be possible to find a second solution $y_2(t)$ of the form

$$y_2(t)=t^{r_2}\sum_{n=0}^{\infty}b_nt^n.$$

This is because $F(r_2+n)=0$ when $n=N$. Thus, the left hand side of (14) becomes

$$0\cdot a_N=-\sum_{k=0}^{N-1}\left[(k+r_2)p_{N-k}+q_{N-k}\right]a_k \tag{19}$$

when $n=N$. This equation cannot be satisfied for any choice of a_N, if

$$\sum_{k=0}^{N-1}\left[(k+r_2)p_{N-k}+q_{N-k}\right]a_k\neq 0.$$

In this case (see Section 2.8.3), Equation (9) has a second solution of the form

$$y_2(t)=y_1(t)\ln t+t^{r_2}\sum_{n=0}^{\infty}b_nt^n$$

where again, the computation of the b_n is a formidable problem.

On the other hand, if the sum on the right hand side of (19) vanishes, then a_N is arbitrary, and we can obtain a second solution of the form

$$y_2(t) = t^{r_2} \sum_{n=0}^{\infty} b_n t^n.$$

We illustrate this situation with the following example.

Example 6. Find two solutions of Bessel's equation of order $\frac{1}{2}$,

$$t^2 \frac{d^2y}{dt^2} + t\frac{dy}{dt} + (t^2 - 1/4)y = 0, \quad 0 < t < \infty. \tag{20}$$

Solution. This equation has a regular singular point at $t=0$ since

$$tp(t) = 1 \quad \text{and} \quad t^2 q(t) = t^2 - \tfrac{1}{4}$$

are both analytic at $t=0$. Set

$$y(t) = \sum_{n=0}^{\infty} a_n t^{n+r}, \quad a_0 \neq 0.$$

Computing

$$y'(t) = \sum_{n=0}^{\infty} (n+r) a_n t^{n+r-1}$$

and

$$y''(t) = \sum_{n=0}^{\infty} (n+r)(n+r-1) a_n t^{n+r-2}$$

we see that

$$\begin{aligned} L[y] &= \sum_{n=0}^{\infty} (n+r)(n+r-1) a_n t^{n+r} + \sum_{n=0}^{\infty} (n+r) a_n t^{n+r} \\ &\quad + \sum_{n=0}^{\infty} a_n t^{n+r+2} - \tfrac{1}{4} \sum_{n=0}^{\infty} a_n t^{n+r} \\ &= \sum_{n=0}^{\infty} \left[(n+r)(n+r-1) + (n+r) - \tfrac{1}{4} \right] a_n t^{n+r} + \sum_{n=2}^{\infty} a_{n-2} t^{n+r}. \end{aligned}$$

Setting the sum of coefficients of like powers of t equal to zero gives

$$F(r)a_0 = \left[r(r-1) + r - \tfrac{1}{4} \right] a_0 = \left(r^2 - \tfrac{1}{4}\right) a_0 = 0 \tag{i}$$

$$F(1+r)a_1 = \left[(1+r)r + (1+r) - \tfrac{1}{4} \right] a_1 = \left[(1+r)^2 - \tfrac{1}{4} \right] a_1 = 0 \tag{ii}$$

and

$$F(n+r)a_n = \left[(n+r)^2 - \tfrac{1}{4} \right] a_n = -a_{n-2}, \quad n \geq 2 \tag{iii}$$

Equation (i) is the indicial equation, and it implies that $r_1 = \frac{1}{2}$, $r_2 = -\frac{1}{2}$.
$r_1 = \frac{1}{2}$: Set $a_0 = 1$. Equation (ii) forces a_1 to be zero, and the recurrence

relation (iii) implies that

$$a_n = \frac{-a_{n-2}}{F(n+\frac{1}{2})} = \frac{-a_{n-2}}{n(n+1)}, \quad n \geq 2.$$

This, in turn, implies that all the odd coefficients $a_3, a_5, \ldots$, are zero, and the even coefficients are given by

$$a_2 = \frac{-a_0}{2\cdot 3} = \frac{-1}{2\cdot 3} = -\frac{1}{3!}$$

$$a_4 = \frac{-a_2}{4\cdot 5} = \frac{1}{2\cdot 3\cdot 4\cdot 5} = \frac{1}{5!}$$

$$a_6 = \frac{-a_4}{6\cdot 7} = \frac{-1}{2\cdot 3\cdot 4\cdot 5\cdot 6\cdot 7} = -\frac{1}{7!}$$

and so on. Proceeding inductively, we see that

$$a_{2n} = \frac{(-1)^n}{(2n)!(2n+1)}$$

Hence

$$y_1(t) = t^{1/2}\left(1 - \frac{t^2}{3!} + \frac{t^4}{5!} - \frac{t^6}{7!} + \cdots\right)$$

is one solution of (20). This solution can be rewritten in the form

$$y_1(t) = \frac{t^{1/2}}{t}\left(t - \frac{t^3}{3!} + \frac{t^5}{5!} - \frac{t^7}{7!} + \cdots\right)$$

$$= \frac{1}{\sqrt{t}}\sin t.$$

$r_2 = -\frac{1}{2}$: Set $a_0 = 1$. Since $1 + r_2 = \frac{1}{2}$ is also a root of the indicial equation, we could, conceivably, run into trouble when trying to solve for a_1. However, Equation (ii) is automatically satisfied, regardless of the value of a_1. We will set $a_1 = 0$. (A nonzero value of a_1 will just reproduce a multiple of $y_1(t)$). The recurrence relation (iii) becomes

$$a_n = \frac{-a_{n-2}}{(n-\frac{1}{2})^2 - \frac{1}{4}} = \frac{-a_{n-2}}{n^2 - n} = \frac{-a_{n-2}}{n(n-1)}, \quad n \geq 2.$$

All the odd coefficients are again zero, and the even coefficients are

$$a_2 = \frac{-a_0}{2\cdot 1} = -\frac{1}{2!}$$

$$a_4 = \frac{-a_2}{4\cdot 3} = \frac{1}{4!}$$

$$a_6 = \frac{-a_4}{6\cdot 5} = -\frac{1}{6!}$$

and so on. Proceeding inductively, we see that

$$a_{2n}=\frac{(-1)^n}{(2n)!}.$$

Hence,

$$\begin{aligned} y_2(t)&=t^{-1/2}\left(1-\frac{t^2}{2!}+\frac{t^4}{4!}-\frac{t^6}{6!}+\cdots\right)\\ &=\frac{1}{\sqrt{t}}\cos t \end{aligned}$$

is a second solution of (20).

Remark 1. If r is a complex root of the indicial equation, then

$$y(t)=t^r\sum_{n=0}^{\infty}a_nt^n$$

is a complex-valued solution of (9). It is easily verified in this case that both the real and imaginary parts of $y(t)$ are real-valued solutions of (9).

Remark 2. We must set

$$y(t)=|t|^r\sum_{n=0}^{\infty}a_nt^n$$

if we want to solve (9) on an interval where t is negative. The proof is exactly analogous to the proof for the Euler equation in Section 2.8.1, and is left as an exercise for the reader.

We summarize the results of this section in the following theorem.

Theorem 8. *Consider the differential equation (9) where $t=0$ is a regular singular point. Then, the functions $tp(t)$ and $t^2q(t)$ are analytic at $t=0$ with power series expansions*

$$tp(t)=p_0+p_1t+p_2t^2+\cdots,\ t^2q(t)=q_0+q_1t+q_2t^2+\cdots$$

which converge for $|t|<\rho$. Let r_1 and r_2 be the two roots of the indicial equation

$$r(r-1)+p_0r+q_0=0$$

with $r_1\geq r_2$ if they are real. Then, Equation (9) has two linearly independent solutions $y_1(t)$ and $y_2(t)$ on the interval $0<t<\rho$ of the following form:
(a) *If r_1-r_2 is not a positive integer, then*

$$y_1(t)=t^{r_1}\sum_{n=0}^{\infty}a_nt^n,\quad y_2(t)=t^{r_2}\sum_{n=0}^{\infty}b_nt^n.$$

(b) *If* $r_1 = r_2$, *then*

$$y_1(t) = t^{r_1} \sum_{n=0}^{\infty} a_n t^n, \quad y_2(t) = y_1(t)\ln t + t^{r_1} \sum_{n=0}^{\infty} b_n t^n.$$

(c) *If* $r_1 - r_2 = N$, *a positive integer, then*

$$y_1(t) = t^{r_1} \sum_{n=0}^{\infty} a_n t^n, \quad y_2(t) = a y_1(t)\ln t + t^{r_2} \sum_{n=0}^{\infty} b_n t^n$$

where the constant a may turn out to be zero.

Exercises

In each of Problems 1–6, determine whether the specified value of t is a regular singular point of the given differential equation.

1. $t(t-2)^2 y'' + ty' + y = 0$; $t = 0$

2. $t(t-2)^2 y'' + ty' + y = 0$; $t = 2$

3. $(\sin t)y'' + (\cos t)y' + \frac{1}{t}y = 0$; $t = 0$

4. $(e^t - 1)y'' + e^t y' + y = 0$; $t = 0$

5. $(1 - t^2)y'' + \frac{1}{\sin(t+1)}y' + y = 0$; $t = -1$

6. $t^3 y'' + (\sin t^2)y' + ty = 0$; $t = 0$

Find the general solution of each of the following equations.

7. $2t^2 y'' + 3ty' - (1+t)y = 0$

8. $2ty'' + (1-2t)y' - y = 0$

9. $2ty'' + (1+t)y' - 2y = 0$

10. $2t^2 y'' - ty' + (1+t)y = 0$

11. $4ty'' + 3y' - 3y = 0$

12. $2t^2 y'' + (t^2 - t)y' + y = 0$

In each of Problems 13–18, find two independent solutions of the given equation. In each problem, the roots of the indicial equation differ by a positive integer, but two solutions exist of the form $t^r \sum_{n=0}^{\infty} a_n t^n$.

13. $t^2 y'' - ty' - (t^2 + \frac{5}{4})y = 0$

14. $t^2 y'' + (t - t^2)y' - y = 0$

15. $ty'' - (t^2 + 2)y' + ty = 0$

16. $t^2 y'' + (3t - t^2)y' - ty = 0$

17. $t^2 y'' + t(t+1)y' - y = 0$

18. $ty'' - (4+t)y' + 2y = 0$

19. Consider the equation

$$t^2 y'' + (t^2 - 3t)y' + 3y = 0 \qquad (*)$$

(a) Show that $r = 1$ and $r = 3$ are the two roots of the indicial equation of $(*)$.
(b) Find a power series solution of $(*)$ of the form

$$y_1(t) = t^3 \sum_{n=0}^{\infty} a_n t^n, \quad a_0 = 1.$$

(c) Show that $y_1(t)=t^3e^{-t}$.
(d) Show that $(*)$ has no solution of the form

$$t\sum_{n=0}^{\infty} b_n t^n.$$

(e) Find a second solution of $(*)$ using the method of reduction of order. Leave your answer in integral form.

20. Consider the equation

$$t^2y''+ty'-(1+t)y=0.$$

(a) Show that $r=-1$ and $r=1$ are the two roots of the indicial equation.
(b) Find one solution of the form

$$y_1(t)=t\sum_{n=0}^{\infty} a_n t^n.$$

(c) Find a second solution using the method of reduction of order.

21. Consider the equation

$$ty''+ty'+2y=0.$$

(a) Show that $r=0$ and $r=1$ are the two roots of the indicial equation.
(b) Find one solution of the form

$$y_1(t)=t\sum_{n=0}^{\infty} a_n t^n.$$

(c) Find a second solution using the method of reduction of order.

22. Consider the equation

$$ty''+(1-t^2)y'+4ty=0.$$

(a) Show that $r=0$ is a double root of the indicial equation.
(b) Find one solution of the form $y_1(t)=\sum_{n=0}^{\infty} a_n t^n$.
(c) Find a second solution using the method of reduction of order.

23. Consider the Bessel equation of order zero

$$t^2y''+ty'+t^2y=0.$$

(a) Show that $r=0$ is a double root of the indicial equation.
(b) Find one solution of the form

$$y_1(t)=1-\frac{t^2}{2^2}+\frac{t^4}{2^2\cdot 4^2}-\frac{t^6}{2^2\cdot 4^2\cdot 6^2}+\cdots.$$

This solution is known as $J_0(t)$.
(c) Find a second solution using the method of reduction of order.

24. Consider the Bessel equation of order ν

$$t^2y''+ty'+(t^2-\nu^2)y=0$$

where ν is real and positive.
(a) Find a power series solution

$$J_\nu(t)=\frac{t^\nu}{2^\nu \nu!}\sum_{n=0}^{\infty} a_n t^n, \quad a_0=1.$$

This function $J_\nu(t)$ is called the Bessel function of order ν.
(b) Find a second solution if 2ν is not an integer.

25. The differential equation

$$ty''+(1-t)y'+\lambda y=0, \quad \lambda \text{ constant},$$

is called the Laguerre differential equation.
(a) Show that the indicial equation is $r^2=0$.
(b) Find a solution $y(t)$ of the Laguerre equation of the form $\Sigma_{n=0}^{\infty} a_n t^n$.
(c) Show that this solution reduces to a polynomial if $\lambda=n$.

26. The differential equation

$$t(1-t)y''+[\gamma-(1+\alpha+\beta)t]y'-\alpha\beta y=0$$

where α, β, and γ are constants, is known as the hypergeometric equation.
(a) Show that $t=0$ is a regular singular point and that the roots of the indicial equation are 0 and $1-\gamma$.
(b) Show that $t=1$ is also a regular singular point, and that the roots of the indicial equation are now 0 and $\gamma-\alpha-\beta$.
(c) Assume that γ is not an integer. Find two solutions $y_1(t)$ and $y_2(t)$ of the hypergeometric equation of the form

$$y_1(t)=\sum_{n=0}^{\infty} a_n t^n, \quad y_2(t)=t^{1-\gamma}\sum_{n=0}^{\infty} b_n t^n.$$

27. (a) Show that the equation

$$2(\sin t)y''+(1-t)y'-2y=0$$

has two solutions of the form

$$y_1(t)=\sum_{n=0}^{\infty} a_n t^n, \quad y_2(t)=t^{1/2}\sum_{n=0}^{\infty} b_n t^n.$$

(b) Find the first 5 terms in these series expansions assuming that $a_0=b_0=1$.

28. Let $y(t)=u(t)+iv(t)$ be a complex-valued solution of (3) with $p(t)$ and $q(t)$ real. Show that both $u(t)$ and $v(t)$ are real-valued solutions of (3).

29. (a) Show that the indicial equation of

$$t^2y''+ty'+(1+t)y=0 \qquad (*)$$

has complex roots $r=\pm i$.

(b) Show that $(*)$ has 2 linearly independent solutions $y(t)$ of the form

$$y(t)=\sin(\ln t)\sum_{n=0}^{\infty} a_n t^n+\cos(\ln t)\sum_{n=0}^{\infty} b_n t^n.$$

2.8.3 *Equal roots, and roots differing by an integer*

Equal roots.
We run into trouble if the indicial equation has equal roots $r_1=r_2$ because then the differential equation

$$P(t)\frac{d^2y}{dt^2}+Q(t)\frac{dy}{dt}+R(t)y=0 \tag{1}$$

has only one solution of the form

$$y(t)=t^r\sum_{n=0}^{\infty} a_n t^n. \tag{2}$$

The method of finding a second solution is very similar to the method used in finding a second solution of Euler's equation, in the case of equal roots. Let us rewrite (2) in the form

$$y(t)=y(t,r)=t^r\sum_{n=0}^{\infty} a_n(r)t^n$$

to emphasize that the solution $y(t)$ depends on our choice of r. Then (see Section 2.8.2)

$$L[y](t,r)=a_0F(r)t^r$$
$$+\sum_{n=1}^{\infty}\left\{a_n(r)F(n+r)+\sum_{k=0}^{n-1}\left[(k+r)p_{n-k}+q_{n-k}\right]a_k\right\}t^{n+r}.$$

We now think of r as a continuous variable and determine a_n as a function of r by requiring that the coefficient of t^{n+r} be zero for $n\geqslant 1$. Thus

$$a_n(r)=\frac{-\sum_{k=0}^{n-1}\left[(k+r)p_{n-k}+q_{n-k}\right]a_k}{F(n+r)}$$

With this choice of $a_n(r)$, we see that

$$L[y](t,r)=a_0F(r)t^r. \tag{3}$$

In the case of equal roots, $F(r)=(r-r_1)^2$, so that (3) can be written in the form

$$L[y](t,r)=a_0(r-r_1)^2t^r.$$

Since $L[y](t,r_1)=0$, we obtain one solution

$$y_1(t)=t^{r_1}\left[a_0+\sum_{n=1}^{\infty} a_n(r_1)t^n\right].$$

Observe now, that

$$\begin{aligned}\frac{\partial}{\partial r}L[y](t,r) &= L\left[\frac{\partial y}{\partial r}\right](t,r)\\ &= \frac{\partial}{\partial r}a_0(r-r_1)^2t^r\\ &= 2a_0(r-r_1)t^r + a_0(r-r_1)^2(\ln t)t^r\end{aligned}$$

also vanishes when $r=r_1$. Thus

$$\begin{aligned}y_2(t) &= \frac{\partial}{\partial r}y_1(t,r)|_{r=r_1}\\ &= \frac{\partial}{\partial r}\left[\sum_{n=0}^{\infty} a_n(r)t^{n+r}\right]_{r=r_1}\\ &= \sum_{n=0}^{\infty}\left[a_n(r_1)t^{n+r_1}\right]\ln t + \sum_{n=0}^{\infty} a_n'(r_1)t^{n+r_1}\\ &= y_1(t)\ln t + \sum_{n=0}^{\infty} a_n'(r_1)t^{n+r_1}\end{aligned}$$

is a second solution of (1).

Example 1. Find two solutions of Bessel's equation of order zero

$$L[y] = t^2\frac{d^2y}{dt^2} + t\frac{dy}{dt} + t^2y = 0, \quad t>0. \tag{4}$$

Solution. Set

$$y(t) = \sum_{n=0}^{\infty} a_n t^{n+r}.$$

Computing

$$y'(t) = \sum_{n=0}^{\infty}(n+r)a_n t^{n+r-1}$$

and

$$y''(t) = \sum_{n=0}^{\infty}(n+r)(n+r-1)a_n t^{n+r-2}$$

we see that

$$\begin{aligned}L[y] &= \sum_{n=0}^{\infty}(n+r)(n+r-1)a_n t^{n+r} + \sum_{n=0}^{\infty}(n+r)a_n t^{n+r} + \sum_{n=0}^{\infty} a_n t^{n+r+2}\\ &= \sum_{n=0}^{\infty}(n+r)^2 a_n t^{n+r} + \sum_{n=2}^{\infty} a_{n-2}t^{n+r}.\end{aligned}$$

Setting the sums of like powers of t equal to zero gives

(i) $r^2 a_0 = F(r)a_0 = 0$
(ii) $(1+r)^2 a_1 = F(1+r)a_1 = 0$
and
(iii) $(n+r)^2 a_n = F(n+r)a_n = -a_{n-2},\ n \geq 2.$

Equation (i) is the indicial equation, and it has equal roots $r_1 = r_2 = 0$. Equation (ii) forces a_1 to be zero, and the recurrence relation (iii) says that

$$a_n = \frac{-a_{n-2}}{(n+r)^2}.$$

Clearly, $a_3 = a_5 = a_7 = \cdots = 0$. The even coefficients are given by

$$a_2(r) = \frac{-a_0}{(2+r)^2} = \frac{-1}{(2+r)^2}$$

$$a_4(r) = \frac{-a_2}{(4+r)^2} = \frac{1}{(2+r)^2(4+r)^2}$$

and so on. Proceeding inductively, we see that

$$a_{2n}(r) = \frac{(-1)^n}{(2+r)^2(4+r)^2\cdots(2n+r)^2}.$$

To determine $y_1(t)$, we set $r=0$. Then

$$a_2(0) = \frac{-1}{2^2}$$

$$a_4(0) = \frac{1}{2^2 \cdot 4^2} = \frac{1}{2^4}\frac{1}{(2!)^2}$$

$$a_6(0) = \frac{-1}{2^2 \cdot 4^2 \cdot 6^2} = \frac{-1}{2^6(3!)^2}$$

and in general

$$a_{2n}(0) = \frac{(-1)^n}{2^2 \cdot 4^2 \cdots (2n)^2} = \frac{(-1)^n}{2^{2n}(n!)^2}.$$

Hence,

$$y_1(t) = 1 - \frac{t^2}{2^2} + \frac{t^4}{2^4 \cdot (2!)^2} - \frac{t^6}{2^6(3!)^2} + \cdots$$

$$= \sum_{n=0}^{\infty} \frac{(-1)^n t^{2n}}{2^{2n}(n!)^2}$$

is one solution of (4). This solution is often referred to as the Bessel function of the first kind of order zero, and is denoted by $J_0(t)$.

To obtain a second solution of (4) we set

$$y_2(t) = y_1(t)\ln t + \sum_{n=0}^{\infty} a'_{2n}(0)t^{2n}.$$

To compute $a'_{2n}(0)$, observe that

$$\frac{a'_{2n}(r)}{a_{2n}(r)} = \frac{d}{dr}\ln|a_{2n}(r)| = \frac{d}{dr}\ln(2+r)^{-2}\cdots(2n+r)^{-2}$$

$$= -2\frac{d}{dr}[\ln(2+r)+\ln(4+r)+\cdots+\ln(2n+r)]$$

$$= -2\left(\frac{1}{2+r}+\frac{1}{4+r}+\cdots+\frac{1}{2n+r}\right).$$

Hence,

$$a'_{2n}(0) = -2\left(\frac{1}{2}+\frac{1}{4}+\cdots+\frac{1}{2n}\right)a_{2n}(0)$$

$$= -\left(1+\frac{1}{2}+\cdots+\frac{1}{n}\right)a_{2n}(0).$$

Setting

$$H_n = 1+\frac{1}{2}+\frac{1}{3}+\cdots+\frac{1}{n} \tag{5}$$

we see that

$$a'_{2n}(0) = \frac{-H_n(-1)^n}{2^{2n}(n!)^2} = \frac{(-1)^{n+1}H_n}{2^{2n}(n!)^2}$$

and thus

$$y_2(t) = y_1(t)\ln t + \sum_{n=0}^{\infty} \frac{(-1)^{n+1}H_n}{2^{2n}(n!)^2}t^{2n}$$

is a second solution of (4) with H_n given by (5).

Roots differing by a positive integer. Suppose that r_2 and $r_1 = r_2 + N$, N a positive integer, are the roots of the indicial equation. Then we can certainly find one solution of (1) of the form

$$y_1(t) = t^{r_1}\sum_{n=0}^{\infty} a_n(r_1)t^n.$$

As we mentioned previously, it may not be possible to find a second solution of the form

$$t^{r_2}\sum_{n=0}^{\infty} b_n t^n.$$

In this case, Equation (1) will have a second solution of the form

$$y_2(t) = \frac{\partial}{\partial r} y(t,r)\bigg|_{r=r_2}$$
$$= ay_1(t)\ln t + \sum_{n=0}^{\infty} a_n'(r_2)t^{n+r_2}$$

where a is a constant, and

$$y(t,r) = t^r \sum_{n=0}^{\infty} a_n(r)t^n$$

with

$$a_0 = a_0(r) = r - r_2.$$

The proof of this result can be found in more advanced books on differential equations. In Exercise 5, we develop a simple proof, using the method of reduction of order, to show why a logarithm term will be present.

Remark. It is usually very difficult, and quite cumbersome, to obtain the second solution $y_2(t)$ when a logarithm term is present. Beginning and intermediate students are not expected, usually, to perform such calculations. We have included several exercises for the more industrious students. In these problems, and in similar problems which occur in applications, it is often more than sufficient to find just the first few terms in the series expansion of $y_2(t)$, and this can usually be accomplished using the method of reduction of order.

Exercises

In Problems 1 and 2, show that the roots of the indicial equation are equal, and find two independent solutions of the given equation.

1. $ty'' + y' - 4y = 0$

2. $t^2y'' - t(1+t)y' + y = 0$

3. (a) Show that $r = -1$ and $r = 1$ are the roots of the indicial equation for Bessel's equation of order one

$$t^2\frac{d^2y}{dt^2} + t\frac{dy}{dt} + (t^2-1)y = 0.$$

(b) Find a solution:

$$J_1(t) = \tfrac{1}{2}t \sum_{n=0}^{\infty} a_n t^n, a_0 = 1.$$

$J_1(t)$ is called the Bessel function of order one.

(c) Find a second solution:

$$y_2(t) = -J_1(t)\ln t + \frac{1}{t}\left[1 - \sum_{n=1}^{\infty} \frac{(-1)^n(H_n + H_{n-1})}{2^{2n}n!(n-1)!}t^{2n}\right].$$

4. Consider the equation

$$ty''+3y'-3y=0, \qquad t>0.$$

(a) Show that $r=0$ and $r=-2$ are the roots of the indicial equation.
(b) Find a solution

$$y_1(t)=\sum_{n=0}^{\infty} a_n t^n.$$

(c) Find a second solution

$$y_2(t)=y_1(t)\ln t+\frac{1}{t^2}-\frac{1}{t}+\frac{1}{4}+\frac{11}{36}t+\frac{31}{576}t^2+\cdots.$$

5. This exercise gives an alternate proof of some of the results of this section, using the method of reduction of order.
(a) Let $t=0$ be a regular singular point of the equation

$$t^2y''+tp(t)y'+q(t)y=0 \tag{i}$$

Show that the substitution $y=t^r z$ reduces (i) to the equation

$$t^2z''+[2r+p(t)]tz'+[r(r-1)+rp(t)+q(t)]z=0. \tag{ii}$$

(b) Let r be a root of the indicial equation. Show that (ii) has an analytic solution $z_1(t)=\sum_{n=0}^{\infty}a_nt^n$.
(c) Set $z_2(t)=z_1(t)v(t)$. Show that

$$v(t)=\int u(t)\,dt, \qquad \text{where } u(t)=\frac{e^{-\int[2r+p(t)]/t\,dt}}{z_1^2(t)}.$$

(d) Suppose that $r=r_0$ is a double root of the indicial equation. Show that $2r_0+p_0=1$, and conclude therefore that

$$u(t)=\frac{u_0}{t}+u_1+u_2t+\cdots.$$

(e) Use the result in (d) to show that $y_2(t)$ has an $\ln t$ term in the case of equal roots.
(f) Suppose the roots of the indicial equation are r_0 and r_0-N, N a positive integer. Show that $2r_0+p_0=1+N$, and conclude therefore, that

$$u(t)=\frac{1}{t^{1+N}}\hat{u}(t)$$

where $\hat{u}(t)$ is analytic at $t=0$.
(g) Use the result in (f) to show that $y_2(t)$ has an $\ln t$ term if the coefficient of t^N in the expansion of $\hat{u}(t)$ is nonzero. Show, in addition, that if this coefficient is zero, then

$$v(t)=\frac{v_{-N}}{t^N}+\cdots+\frac{v_{-1}}{t}+v_1t+v_2t^2+\cdots$$

and $y_2(t)$ has no $\ln t$ term.

2.9 The method of Laplace transforms

In this section we describe a very different and extremely clever way of solving the initial-value problem

$$a\frac{d^2y}{dt^2}+b\frac{dy}{dt}+cy=f(t); \qquad y(0)=y_0, \;\; y'(0)=y_0' \tag{1}$$

where a, b and c are constants. This method, which is known as the method of Laplace transforms, is especially useful in two cases which arise quite often in applications. The first case is when $f(t)$ is a discontinuous function of time. The second case is when $f(t)$ is zero except for a very short time interval in which it is very large.

To put the method of Laplace transforms into proper perspective, we consider the following hypothetical situation. Suppose that we want to multiply the numbers 3.163 and 16.38 together, but that we have forgotten completely how to multiply. We only remember how to add. Being good mathematicians, we ask ourselves the following question.

Question: Is it possible to reduce the problem of multiplying the two numbers 3.163 and 16.38 together to the simpler problem of adding two numbers together?

The answer to this question, of course, is yes, and is obtained as follows. First, we consult our logarithm tables and find that $\ln 3.163 = 1.15152094$, and $\ln 16.38 = 2.79606108$. Then, we add these two numbers together to yield 3.94758202. Finally, we consult our anti-logarithm tables and find that $3.94758202 = \ln 51.80994$. Hence, we conclude that $3.163 \times 16.38 = 51.80994$.

The key point in this analysis is that the operation of multiplication is replaced by the simpler operation of addition when we work with the logarithms of numbers, rather than with the numbers themselves. We represent this schematically in Table 1. In the method to be discussed below, the unknown function $y(t)$ will be replaced by a new function $Y(s)$, known as the Laplace transform of $y(t)$. This association will have the property that $y'(t)$ will be replaced by $sY(s)-y(0)$. Thus, the operation of differentiation with respect to t will be replaced, essentially, by the operation of multiplication with respect to s. In this manner, we will replace the initial-value problem (1) by an algebraic equation which can be solved explicitly for $Y(s)$. Once we know $Y(s)$, we can consult our "anti-Laplace transform" tables and recover $y(t)$.

Table 1

a	$\longrightarrow$	$\ln a$
b	$\longrightarrow$	$\ln b$
$a \cdot b$	$\longrightarrow$	$\ln a + \ln b$

We begin with the definition of the Laplace transform.

Definition. Let $f(t)$ be defined for $0 \leqslant t < \infty$. The Laplace transform of $f(t)$, which is denoted by $F(s)$, or $\mathcal{L}\{f(t)\}$, is given by the formula

$$F(s) = \mathcal{L}\{f(t)\} = \int_0^\infty e^{-st} f(t)\,dt \tag{2}$$

where

$$\int_0^\infty e^{-st} f(t)\,dt = \lim_{A\to\infty} \int_0^A e^{-st} f(t)\,dt.$$

Example 1. Compute the Laplace transform of the function $f(t)=1$.
Solution. From (2),

$$\mathcal{L}\{f(t)\} = \lim_{A\to\infty} \int_0^A e^{-st}\,dt = \lim_{A\to\infty} \frac{1-e^{-sA}}{s}$$

$$= \begin{cases} \dfrac{1}{s}, & s>0 \\ \infty, & s\leqslant 0 \end{cases}.$$

Example 2. Compute the Laplace transform of the function e^{at}.
Solution. From (2),

$$\mathcal{L}\{e^{at}\} = \lim_{A\to\infty} \int_0^A e^{-st} e^{at}\,dt = \lim_{A\to\infty} \frac{e^{(a-s)A}-1}{a-s}$$

$$= \begin{cases} \dfrac{1}{s-a}, & s>a \\ \infty, & s\leqslant a \end{cases}.$$

Example 3. Compute the Laplace transform of the functions $\cos\omega t$ and $\sin\omega t$.
Solution. From (2),

$$\mathcal{L}\{\cos\omega t\} = \int_0^\infty e^{-st}\cos\omega t\,dt \quad \text{and} \quad \mathcal{L}\{\sin\omega t\} = \int_0^\infty e^{-st}\sin\omega t\,dt.$$

Now, observe that

$$\mathcal{L}\{\cos\omega t\} + i\mathcal{L}\{\sin\omega t\} = \int_0^\infty e^{-st} e^{i\omega t}\,dt = \lim_{A\to\infty} \int_0^A e^{(i\omega - s)t}\,dt$$

$$= \lim_{A\to\infty} \frac{e^{(i\omega-s)A}-1}{i\omega - s}$$

$$= \begin{cases} \dfrac{1}{s-i\omega} = \dfrac{s+i\omega}{s^2+\omega^2}, & s>0 \\ \text{undefined}, & s\leqslant 0 \end{cases}.$$

Equating real and imaginary parts in this equation gives

$$\mathcal{L}\{\cos\omega t\}=\frac{s}{s^2+\omega^2}\quad\text{and}\quad\mathcal{L}\{\sin\omega t\}=\frac{\omega}{s^2+\omega^2},\qquad s>0.$$

Equation (2) associates with every function $f(t)$ a new function, which we call $F(s)$. As the notation $\mathcal{L}\{f(t)\}$ suggests, the Laplace transform is an operator acting on functions. It is also a linear operator, since

$$\begin{aligned}\mathcal{L}\{c_1f_1(t)+c_2f_2(t)\}&=\int_0^\infty e^{-st}\left[c_1f_1(t)+c_2f_2(t)\right]dt\\&=c_1\int_0^\infty e^{-st}f_1(t)\,dt+c_2\int_0^\infty e^{-st}f_2(t)\,dt\\&=c_1\mathcal{L}\{f_1(t)\}+c_2\mathcal{L}\{f_2(t)\}.\end{aligned}$$

It is to be noted, though, that whereas $f(t)$ is defined for $0\leqslant t<\infty$, its Laplace transform is usually defined in a different interval. For example, the Laplace transform of e^{2t} is only defined for $2<s<\infty$, and the Laplace transform of e^{8t} is only defined for $8<s<\infty$. This is because the integral (2) will only exist, in general, if s is sufficiently large.

One very serious difficulty with the definition (2) is that this integral may fail to exist for every value of s. This is the case, for example, if $f(t)=e^{t^2}$ (see Exercise 13). To guarantee that the Laplace transform of $f(t)$ exists at least in some interval $s>s_0$, we impose the following conditions on $f(t)$.

(i) The function $f(t)$ is piecewise continuous. This means that $f(t)$ has at most a finite number of discontinuities on any interval $0\leqslant t\leqslant A$, and both the limit from the right and the limit from the left of f exist at every point of discontinuity. In other words, $f(t)$ has only a finite number of "jump discontinuities" in any finite interval. The graph of a typical piecewise continuous function $f(t)$ is described in Figure 1.

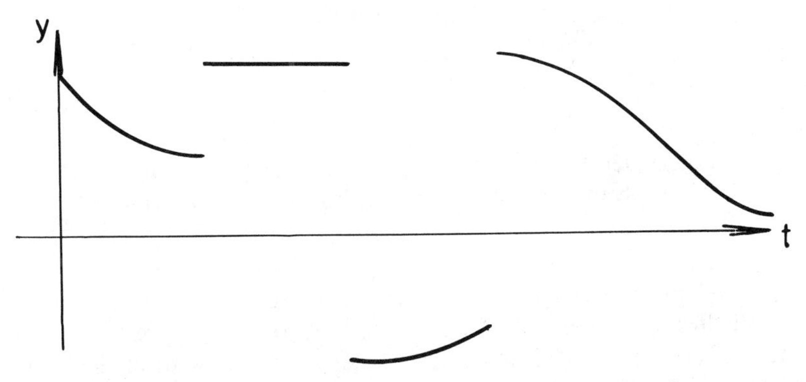

Figure 1. Graph of a typical piecewise continuous function

(ii) The function $f(t)$ is of exponential order, that is, there exist constants M and c such that

$$|f(t)| \leqslant Me^{ct}, \qquad 0 \leqslant t < \infty.$$

Lemma 1. *If $f(t)$ is piecewise continuous and of exponential order, then its Laplace transform exists for all s sufficiently large. Specifically, if $f(t)$ is piecewise continuous, and $|f(t)| \leqslant Me^{ct}$, then $F(s)$ exists for $s > c$.*

We prove Lemma 1 with the aid of the following lemma from integral calculus, which we quote without proof.

Lemma 2. *Let $g(t)$ be piecewise continuous. Then, the improper integral $\int_0^\infty g(t)\,dt$ exists if $\int_0^\infty |g(t)|\,dt$ exists. To prove that this latter integral exists, it suffices to show that there exists a constant K such that*

$$\int_0^A |g(t)|\,dt \leqslant K$$

for all A.

Remark. Notice the similarity of Lemma 2 with the theorem of infinite series (see Appendix B) which states that the infinite series $\sum a_n$ converges if $\sum |a_n|$ converges, and that $\sum |a_n|$ converges if there exists a constant K such that $|a_1| + \ldots + |a_n| \leqslant K$ for all n.

We are now in a position to prove Lemma 1.

PROOF OF LEMMA 1. Since $f(t)$ is piecewise continuous, the integral $\int_0^A e^{-st} f(t)\,dt$ exists for all A. To prove that this integral has a limit for all s sufficiently large, observe that

$$\int_0^A |e^{-st} f(t)|\,dt \leqslant M \int_0^A e^{-st} e^{ct}\,dt$$

$$= \frac{M}{c-s}\left[e^{(c-s)A} - 1\right] \leqslant \frac{M}{s-c}$$

for $s > c$. Consequently, by Lemma 2, the Laplace transform of $f(t)$ exists for $s > c$. Thus, from here on, we tacitly assume that $|f(t)| \leqslant Me^{ct}$, and $s > c$. □

The real usefulness of the Laplace transform in solving differential equations lies in the fact that the Laplace transform of $f'(t)$ is very closely related to the Laplace transform of $f(t)$. This is the content of the following important lemma.

Lemma 3. *Let* $F(s)=\mathcal{L}\{f(t)\}$. *Then*

$$\mathcal{L}\{f'(t)\}=s\mathcal{L}\{f(t)\}-f(0)=sF(s)-f(0).$$

PROOF. The proof of Lemma 3 is very elementary; we just write down the formula for the Laplace transform of $f'(t)$ and integrate by parts. To wit,

$$\begin{aligned}
\mathcal{L}\{f'(t)\} &= \lim_{A\to\infty}\int_0^A e^{-st}f'(t)\,dt \\
&= \lim_{A\to\infty} e^{-st}f(t)\Big|_0^A + \lim_{A\to\infty} s\int_0^A e^{-st}f(t)\,dt \\
&= -f(0)+s\lim_{A\to\infty}\int_0^A e^{-st}f(t),dt \\
&= -f(0)+sF(s).
\end{aligned}$$

□

Our next step is to relate the Laplace transform of $f''(t)$ to the Laplace transform of $f(t)$. This is the content of Lemma 4.

Lemma 4. *Let* $F(s)=\mathcal{L}\{f(t)\}$. *Then,*

$$\mathcal{L}\{f''(t)\}=s^2F(s)-sf(0)-f'(0).$$

PROOF. Using Lemma 3 twice, we see that

$$\begin{aligned}
\mathcal{L}\{f''(t)\} &= s\mathcal{L}\{f'(t)\}-f'(0) \\
&= s\left[sF(s)-f(0)\right]-f'(0) \\
&= s^2F(s)-sf(0)-f'(0).
\end{aligned}$$

□

We have now developed all the machinery necessary to reduce the problem of solving the initial-value problem

$$a\frac{d^2y}{dt^2}+b\frac{dy}{dt}+cy=f(t);\qquad y(0)=y_0,\quad y'(0)=y_0' \tag{3}$$

to that of solving an algebraic equation. Let $Y(s)$ and $F(s)$ be the Laplace transforms of $y(t)$ and $f(t)$ respectively. Taking Laplace transforms of both sides of the differential equation gives

$$\mathcal{L}\{ay''(t)+by'(t)+cy(t)\}=F(s).$$

By the linearity of the Laplace transform operator,

$$\mathcal{L}\{ay''(t)+by'(t)+cy(t)\}=a\mathcal{L}\{y''(t)\}+b\mathcal{L}\{y'(t)\}+c\mathcal{L}\{y(t)\},$$

and from Lemmas 3 and 4

$$\mathcal{L}\{y'(t)\}=sY(s)-y_0, \qquad \mathcal{L}\{y''(t)\}=s^2Y(s)-sy_0-y_0'.$$

Hence,

$$a\left[s^2Y(s)-sy_0-y_0'\right]+b\left[sY(s)-y_0\right]+cY(s)=F(s)$$

and this *algebraic equation* implies that

$$Y(s)=\frac{(as+b)y_0}{as^2+bs+c}+\frac{ay_0'}{as^2+bs+c}+\frac{F(s)}{as^2+bs+c}. \tag{4}$$

Equation (4) tells us the Laplace transform of the solution $y(t)$ of (3). To find $y(t)$, we must consult our anti, or inverse, Laplace transform tables. Now, just as $Y(s)$ is expressed explicitly in terms of $y(t)$; that is, $Y(s) = \int_0^\infty e^{-st}y(t)\,dt$, we can write down an explicit formula for $y(t)$. However, this formula, which is written symbolically as $y(t)=\mathcal{L}^{-1}\{Y(s)\}$, involves an integration with respect to a complex variable, and this is beyond the scope of this book. Therefore, instead of using this formula, we will derive several elegant properties of the Laplace transform operator in the next section. These properties will enable us to invert many Laplace transforms by inspection; that is, by recognizing "which functions they are the Laplace transform of".

Example 4. Solve the initial-value problem

$$\frac{d^2y}{dt^2}-3\frac{dy}{dt}+2y=e^{3t}; \qquad y(0)=1, \quad y'(0)=0.$$

Solution. Let $Y(s)=\mathcal{L}\{y(t)\}$. Taking Laplace transforms of both sides of the differential equation gives

$$s^2Y(s)-s-3\left[sY(s)-1\right]+2Y(s)=\frac{1}{s-3}$$

and this implies that

$$\begin{aligned} Y(s)&=\frac{1}{(s-3)(s^2-3s+2)}+\frac{s-3}{s^2-3s+2}\\ &=\frac{1}{(s-1)(s-2)(s-3)}+\frac{s-3}{(s-1)(s-2)}. \end{aligned} \tag{5}$$

To find $y(t)$, we expand each term on the right-hand side of (5) in partial fractions. Thus, we write

$$\frac{1}{(s-1)(s-2)(s-3)}=\frac{A}{s-1}+\frac{B}{s-2}+\frac{C}{s-3}.$$

This implies that

$$A(s-2)(s-3)+B(s-1)(s-3)+C(s-1)(s-2)=1. \tag{6}$$

Setting $s=1$ in (6) gives $A=\frac{1}{2}$; setting $s=2$ gives $B=-1$; and setting $s=3$ gives $C=\frac{1}{2}$. Hence,

$$\frac{1}{(s-1)(s-2)(s-3)}=\frac{1}{2}\frac{1}{s-1}-\frac{1}{s-2}+\frac{1}{2}\frac{1}{s-3}.$$

Similarly, we write

$$\frac{s-3}{(s-1)(s-2)}=\frac{D}{s-1}+\frac{E}{s-2}$$

and this implies that

$$D(s-2)+E(s-1)=s-3. \tag{7}$$

Setting $s=1$ in (7) gives $D=2$, while setting $s=2$ gives $E=-1$. Hence,

$$\begin{aligned} Y(s)&=\frac{1}{2}\frac{1}{s-1}-\frac{1}{s-2}+\frac{1}{2}\frac{1}{s-3}+\frac{2}{s-1}-\frac{1}{s-2}\\ &=\frac{5}{2}\frac{1}{s-1}-\frac{2}{s-2}+\frac{1}{2}\frac{1}{s-3}. \end{aligned}$$

Now, we recognize the first term as being the Laplace transform of $\frac{5}{2}e^{t}$. Similarly, we recognize the second and third terms as being the Laplace transforms of $-2e^{2t}$ and $\frac{1}{2}e^{3t}$, respectively. Therefore,

$$Y(s)=\mathcal{L}\left\{\tfrac{5}{2}e^{t}-2e^{2t}+\tfrac{1}{2}e^{3t}\right\}$$

so that

$$y(t)=\tfrac{5}{2}e^{t}-2e^{2t}+\tfrac{1}{2}e^{3t}.$$

Remark. We have cheated a little bit in this problem because there are actually infinitely many functions whose Laplace transform is a given function. For example, the Laplace transform of the function

$$z(t)=\begin{cases} \frac{5}{2}e^{t}-2e^{2t}+\frac{1}{2}e^{3t}, & t\neq 1,\ 2, \text{ and } 3\\ 0, & t=1,2,3 \end{cases}$$

is also $Y(s)$, since $z(t)$ differs from $y(t)$ at only three points.* However, there is only one *continuous* function $y(t)$ whose Laplace transform is a given function $Y(s)$, and it is in this sense that we write $y(t)=\mathcal{L}^{-1}\{Y(s)\}$.

We wish to emphasize that Example 4 is just by way of illustrating the method of Laplace transforms for solving initial-value problems. The best way of solving this particular initial-value problem is by the method of

*If $f(t)=g(t)$ except at a finite number of points, then $\int_a^b f(t)\,dt=\int_a^b g(t)\,dt$.

judicious guessing. However, even though it is longer to solve this particular initial-value problem by the method of Laplace transforms, there is still something "nice and satisfying" about this method. If we had done this problem by the method of judicious guessing, we would have first computed a particular solution $\psi(t)=\frac{1}{2}e^{3t}$. Then, we would have found two independent solutions e^t and e^{2t} of the homogeneous equation, and we would have written

$$y(t)=c_1e^t+c_2e^{2t}+\tfrac{1}{2}e^{3t}$$

as the general solution of the differential equation. Finally, we would have computed $c_1=\frac{5}{2}$ and $c_2=-2$ from the initial conditions. What is unsatisfying about this method is that we first had to find *all* the solutions of the differential equation before we could find the specific solution $y(t)$ which we were interested in. The method of Laplace transforms, on the other hand, enables us to find $y(t)$ directly, without first finding all solutions of the differential equation.

Exercises

Determine the Laplace transform of each of the following functions.

1. t

2. t^n

3. $e^{at}\cos bt$

4. $e^{at}\sin bt$

5. $\cos^2 at$

6. $\sin^2 at$

7. $\sin at\cos bt$

8. $t^2\sin t$

9. Given that $\int_0^\infty e^{-x^2}dx=\sqrt{\pi}\,/2$, find $\mathcal{L}\{t^{-1/2}\}$. *Hint*: Make the change of variable $u=\sqrt{t}$ in (2).

Show that each of the following functions are of exponential order.

10. t^n

11. $\sin at$

12. $e^{\sqrt{t}}$

13. Show that e^{t^2} does not possess a Laplace transform. *Hint*: Show that $e^{t^2-st}>e^t$ for $t>s+1$.

14. Suppose that $f(t)$ is of exponential order. Show that $F(s)=\mathcal{L}\{f(t)\}$ approaches 0 as $s\to\infty$.

Solve each of the following initial-value problems.

15. $y''-5y'+4y=e^{2t};\quad y(0)=1,\ y'(0)=-1$

16. $2y''+y'-y=e^{3t};\quad y(0)=2,\ y'(0)=0$

Find the Laplace transform of the solution of each of the following initial-value problems.

17. $y''+2y'+y=e^{-t};\quad y(0)=1,\ y'(0)=3$

18. $y''+y=t^2\sin t;\quad y(0)=y'(0)=0$

19. $y''+3y'+7y=\cos t;\quad y(0)=0,\ y'(0)=2$

20. $y''+y'+y=t^3;\quad y(0)=2,\ y'(0)=0$

21. Prove that all solutions $y(t)$ of $ay''+by'+cy=f(t)$ are of exponential order if $f(t)$ is of exponential order. *Hint*: Show that all solutions of the homogeneous equation are of exponential order. Obtain a particular solution using the method of variation of parameters, and show that it, too, is of exponential order.

22. Let $F(s)=\mathcal{L}\{f(t)\}$. Prove that

$$\mathcal{L}\left\{\frac{d^nf(t)}{dt^n}\right\}=s^nF(s)-s^{n-1}f(0)-\ldots-\frac{df^{(n-1)}(0)}{dt^{n-1}}.$$

Hint: Try induction.

23. Solve the initial-value problem

$$y'''-6y''+11y'-6y=e^{4t};\qquad y(0)=y'(0)=y''(0)=0$$

24. Solve the initial-value problem

$$y''-3y'+2y=e^{-t};\qquad y(t_0)=1,\quad y'(t_0)=0$$

by the method of Laplace transforms. *Hint*: Let $\phi(t)=y(t+t_0)$.

2.10 Some useful properties of Laplace transforms

In this section we derive several important properties of Laplace transforms. Using these properties, we will be able to compute the Laplace transform of most functions without performing tedious integrations, and to invert many Laplace transforms by inspection.

Property 1. If $\mathcal{L}\{f(t)\}=F(s)$, then

$$\mathcal{L}\{-tf(t)\}=\frac{d}{ds}F(s).$$

PROOF. By definition, $F(s)=\int_0^\infty e^{-st}f(t)\,dt$. Differentiating both sides of this equation with respect to s gives

$$\begin{aligned}\frac{d}{ds}F(s)&=\frac{d}{ds}\int_0^\infty e^{-st}f(t)\,dt\\&=\int_0^\infty\frac{\partial}{\partial s}(e^{-st})f(t)\,dt=\int_0^\infty -te^{-st}f(t)\,dt\\&=\mathcal{L}\{-tf(t)\}.\end{aligned}$$

□

Property 1 states that the Laplace transform of the function $-tf(t)$ is the derivative of the Laplace transform of $f(t)$. Thus, if we know the Laplace transform $F(s)$ of $f(t)$, then, we don't have to perform a tedious integration to find the Laplace transform of $tf(t)$; we need only differentiate $F(s)$ and multiply by -1.

Example 1. Compute the Laplace transform of te^t.
Solution. The Laplace transform of e^t is $1/(s-1)$. Hence, by Property 1, the Laplace transform of te^t is

$$\mathcal{L}\{te^t\} = -\frac{d}{ds}\frac{1}{s-1} = \frac{1}{(s-1)^2}.$$

Example 2. Compute the Laplace transform of t^{13}.
Solution. Using Property 1 thirteen times gives

$$\mathcal{L}\{t^{13}\} = (-1)^{13}\frac{d^{13}}{ds^{13}}\mathcal{L}\{1\} = (-1)^{13}\frac{d^{13}}{ds^{13}}\frac{1}{s} = \frac{(13)!}{s^{14}}.$$

The main usefulness of Property 1 is in inverting Laplace transforms, as the following examples illustrate.

Example 3. What function has Laplace transform $-1/(s-2)^2$?
Solution. Observe that

$$-\frac{1}{(s-2)^2} = \frac{d}{ds}\frac{1}{s-2} \quad \text{and} \quad \frac{1}{s-2} = \mathcal{L}\{e^{2t}\}.$$

Hence, by Property 1,

$$\mathcal{L}^{-1}\left\{-\frac{1}{(s-2)^2}\right\} = -te^{2t}.$$

Example 4. What function has Laplace transform $-4s/(s^2+4)^2$?
Solution. Observe that

$$-\frac{4s}{(s^2+4)^2} = \frac{d}{ds}\frac{2}{s^2+4} \quad \text{and} \quad \frac{2}{s^2+4} = \mathcal{L}\{\sin 2t\}.$$

Hence, by Property 1,

$$\mathcal{L}^{-1}\left\{-\frac{4s}{(s^2+4)^2}\right\} = -t\sin 2t.$$

Example 5. What function has Laplace transform $1/(s-4)^3$?

Solution. We recognize that

$$\frac{1}{(s-4)^3}=\frac{d^2}{ds^2}\frac{1}{2}\frac{1}{s-4}.$$

Hence, using Property 1 twice, we see that

$$\frac{1}{(s-4)^3}=\mathcal{L}\left\{\frac{1}{2}t^2e^{4t}\right\}.$$

Property 2. If $F(s)=\mathcal{L}\{f(t)\}$, then

$$\mathcal{L}\left\{e^{at}f(t)\right\}=F(s-a).$$

PROOF. By definition,

$$\mathcal{L}\left\{e^{at}f(t)\right\}=\int_0^\infty e^{-st}e^{at}f(t)\,dt=\int_0^\infty e^{(a-s)t}f(t)\,dt$$

$$=\int_0^\infty e^{-(s-a)t}f(t)\,dt\equiv F(s-a).$$ □

Property 2 states that the Laplace transform of $e^{at}f(t)$ evaluated at the point s equals the Laplace transform of $f(t)$ evaluated at the point $(s-a)$. Thus, if we know the Laplace transform $F(s)$ of $f(t)$, then we don't have to perform an integration to find the Laplace transform of $e^{at}f(t)$; we need only replace every s in $F(s)$ by $s-a$.

Example 6. Compute the Laplace transform of $e^{3t}\sin t$.
Solution. The Laplace transform of $\sin t$ is $1/(s^2+1)$. Therefore, to compute the Laplace transform of $e^{3t}\sin t$, we need only replace every s by $s-3$; that is,

$$\mathcal{L}\{e^{3t}\sin t\}=\frac{1}{(s-3)^2+1}.$$

The real usefulness of Property 2 is in inverting Laplace transforms, as the following examples illustrate.

Example 7. What function $g(t)$ has Laplace transform

$$G(s)=\frac{s-7}{25+(s-7)^2}?$$

Solution. Observe that

$$F(s)=\frac{s}{s^2+5^2}=\mathcal{L}\{\cos 5t\}$$

and that $G(s)$ is obtained from $F(s)$ by replacing every s by $s-7$. Hence, by Property 2,

$$\frac{s-7}{(s-7)^2+25}=\mathcal{L}\{e^{7t}\cos 5t\}.$$

Example 8. What function has Laplace transform $1/(s^2-4s+9)$?
Solution. One way of solving this problem is to expand $1/(s^2-4s+9)$ in partial fractions. A much better way is to complete the square of s^2-4s+9. Thus, we write

$$\frac{1}{s^2-4s+9}=\frac{1}{s^2-4s+4+(9-4)}=\frac{1}{(s-2)^2+5}.$$

Now,

$$\frac{1}{s^2+5}=\mathcal{L}\left\{\frac{1}{\sqrt{5}}\sin\sqrt{5}\,t\right\}.$$

Hence, by Property 2,

$$\frac{1}{s^2-4s+9}=\frac{1}{(s-2)^2+5}=\mathcal{L}\left\{\frac{1}{\sqrt{5}}e^{2t}\sin\sqrt{5}\,t\right\}.$$

Example 9. What function has Laplace transform $s/(s^2-4s+9)$?
Solution. Observe that

$$\frac{s}{s^2-4s+9}=\frac{s-2}{(s-2)^2+5}+\frac{2}{(s-2)^2+5}.$$

The function $s/(s^2+5)$ is the Laplace transform of $\cos\sqrt{5}\,t$. Therefore, by Property 2,

$$\frac{s-2}{(s-2)^2+5}=\mathcal{L}\{e^{2t}\cos\sqrt{5}\,t\},$$

and

$$\frac{s}{s^2-4s+9}=\mathcal{L}\left\{e^{2t}\cos\sqrt{5}\,t+\frac{2}{\sqrt{5}}e^{2t}\sin\sqrt{5}\,t\right\}.$$

In the previous section we showed that the Laplace transform is a linear operator; that is

$$\mathcal{L}\{c_1f_1(t)+c_2f_2(t)\}=c_1\mathcal{L}\{f_1(t)\}+c_2\mathcal{L}\{f_2(t)\}.$$

Thus, if we know the Laplace transforms $F_1(s)$ and $F_2(s)$, of $f_1(t)$ and $f_2(t)$, then we don't have to perform any integrations to find the Laplace transform of a linear combination of $f_1(t)$ and $f_2(t)$; we need only take the same linear combination of $F_1(s)$ and $F_2(s)$. For example, two functions which appear quite often in the study of differential equations are the hyperbolic cosine and hyperbolic sine functions. These functions are defined by the equations

$$\cosh at=\frac{e^{at}+e^{-at}}{2},\qquad \sinh at=\frac{e^{at}-e^{-at}}{2}.$$

Therefore, by the linearity of the Laplace transform,

$$\mathcal{L}\{\cosh at\} = \frac{1}{2}\mathcal{L}\{e^{at}\} + \frac{1}{2}\mathcal{L}\{e^{-at}\}$$
$$= \frac{1}{2}\left[\frac{1}{s-a} + \frac{1}{s+a}\right] = \frac{s}{s^2-a^2}$$

and

$$\mathcal{L}\{\sinh at\} = \frac{1}{2}\mathcal{L}\{e^{at}\} - \frac{1}{2}\mathcal{L}\{e^{-at}\}$$
$$= \frac{1}{2}\left[\frac{1}{s-a} - \frac{1}{s+a}\right] = \frac{a}{s^2-a^2}.$$

Exercises

Use Properties 1 and 2 to find the Laplace transform of each of the following functions.

1. t^n **2.** $t^n e^{at}$ **3.** $t\sin at$ **4.** $t^2\cos at$

5. $t^{5/2}$ (see Exercise 9, Section 2.9)

6. Let $F(s) = \mathcal{L}\{f(t)\}$, and suppose that $f(t)/t$ has a limit as t approaches zero. Prove that

$$\mathcal{L}\{f(t)/t\} = \int_s^\infty F(u)\,du. \qquad (*)$$

(The assumption that $f(t)/t$ has a limit as $t \to 0$ guarantees that the integral on the right-hand side of (*) exists.)

7. Use Equation (*) of Problem 6 to find the Laplace transform of each of the following functions:

(a) $\dfrac{\sin t}{t}$ (b) $\dfrac{\cos at - 1}{t}$ (c) $\dfrac{e^{at}-e^{bt}}{t}$

Find the inverse Laplace transform of each of the following functions. In several of these problems, it will be helpful to write the functions

$$p_1(s) = \frac{\alpha_1 s^3 + \beta_1 s^2 + \gamma_1 s + \delta_1}{(as^2+bs+c)(ds^2+es+f)} \quad \text{and} \quad p_2(s) = \frac{\alpha_1 s^2 + \beta_1 s + \gamma}{(as+b)(cs^2+ds+e)}$$

in the simpler form

$$p_1(s) = \frac{As+B}{as^2+bs+c} + \frac{Cs+D}{ds^2+es+f} \quad \text{and} \quad p_2(s) = \frac{A}{as+b} + \frac{Cs+D}{cs^2+ds+e}.$$

8. $\dfrac{s}{(s+a)^2+b^2}$ **9.** $\dfrac{s^2-5}{s^3+4s^2+3s}$

10. $\dfrac{1}{s(s^2+4)}$ **11.** $\dfrac{s}{s^2-3s-12}$

12. $\dfrac{1}{(s^2+a^2)(s^2+b^2)}$ **13.** $\dfrac{3s}{(s+1)^4}$

14. $\dfrac{1}{s(s+4)^2}$

15. $\dfrac{s}{(s+1)^2(s^2+1)}$

16. $\dfrac{1}{(s^2+1)^2}$

17. Let $F(s)=\mathcal{L}\{f(t)\}$. Show that

$$f(t)=-\frac{1}{t}\mathcal{L}^{-1}\{F'(s)\}.$$

Thus, if we know how to invert $F'(s)$, then we can also invert $F(s)$.

18. Use the result of Problem 17 to invert each of the following Laplace transforms

(a) $\ln\left(\dfrac{s+a}{s-a}\right)$ (b) $\arctan\dfrac{a}{s}$ (c) $\ln\left(1-\dfrac{a^2}{s^2}\right)$

Solve each of the following initial-value problems by the method of Laplace transforms.

19. $y''+y=\sin t;\quad y(0)=1, y'(0)=2$

20. $y''+y=t\sin t;\quad y(0)=1, y'(0)=2$

21. $y''-2y'+y=te^t;\quad y(0)=0, y'(0)=0$

22. $y''-2y'+7y=\sin t;\quad y(0)=0, y'(0)=0$

23. $y''+y'+y=1+e^{-t};\quad y(0)=3, y'(0)=-5$

24. $y''+y=\begin{cases} 2, & 0\leqslant t\leqslant 3 \\ 3t-7, & 3<t<\infty \end{cases};\quad y(0)=0, y'(0)=0$

2.11 Differential equations with discontinuous right-hand sides

In many applications, the right-hand side of the differential equation $ay''+by'+cy=f(t)$ has a jump discontinuity at one or more points. For example, a particle may be moving under the influence of a force $f_1(t)$, and suddenly, at time t_1, an additional force $f_2(t)$ is applied to the particle. Such equations are often quite tedious and cumbersome to solve, using the methods developed in Sections 2.4 and 2.5. In this section we show how to handle such problems by the method of Laplace transforms. We begin by computing the Laplace transform of several simple discontinuous functions.

The simplest example of a function with a single jump discontinuity is the function

$$H_c(t)=\begin{cases} 0, & 0\leqslant t<c \\ 1, & t\geqslant c \end{cases}.$$

This function, whose graph is given in Figure 1, is often called the unit step

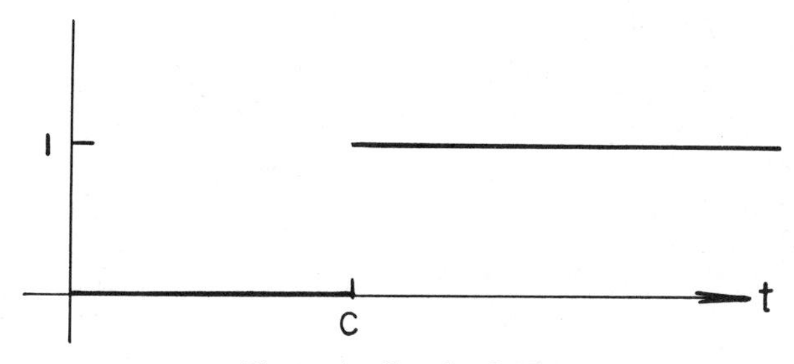

Figure 1. Graph of $H_c(t)$

function, or the Heaviside function. Its Laplace transform is

$$\begin{aligned}\mathcal{L}\{H_c(t)\} &= \int_0^\infty e^{-st}H_c(t)\,dt = \int_c^\infty e^{-st}\,dt\\ &= \lim_{A\to\infty}\int_c^A e^{-st}\,dt = \lim_{A\to\infty}\frac{e^{-cs}-e^{-sA}}{s}\\ &= \frac{e^{-cs}}{s}, \qquad s>0.\end{aligned}$$

Next, let f be any function defined on the interval $0\leqslant t<\infty$, and let g be the function obtained from f by moving the graph of f over c units to the right, as shown in Figure 2. More precisely, $g(t)=0$ for $0\leqslant t<c$, and $g(t)=f(t-c)$ for $t\geqslant c$. For example, if $c=2$ then the value of g at $t=7$ is the value of f at $t=5$. A convenient analytical expression for $g(t)$ is

$$g(t)=H_c(t)f(t-c).$$

The factor $H_c(t)$ makes g zero for $0\leqslant t<c$, and replacing the argument t of f by $t-c$ moves f over c units to the right. Since $g(t)$ is obtained in a simple manner from $f(t)$, we would expect that its Laplace transform can also be obtained in a simple manner from the Laplace transform of $f(t)$. This is indeed the case, as we now show.

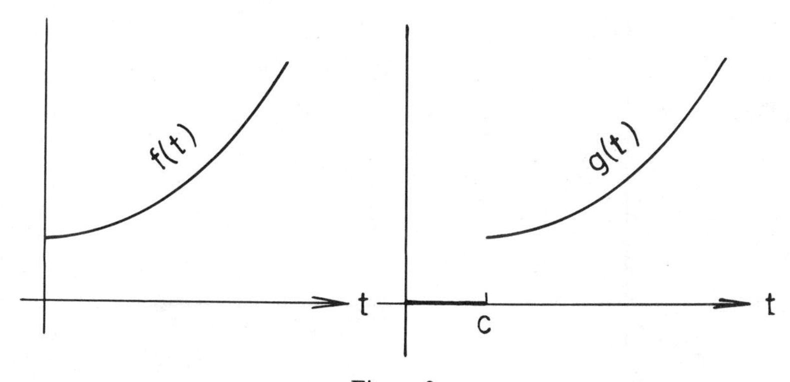

Figure 2

Property 3. Let $F(s)=\mathcal{L}\{f(t)\}$. Then,

$$\mathcal{L}\{H_c(t)f(t-c)\}=e^{-cs}F(s).$$

PROOF. By definition,

$$\begin{aligned}\mathcal{L}\{H_c(t)f(t-c)\}&=\int_0^\infty e^{-st}H_c(t)f(t-c)\,dt\\&=\int_c^\infty e^{-st}f(t-c)\,dt.\end{aligned}$$

This integral suggests the substitution

$$\xi=t-c.$$

Then,

$$\begin{aligned}\int_c^\infty e^{-st}f(t-c)\,dt&=\int_0^\infty e^{-s(\xi+c)}f(\xi)\,d\xi\\&=e^{-cs}\int_0^\infty e^{-s\xi}f(\xi)\,d\xi\\&=e^{-cs}F(s).\end{aligned}$$

Hence, $\mathcal{L}\{H_c(t)f(t-c)\}=e^{-cs}\mathcal{L}\{f(t)\}$. □

Example 1. What function has Laplace transform e^{-s}/s^2?
Solution. We know that $1/s^2$ is the Laplace transform of the function t. Hence, by Property 3

$$\frac{e^{-s}}{s^2}=\mathcal{L}\{H_1(t)(t-1)\}.$$

The graph of $H_1(t)(t-1)$ is given in Figure 3.

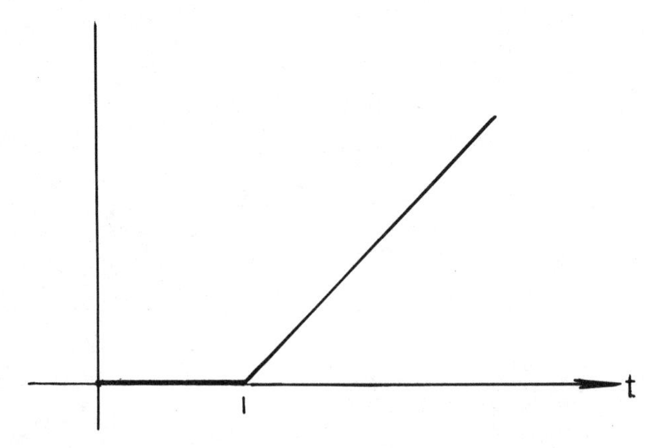

Figure 3. Graph of $H_1(t)\,(t-1)$

Example 2. What function has Laplace transform $e^{-3s}/(s^2-2s-3)$?
Solution. Observe that

$$\frac{1}{s^2-2s-3}=\frac{1}{s^2-2s+1-4}=\frac{1}{(s-1)^2-2^2}.$$

Since $1/(s^2-2^2)=\mathcal{L}\{\frac{1}{2}\sinh 2t\}$, we conclude from Property 2 that

$$\frac{1}{(s-1)^2-2^2}=\mathcal{L}\left\{\frac{1}{2}e^t\sinh 2t\right\}.$$

Consequently, from Property 3,

$$\frac{e^{-3s}}{s^2-2s-3}=\mathcal{L}\left\{\frac{1}{2}H_3(t)e^{t-3}\sinh 2(t-3)\right\}.$$

Example 3. Let $f(t)$ be the function which is t for $0\leqslant t<1$, and 0 for $t\geqslant 1$. Find the Laplace transform of f without performing any integrations.
Solution. Observe that $f(t)$ can be written in the form

$$f(t)=t[H_0(t)-H_1(t)]=t-tH_1(t).$$

Hence, from Property 1,

$$\begin{aligned}\mathcal{L}\{f(t)\}&=\mathcal{L}\{t\}-\mathcal{L}\{tH_1(t)\}\\&=\frac{1}{s^2}+\frac{d}{ds}\frac{e^{-s}}{s}=\frac{1}{s^2}-\frac{e^{-s}}{s}-\frac{e^{-s}}{s^2}.\end{aligned}$$

Example 4. Solve the initial-value problem

$$\frac{d^2y}{dt^2}-3\frac{dy}{dt}+2y=f(t)=\begin{cases}1, & 0\leqslant t<1; \quad 0, \ 1\leqslant t<2;\\ 1, & 2\leqslant t<3; \quad 0, \ 3\leqslant t<4;\\ 1, & 4\leqslant t<5; \quad 0, \ 5\leqslant t<\infty.\end{cases} \qquad y(0)=0, \quad y'(0)=0$$

Solution. Let $Y(s)=\mathcal{L}\{y(t)\}$ and $F(s)=\mathcal{L}\{f(t)\}$. Taking Laplace transforms of both sides of the differential equation gives $(s^2-3s+2)Y(s)=F(s)$, so that

$$Y(s)=\frac{F(s)}{s^2-3s+2}=\frac{F(s)}{(s-1)(s-2)}.$$

One way of computing $F(s)$ is to write $f(t)$ in the form

$$f(t)=[H_0(t)-H_1(t)]+[H_2(t)-H_3(t)]+[H_4(t)-H_5(t)].$$

Hence, by the linearity of the Laplace transform

$$F(s)=\frac{1}{s}-\frac{e^{-s}}{s}+\frac{e^{-2s}}{s}-\frac{e^{-3s}}{s}+\frac{e^{-4s}}{s}-\frac{e^{-5s}}{s}.$$

A second way of computing $F(s)$ is to evaluate the integral

$$\int_0^\infty e^{-st}f(t)\,dt=\int_0^1 e^{-st}\,dt+\int_2^3 e^{-st}\,dt+\int_4^5 e^{-st}\,dt$$
$$=\frac{1-e^{-s}}{s}+\frac{e^{-2s}-e^{-3s}}{s}+\frac{e^{-4s}-e^{-5s}}{s}.$$

Consequently,

$$Y(s)=\frac{1-e^{-s}+e^{-2s}-e^{-3s}+e^{-4s}-e^{-5s}}{s(s-1)(s-2)}.$$

Our next step is to expand $1/s(s-1)(s-2)$ in partial fractions; i.e., we write

$$\frac{1}{s(s-1)(s-2)}=\frac{A}{s}+\frac{B}{s-1}+\frac{C}{s-2}.$$

This implies that

$$A(s-1)(s-2)+Bs(s-2)+Cs(s-1)=1. \tag{1}$$

Setting $s=0$ in (1) gives $A=\frac{1}{2}$; setting $s=1$ gives $B=-1$; and setting $s=2$ gives $C=\frac{1}{2}$. Thus,

$$\frac{1}{s(s-1)(s-2)}=\frac{1}{2}\frac{1}{s}-\frac{1}{s-1}+\frac{1}{2}\frac{1}{s-2}$$
$$=\mathcal{L}\left\{\frac{1}{2}-e^t+\frac{1}{2}e^{2t}\right\}.$$

Consequently, from Property 3,

$$\begin{aligned}y(t)=&\left[\tfrac{1}{2}-e^t+\tfrac{1}{2}e^{2t}\right]-H_1(t)\left[\tfrac{1}{2}-e^{(t-1)}+\tfrac{1}{2}e^{2(t-1)}\right]\\&+H_2(t)\left[\tfrac{1}{2}-e^{(t-2)}+\tfrac{1}{2}e^{2(t-2)}\right]-H_3(t)\left[\tfrac{1}{2}-e^{(t-3)}+\tfrac{1}{2}e^{2(t-3)}\right]\\&+H_4(t)\left[\tfrac{1}{2}-e^{(t-4)}+\tfrac{1}{2}e^{2(t-4)}\right]-H_5(t)\left[\tfrac{1}{2}-e^{(t-5)}+\tfrac{1}{2}e^{2(t-5)}\right].\end{aligned}$$

Remark. It is easily verified that the function

$$\tfrac{1}{2}-e^{(t-n)}+\tfrac{1}{2}e^{2(t-n)}$$

and its derivative are both zero at $t=n$. Hence, both $y(t)$ and $y'(t)$ are continuous functions of time, even though $f(t)$ is discontinuous at $t=1, 2, 3, 4$, and 5. More generally, both the solution $y(t)$ of the initial-value problem

$$a\frac{d^2y}{dt^2}+b\frac{dy}{dt}+cy=f(t);\qquad y(t_0)=y_0,\quad y'(t_0)=y_0'$$

and its derivative $y'(t)$ are always continuous functions of time, if $f(t)$ is piecewise continuous. We will indicate the proof of this result in Section 2.12.

Exercises

Find the solution of each of the following initial-value problems.

1. $y''+2y'+y=2(t-3)H_3(t);\quad y(0)=2, y'(0)=1$

2. $y''+y'+y=H_\pi(t)-H_{2\pi}(t);\quad y(0)=1, y'(0)=0$

3. $y''+4y=\begin{cases}1, & 0\leqslant t<4\\ 0, & t>4\end{cases};\quad y(0)=3, y'(0)=-2$

4. $y''+y=\begin{cases}\sin t, & 0\leqslant t<\pi\\ \cos t, & \pi\leqslant t<\infty\end{cases};\quad y(0)=1, y'(0)=0$

5. $y''+y=\begin{cases}\cos t, & 0\leqslant t<\pi/2\\ 0, & \pi/2\leqslant t<\infty\end{cases};\quad y(0)=3, y'(0)=-1$

6. $y''+2y'+y=\begin{cases}\sin 2t, & 0\leqslant t<\pi/2\\ 0, & \pi/2\leqslant t<\infty\end{cases};\quad y(0)=1, y'(0)=0$

7. $y''+y'+7y=\begin{cases}t, & 0\leqslant t<2\\ 0, & 2\leqslant t<\infty\end{cases};\quad y(0)=0, y'(0)=0$

8. $y''+y=\begin{cases}t^2, & 0\leqslant t<1\\ 0, & 1\leqslant t<\infty\end{cases};\quad y(0)=0, y'(0)=0$

9. $y''-2y'+y=\begin{cases}0, & 0\leqslant t<1\\ t, & 1\leqslant t<2\\ 0, & 2\leqslant t<\infty\end{cases};\quad y(0)=0, y'(0)=1$

10. Find the Laplace transform of $|\sin t|$. *Hint*: Observe that

$$|\sin t|=\sin t+2\sum_{n=1}^{\infty} H_{n\pi}(t)\sin(t-n\pi).$$

11. Solve the initial-value problem of Example 4 by the method of judicious guessing. *Hint*: Find the general solution of the differential equation in each of the intervals $0<t<1$, $1<t<2$, $2<t<3$, $3<t<4$, $4<t<5$, $5<t<\infty$, and choose the arbitrary constants so that $y(t)$ and $y'(t)$ are continuous at the points $t=1$, 2, 3, 4, and 5.

2.12 The Dirac delta function

In many physical and biological applications we are often confronted with an initial-value problem

$$a\frac{d^2y}{dt^2}+b\frac{dy}{dt}+cy=f(t);\qquad y(0)=y_0,\quad y'(0)=y_0' \tag{1}$$

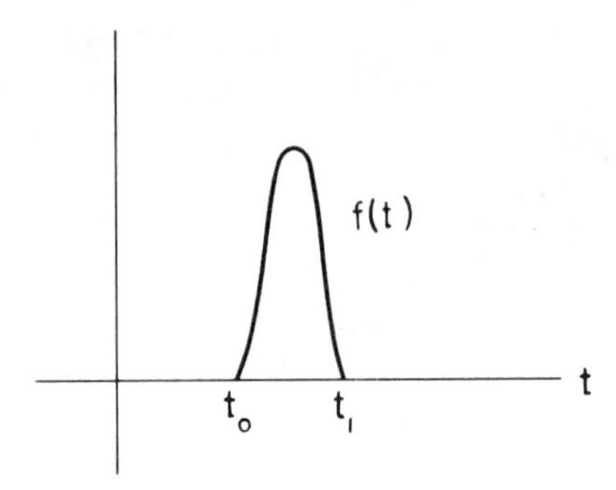

Figure 1. The graph of a typical impulsive function $f(t)$

where we do not know $f(t)$ explicitly. Such problems usually arise when we are dealing with phenomena of an impulsive nature. In these situations, the only information we have about $f(t)$ is that it is identically zero except for a very short time interval $t_0 \leqslant t \leqslant t_1$, and that its integral over this time interval is a given number $I_0 \neq 0$. If I_0 is not very small, then $f(t)$ will be quite large in the interval $t_0 \leqslant t \leqslant t_1$. Such functions are called impulsive functions, and the graph of a typical $f(t)$ is given in Figure 1.

In the early 1930's the Nobel Prize winning physicist P. A. M. Dirac developed a very controversial method for dealing with impulsive functions. His method is based on the following argument. Let t_1 get closer and closer to t_0. Then the function $f(t)/I_0$ approaches the function which is 0 for $t \neq t_0$, and ∞ for $t = t_0$, and whose integral over any interval containing t_0 is 1. We will denote this function, which is known as the Dirac delta function, by $\delta(t-t_0)$. Of course, $\delta(t-t_0)$ is not an ordinary function. However, says Dirac, let us formally operate with $\delta(t-t_0)$ as if it really were an ordinary function. Then, if we set $f(t) = I_0\delta(t-t_0)$ in (1) and impose the condition

$$\int_a^b g(t)\delta(t-t_0)\,dt = \begin{cases} g(t_0) & \text{if } a \leqslant t_0 \leqslant b \\ 0 & \text{otherwise} \end{cases} \tag{2}$$

for any continuous function $g(t)$, we will always obtain the correct solution $y(t)$.

Remark. Equation (2) is certainly a very reasonable condition to impose on $\delta(t-t_0)$. To see this, suppose that $f(t)$ is an impulsive function which is positive for $t_0 < t < t_1$, zero otherwise, and whose integral over the interval $[t_0, t_1]$ is 1. For any continuous function $g(t)$,

$$\Big[\min_{t_0 \leqslant t \leqslant t_1} g(t)\Big] f(t) \leqslant g(t) f(t) \leqslant \Big[\max_{t_0 \leqslant t \leqslant t_1} g(t)\Big] f(t).$$

Consequently,

$$\int_{t_0}^{t_1} \Big[\min_{t_0 \leqslant t \leqslant t_1} g(t)\Big] f(t)\,dt \leqslant \int_{t_0}^{t_1} g(t) f(t)\,dt \leqslant \int_{t_0}^{t_1} \Big[\max_{t_0 \leqslant t \leqslant t_1} g(t)\Big] f(t)\,dt,$$

or

$$\min_{t_0 \leqslant t \leqslant t_1} g(t) \leqslant \int_{t_0}^{t_1} g(t) f(t)\, dt \leqslant \max_{t_0 \leqslant t \leqslant t_1} g(t).$$

Thus, as $t_1 \to t_0$, $\int_{t_0}^{t_1} g(t) f(t)\, dt \to g(t_0)$.

Now, most mathematicians, of course, usually ridiculed this method. "How can you make believe that $\delta(t-t_0)$ is an ordinary function if it is obviously not," they asked. However, they never laughed too loud since Dirac and his followers always obtained the right answer. In the late 1940's, in one of the great success stories of mathematics, the French mathematician Laurent Schwartz succeeded in placing the delta function on a firm mathematical foundation. He accomplished this by enlarging the class of all functions so as to include the delta function. In this section we will first present a physical justification of the method of Dirac. Then we will illustrate how to solve the initial-value problem (1) by the method of Laplace transforms. Finally, we will indicate very briefly the "germ" of Laurent Schwartz's brilliant idea.

Physical justification of the method of Dirac. Newton's second law of motion is usually written in the form

$$\frac{d}{dt} mv(t) = f(t) \tag{3}$$

where m is the mass of the particle, v is its velocity, and $f(t)$ is the total force acting on the particle. The quantity mv is called the momentum of the particle. Integrating Equation (3) between t_0 and t_1 gives

$$mv(t_1) - mv(t_0) = \int_{t_0}^{t_1} f(t)\, dt.$$

This equation says that the change in momentum of the particle from time t_0 to time t_1 equals $\int_{t_0}^{t_1} f(t)\, dt$. Thus, the physically important quantity is the integral of the force, which is known as the impulse imparted by the force, rather than the force itself. Now, we may assume that $a > 0$ in Equation (1), for otherwise we can multiply both sides of the equation by -1 to obtain $a > 0$. In this case (see Section 2.6) we can view $y(t)$, for $t \leqslant t_0$, as the position at time t of a particle of mass a moving under the influence of the force $-b(dy/dt) - cy$. At time t_0 a force $f(t)$ is applied to the particle, and this force acts over an extremely short time interval $t_0 \leqslant t \leqslant t_1$. Since the time interval is extremely small, we may assume that the position of the particle does not change while the force $f(t)$ acts. Thus the sum result of the impulsive force $f(t)$ is that the velocity of the particle jumps by an amount I_0/a at time t_0. In other words, $y(t)$ satisfies the initial-value problem

$$a\frac{d^2y}{dt^2} + b\frac{dy}{dt} + cy = 0; \qquad y(0) = y_0, \quad y'(0) = y_0'$$

for $0 \leqslant t < t_0$, and

$$a\frac{d^2y}{dt^2}+b\frac{dy}{dt}+cy=0; \qquad y(t_0)=z_0, \quad y'(t_0)=z_0'+\frac{I_0}{a} \tag{4}$$

for $t \geqslant t_0$, where z_0 and z_0' are the position and velocity of the particle just before the impulsive force acts. It is clear, therefore, that any method which correctly takes into account the momentum I_0 transferred to the particle at time t_0 by the impulsive force $f(t)$ must yield the correct answer. It is also clear that we always keep track of the momentum I_0 transferred to the particle by $f(t)$ if we replace $f(t)$ by $I_0\delta(t-t_0)$ and obey Equation (2). Hence the method of Dirac will always yield the correct answer.

Remark. We can now understand why any solution $y(t)$ of the differential equation

$$a\frac{d^2y}{dt^2}+b\frac{dy}{dt}+cy=f(t), \qquad f(t) \text{ a piecewise continuous function,}$$

is a continuous function of time even though $f(t)$ is discontinuous. To wit, since the integral of a piecewise continuous function is continuous, we see that $y'(t)$, must vary continuously with time. Consequently, $y(t)$ must also vary continuously with time.

Solution of Equation (1) *by the method of Laplace transforms*. In order to solve the initial-value problem (1) by the method of Laplace transforms, we need only know the Laplace transform of $\delta(t-t_0)$. This is obtained directly from the definition of the Laplace transform and Equation (2), for

$$\mathcal{L}\{\delta(t-t_0)\} \equiv \int_0^\infty e^{-st}\delta(t-t_0)\,dt = e^{-st_0} \qquad (\text{for } t_0 \geqslant 0).$$

Example 1. Find the solution of the initial-value problem

$$\frac{d^2y}{dt^2}-4\frac{dy}{dt}+4y=3\delta(t-1)+\delta(t-2); \qquad y(0)=1, \quad y'(0)=1.$$

Solution. Let $Y(s)=\mathcal{L}\{y(t)\}$. Taking Laplace transforms of both sides of the differential equation gives

$$s^2Y-s-1-4(sY-1)+4Y=3e^{-s}+e^{-2s}$$

or

$$(s^2-4s+4)Y(s)=s-3+3e^{-s}+e^{-2s}.$$

Consequently,

$$Y(s)=\frac{s-3}{(s-2)^2}+\frac{3e^{-s}}{(s-2)^2}+\frac{e^{-2s}}{(s-2)^2}.$$

Now, $1/(s-2)^2=\mathcal{L}\{te^{2t}\}$. Hence,

$$\frac{3e^{-s}}{(s-2)^2}+\frac{e^{-2s}}{(s-2)^2}=\mathcal{L}\{3H_1(t)(t-1)e^{2(t-1)}+H_2(t)(t-2)e^{2(t-2)}\}.$$

To invert the first term of $Y(s)$, observe that

$$\frac{s-3}{(s-2)^2}=\frac{s-2}{(s-2)^2}-\frac{1}{(s-2)^2}=\mathcal{L}\{e^{2t}\}-\mathcal{L}\{te^{2t}\}.$$

Thus, $y(t)=(1-t)e^{2t}+3H_1(t)(t-1)e^{2(t-1)}+H_2(t)(t-2)e^{2(t-2)}$.

It is instructive to do this problem the long way, that is, to find $y(t)$ separately in each of the intervals $0\leqslant t<1$, $1\leqslant t<2$ and $2\leqslant t<\infty$. For $0\leqslant t<1$, $y(t)$ satisfies the initial-value problem

$$\frac{d^2y}{dt^2}-4\frac{dy}{dt}+4y=0;\qquad y(0)=1,\quad y'(0)=1.$$

The characteristic equation of this differential equation is $r^2-4r+4=0$, whose roots are $r_1=r_2=2$. Hence, any solution $y(t)$ must be of the form $y(t)=(a_1+a_2t)e^{2t}$. The constants a_1 and a_2 are determined from the initial conditions

$$1=y(0)=a_1\quad\text{and}\quad 1=y'(0)=2a_1+a_2.$$

Hence, $a_1=1$, $a_2=-1$ and $y(t)=(1-t)e^{2t}$ for $0\leqslant t<1$. Now $y(1)=0$ and $y'(1)=-e^2$. At time $t=1$ the derivative of $y(t)$ is suddenly increased by 3. Consequently, for $1\leqslant t<2$, $y(t)$ satisfies the initial-value problem

$$\frac{d^2y}{dt^2}-4\frac{dy}{dt}+4y=0;\qquad y(1)=0,\quad y'(1)=3-e^2.$$

Since the initial conditions are given at $t=1$, we write this solution in the form $y(t)=[b_1+b_2(t-1)]e^{2(t-1)}$ (see Exercise 1). The constants b_1 and b_2 are determined from the initial conditions

$$0=y(1)=b_1\quad\text{and}\quad 3-e^2=y'(1)=2b_1+b_2.$$

Thus, $b_1=0$, $b_2=3-e^2$ and $y(t)=(3-e^2)(t-1)e^{2(t-1)}$, $1\leqslant t<2$. Now, $y(2)=(3-e^2)e^2$ and $y'(2)=3(3-e^2)e^2$. At time $t=2$ the derivative of $y(t)$ is suddenly increased by 1. Consequently, for $2\leqslant t<\infty$, $y(t)$ satisfies the initial-value problem

$$\frac{d^2y}{dt^2}-4\frac{dy}{dt}+4y=0;\qquad y(2)=e^2(3-e^2),\quad y'(2)=1+3e^2(3-e^2).$$

Hence $y(t)=[c_1+c_2(t-2)]e^{2(t-2)}$. The constants c_1 and c_2 are determined from the equations

$$e^2(3-e^2)=c_1\quad\text{and}\quad 1+3e^2(3-e^2)=2c_1+c_2.$$

Thus,

$$c_1=e^2(3-e^2), \qquad c_2=1+3e^2(3-e^2)-2e^2(3-e^2)=1+e^2(3-e^2)$$

and $y(t)=[e^2(3-e^2)+(1+e^2(3-e^2))(t-2)]e^{2(t-2)}$, $t\geqslant 2$. The reader should verify that this expression agrees with the expression obtained for $y(t)$ by the method of Laplace transforms.

Example 2. A particle of mass 1 is attached to a spring dashpot mechanism. The stiffness constant of the spring is 1 N/ft and the drag force exerted by the dashpot mechanism on the particle is twice its velocity. At time $t=0$, when the particle is at rest, an external force e^{-t} is applied to the system. At time $t=1$, an additional force $f(t)$ of very short duration is applied to the particle. This force imparts an impulse of 3 N·s to the particle. Find the position of the particle at any time t greater than 1.

Solution. Let $y(t)$ be the distance of the particle from its equilibrium position. Then, $y(t)$ satisfies the initial-value problem

$$\frac{d^2y}{dt^2}+2\frac{dy}{dt}+y=e^{-t}+3\delta(t-1); \qquad y(0)=0, \quad y'(0)=0.$$

Let $Y(s)=\mathcal{L}\{y(t)\}$. Taking Laplace transforms of both sides of the differential equation gives

$$(s^2+2s+1)Y(s)=\frac{1}{s+1}+3e^{-s}, \quad \text{or} \quad Y(s)=\frac{1}{(s+1)^3}+\frac{3e^{-s}}{(s+1)^2}.$$

Since

$$\frac{1}{(s+1)^3}=\mathcal{L}\left\{\frac{t^2e^{-t}}{2}\right\} \quad \text{and} \quad \frac{3e^{-s}}{(s+1)^2}=3\mathcal{L}\{H_1(t)(t-1)e^{-(t-1)}\}$$

we see that

$$y(t)=\frac{t^2e^{-t}}{2}+3H_1(t)(t-1)e^{-(t-1)}.$$

Consequently, $y(t)=\frac{1}{2}t^2e^{-t}+3(t-1)e^{-(t-1)}$ for $t>1$.

We conclude this section with a very brief description of Laurent Schwartz's method for placing the delta function on a rigorous mathematical foundation. The main step in his method is to rethink our notion of "function." In Calculus, we are taught to recognize a function by its value at each time t. A much more subtle (and much more difficult) way of recognizing a function is by what it does to other functions. More precisely, let f be a piecewise continuous function defined for $-\infty<t<\infty$. To each function ϕ which is infinitely often differentiable and which vanishes for $|t|$

sufficiently large, we assign a number $K[\phi]$ according to the formula

$$K[\phi]=\int_{-\infty}^{\infty}\phi(t)f(t)\,dt. \tag{5}$$

As the notation suggests, K is an operator acting on functions. However, it differs from the operators introduced previously in that it associates a number, rather than a function, with ϕ. For this reason, we say that $K[\phi]$ is a functional, rather than a function. Now, observe that the association $\phi\to K[\phi]$ is a linear association, since

$$\begin{aligned}K[c_1\phi_1+c_2\phi_2]&=\int_{-\infty}^{\infty}(c_1\phi_1+c_2\phi_2)(t)f(t)\,dt\\&=c_1\int_{-\infty}^{\infty}\phi_1(t)f(t)\,dt+c_2\int_{-\infty}^{\infty}\phi_2(t)f(t)\,dt\\&=c_1K[\phi_1]+c_2K[\phi_2].\end{aligned}$$

Hence every piecewise continuous function defines, through (5), a linear functional on the space of all infinitely often differentiable functions which vanish for $|t|$ sufficiently large.

Now consider the functional $K[\phi]$ defined by the relation $K[\phi]=\phi(t_0)$. K is a linear functional since

$$K[c_1\phi_1+c_2\phi_2]=c_1\phi_1(t_0)+c_2\phi_2(t_0)=c_1K[\phi_1]+c_2K[\phi_2].$$

To mimic (5), we write K symbolically in the form

$$K[\phi]=\int_{-\infty}^{\infty}\phi(t)\delta(t-t_0)\,dt. \tag{6}$$

In this sense, $\delta(t-t_0)$ is a "generalized function." It is important to realize though, that we cannot speak of the value of $\delta(t-t_0)$ at any time t. The only meaningful quantity is the expression $\int_{-\infty}^{\infty}\phi(t)\delta(t-t_0)\,dt$, and we must always assign the value $\phi(t_0)$ to this expression.

Admittedly, it is very difficult to think of a function in terms of the linear functional (5) that it induces. The advantage to this way of thinking, though, is that it is now possible to assign a derivative to every piecewise continuous function and to every "generalized function." To wit, suppose that $f(t)$ is a differentiable function. Then $f'(t)$ induces the linear functional

$$K'[\phi]=\int_{-\infty}^{\infty}\phi(t)f'(t)\,dt. \tag{7}$$

Integrating by parts and using the fact that $\phi(t)$ vanishes for $|t|$ sufficiently large, we see that

$$K'[\phi]=\int_{-\infty}^{\infty}[-\phi'(t)]f(t)\,dt=K[-\phi']. \tag{8}$$

Now, notice that the formula $K'[\phi]=K[-\phi']$ makes sense even if $f(t)$ is not differentiable. This motivates the following definition.

Definition. To every linear functional $K[\phi]$ we assign the new linear functional $K'[\phi]$ by the formula $K'[\phi]=K[-\phi']$. The linear functional $K'[\phi]$ is called the derivative of $K[\phi]$ since if $K[\phi]$ is induced by a differentiable function $f(t)$ then $K'[\phi]$ is induced by $f'(t)$.

Finally, we observe from (8) that the derivative of the delta function $\delta(t-t_0)$ is the linear functional which assigns to each function ϕ the number $-\phi'(t_0)$, for if $K[\phi]=\phi(t_0)$ then $K'[\phi]=K[-\phi']=-\phi'(t_0)$. Thus,

$$\int_{-\infty}^{\infty} \phi(t)\delta'(t-t_0)\,dt = -\phi'(t_0)$$

for all differentiable functions $\phi(t)$.

Exercises

1. Let a be a fixed constant. Show that every solution of the differential equation $(d^2y/dt^2)+2\alpha(dy/dt)+\alpha^2y=0$ can be written in the form

$$y(t)=[c_1+c_2(t-a)]e^{-\alpha(t-a)}.$$

2. Solve the initial-value problem $(d^2y/dt^2)+4(dy/dt)+5y=f(t)$; $y(0)=1, y'(0)=0$, where $f(t)$ is an impulsive force which acts on the extremely short time interval $1\leq t\leq 1+\tau$, and $\int_1^{1+\tau} f(t)\,dt=2$.

3. (a) Solve the initial-value problem $(d^2y/dt^2)-3(dy/dt)+2y=f(t)$; $y(0)=1$, $y'(0)=0$, where $f(t)$ is an impulsive function which acts on the extremely short time interval $2\leq t\leq 2+\tau$, and $\int_2^{2+\tau} f(t)\,dt=-1$.

(b) Solve the initial-value problem $(d^2y/dt^2)-3(dy/dt)+2y=0$; $y(0)=1$, $y'(0)=0$, on the interval $0\leq t\leq 2$. Compute $z_0=y(2)$ and $z_0'=y'(2)$. Then solve the initial-value problem

$$\frac{d^2y}{dt^2}-3\frac{dy}{dt}+2y=0;\qquad y(2)=z_0,\ \ y'(2)=z_0'-1,\qquad 2\leq t<\infty.$$

Compare this solution with the solution of part (a).

4. A particle of mass 1 is attached to a spring dashpot mechanism. The stiffness constant of the spring is 3 N/m and the drag force exerted on the particle by the dashpot mechanism is 4 times its velocity. At time $t=0$, the particle is stretched $\frac{1}{4}$ m from its equilibrium position. At time $t=3$ seconds, an impulsive force of very short duration is applied to the system. This force imparts an impulse of 2 N·s to the particle. Find the displacement of the particle from its equilibrium position.

In Exercises 5–7 solve the given initial-value problem.

5. $\dfrac{d^2y}{dt^2}+y=\sin t+\delta(t-\pi);\quad y(0)=0,\ y'(0)=0$

6. $\dfrac{d^2y}{dt^2}+\dfrac{dy}{dt}+y=2\delta(t-1)-\delta(t-2);\quad y(0)=1,\ y'(0)=0$

7. $\dfrac{d^2y}{dt^2}+2\dfrac{dy}{dt}+y=e^{-t}+3\delta(t-3);\quad y(0)=0,\ y'(0)=3$

8. (a) Solve the initial-value problem

$$\frac{d^2y}{dt^2}+y=\sum_{j=0}^{\infty}\delta(t-j\pi),\qquad y(0)=y'(0)=0,$$

and show that

$$y(t)=\begin{cases}\sin t, & n \text{ even}\\ 0, & n \text{ odd}\end{cases}$$

in the interval $n\pi<t<(n+1)\pi$.

(b) Solve the initial-value problem

$$\frac{d^2y}{dt^2}+y=\sum_{j=0}^{\infty}\delta(t-2j\pi),\qquad y(0)=y'(0)=0,$$

and show that $y(t)=(n+1)\sin t$ in the interval $2n\pi<t<2(n+1)\pi$.

This example indicates why soldiers are instructed to break cadence when marching across a bridge. To wit, if the soldiers are in step with the natural frequency of the steel in the bridge, then a resonance situation of the type (b) may be set up.

9. Let $f(t)$ be the function which is $\frac{1}{2}$ for $t>t_0$, 0 for $t=t_0$, and $-\frac{1}{2}$ for $t<t_0$. Let $K[\phi]$ be the linear functional

$$K[\phi]=\int_{-\infty}^{\infty}\phi(t)f(t)\,dt.$$

Show that $K'[\phi]\equiv K[-\phi']=\phi(t_0)$. Thus, $\delta(t-t_0)$ may be viewed as the derivative of $f(t)$.

2.13 The convolution integral

Consider the initial-value problem

$$a\frac{d^2y}{dt^2}+b\frac{dy}{dt}+cy=f(t);\qquad y(0)=y_0,\quad y'(0)=y_0'. \tag{1}$$

Let $Y(s)=\mathcal{L}\{y(t)\}$ and $F(s)=\mathcal{L}\{f(t)\}$. Taking Laplace transforms of both sides of the differential equation gives

$$a\left[s^2Y(s)-sy_0-y_0'\right]+b\left[sY(s)-y_0\right]+cY(s)=F(s)$$

and this implies that

$$Y(s)=\frac{as+b}{as^2+bs+c}y_0+\frac{a}{as^2+bs+c}y_0'+\frac{F(s)}{as^2+bs+c}.$$

Now, let

$$y_1(t)=\mathcal{L}^{-1}\left\{\frac{as+b}{as^2+bs+c}\right\}$$

and

$$y_2(t)=\mathcal{L}^{-1}\left\{\frac{a}{as^2+bs+c}\right\}.$$

Setting $f(t)=0$, $y_0=1$ and $y_0'=0$, we see that $y_1(t)$ is the solution of the homogeneous equation which satisfies the initial conditions $y_1(0)=1$, $y_1'(0)=0$. Similarly, by setting $f(t)=0$, $y_0=0$ and $y_0'=1$, we see that $y_2(t)$ is the solution of the homogeneous equation which satisfies the initial conditions $y_2(0)=0$, $y_2'(0)=1$. This implies that

$$\psi(t)=\mathcal{L}^{-1}\left\{\frac{F(s)}{as^2+bs+c}\right\}$$

is the particular solution of the nonhomogeneous equation which satisfies the initial conditions $\psi(0)=0$, $\psi'(0)=0$. Thus, the problem of finding a particular solution $\psi(t)$ of the nonhomogeneous equation is now reduced to the problem of finding the inverse Laplace transform of the function $F(s)/(as^2+bs+c)$. If we look carefully at this function, we see that it is the product of two Laplace transforms; that is

$$\frac{F(s)}{as^2+bs+c}=\mathcal{L}\{f(t)\}\times\mathcal{L}\left\{\frac{y_2(t)}{a}\right\}.$$

It is natural to ask whether there is any simple relationship between $\psi(t)$ and the functions $f(t)$ and $y_2(t)/a$. It would be nice, of course, if $\psi(t)$ were the product of $f(t)$ with $y_2(t)/a$, but this is obviously false. However, there is an extremely interesting way of combining two functions f and g together to form a new function $f*g$, which resembles multiplication, and for which

$$\mathcal{L}\{(f*g)(t)\}=\mathcal{L}\{f(t)\}\times\mathcal{L}\{g(t)\}.$$

This combination of f and g appears quite often in applications, and is known as the *convolution* of f with g.

Definition. The *convolution* $(f*g)(t)$ of f with g is defined by the equation

$$(f*g)(t)=\int_0^t f(t-u)\,g(u)\,du. \tag{2}$$

For example, if $f(t)=\sin 2t$ and $g(t)=e^{t^2}$, then

$$(f*g)(t)=\int_0^t \sin 2(t-u)e^{u^2}\,du.$$

The convolution operator $*$ clearly bears some resemblance to the multiplication operator since we multiply the value of f at the point $t-u$ by the value of g at the point u, and then integrate this product with respect to u. Therefore, it should not be too surprising to us that the convolution operator satisfies the following properties.

Property 1. The convolution operator obeys the commutative law of multiplication; that is, $(f*g)(t)=(g*f)(t)$.

PROOF. By definition,

$$(f*g)(t)=\int_0^t f(t-u)\,g(u)\,du.$$

Let us make the substitution $t-u=s$ in this integral. Then,

$$\begin{aligned}(f*g)(t)&=-\int_t^0 f(s)\,g(t-s)\,ds\\&=\int_0^t g(t-s)f(s)\,ds\equiv(g*f)(t).\end{aligned}$$ □

Property 2. The convolution operator satisfies the distributive law of multiplication; that is,

$$f*(g+h)=f*g+f*h.$$

PROOF. See Exercise 19. □

Property 3. The convolution operator satisfies the associative law of multiplication; that is, $(f*g)*h=f*(g*h)$.

PROOF. See Exercise 20. □

Property 4. The convolution of any function f with the zero function is zero.

PROOF. Obvious. □

On the other hand, the convolution operator differs from the multiplication operator in that $f*1\neq f$ and $f*f\neq f^2$. Indeed, the convolution of a function f with itself may even be negative.

Example 1. Compute the convolution of $f(t)=t^2$ with $g(t)=1$.

Solution. From Property 1,

$$(f*g)(t)=(g*f)(t)=\int_0^t 1\cdot u^2\,du=\frac{t^3}{3}.$$

Example 2. Compute the convolution of $f(t)=\cos t$ with itself, and show that it is not always positive.
Solution. By definition,

$$\begin{aligned}(f*f)(t)&=\int_0^t \cos(t-u)\cos u\,du\\&=\int_0^t(\cos t\cos^2 u+\sin t\sin u\cos u)\,du\\&=\cos t\int_0^t\frac{1+\cos 2u}{2}\,du+\sin t\int_0^t\sin u\cos u\,du\\&=\cos t\left[\frac{t}{2}+\frac{\sin 2t}{4}\right]+\frac{\sin^3 t}{2}\\&=\frac{t\cos t+\sin t\cos^2 t+\sin^3 t}{2}\\&=\frac{t\cos t+\sin t(\cos^2 t+\sin^2 t)}{2}\\&=\frac{t\cos t+\sin t}{2}.\end{aligned}$$

This function, clearly, is negative for

$$(2n+1)\pi\leqslant t\leqslant(2n+1)\pi+\tfrac{1}{2}\pi,\qquad n=0,1,2,\ldots.$$

We now show that the Laplace transform of $f*g$ is the product of the Laplace transform of f with the Laplace transform of g.

Theorem 9. $\mathcal{L}\{(f*g)(t)\}=\mathcal{L}\{f(t)\}\times\mathcal{L}\{g(t)\}$.

PROOF. By definition,

$$\mathcal{L}\{(f*g)(t)\}=\int_0^\infty e^{-st}\left[\int_0^t f(t-u)\,g(u)\,du\right]dt.$$

This iterated integral equals the double integral

$$\iint_R e^{-st}f(t-u)\,g(u)\,du\,dt$$

where R is the triangular region described in Figure 1. Integrating first

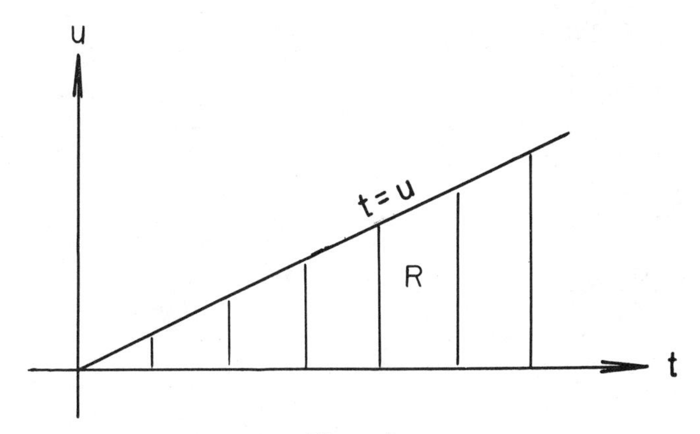

Figure 1

with respect to t, instead of u, gives

$$\mathcal{L}\{(f*g)(t)\} = \int_0^\infty g(u)\left[\int_u^\infty e^{-st} f(t-u)\,dt\right] du.$$

Setting $t-u=\xi$, we see that

$$\int_u^\infty e^{-st} f(t-u)\,dt = \int_0^\infty e^{-s(u+\xi)} f(\xi)\,d\xi.$$

Hence,

$$\begin{aligned}\mathcal{L}\{(f*g)(t)\} &= \int_0^\infty g(u)\left[\int_0^\infty e^{-su} e^{-s\xi} f(\xi)\,d\xi\right] du \\ &= \left[\int_0^\infty g(u) e^{-su}\,du\right]\left[\int_0^\infty e^{-s\xi} f(\xi)\,d\xi\right] \\ &\equiv \mathcal{L}\{f(t)\} \times \mathcal{L}\{g(t)\}.\end{aligned}$$

□

Example 3. Find the inverse Laplace transform of the function

$$\frac{a}{s^2(s^2+a^2)}.$$

Solution. Observe that

$$\frac{1}{s^2} = \mathcal{L}\{t\} \quad \text{and} \quad \frac{a}{s^2+a^2} = \mathcal{L}\{\sin at\}.$$

Hence, by Theorem 9

$$\begin{aligned}\mathcal{L}^{-1}\left\{\frac{a}{s^2(s^2+a^2)}\right\} &= \int_0^t (t-u)\sin au\,du \\ &= \frac{at - \sin at}{a^2}.\end{aligned}$$

Example 4. Find the inverse Laplace transform of the function

$$\frac{1}{s(s^2+2s+2)}.$$

Solution. Observe that

$$\frac{1}{s}=\mathcal{L}\{1\} \quad \text{and} \quad \frac{1}{s^2+2s+2}=\frac{1}{(s+1)^2+1}=\mathcal{L}\{e^{-t}\sin t\}.$$

Hence, by Theorem 9,

$$\mathcal{L}^{-1}\left\{\frac{1}{s(s^2+2s+2)}\right\}=\int_0^t e^{-u}\sin u\,du$$
$$=\frac{1}{2}\left[1-e^{-t}(\cos t+\sin t)\right].$$

Remark. Let $y_2(t)$ be the solution of the homogeneous equation $ay''+by'+cy=0$ which satisfies the initial conditions $y_2(0)=0$, $y_2'(0)=1$. Then,

$$\psi(t)=f(t)*\frac{y_2(t)}{a} \tag{3}$$

is the particular solution of the nonhomogeneous equation $ay''+by'+cy=f(t)$ which satisfies the initial conditions $\psi(0)=\psi'(0)=0$. Equation (3) is often much simpler to use than the variation of parameters formula derived in Section 2.4.

Exercises

Compute the convolution of each of the following pairs of functions.

1. $e^{at}, e^{bt}, a\neq b$

2. e^{at}, e^{at}

3. $\cos at, \cos bt$

4. $\sin at, \sin bt, a\neq b$

5. $\sin at, \sin at$

6. $t, \sin t$

Use Theorem 9 to invert each of the following Laplace transforms.

7. $\dfrac{1}{s^2(s^2+1)}$

8. $\dfrac{s}{(s+1)(s^2+4)}$

9. $\dfrac{s}{(s^2+1)^2}$

10. $\dfrac{1}{s(s^2+1)}$

11. $\dfrac{1}{s^2(s+1)^2}$

12. $\dfrac{1}{(s^2+1)^2}$

Use Theorem 9 to find the solution $y(t)$ of each of the following integro-differential equations.

13. $y(t)=4t-3\int_0^t y(u)\sin(t-u)\,du$

14. $y(t)=4t-3\int_0^t y(t-u)\sin u\,du$

15. $y'(t)=\sin t+\int_0^t y(t-u)\cos u\,du,\ y(0)=0$

16. $y(t)=4t^2-\int_0^t y(u)e^{-(t-u)}\,du$

17. $y'(t)+2y+\int_0^t y(u)\,du=\sin t,\ y(0)=1$

18. $y(t)=t-e^t\int_0^t y(u)e^{-u}\,du$

19. Prove that $f*(g+h)=f*g+f*h$.

20. Prove that $(f*g)*h=f*(g*h)$.

2.14 The method of elimination for systems

The theory of second-order linear differential equations can also be used to find the solutions of two simultaneous first-order equations of the form

$$\begin{aligned} x'=\frac{dx}{dt}&=a(t)x+b(t)y+f(t)\\ y'=\frac{dy}{dt}&=c(t)x+d(t)y+g(t). \end{aligned} \tag{1}$$

The key idea is to eliminate one of the variables, say y, and then find x as the solution of a second-order linear differential equation. This technique is known as the *method of elimination*, and we illustrate it with the following two examples.

Example 1. Find all solutions of the simultaneous equations

$$\begin{aligned} x'&=2x+y+t\\ y'&=x+3y+1. \end{aligned} \tag{2}$$

Solution. First, we solve for

$$y=x'-2x-t \tag{3}$$

from the first equation of (2). Differentiating this equation gives

$$y'=x''-2x'-1=x+3y+1.$$

Then, substituting for y from (3) gives

$$x''-2x'-1=x+3(x'-2x-t)+1$$

so that

$$x''-5x'+5x=2-3t. \tag{4}$$

Equation (4) is a second-order linear equation and its solution is

$$x(t)=e^{5t/2}\left[c_1e^{\sqrt{5}\,t/2}+c_2e^{-\sqrt{5}\,t/2}\right]-\frac{(1+3t)}{5}$$

for some constants c_1 and c_2. Finally, plugging this expression into (3) gives

$$y(t)=e^{5t/2}\left[\frac{1+\sqrt{5}}{2}c_1e^{\sqrt{5}\,t/2}+\frac{1-\sqrt{5}}{2}c_2e^{-\sqrt{5}\,t/2}\right]+\frac{t-1}{5}.$$

Example 2. Find the solution of the initial-value problem

$$\begin{aligned} x'&=3x-y, \qquad x(0)=3\\ y'&=x+y, \qquad y(0)=0. \end{aligned} \tag{5}$$

Solution. From the first equation of (5),

$$y=3x-x'. \tag{6}$$

Differentiating this equation gives

$$y'=3x'-x''=x+y.$$

Then, substituting for y from (6) gives

$$3x'-x''=x+3x-x'$$

so that

$$x''-4x'+4x=0.$$

This implies that

$$x(t)=(c_1+c_2t)e^{2t}$$

for some constants c_1, c_2, and plugging this expression into (6) gives

$$y(t)=(c_1-c_2+c_2t)e^{2t}.$$

The constants c_1 and c_2 are determined from the initial conditions

$$\begin{aligned} x(0)&=3=c_1\\ y(0)&=0=c_1-c_2. \end{aligned}$$

Hence $c_1=3$, $c_2=3$ and

$$x(t)=3(1+t)e^{2t}, y(t)=3te^{2t}$$

is the solution of (5).

Remark. The simultaneous equations (1) are usually referred to as a first-order *system* of equations. Systems of equations are treated fully in Chapters 3 and 4.

Exercises

Find all solutions of each of the following systems of equations.

1. $x'=6x-3y$
$y'=2x+y$

2. $x'=-2x+y+t$
$y'=-4x+3y-1$

3. $x' = -3x + 2y$
$y' = -x - y$

4. $x' = x + y + e^t$
$y' = x - y - e^t$

Find the solution of each of the following initial-value problems.

5. $x' = x + y, \quad x(0) = 2$
$y' = 4x + y, \quad y(0) = 3$

6. $x' = x - 3y, \quad x(0) = 0$
$y' = -2x + 2y, \quad y(0) = 5$

7. $x' = x - y, \quad x(0) = 1$
$y' = 5x - 3y, \quad y(0) = 2$

8. $x' = 3x - 2y, \quad x(0) = 1$
$y' = 4x - y, \quad y(0) = 5$

9. $x' = 4x + 5y + 4e^t \cos t, \quad x(0) = 0$
$y' = -2x - 2y, \quad y(0) = 0$

10. $x' = 3x - 4y + e^t, \quad x(0) = 1$
$y' = x - y + e^t, \quad y(0) = 1$

11. $x' = 2x - 5y + \sin t, \quad x(0) = 0$
$y' = x - 2y + \tan t, \quad y(0) = 0$

12. $x' = y + f_1(t), \quad x(0) = 0$
$y' = -x + f_2(t), \quad y(0) = 0$

2.15 Higher-order equations

In this section we briefly discuss higher-order linear differential equations.

Definition. The equation

$$L[y] = a_n(t)\frac{d^n y}{dt^n} + a_{n-1}(t)\frac{d^{n-1}y}{dt^{n-1}} + \dots + a_0(t)y = 0, \quad a_n(t) \neq 0 \tag{1}$$

is called the general nth order homogeneous linear equation. The differential equation (1) together with the initial conditions

$$y(t_0) = y_0, \quad y'(t_0) = y_0', \dots, y^{(n-1)}(t_0) = y_0^{(n-1)} \tag{1'}$$

is called an initial-value problem. The theory for Equation (1) is completely analogous to the theory for the second-order linear homogeneous equation which we studied in Sections 2.1 and 2.2. Therefore, we will state the relevant theorems without proof. Complete proofs can be obtained by generalizing the methods used in Sections 2.1 and 2.2, or by using the methods to be developed in Chapter 3.

Theorem 10. *Let $y_1(t), \dots, y_n(t)$ be n independent solutions of (1); that is, no solution $y_j(t)$ is a linear combination of the other solutions. Then, every solution $y(t)$ of (1) is of the form*

$$y(t) = c_1 y_1(t) + \dots + c_n y_n(t) \tag{2}$$

for some choice of constants $c_1, \dots, c_n$. For this reason, we say that (2) is the general solution of (1).

To find n independent solutions of (1) when the coefficients $a_0, a_1, \dots, a_n$ do not depend on t, we compute

$$L[e^{rt}] = (a_n r^n + a_{n-1} r^{n-1} + \dots + a_0)e^{rt}. \tag{3}$$

This implies that e^{rt} is a solution of (1) if, and only if, r is a root of the characteristic equation

$$a_n r^n + a_{n-1} r^{n-1} + \dots + a_0 = 0. \tag{4}$$

Thus, if Equation (4) has n distinct roots $r_1, \dots, r_n$, then the general solution of (1) is $y(t) = c_1 e^{r_1 t} + \dots + c_n e^{r_n t}$. If $r_j = \alpha_j + i\beta_j$ is a complex root of (4), then

$$u(t) = \mathrm{Re}\{e^{r_j t}\} = e^{\alpha_j t}\cos\beta_j t$$

and

$$v(t) = \mathrm{Im}\{e^{r_j t}\} = e^{\alpha_j t}\sin\beta_j t$$

are two real-valued solutions of (1). Finally, if r_1 is a root of multiplicity k; that is, if

$$a_n r^n + \dots + a_0 = (r - r_1)^k q(r)$$

where $q(r_1) \neq 0$, then $e^{r_1 t}, te^{r_1 t}, \dots, t^{k-1} e^{r_1 t}$ are k independent solutions of (1). We prove this last assertion in the following manner. Observe from (3) that

$$L[e^{rt}] = (r - r_1)^k q(r) e^{rt}$$

if r_1 is a root of multiplicity k. Therefore,

$$\begin{aligned} L[t^j e^{r_1 t}] &= L\left[\frac{\partial^j}{\partial r^j} e^{rt}\right]\Bigg|_{r=r_1} \\ &= \frac{\partial^j}{\partial r^j} L[e^{rt}]\Bigg|_{r=r_1} \\ &= \frac{\partial^j}{\partial r^j}(r - r_1)^k q(r) e^{rt}\Bigg|_{r=r_1} \\ &= 0, \text{ for } 1 \leqslant j < k. \end{aligned}$$

Example 1. Find the general solution of the equation

$$\frac{d^4 y}{dt^4} + y = 0. \tag{5}$$

Solution. The characteristic equation of (5) is $r^4 + 1 = 0$. We find the roots of this equation by noting that

$$-1 = e^{i\pi} = e^{3\pi i} = e^{5\pi i} = e^{7\pi i}.$$

Hence,

$$r_1 = e^{i\pi/4} = \cos\frac{\pi}{4} + i\sin\frac{\pi}{4} = \frac{1}{\sqrt{2}}(1 + i),$$

$$r_2 = e^{3\pi i/4} = \cos\frac{3\pi}{4} + i\sin\frac{3\pi}{4} = -\frac{1}{\sqrt{2}}(1 - i),$$

$$r_3 = e^{5\pi i/4} = \cos\frac{5\pi}{4} + i\sin\frac{5\pi}{4} = -\frac{1}{\sqrt{2}}(1 + i),$$

and

$$r_4 = e^{7\pi i/4} = \cos\frac{7\pi}{4} + i\sin\frac{7\pi}{4} = \frac{1}{\sqrt{2}}(1-i)$$

are 4 roots of the equation $r^4+1=0$. The roots r_3 and r_4 are the complex conjugates of r_2 and r_1, respectively. Thus,

$$e^{r_1 t} = e^{t/\sqrt{2}}\left[\cos\frac{t}{\sqrt{2}} + i\sin\frac{t}{\sqrt{2}}\right]$$

and

$$e^{r_2 t} = e^{-t/\sqrt{2}}\left[\cos\frac{t}{\sqrt{2}} + i\sin\frac{t}{\sqrt{2}}\right]$$

are 2 complex-valued solutions of (5), and this implies that

$$y_1(t) = e^{t/\sqrt{2}}\cos\frac{t}{\sqrt{2}}, \qquad y_2(t) = e^{t/\sqrt{2}}\sin\frac{t}{\sqrt{2}},$$
$$y_3(t) = e^{-t/\sqrt{2}}\cos\frac{t}{\sqrt{2}}, \quad \text{and} \quad y_4(t) = e^{-t/\sqrt{2}}\sin\frac{t}{\sqrt{2}}$$

are 4 real-valued solutions of (5). These solutions are clearly independent. Hence, the general solution of (5) is

$$y(t) = e^{t/\sqrt{2}}\left[a_1\cos\frac{t}{\sqrt{2}} + b_1\sin\frac{t}{\sqrt{2}}\right] + e^{-t/\sqrt{2}}\left[a_2\cos\frac{t}{\sqrt{2}} + b_2\sin\frac{t}{\sqrt{2}}\right].$$

Example 2. Find the general solution of the equation

$$\frac{d^4y}{dt^4} - 3\frac{d^3y}{dt^3} + 3\frac{d^2y}{dt^2} - \frac{dy}{dt} = 0. \tag{6}$$

Solution. The characteristic equation of (6) is

$$0 = r^4 - 3r^3 + 3r^2 - r = r(r^3 - 3r^2 + 3r - 1)$$
$$= r(r-1)^3.$$

Its roots are $r_1=0$ and $r_2=1$, with $r_2=1$ a root of multiplicity three. Hence, the general solution of (6) is

$$y(t) = c_1 + (c_2 + c_3 t + c_4 t^2)e^t.$$

The theory for the nonhomogeneous equation

$$L[y] = a_n(t)\frac{d^ny}{dt^n} + \dots + a_0(t)y = f(t), \quad a_n(t) \neq 0 \tag{7}$$

is also completely analogous to the theory for the second-order nonhomogeneous equation. The following results are the analogs of Lemma 1 and Theorem 5 of Section 2.3.

Lemma 1. *The difference of any two solutions of the nonhomogeneous equation* (7) *is a solution of the homogeneous equation* (1).

Theorem 11. *Let $\psi(t)$ be a particular solution of the nonhomogeneous equation* (7), *and let $y_1(t),\dots,y_n(t)$ be n independent solutions of the homogeneous equation* (1). *Then, every solution $y(t)$ of* (7) *is of the form*

$$y(t)=\psi(t)+c_1y_1(t)+\dots+c_ny_n(t)$$

for some choice of constants $c_1, c_2,\dots,c_n$.

The method of judicious guessing also applies to the nth-order equation

$$a_n\frac{d^ny}{dt^n}+\dots+a_0y=\left[b_0+b_1t+\dots+b_kt^k\right]e^{\alpha t}. \tag{8}$$

It is easily verified that Equation (8) has a particular solution $\psi(t)$ of the form

$$\psi(t)=\left[c_0+c_1t+\dots+c_kt^k\right]e^{\alpha t}$$

if $e^{\alpha t}$ is not a solution of the homogeneous equation, and

$$\psi(t)=t^j\left[c_0+c_1t+\dots+c_kt^k\right]e^{\alpha t}$$

if $t^{j-1}e^{\alpha t}$ is a solution of the homogeneous equation, but $t^je^{\alpha t}$ is not.

Example 3. Find a particular solution $\psi(t)$ of the equation

$$L[y]=\frac{d^3y}{dt^3}+3\frac{d^2y}{dt^2}+3\frac{dy}{dt}+y=e^t. \tag{9}$$

Solution. The characteristic equation

$$r^3+3r^2+3r+1=(r+1)^3$$

has $r=-1$ as a triple root. Hence, e^t is not a solution of the homogeneous equation, and Equation (9) has a particular solution $\psi(t)$ of the form

$$\psi(t)=Ae^t.$$

Computing $L[\psi](t)=8Ae^t$, we see that $A=\frac{1}{8}$. Consequently, $\psi(t)=\frac{1}{8}e^t$ is a particular solution of (9).

There is also a variation of parameters formula for the nonhomogeneous equation (7). Let $v(t)$ be the solution of the homogeneous equation

(1) which satisfies the initial conditions $v(t_0)=0$, $v'(t_0)=0,\ldots,v^{(n-2)}(t_0)=0$, $v^{(n-1)}(t_0)=1$. Then,

$$\psi(t)=\int_{t_0}^{t}\frac{v(t-s)}{a_n(s)}f(s)\,ds$$

is a particular solution of the nonhomogeneous equation (7). We will prove this assertion in Section 3.12. (This can also be proven using the method of Laplace transforms; see Section 2.13.)

Exercises

Find the general solution of each of the following equations.

1. $y'''-2y''-y'+2y=0$

2. $y'''-6y''+5y'+12y=0$

3. $y^{(\text{iv})}-5y'''+6y''+4y'-8y=0$

4. $y'''-y''+y'-y=0$

Solve each of the following initial-value problems.

5. $y^{(\text{iv})}+4y'''+14y''-20y'+25y=0;\quad y(0)=y'(0)=y''(0)=0,\ y'''(0)=0$

6. $y^{(\text{iv})}-y=0;\quad y(0)=1,\ y'(0)=y''(0)=0,\ y'''(0)=-1$

7. $y^{(\text{v})}-2y^{(\text{iv})}+y'''=0;\quad y(0)=y'(0)=y''(0)=y'''(0)=0,\ y^{(\text{iv})}(0)=-1$

8. Given that $y_1(t)=e^t\cos t$ is a solution of

$$y^{(\text{iv})}-2y'''+y''+2y'-2y=0, \qquad (*)$$

find the general solution of (*). *Hint*: Use this information to find the roots of the characteristic equation of (*).

Find a particular solution of each of the following equations.

9. $y'''+y'=\tan t$

10. $y^{(\text{iv})}-y=g(t)$

11. $y^{(\text{iv})}+y=g(t)$

12. $y'''+y'=2t^2+4\sin t$

13. $y'''-4y'=t+\cos t+2e^{-2t}$

14. $y^{(\text{iv})}-y=t+\sin t$

15. $y^{(\text{iv})}+2y''+y=t^2\sin t$

16. $y^{(\text{vi})}+y''=t^2$

17. $y'''+y''+y'+y=t+e^{-t}$

18. $y^{(\text{iv})}+4y'''+6y''+4y'+y=t^3e^{-t}$

Hint for (18): Make the substitution $y=e^{-t}v$ and solve for v. Otherwise, it will take an awfully long time to do this problem.

3 Systems of differential equations

3.1 Algebraic properties of solutions of linear systems

In this chapter we will consider simultaneous first-order differential equations in several variables, that is, equations of the form

$$\begin{aligned}\frac{dx_1}{dt} &= f_1(t,x_1,\dots,x_n),\\ \frac{dx_2}{dt} &= f_2(t,x_1,\dots,x_n),\\ &\vdots\\ \frac{dx_n}{dt} &= f_n(t,x_1,\dots,x_n).\end{aligned} \tag{1}$$

A solution of (1) is n functions $x_1(t),\dots,x_n(t)$ such that $dx_j(t)/dt = f_j(t,x_1(t),\dots,x_n(t))$, $j=1,2,\dots,n$. For example, $x_1(t)=t$ and $x_2(t)=t^2$ is a solution of the simultaneous first-order differential equations

$$\frac{dx_1}{dt}=1 \quad\text{and}\quad \frac{dx_2}{dt}=2x_1$$

since $dx_1(t)/dt=1$ and $dx_2(t)/dt=2t=2x_1(t)$.

In addition to Equation (1), we will often impose initial conditions on the functions $x_1(t),\dots,x_n(t)$. These will be of the form

$$x_1(t_0)=x_1^0,\quad x_2(t_0)=x_2^0,\dots,x_n(t_0)=x_n^0. \tag{1'}$$

Equation (1), together with the initial conditions (1)′, is referred to as an initial-value problem. A solution of this initial-value problem is n functions $x_1(t),\dots,x_n(t)$ which satisfy (1) and the initial conditions

$$x_1(t_0)=x_1^0,\dots,x_n(t_0)=x_n^0.$$

For example, $x_1(t)=e^t$ and $x_2(t)=1+e^{2t}/2$ is a solution of the initial-value problem

$$\frac{dx_1}{dt} = x_1, \qquad x_1(0) = 1,$$

$$\frac{dx_2}{dt} = x_1^2, \qquad x_2(0) = \frac{3}{2},$$

since $dx_1(t)/dt = e^t = x_1(t)$, $dx_2(t)/dt = e^{2t} = x_1^2(t)$, $x_1(0) = 1$ and $x_2(0) = \frac{3}{2}$.

Equation (1) is usually referred to as a system of n first-order differential equations. Equations of this type arise quite often in biological and physical applications and frequently describe very complicated systems since the rate of change of the variable x_j depends not only on t and x_j, but on the value of all the other variables as well. One particular example is the blood glucose model we studied in Section 2.7. In this model, the rates of change of g and h (respectively, the deviations of the blood glucose and net hormonal concentrations from their optimal values) are given by the equations

$$\frac{dg}{dt} = -m_1 g - m_2 h + J(t), \qquad \frac{dh}{dt} = -m_3 h + m_4 g.$$

This is a system of two first-order equations for the functions $g(t)$ and $h(t)$.

First-order systems of differential equations also arise from higher-order equations for a single variable $y(t)$. Every nth-order differential equation for the single variable y can be converted into a system of n first-order equations for the variables

$$x_1(t) = y, \quad x_2(t) = \frac{dy}{dt}, \ldots, x_n(t) = \frac{d^{n-1}y}{dt^{n-1}}.$$

Examples 1 and 2 illustrate how this works.

Example 1. Convert the differential equation

$$a_n(t)\frac{d^n y}{dt^n} + a_{n-1}(t)\frac{d^{n-1}y}{dt^{n-1}} + \ldots + a_0 y = 0$$

into a system of n first-order equations.

Solution. Let $x_1(t) = y$, $x_2(t) = dy/dt, \ldots,$ and $x_n(t) = d^{n-1}y/dt^{n-1}$. Then,

$$\frac{dx_1}{dt} = x_2, \quad \frac{dx_2}{dt} = x_3, \ldots, \quad \frac{dx_{n-1}}{dt} = x_n,$$

and

$$\frac{dx_n}{dt} = -\frac{a_{n-1}(t)x_n + a_{n-2}(t)x_{n-1} + \ldots + a_0 x_1}{a_n(t)}.$$

Example 2. Convert the initial-value problem

$$\frac{d^3y}{dt^3} + \left(\frac{dy}{dt}\right)^2 + 3y = e^t; \qquad y(0) = 1, \quad y'(0) = 0, \quad y''(0) = 0$$

into an initial-value problem for the variables y, dy/dt, and d^2y/dt^2.

Solution. Set $x_1(t)=y$, $x_2(t)=dy/dt$, and $x_3(t)=d^2y/dt^2$. Then,

$$\frac{dx_1}{dt}=x_2,\qquad \frac{dx_2}{dt}=x_3,\qquad \frac{dx_3}{dt}=e^t-x_2^2-3x_1.$$

Moreover, the functions x_1, x_2, and x_3 satisfy the initial conditions $x_1(0)=1$, $x_2(0)=0$, and $x_3(0)=0$.

If each of the functions $f_1,\dots,f_n$ in (1) is a linear function of the dependent variables $x_1,\dots,x_n$, then the system of equations is said to be linear. The most general system of n first-order linear equations has the form

$$\begin{aligned}\frac{dx_1}{dt}&=a_{11}(t)x_1+\dots+a_{1n}(t)x_n+g_1(t)\\ &\vdots\\ \frac{dx_n}{dt}&=a_{n1}(t)x_1+\dots+a_{nn}(t)x_n+g_n(t).\end{aligned}\tag{2}$$

If each of the functions $g_1,\dots,g_n$ is identically zero, then the system (2) is said to be homogeneous; otherwise it is nonhomogeneous. In this chapter, we only consider the case where the coefficients a_{ij} do not depend on t.

Now, even the homogeneous linear system with constant coefficients

$$\begin{aligned}\frac{dx_1}{dt}&=a_{11}x_1+\dots+a_{1n}x_n\\ &\vdots\\ \frac{dx_n}{dt}&=a_{n1}x_1+\dots+a_{nn}x_n\end{aligned}\tag{3}$$

is quite cumbersome to handle. This is especially true if n is large. Therefore, we seek to write these equations in as concise a manner as possible. To this end we introduce the concepts of *vectors* and *matrices*.

Definition. A *vector*

$$\mathbf{x}=\begin{bmatrix}x_1\\ x_2\\ \vdots\\ x_n\end{bmatrix}$$

is a shorthand notation for the sequence of numbers $x_1,\dots,x_n$. The numbers $x_1,\dots,x_n$ are called the *components* of $\mathbf{x}$. If $x_1=x_1(t),\dots,$ and $x_n=x_n(t)$, then

$$\mathbf{x}(t)=\begin{bmatrix}x_1(t)\\ \vdots\\ x_n(t)\end{bmatrix}$$

is called a vector-valued function. Its derivative $d\mathbf{x}(t)/dt$ is the vector-valued function

$$\begin{bmatrix} \dfrac{dx_1(t)}{dt} \\ \vdots \\ \dfrac{dx_n(t)}{dt} \end{bmatrix}.$$

Definition. A *matrix*

$$\mathbf{A}=\begin{bmatrix} a_{11} & a_{12} & \dots & a_{1n} \\ a_{21} & a_{22} & \dots & a_{2n} \\ \vdots & \vdots & & \vdots \\ a_{m1} & a_{m2} & \dots & a_{mn} \end{bmatrix}$$

is a shorthand notation for the array of numbers a_{ij} arranged in m rows and n columns. The element lying in the ith row and jth column is denoted by a_{ij}, the first subscript identifying its row and the second subscript identifying its column. $\mathbf{A}$ is said to be a square matrix if $m=n$.

Next, we define the product of a matrix $\mathbf{A}$ with a vector $\mathbf{x}$.

Definition. Let $\mathbf{A}$ be an $n\times n$ matrix with elements a_{ij} and let $\mathbf{x}$ be a vector with components $x_1,\dots,x_n$. We define the product of $\mathbf{A}$ with $\mathbf{x}$, denoted by $\mathbf{Ax}$, as the vector whose ith component is

$$a_{i1}x_1+a_{i2}x_2+\dots+a_{in}x_n, \qquad i=1,2,\dots,n.$$

In other words, the ith component of $\mathbf{Ax}$ is the sum of the product of corresponding terms of the ith row of $\mathbf{A}$ with the vector $\mathbf{x}$. Thus,

$$\mathbf{Ax}=\begin{bmatrix} a_{11} & a_{12} & \dots & a_{1n} \\ a_{21} & a_{22} & \dots & a_{2n} \\ \vdots & \vdots & & \vdots \\ a_{n1} & a_{n2} & \dots & a_{nn} \end{bmatrix}\begin{bmatrix} x_1 \\ x_2 \\ \vdots \\ x_n \end{bmatrix}$$

$$=\begin{bmatrix} a_{11}x_1+a_{12}x_2+\dots+a_{1n}x_n \\ a_{21}x_1+a_{22}x_2+\dots+a_{2n}x_n \\ \vdots \\ a_{n1}x_1+a_{n2}x_2+\dots+a_{nn}x_n \end{bmatrix}.$$

For example,

$$\begin{bmatrix} 1 & 2 & 4 \\ -1 & 0 & 6 \\ 1 & 1 & 1 \end{bmatrix}\begin{bmatrix} 3 \\ 2 \\ 1 \end{bmatrix}=\begin{bmatrix} 3+4+4 \\ -3+0+6 \\ 3+2+1 \end{bmatrix}=\begin{bmatrix} 11 \\ 3 \\ 6 \end{bmatrix}.$$

Finally, we observe that the left-hand sides of (3) are the components of the vector $d\mathbf{x}/dt$, while the right-hand sides of (3) are the components of the vector $\mathbf{Ax}$. Hence, we can write (3) in the concise form

$$\dot{\mathbf{x}}=\frac{d\mathbf{x}}{dt}=\mathbf{Ax}, \quad \text{where} \quad \mathbf{x}=\begin{bmatrix} x_1 \\ \vdots \\ x_n \end{bmatrix} \quad \text{and} \quad \mathbf{A}=\begin{bmatrix} a_{11} & a_{12} & \dots & a_{1n} \\ a_{21} & a_{22} & \dots & a_{2n} \\ \vdots & \vdots & & \vdots \\ a_{n1} & a_{n2} & \dots & a_{nn} \end{bmatrix}. \tag{4}$$

Moreover, if $x_1(t),\dots,x_n(t)$ satisfy the initial conditions

$$x_1(t_0)=x_1^0,\dots,x_n(t_0)=x_n^0,$$

then $\mathbf{x}(t)$ satisfies the initial-value problem

$$\dot{\mathbf{x}}=\mathbf{Ax}, \qquad \mathbf{x}(t_0)=\mathbf{x}^0, \quad \text{where} \quad \mathbf{x}^0=\begin{bmatrix} x_1^0 \\ \vdots \\ x_n^0 \end{bmatrix}. \tag{5}$$

For example, the system of equations

$$\frac{dx_1}{dt}=3x_1-7x_2+9x_3$$

$$\frac{dx_2}{dt}=15x_1+x_2-x_3$$

$$\frac{dx_3}{dt}=7x_1+6x_3$$

can be written in the concise form

$$\dot{\mathbf{x}}=\begin{bmatrix} 3 & -7 & 9 \\ 15 & 1 & -1 \\ 7 & 0 & 6 \end{bmatrix}\mathbf{x}, \qquad \mathbf{x}=\begin{bmatrix} x_1 \\ x_2 \\ x_3 \end{bmatrix},$$

and the initial-value problem

$$\frac{dx_1}{dt}=x_1-x_2+x_3, \qquad x_1(0)=1$$

$$\frac{dx_2}{dt}=3x_2-x_3, \qquad x_2(0)=0$$

$$\frac{dx_3}{dt}=x_1+7x_3, \qquad x_3(0)=-1$$

can be written in the concise form

$$\dot{\mathbf{x}}=\begin{bmatrix} 1 & -1 & 1 \\ 0 & 3 & -1 \\ 1 & 0 & 7 \end{bmatrix}\mathbf{x}, \qquad \mathbf{x}(0)=\begin{bmatrix} 1 \\ 0 \\ -1 \end{bmatrix}.$$

Now that we have succeeded in writing (3) in the more manageable form (4), we can tackle the problem of finding all of its solutions. Since these equations are linear, we will try and play the same game that we played, with so much success, with the second-order linear homogeneous equation. To wit, we will show that a constant times a solution and the sum of two solutions are again solutions of (4). Then, we will try and show that we can find every solution of (4) by taking all linear combinations of a finite number of solutions. Of course, we must first define what we mean by a constant times $\mathbf{x}$ and the sum of $\mathbf{x}$ and $\mathbf{y}$ if $\mathbf{x}$ and $\mathbf{y}$ are vectors with n components.

Definition. Let c be a number and $\mathbf{x}$ a vector with n components $x_1,\ldots,x_n$. We define $c\mathbf{x}$ to be the vector whose components are $cx_1,\ldots,cx_n$, that is

$$c\mathbf{x}=c\begin{bmatrix} x_1\\ x_2\\ \vdots\\ x_n\end{bmatrix}=\begin{bmatrix} cx_1\\ cx_2\\ \vdots\\ cx_n\end{bmatrix}.$$

For example, if

$$c=2 \quad\text{and}\quad \mathbf{x}=\begin{bmatrix}3\\1\\7\end{bmatrix}, \quad\text{then}\quad 2\mathbf{x}=2\begin{bmatrix}3\\1\\7\end{bmatrix}=\begin{bmatrix}6\\2\\14\end{bmatrix}.$$

This process of multiplying a vector $\mathbf{x}$ by a number c is called scalar multiplication.

Definition. Let $\mathbf{x}$ and $\mathbf{y}$ be vectors with components $x_1,\ldots,x_n$ and $y_1,\ldots,y_n$ respectively. We define $\mathbf{x}+\mathbf{y}$ to be the vector whose components are $x_1+y_1,\ldots,x_n+y_n$, that is

$$\mathbf{x}+\mathbf{y}=\begin{bmatrix} x_1\\ x_2\\ \vdots\\ x_n\end{bmatrix}+\begin{bmatrix} y_1\\ y_2\\ \vdots\\ y_n\end{bmatrix}=\begin{bmatrix} x_1+y_1\\ x_2+y_2\\ \vdots\\ x_n+y_n\end{bmatrix}.$$

For example, if

$$\mathbf{x}=\begin{bmatrix}1\\6\\3\\2\end{bmatrix} \quad\text{and}\quad \mathbf{y}=\begin{bmatrix}-1\\-6\\7\\9\end{bmatrix},$$

then

$$\mathbf{x}+\mathbf{y}=\begin{bmatrix}1\\6\\3\\2\end{bmatrix}+\begin{bmatrix}-1\\-6\\7\\9\end{bmatrix}=\begin{bmatrix}0\\0\\10\\11\end{bmatrix}.$$

This process of adding two vectors together is called vector addition.

Having defined the processes of scalar multiplication and vector addition, we can now state the following theorem.

Theorem 1. *Let* $\mathbf{x}(t)$ *and* $\mathbf{y}(t)$ *be two solutions of* (4). *Then* (*a*) $c\mathbf{x}(t)$ *is a solution, for any constant c, and* (*b*) $\mathbf{x}(t)+\mathbf{y}(t)$ *is again a solution.*

Theorem 1 can be proven quite easily with the aid of the following lemma.

Lemma. *Let* $\mathbf{A}$ *be an* $n\times n$ *matrix. For any vectors* $\mathbf{x}$ *and* $\mathbf{y}$ *and constant c,* (*a*) $\mathbf{A}(c\mathbf{x})=c\mathbf{A}\mathbf{x}$ *and* (*b*) $\mathbf{A}(\mathbf{x}+\mathbf{y})=\mathbf{A}\mathbf{x}+\mathbf{A}\mathbf{y}$.

PROOF OF LEMMA.
(a) We prove that two vectors are equal by showing that they have the same components. To this end, observe that the ith component of the vector $c\mathbf{A}\mathbf{x}$ is

$$ca_{i1}x_1+ca_{i2}x_2+\ldots+ca_{in}x_n=c(a_{i1}x_1+\ldots+a_{in}x_n),$$

and the ith component of the vector $\mathbf{A}(c\mathbf{x})$ is

$$a_{i1}(cx_1)+a_{i2}(cx_2)+\ldots+a_{in}(cx_n)=c(a_{i1}x_1+\ldots+a_{in}x_n).$$

Hence $\mathbf{A}(c\mathbf{x})=c\mathbf{A}\mathbf{x}$.
(b) The ith component of the vector $\mathbf{A}(\mathbf{x}+\mathbf{y})$ is

$$a_{i1}(x_1+y_1)+\ldots+a_{in}(x_n+y_n)=(a_{i1}x_1+\ldots+a_{in}x_n)+(a_{i1}y_1+\ldots+a_{in}y_n).$$

But this is also the ith component of the vector $\mathbf{A}\mathbf{x}+\mathbf{A}\mathbf{y}$ since the ith component of $\mathbf{A}\mathbf{x}$ is $a_{i1}x_1+\ldots+a_{in}x_n$ and the ith component of $\mathbf{A}\mathbf{y}$ is $a_{i1}y_1+\ldots+a_{in}y_n$. Hence $\mathbf{A}(\mathbf{x}+\mathbf{y})=\mathbf{A}\mathbf{x}+\mathbf{A}\mathbf{y}$. □

PROOF OF THEOREM 1.
(a). If $\mathbf{x}(t)$ is a solution of (4), then

$$\frac{d}{dt}c\mathbf{x}(t)=c\frac{d\mathbf{x}(t)}{dt}=c\mathbf{A}\mathbf{x}(t)=\mathbf{A}(c\mathbf{x}(t)).$$

Hence, $c\mathbf{x}(t)$ is also a solution of (4).
(b). If $\mathbf{x}(t)$ and $\mathbf{y}(t)$ are solutions of (4) then

$$\frac{d}{dt}(\mathbf{x}(t)+\mathbf{y}(t))=\frac{d\mathbf{x}(t)}{dt}+\frac{d\mathbf{y}(t)}{dt}=\mathbf{A}\mathbf{x}(t)+\mathbf{A}\mathbf{y}(t)=\mathbf{A}(\mathbf{x}(t)+\mathbf{y}(t)).$$

Hence, $\mathbf{x}(t)+\mathbf{y}(t)$ is also a solution of (4). □

An immediate corollary of Theorem 1 is that any linear combination of solutions of (4) is again a solution of (4). That is to say, if $\mathbf{x}^1(t),\ldots,\mathbf{x}^j(t)$ are j solutions of (4), then $c_1\mathbf{x}^1(t)+\ldots+c_j\mathbf{x}^j(t)$ is again a solution for any choice of constants $c_1, c_2, \ldots, c_j$. For example, consider the system of equations

$$\frac{dx_1}{dt} = x_2, \quad \frac{dx_2}{dt} = -4x_1, \quad \text{or} \quad \frac{d\mathbf{x}}{dt} = \begin{pmatrix} 0 & 1 \\ -4 & 0 \end{pmatrix}\mathbf{x}, \quad \mathbf{x} = \begin{pmatrix} x_1 \\ x_2 \end{pmatrix}. \tag{6}$$

This system of equations was derived from the second-order scalar equation $(d^2y/dt^2) + 4y = 0$ by setting $x_1 = y$ and $x_2 = dy/dt$. Since $y_1(t) = \cos 2t$ and $y_2(t) = \sin 2t$ are two solutions of the scalar equation, we know that

$$\mathbf{x}(t) = \begin{pmatrix} x_1(t) \\ x_2(t) \end{pmatrix} = c_1 \begin{pmatrix} \cos 2t \\ -2\sin 2t \end{pmatrix} + c_2 \begin{pmatrix} \sin 2t \\ 2\cos 2t \end{pmatrix}$$
$$= \begin{pmatrix} c_1 \cos 2t + c_2 \sin 2t \\ -2c_1 \sin 2t + 2c_2 \cos 2t \end{pmatrix}$$

is a solution of (6) for any choice of constants c_1 and c_2.

The next step in our game plan is to show that every solution of (4) can be expressed as a linear combination of finitely many solutions. Equivalently, we seek to determine how many solutions we must find before we can generate all the solutions of (4). There is a branch of mathematics known as linear algebra, which addresses itself to exactly this question, and it is to this area that we now turn our attention.

Exercises

In each of Exercises 1–3 convert the given differential equation for the single variable y into a system of first-order equations.

1. $\dfrac{d^3y}{dt^3} + \left(\dfrac{dy}{dt}\right)^2 = 0$ **2.** $\dfrac{d^3y}{dt^3} + \cos y = e^t$ **3.** $\dfrac{d^4y}{dt^4} + \dfrac{d^2y}{dt^2} = 1$

4. Convert the pair of second-order equations

$$\frac{d^2y}{dt^2} + 3\frac{dz}{dt} + 2y = 0, \qquad \frac{d^2z}{dt^2} + 3\frac{dy}{dt} + 2z = 0$$

into a system of 4 first-order equations for the variables

$$x_1 = y, \qquad x_2 = y', \qquad x_3 = z, \quad \text{and} \quad x_4 = z'.$$

5. (a) Let $y(t)$ be a solution of the equation $y'' + y' + y = 0$. Show that

$$\mathbf{x}(t) = \begin{pmatrix} y(t) \\ y'(t) \end{pmatrix}$$

is a solution of the system of equations

$$\dot{\mathbf{x}} = \begin{pmatrix} 0 & 1 \\ -1 & -1 \end{pmatrix}\mathbf{x}.$$

(b) Let

$$\mathbf{x}(t)=\begin{pmatrix} x_1(t) \\ x_2(t) \end{pmatrix}$$

be a solution of the system of equations

$$\dot{\mathbf{x}}=\begin{pmatrix} 0 & 1 \\ -1 & -1 \end{pmatrix}\mathbf{x}.$$

Show that $y=x_1(t)$ is a solution of the equation $y''+y'+y=0$.

In each of Exercises 6–9, write the given system of differential equations and initial values in the form $\dot{\mathbf{x}}=\mathbf{A}\mathbf{x}, \mathbf{x}(t_0)=\mathbf{x}^0$.

6. $\dot{x}_1=3x_1-7x_2, \quad x_1(0)=1$
$\dot{x}_2=4x_1, \quad x_2(0)=1$

7. $\dot{x}_1=5x_1+5x_2, \quad x_1(3)=0$
$\dot{x}_2=-x_1+7x_2, \quad x_2(3)=6$

8. $\dot{x}_1=x_1+x_2-x_3, \quad x_1(0)=0$
$\dot{x}_2=3x_1-x_2+4x_3, \quad x_2(0)=1$
$\dot{x}_3=-x_1-x_2, \quad x_3(0)=-1$

9. $\dot{x}_1=-x_3, \quad x_1(-1)=2$
$\dot{x}_2=x_1, \quad x_2(-1)=3$
$\dot{x}_3=-x_2, \quad x_3(-1)=4$

10. Let

$$\mathbf{x}=\begin{pmatrix} 1 \\ 3 \\ 2 \end{pmatrix} \quad\text{and}\quad \mathbf{y}=\begin{pmatrix} -1 \\ 0 \\ 4 \end{pmatrix}.$$

Compute $\mathbf{x}+\mathbf{y}$ and $3\mathbf{x}-2\mathbf{y}$.

11. Let

$$\mathbf{A}=\begin{pmatrix} 1 & 2 & -1 \\ 3 & 0 & 4 \\ -1 & -1 & 2 \end{pmatrix}.$$

Compute $\mathbf{A}\mathbf{x}$ if

$$\text{(a) } \mathbf{x}=\begin{pmatrix} 1 \\ 0 \\ 0 \end{pmatrix}, \qquad \text{(b) } \mathbf{x}=\begin{pmatrix} 0 \\ 1 \\ 0 \end{pmatrix}, \qquad \text{(c) } \mathbf{x}=\begin{pmatrix} 0 \\ 0 \\ 1 \end{pmatrix}.$$

12. Let $\mathbf{A}$ be any $n\times n$ matrix and let $\mathbf{e}^j$ be the vector whose jth component is 1 and whose remaining components are zero. Verify that the vector $\mathbf{A}\mathbf{e}^j$ is the jth column of $\mathbf{A}$.

13. Let

$$\mathbf{A}=\begin{pmatrix} -1 & 0 & 0 \\ 2 & 1 & 3 \\ -1 & 0 & 2 \end{pmatrix}.$$

Compute $\mathbf{A}\mathbf{x}$ if

$$\text{(a) } \mathbf{x}=\begin{pmatrix} 1 \\ 2 \\ 4 \end{pmatrix}, \qquad \text{(b) } \mathbf{x}=\begin{pmatrix} 1 \\ -1 \\ -1 \end{pmatrix}, \qquad \text{(c) } \mathbf{x}=\begin{pmatrix} 1 \\ 1 \\ 1 \end{pmatrix}, \qquad \text{(d) } \mathbf{x}=\begin{pmatrix} 1 \\ 0 \\ 1 \end{pmatrix}.$$

14. Let $\mathbf{A}$ be a 3×3 matrix with the property that

$$\mathbf{A}\begin{pmatrix} 1 \\ 1 \\ 1 \end{pmatrix}=\begin{pmatrix} 3 \\ 2 \\ 6 \end{pmatrix} \quad\text{and}\quad \mathbf{A}\begin{pmatrix} 1 \\ -1 \\ -1 \end{pmatrix}=\begin{pmatrix} 1 \\ 2 \\ 3 \end{pmatrix}.$$

Compute

$$\mathbf{A}\begin{pmatrix} 3 \\ -1 \\ -1 \end{pmatrix}.$$

Hint: Write

$$\begin{pmatrix} 3 \\ -1 \\ -1 \end{pmatrix}$$

as a linear combination of

$$\begin{pmatrix} 1 \\ 1 \\ 1 \end{pmatrix} \quad \text{and} \quad \begin{pmatrix} 1 \\ -1 \\ -1 \end{pmatrix}.$$

15. Let $\mathbf{A}$ be a 2×2 matrix with the property that

$$\mathbf{A}\begin{pmatrix} 1 \\ 1 \end{pmatrix} = \begin{pmatrix} 4 \\ 2 \end{pmatrix} \quad \text{and} \quad \mathbf{A}\begin{pmatrix} 1 \\ -1 \end{pmatrix} = \begin{pmatrix} -3 \\ -6 \end{pmatrix}.$$

Find $\mathbf{A}$. *Hint*: The easy way is to use Exercise 12.

3.2 Vector spaces

In the previous section we defined, in a natural manner, a process of adding two vectors $\mathbf{x}$ and $\mathbf{y}$ together to form a new vector $\mathbf{z}=\mathbf{x}+\mathbf{y}$, and a process of multiplying a vector $\mathbf{x}$ by a scalar c to form a new vector $\mathbf{u}=c\mathbf{x}$. The former process was called vector addition and the latter process was called scalar multiplication. Our study of linear algebra begins with the more general premise that we have a set $\mathbf{V}$ of elements $\mathbf{x},\mathbf{y},\mathbf{z},\ldots$ and that we have one process that combines two elements $\mathbf{x}$ and $\mathbf{y}$ of $\mathbf{V}$ to form a third element $\mathbf{z}$ in $\mathbf{V}$ and a second process that combines a number c and an element $\mathbf{x}$ in $\mathbf{V}$ to form a new element $\mathbf{u}$ in $\mathbf{V}$. We will denote the first process by addition; that is, we will write $\mathbf{z}=\mathbf{x}+\mathbf{y}$, and the second process by scalar multiplication; that is, we will write $\mathbf{u}=c\mathbf{x}$, if they satisfy the usual axioms of addition and multiplication. These axioms are:

(i) $\mathbf{x}+\mathbf{y}=\mathbf{y}+\mathbf{x}$ (commutative law)
(ii) $\mathbf{x}+(\mathbf{y}+\mathbf{z})=(\mathbf{x}+\mathbf{y})+\mathbf{z}$ (associative law)
(iii) There is a unique element in $\mathbf{V}$, called the zero element, and denoted by $\mathbf{0}$, having the property that $\mathbf{x}+\mathbf{0}=\mathbf{x}$ for all $\mathbf{x}$ in $\mathbf{V}$.
(iv) For each element $\mathbf{x}$ in $\mathbf{V}$ there is a unique element, denoted by $-\mathbf{x}$ and called minus $\mathbf{x}$, such that $\mathbf{x}+(-\mathbf{x})=\mathbf{0}$.
(v) $1\cdot\mathbf{x}=\mathbf{x}$ for all $\mathbf{x}$ in $\mathbf{V}$.
(vi) $(ab)\mathbf{x}=a(b\mathbf{x})$ for any numbers a,b and any element $\mathbf{x}$ in $\mathbf{V}$.
(vii) $a(\mathbf{x}+\mathbf{y})=a\mathbf{x}+a\mathbf{y}$
(viii) $(a+b)\mathbf{x}=a\mathbf{x}+b\mathbf{x}$.

A set $\mathbf{V}$, together with processes addition and multiplication satisfying (i)–(viii) is said to be a *vector space* and its elements are called *vectors*. The numbers a, b will usually be real numbers, except in certain special cases where they will be complex numbers.

Remark 1. Implicit in axioms (i)–(viii) is the fact that if $\mathbf{x}$ and $\mathbf{y}$ are in $\mathbf{V}$, then the linear combination $a\mathbf{x}+b\mathbf{y}$ is again in $\mathbf{V}$ for any choice of constants a and b.

Remark 2. In the previous section we defined a vector $\mathbf{x}$ as a sequence of n numbers. In the more general context of this section, a quantity $\mathbf{x}$ is a vector by dint of its being in a vector space. That is to say, a quantity $\mathbf{x}$ is a vector if it belongs to a set of elements $\mathbf{V}$ which is equipped with two processes (addition and scalar multiplication) which satisfy (i)–(viii). As we shall see in Example 3 below, the set of all sequences

$$\mathbf{x}=\begin{pmatrix} x_1 \\ x_2 \\ \vdots \\ x_n \end{pmatrix}$$

of n real numbers is a vector space (with the usual operations of vector addition and scalar multiplication defined in Section 3.1). Thus, our two definitions are consistent.

Example 1. Let $\mathbf{V}$ be the set of all functions $x(t)$ which satisfy the differential equation

$$\frac{d^2x}{dt^2}-x=0 \tag{1}$$

with the sum of two functions and the product of a function by a number being defined in the usual manner. That is to say,

$$(f_1+f_2)(t)=f_1(t)+f_2(t)$$

and

$$(cf)(t)=cf(t).$$

It is trivial to verify that $\mathbf{V}$ is a vector space. Observe first that if x^1 and x^2 are in $\mathbf{V}$, then every linear combination $c_1x^1+c_2x^2$ is in $\mathbf{V}$, since the differential equation (1) is linear. Moreover, axioms (i), (ii), and (v)–(viii) are automatically satisfied since all we are doing at any time t in function addition and multiplication of a function by a number is adding or multiplying two numbers together. The zero vector in $\mathbf{V}$ is the function whose value at any time t is zero; this function is in $\mathbf{V}$ since $x(t)\equiv 0$ is a solution of (1). Finally, the negative of any function in $\mathbf{V}$ is again in $\mathbf{V}$, since the negative of any solution of (1) is again a solution of (1).

Example 2. Let **V** be the set of all solutions $x(t)$ of the differential equation $(d^2x/dt^2)-6x^2=0$, with the sum of two functions and the product of a function by a number being defined in the usual manner. **V** is not a vector space since the sum of any two solutions, while being defined, is not necessarily in **V**. Similarly, the product of a solution by a constant is not necessarily in **V**. For example, the function $x(t)=1/t^2$ is in **V** since it satisfies the differential equation, but the function $2x(t)=2/t^2$ is not in **V** since it does not satisfy the differential equation.

Example 3. Let **V** be the set of all sequences

$$\mathbf{x}=\begin{pmatrix} x_1 \\ \vdots \\ x_n \end{pmatrix}$$

of n real numbers. Define $\mathbf{x}+\mathbf{y}$ and $c\mathbf{x}$ as the vector addition and scalar multiplication defined in Section 3.1. It is trivial to verify that **V** is a vector space under these operations. The zero vector is the sequence

$$\begin{pmatrix} 0 \\ 0 \\ \vdots \\ 0 \end{pmatrix}$$

and the vector $-\mathbf{x}$ is the vector

$$\begin{pmatrix} -x_1 \\ \vdots \\ -x_n \end{pmatrix}.$$

This space is usually called n dimensional Euclidean space and is denoted by R^n.

Example 4. Let **V** be the set of all sequences

$$\mathbf{x}=\begin{pmatrix} x_1 \\ \vdots \\ x_n \end{pmatrix}$$

of n complex numbers $x_1,\ldots,x_n$. Define $\mathbf{x}+\mathbf{y}$ and $c\mathbf{x}$, for any complex number c, as the vector addition and scalar multiplication defined in Section 3.1. Again, it is trivial to verify that **V** is a vector space under these operations. This space is usually called complex n dimensional space and is denoted by C^n.

Example 5. Let $\mathbf{V}$ be the set of all $n\times n$ matrices $\mathbf{A}$. Define the sum of two matrices $\mathbf{A}$ and $\mathbf{B}$ to be the matrix obtained by adding together corresponding elements of $\mathbf{A}$ and $\mathbf{B}$, and define the matrix $c\mathbf{A}$ to be the matrix obtained by multiplying every element of $\mathbf{A}$ by c. In other words,

$$\begin{bmatrix} a_{11} & a_{12} & \dots & a_{1n} \\ a_{21} & a_{22} & \dots & a_{2n} \\ \vdots & \vdots & & \vdots \\ a_{n1} & a_{n2} & \dots & a_{nn} \end{bmatrix} + \begin{bmatrix} b_{11} & b_{12} & \dots & b_{1n} \\ b_{21} & b_{22} & \dots & b_{2n} \\ \vdots & \vdots & & \vdots \\ b_{n1} & b_{n2} & \dots & b_{nn} \end{bmatrix}$$

$$= \begin{bmatrix} a_{11}+b_{11} & a_{12}+b_{12} & \dots & a_{1n}+b_{1n} \\ a_{21}+b_{21} & a_{22}+b_{22} & \dots & a_{2n}+b_{2n} \\ \vdots & \vdots & & \vdots \\ a_{n1}+b_{n1} & a_{n2}+b_{n2} & \dots & a_{nn}+b_{nn} \end{bmatrix}$$

and

$$c\begin{bmatrix} a_{11} & a_{12} & \dots & a_{1n} \\ a_{21} & a_{22} & \dots & a_{2n} \\ \vdots & \vdots & & \vdots \\ a_{n1} & a_{n2} & \dots & a_{nn} \end{bmatrix} = \begin{bmatrix} ca_{11} & ca_{12} & \dots & ca_{1n} \\ ca_{21} & ca_{22} & \dots & ca_{2n} \\ \vdots & \vdots & & \vdots \\ ca_{n1} & ca_{n2} & \dots & ca_{nn} \end{bmatrix}.$$

Axioms (i), (ii), and (v)–(viii) are automatically satisfied since all we are doing in adding two matrices together or multiplying a matrix by a number is adding or multiplying two numbers together. The zero vector, or the matrix $\mathbf{0}$, is the matrix whose every element is the number zero, and the negative of any matrix $\mathbf{A}$ is the matrix

$$\begin{bmatrix} -a_{11} & \dots & -a_{1n} \\ \vdots & & \vdots \\ -a_{n1} & \dots & -a_{nn} \end{bmatrix}.$$

Hence $\mathbf{V}$ is a vector space under these operations of matrix addition and scalar multiplication.

Example 6. We now present an example of a set of elements which comes close to being a vector space, but which doesn't quite make it. The purpose of this example is to show that the elements of $\mathbf{V}$ can be just about anything, and the operation of addition can be a rather strange process. Let $\mathbf{V}$ be the set consisting of three animals, a cat, a dog, and a mouse. Whenever any two of these animals meet, one eats up the other and changes into a different animal. The rules of eating are as follows.

(1) If a dog meets a cat, then the dog eats up the cat and changes into a mouse.
(2) If a dog meets another dog, then one dog eats up the other and changes into a cat.
(3) If a dog meets a mouse, then the dog eats up the mouse and remains unchanged.
(4) If a cat meets another cat, then one cat eats up the other and changes into a dog.
(5) If a cat meets a mouse, then the cat eats up the mouse and remains unchanged.
(6) If a mouse meets another mouse, then one mouse eats up the other and remains unchanged.

Clearly, "eating" is a process which combines two elements of **V** to form a third element in **V**. If we call this eating process addition, and denote it by $+$, then rules 1–6 can be written concisely in the form

1. $D+C=M$ 2. $D+D=C$ 3. $D+M=D$
4. $C+C=D$ 5. $C+M=C$ 6. $M+M=M$.

This operation of eating satisfies all the axioms of addition. To see this, note that axiom (i) is satisfied since the eating formulae do not depend on the order of the two animals involved. This is to say, $D+C=C+D$, etc. Moreover, the result of any addition is again an animal in **V**. This would not be the case, for example, if a dog ate up a cat and changed into a hippopotamus. The associative law (ii) is also satisfied, but it has to be verified explicitly. For example, suppose that we have an encounter between two cats and a dog. It is not obvious, a priori, that it does not make a difference whether the two cats meet first and their resultant meets the dog or whether one cat meets the dog and their resultant meets the other cat. To check that this is so we compute

$$(C+C)+D=D+D=C$$

and

$$(C+D)+C=M+C=C.$$

In a similar manner we can show that the result of any encounter between three animals is independent of the order in which they meet. Next, observe that the zero element in **V** is the mouse, since every animal is unchanged after eating a mouse. Finally, "minus a dog" is a cat (since $D+C=M$), "minus a cat" is a dog and "minus a mouse" is a mouse. However, **V** is *not* a vector space since there is no operation of scalar multiplication defined. Moreover, it is clearly impossible to define the quantities aC and aD, for all real numbers a, so as to satisfy axioms (v)–(viii).

Example 7. Let $\mathbf{V}$ be the set of all vector-valued solutions

$$\mathbf{x}(t)=\begin{bmatrix} x_1(t) \\ \vdots \\ x_n(t) \end{bmatrix}$$

of the vector differential equation

$$\dot{\mathbf{x}}=\mathbf{A}\mathbf{x}, \qquad \mathbf{A}=\begin{bmatrix} a_{11} & \dots & a_{1n} \\ \vdots & & \vdots \\ a_{n1} & \dots & a_{nn} \end{bmatrix}. \tag{2}$$

$\mathbf{V}$ is a vector space under the usual operations of vector addition and scalar multiplication. To wit, observe that axioms (i), (ii), and (v)–(viii) are automatically satisfied. Hence, we need only verify that

(a) The sum of any two solutions of (2) is again a solution.
(b) A constant times a solution of (2) is again a solution.
(c) The vector-valued function

$$\mathbf{x}(t)=\begin{bmatrix} x_1(t) \\ \vdots \\ x_n(t) \end{bmatrix}=\begin{bmatrix} 0 \\ \vdots \\ 0 \end{bmatrix}$$

is a solution of (2) (axiom (iii)).
(d) The negative of any solution of (2) is again a solution (axiom (iv)).

Now (a) and (b) are exactly Theorem 1 of the previous section, while (d) is a special case of (b). To verify (c) we observe that

$$\frac{d}{dt}\begin{bmatrix} 0 \\ \vdots \\ 0 \end{bmatrix}=\begin{bmatrix} 0 \\ \vdots \\ 0 \end{bmatrix} \quad \text{and} \quad \mathbf{A}\begin{bmatrix} 0 \\ \vdots \\ 0 \end{bmatrix}=\begin{bmatrix} 0 \\ \vdots \\ 0 \end{bmatrix}.$$

Hence the vector-valued function $\mathbf{x}(t)\equiv\mathbf{0}$ is always a solution of the differential equation (2).

Exercises

In each of Problems 1–6, determine whether the given set of elements

$$\mathbf{x}=\begin{bmatrix} x_1 \\ x_2 \\ x_3 \end{bmatrix}$$

form a vector space under the properties of vector addition and scalar multiplication defined in Section 3.1.

1. The set of all elements $\mathbf{x}=\begin{pmatrix} x_1 \\ x_2 \\ x_3 \end{pmatrix}$ where $3x_1-2x_2=0$

2. The set of all elements $\mathbf{x}=\begin{pmatrix} x_1 \\ x_2 \\ x_3 \end{pmatrix}$ where $x_1+x_2+x_3=0$

3. The set of all elements $\mathbf{x}=\begin{pmatrix} x_1 \\ x_2 \\ x_3 \end{pmatrix}$ where $x_1^2+x_2^2+x_3^2=1$

4. The set of all elements $\mathbf{x}=\begin{pmatrix} x_1 \\ x_2 \\ x_3 \end{pmatrix}$ where $x_1+x_2+x_3=1$

5. The set of elements $\mathbf{x}=\begin{pmatrix} 1 \\ a \\ b \end{pmatrix}$ for all real numbers a and b

6. The set of all elements $\mathbf{x}=\begin{pmatrix} x_1 \\ x_2 \\ x_3 \end{pmatrix}$ where

$$x_1+x_2+x_3=0, \quad x_1-x_2+2x_3=0, \quad 3x_1-x_2+5x_3=0$$

In each of Problems 7–11 determine whether the given set of functions form a vector space under the usual operations of function addition and multiplication of a function by a constant.

7. The set of all polynomials of degree $\leqslant 4$

8. The set of all differentiable functions

9. The set of all differentiable functions whose derivative at $t=1$ is three

10. The set of all solutions of the differential equation $y''+y=\cos t$

11. The set of all functions $y(t)$ which have period 2π, that is $y(t+2\pi)=y(t)$

12. Show that the set of all vector-valued solutions

$$\mathbf{x}(t)=\begin{pmatrix} x_1(t) \\ x_2(t) \end{pmatrix}$$

of the system of differential equations

$$\frac{dx_1}{dt}=x_2+1, \qquad \frac{dx_2}{dt}=x_1+1$$

is not a vector space.

3.3 Dimension of a vector space

Let $\mathbf{V}$ be the set of all solutions $y(t)$ of the second-order linear homogeneous equation $(d^2y/dt^2)+p(t)(dy/dt)+q(t)y=0$. Recall that every solution $y(t)$ can be expressed as a linear combination of any two linearly independent solutions. Thus, if we knew two "independent" functions $y^1(t)$ and

$y^2(t)$ in **V**, then we could find every function in **V** by taking all linear combinations $c_1y^1(t)+c_2y^2(t)$ of y^1 and y^2. We would like to derive a similar property for solutions of the equation $\dot{\mathbf{x}}=\mathbf{A}\mathbf{x}$. To this end, we define the notion of a finite set of vectors generating the whole space, and the notion of independence of vectors in an arbitrary vector space **V**.

Definition. A set of vectors $\mathbf{x}^1,\mathbf{x}^2,\ldots,\mathbf{x}^n$ is said to *span* **V** if the set of all linear combinations $c_1\mathbf{x}^1+c_2\mathbf{x}^2+\ldots+c_n\mathbf{x}^n$ exhausts **V**. That is to say, the vectors $\mathbf{x}^1,\mathbf{x}^2,\ldots,\mathbf{x}^n$ span **V** if every element of **V** can be expressed as a linear combination of $\mathbf{x}^1,\mathbf{x}^2,\ldots,\mathbf{x}^n$.

Example 1. Let **V** be the set of all solutions of the differential equation $(d^2x/dt^2)-x=0$. Let x^1 be the function whose value at any time t is e^t and let x^2 be the function whose value at any time t is e^{-t}. The functions x^1 and x^2 are in **V** since they satisfy the differential equation. Moreover, these functions also span **V** since every solution $x(t)$ of the differential equation can be written in the form

$$x(t)=c_1e^t+c_2e^{-t}$$

so that

$$x=c_1x^1+c_2x^2.$$

Example 2. Let $\mathbf{V}=\mathsf{R}^n$ and let $\mathbf{e}^j$ denote the vector with a 1 in the jth place and zeros everywhere else, that is,

$$\mathbf{e}^1=\begin{bmatrix}1\\0\\0\\ \vdots\\0\end{bmatrix},\qquad \mathbf{e}^2=\begin{bmatrix}0\\1\\0\\ \vdots\\0\end{bmatrix},\ldots,\mathbf{e}^n=\begin{bmatrix}0\\0\\ \vdots\\0\\1\end{bmatrix}.$$

The set of vectors $\mathbf{e}^1,\mathbf{e}^2,\ldots,\mathbf{e}^n$ span R^n since any vector

$$\mathbf{x}=\begin{bmatrix}x_1\\x_2\\ \vdots\\x_n\end{bmatrix}$$

can be written in the form

$$\mathbf{x}=\begin{bmatrix}x_1\\0\\ \vdots\\0\end{bmatrix}+\begin{bmatrix}0\\x_2\\ \vdots\\0\end{bmatrix}+\ldots+\begin{bmatrix}0\\0\\ \vdots\\x_n\end{bmatrix}=x_1\mathbf{e}^1+x_2\mathbf{e}^2+\ldots+x_n\mathbf{e}^n.$$

Definition. A set of vectors $\mathbf{x}^1, \mathbf{x}^2, \ldots, \mathbf{x}^n$ in $\mathbf{V}$ is said to be *linearly dependent* if one of these vectors is a linear combination of the others. A very precise mathematical way of saying this is as follows. A set of vectors $\mathbf{x}^1, \mathbf{x}^2, \ldots, \mathbf{x}^n$ is said to be linearly dependent if there exist constants $c_1, c_2, \ldots, c_n$, *not all zero* such that

$$c_1\mathbf{x}^1 + c_2\mathbf{x}^2 + \ldots + c_n\mathbf{x}^n = \mathbf{0}.$$

These two definitions are equivalent, for if $\mathbf{x}^j$ is a linear combination of $\mathbf{x}^1, \ldots, \mathbf{x}^{j-1}, \mathbf{x}^{j+1}, \ldots, \mathbf{x}^n$, that is

$$\mathbf{x}^j = c_1\mathbf{x}^1 + \ldots + c_{j-1}\mathbf{x}^{j-1} + c_{j+1}\mathbf{x}^{j+1} + \ldots + c_n\mathbf{x}^n,$$

then the linear combination

$$c_1\mathbf{x}^1 + \ldots + c_{j-1}\mathbf{x}^{j-1} - \mathbf{x}^j + c_{j+1}\mathbf{x}^{j+1} + \ldots + c_n\mathbf{x}^n$$

equals zero and not all the constants are zero. Conversely, if $c_1\mathbf{x}^1 + c_2\mathbf{x}^2 + \ldots + c_n\mathbf{x}^n = \mathbf{0}$ and $c_j \neq 0$ for some j, then we can divide by c_j and solve for $\mathbf{x}^j$ as a linear combination of $\mathbf{x}^1, \ldots, \mathbf{x}^{j-1}, \mathbf{x}^{j+1}, \ldots, \mathbf{x}^n$. For example, if $c_1 \neq 0$ then we can divide by c_1 to obtain that

$$\mathbf{x}^1 = -\frac{c_2}{c_1}\mathbf{x}^2 - \frac{c_3}{c_1}\mathbf{x}^3 - \ldots - \frac{c_n}{c_1}\mathbf{x}^n.$$

Definition. If the vectors $\mathbf{x}^1, \mathbf{x}^2, \ldots, \mathbf{x}^n$ are not linearly dependent, that is, none of these vectors can be expressed as a linear combination of the others, then they are said to be *linearly independent*. The precise mathematical way of saying this is that the vectors $\mathbf{x}^1, \mathbf{x}^2, \ldots, \mathbf{x}^n$ are linearly independent if the equation

$$c_1\mathbf{x}^1 + c_2\mathbf{x}^2 + \ldots + c_n\mathbf{x}^n = \mathbf{0}$$

implies, of necessity, that all the constants $c_1, c_2, \ldots, c_n$ are zero.

In order to determine whether a set of vectors $\mathbf{x}^1, \mathbf{x}^2, \ldots, \mathbf{x}^n$ is linearly dependent or linearly independent, we write down the equation $c_1\mathbf{x}^1 + c_2\mathbf{x}^2 + \ldots + c_n\mathbf{x}^n = \mathbf{0}$ and see what this implies about the constants $c_1, c_2, \ldots, c_n$. If all these constants must be zero, then $\mathbf{x}^1, \mathbf{x}^2, \ldots, \mathbf{x}^n$ are linearly independent. On the other hand, if not all the constants $c_1, c_2, \ldots, c_n$ must be zero, then $\mathbf{x}^1, \mathbf{x}^2, \ldots, \mathbf{x}^n$ are linearly dependent.

Example 3. Let $\mathbf{V} = \mathsf{R}^3$ and let $\mathbf{x}^1$, $\mathbf{x}^2$, and $\mathbf{x}^3$ be the vectors

$$\mathbf{x}^1 = \begin{bmatrix} 1 \\ -1 \\ 1 \end{bmatrix}, \quad \mathbf{x}^2 = \begin{bmatrix} 1 \\ 2 \\ 3 \end{bmatrix} \quad \text{and} \quad \mathbf{x}^3 = \begin{bmatrix} 3 \\ 0 \\ 5 \end{bmatrix}.$$

To determine whether these vectors are linearly dependent or linearly independent, we write down the equation $c_1\mathbf{x}^1+c_2\mathbf{x}^2+c_3\mathbf{x}^3=\mathbf{0}$, that is

$$c_1\begin{bmatrix}1\\-1\\1\end{bmatrix}+c_2\begin{bmatrix}1\\2\\3\end{bmatrix}+c_3\begin{bmatrix}3\\0\\5\end{bmatrix}=\begin{bmatrix}0\\0\\0\end{bmatrix}.$$

The left-hand side of this equation is the vector

$$\begin{bmatrix}c_1+c_2+3c_3\\-c_1+2c_2\\c_1+3c_2+5c_3\end{bmatrix}.$$

Hence the constants c_1, c_2, and c_3 must satisfy the equations

$$c_1+c_2+3c_3=0, \tag{i}$$

$$-c_1+2c_2=0, \tag{ii}$$

$$c_1+3c_2+5c_3=0. \tag{iii}$$

Equation (ii) says that $c_1=2c_2$. Substituting this into Equations (i) and (iii) gives

$$3c_2+3c_3=0 \quad\text{and}\quad 5c_2+5c_3=0.$$

These equations have infinitely many solutions c_2,c_3 since they both reduce to the single equation $c_2+c_3=0$. One solution, in particular, is $c_2=-1$, $c_3=1$. Then, from Equation (ii), $c_1=-2$. Hence,

$$-2\begin{bmatrix}1\\-1\\1\end{bmatrix}-\begin{bmatrix}1\\2\\3\end{bmatrix}+\begin{bmatrix}3\\0\\5\end{bmatrix}=\begin{bmatrix}0\\0\\0\end{bmatrix}$$

and $\mathbf{x}^1$, $\mathbf{x}^2$ and $\mathbf{x}^3$ are linearly dependent vectors in R^3.

Example 4. Let $\mathbf{V}=\mathsf{R}^n$ and let $\mathbf{e}^1,\mathbf{e}^2,\ldots,\mathbf{e}^n$ be the vectors

$$\mathbf{e}^1=\begin{bmatrix}1\\0\\0\\\vdots\\0\end{bmatrix},\qquad \mathbf{e}^2=\begin{bmatrix}0\\1\\0\\\vdots\\0\end{bmatrix},\ldots,\mathbf{e}^n=\begin{bmatrix}0\\0\\\vdots\\0\\1\end{bmatrix}.$$

To determine whether $\mathbf{e}^1,\mathbf{e}^2,\ldots,\mathbf{e}^n$ are linearly dependent or linearly independent, we write down the equation $c_1\mathbf{e}^1+\ldots+c_n\mathbf{e}^n=\mathbf{0}$, that is

$$c_1\begin{bmatrix}1\\0\\0\\\vdots\\0\end{bmatrix}+c_2\begin{bmatrix}0\\1\\0\\\vdots\\0\end{bmatrix}+\ldots+c_n\begin{bmatrix}0\\0\\\vdots\\0\\1\end{bmatrix}=\begin{bmatrix}0\\0\\0\\\vdots\\0\end{bmatrix}.$$

The left-hand side of this equation is the vector

$$\begin{bmatrix} c_1 \\ c_2 \\ \vdots \\ c_n \end{bmatrix}.$$

Hence $c_1=0, c_2=0, \ldots,$ and $c_n=0$. Consequently, $\mathbf{e}^1, \mathbf{e}^2, \ldots, \mathbf{e}^n$ are linearly independent vectors in R^n.

Definition. The *dimension* of a vector space $\mathbf{V}$, denoted by $\dim \mathbf{V}$, is the fewest number of linearly independent vectors which span $\mathbf{V}$. $\mathbf{V}$ is said to be a finite dimensional space if its dimension is finite. On the other hand, $\mathbf{V}$ is said to be an infinite dimensional space if no set of finitely many elements span $\mathbf{V}$.

The dimension of a space $\mathbf{V}$ can be characterized as the fewest number of elements that we have to find in order to know all the elements of $\mathbf{V}$. In this sense, the definition of dimension captures our intuitive feeling. However, it is extremely difficult to compute the dimension of a space $\mathbf{V}$ from this definition alone. For example, let $\mathbf{V}=\mathsf{R}^n$. We have shown in Examples 2 and 4 that the vectors $\mathbf{e}^1, \mathbf{e}^2, \ldots, \mathbf{e}^n$ are linearly independent and span $\mathbf{V}$. Moreover, it seems intuitively obvious to us that we cannot generate R^n from fewer than n vectors. Thus, the dimension of R^n should be n. But how can we prove this rigorously? To wit, how can we prove that it is impossible to find a set of $(n-1)$ linearly independent vectors that span R^n? Thus, our definition of dimension is not, as yet, a very useful one. However, it will become extremely useful after we prove the following theorem.

Theorem 2. *If n linearly independent vectors span $\mathbf{V}$, then* $\dim \mathbf{V}=n$.

We will need two lemmas to prove Theorem 2. The first lemma concerns itself with the solutions of simultaneous linear equations and can be motivated as follows. Suppose that we are interested in determining n unknown numbers $x_1, x_2, \ldots, x_n$ uniquely. It seems pretty reasonable that we should be given n equations satisfied by these unknowns. If we are given too few equations then there may be many different solutions, that is, many different sets of values for $x_1, x_2, \ldots, x_n$ which satisfy the given equations. Lemma 1 proves this in the special case that we have m homogeneous linear equations for $n>m$ unknowns.

Lemma 1. *A set of m homogeneous linear equations for n unknowns $x_1, x_2, \ldots, x_n$ always admits a nontrivial solution if $m<n$. That is to say,*

the set of m equations in n unknowns

$$
\begin{aligned}
a_{11}x_1 + a_{12}x_2 + \dots + a_{1n}x_n &= 0\\
a_{21}x_1 + a_{22}x_2 + \dots + a_{2n}x_n &= 0\\
&\vdots\\
a_{m1}x_1 + a_{m2}x_2 + \dots + a_{mn}x_n &= 0
\end{aligned}
\tag{1}
$$

always has a solution $x_1, x_2, \dots, x_n$, *other than* $x_1 = \dots = x_n = 0$, *if* $m < n$.

Remark. Notice that $x_1 = 0, x_2 = 0, \dots, x_n = 0$ is certainly one solution of the system of equations (1). Thus, Lemma 1 is telling us that these equations have more than one solution.

PROOF OF LEMMA 1. We will prove Lemma 1 by induction on m. To this end, observe that the lemma is certainly true if $m = 1$, for in this case we have a single equation of the form $a_{11}x_1 + a_{12}x_2 + \dots + a_{1n}x_n = 0$, with $n \geqslant 2$. We can find a nontrivial solution of this equation, if $a_{11} = 0$, by taking $x_1 = 1, x_2 = 0, \dots, x_n = 0$. We can find a nontrivial solution of this equation, if $a_{11} \neq 0$, by taking $x_2 = 1, \dots, x_n = 1$ and $x_1 = -(a_{12} + \dots + a_{1n})/a_{11}$.

For the next step in our induction proof, we assume that Lemma 1 is true for some integer $m = k$ and show that this implies that Lemma 1 is true for $m = k+1$, and $k+1 < n$. To this end, consider the $k+1$ equations for the n unknowns $x_1, x_2, \dots, x_n$

$$
\begin{aligned}
a_{11}x_1 \ + \ a_{12}x_2 \ + \dots + \ a_{1n}x_n &= 0\\
a_{21}x_1 \ + \ a_{22}x_2 \ + \dots + \ a_{2n}x_n &= 0\\
&\vdots\\
a_{k+1,1}x_1 + a_{k+1,2}x_2 + \dots + a_{k+1,n}x_n &= 0
\end{aligned}
\tag{2}
$$

with $k+1 < n$. If $a_{11}, a_{21}, \dots, a_{k+1,1}$ are all zero, then $x_1 = 1, x_2 = 0, \dots, x_n = 0$ is clearly a non-trivial solution. Hence, we may assume that at least one of these coefficients is not zero. Without any loss of generality, we may assume that $a_{11} \neq 0$, for otherwise we can take the equation with the nonzero coefficient of x_1 and relabel it as the first equation. Then

$$
x_1 = -\frac{a_{12}}{a_{11}}x_2 - \frac{a_{13}}{a_{11}}x_3 - \dots - \frac{a_{1n}}{a_{11}}x_n.
$$

Substituting this value of x_1 in the second through the $(k+1)$st equations, we obtain the equivalent equations

$$\begin{aligned}
a_{11}x_1 + a_{12}x_2 + a_{13}x_3 + \dots + a_{1n}x_n &= 0\\
b_{22}x_2 + b_{23}x_3 + \dots + b_{2n}x_n &= 0\\
&\vdots\\
b_{k2}x_2 + b_{k3}x_3 + \dots + b_{kn}x_n &= 0\\
b_{k+1,2}x_2 + b_{k+1,3}x_3 + \dots + b_{k+1,n}x_n &= 0
\end{aligned} \tag{3}$$

where $b_{ij} = a_{ij} - a_{i1}a_{1j}/a_{11}$. Now, the last k equations of (3) are k homogeneous linear equations for the $(n-1)$ unknowns $x_2,\dots,x_n$. Moreover, k is less than $n-1$ since $k+1$ is less than n. Hence, by the induction hypothesis, these equations have a nontrivial solution $x_2,\dots,x_n$. Once $x_2,\dots,x_n$ are known, we have as before $x_1 = -(a_{12}x_2 + \dots + a_{1n}x_n)/a_{11}$ from the first equation of (3). This establishes Lemma 1 for $m = k+1$, and therefore for all m, by induction. □

If a vector space **V** has dimension m, then it has m linearly independent vectors $\mathbf{x}^1,\dots,\mathbf{x}^m$ and every vector in the space can be written as a linear combination of the m vectors $\mathbf{x}^1,\mathbf{x}^2,\dots,\mathbf{x}^m$. It seems intuitively obvious to us in this case that there cannot be more than m linear independent vectors in **V**. This is the content of Lemma 2.

Lemma 2. *In an m dimensional space, any set of $n > m$ vectors must be linearly dependent. In other words, the maximum number of linearly independent vectors in a finite dimensional space is the dimension of the space.*

Proof. Since **V** has dimension m, there exist m linearly independent vectors $\mathbf{x}^1,\mathbf{x}^2,\dots,\mathbf{x}^m$ which span **V**. Let $\mathbf{y}^1,\mathbf{y}^2,\dots,\mathbf{y}^n$ be a set of n vectors in **V**, with $n > m$. Since $\mathbf{x}^1,\mathbf{x}^2,\dots,\mathbf{x}^m$ span **V**, all the $\mathbf{y}^j$ can be written as linear combinations of these vectors. That is to say, there exist constants a_{ij}, $1 \leqslant i \leqslant n$, $1 \leqslant j \leqslant m$ such that

$$\begin{aligned}
\mathbf{y}^1 &= a_{11}\mathbf{x}^1 + a_{12}\mathbf{x}^2 + \dots + a_{1m}\mathbf{x}^m\\
\mathbf{y}^2 &= a_{21}\mathbf{x}^1 + a_{22}\mathbf{x}^2 + \dots + a_{2m}\mathbf{x}^m\\
&\vdots\\
\mathbf{y}^n &= a_{n1}\mathbf{x}^1 + a_{n2}\mathbf{x}^2 + \dots + a_{nm}\mathbf{x}^m.
\end{aligned} \tag{4}$$

To determine whether $\mathbf{y}^1,\mathbf{y}^2,\ldots,\mathbf{y}^n$ are linearly dependent or linearly independent, we consider the equation

$$c_1\mathbf{y}^1+c_2\mathbf{y}^2+\ldots+c_n\mathbf{y}^n=\mathbf{0}. \tag{5}$$

Using (4) we can rewrite (5) in the form

$$\begin{aligned}\mathbf{0}&=c_1\mathbf{y}^1+c_2\mathbf{y}^2+\ldots+c_n\mathbf{y}^n\\&=(c_1a_{11}+\ldots+c_na_{n1})\mathbf{x}^1+(c_1a_{12}+\ldots+c_na_{n2})\mathbf{x}^2\\&\quad+\ldots+(c_1a_{1m}+\ldots+c_na_{nm})\mathbf{x}^m.\end{aligned}$$

This equation states that a linear combination of $\mathbf{x}^1,\mathbf{x}^2,\ldots,\mathbf{x}^m$ is zero. Since $\mathbf{x}^1,\mathbf{x}^2,\ldots,\mathbf{x}^m$ are linearly independent, all these coefficients must be zero. Hence,

$$\begin{aligned}c_1a_{11}+c_2a_{21}+\ldots+c_na_{n1}&=0\\c_1a_{12}+c_2a_{22}+\ldots+c_na_{n2}&=0\\&\vdots\\c_1a_{1m}+c_2a_{2m}+\ldots+c_na_{nm}&=0.\end{aligned} \tag{6}$$

Now, observe that the system of Equations (6) is a set of m homogeneous linear equations for n unknowns $c_1,c_2,\ldots,c_n$, with $n>m$. By Lemma 1, these equations have a nontrivial solution. Thus, there exist constants $c_1,c_2,\ldots,c_n$, not all zero, such that $c_1\mathbf{y}^1+c_2\mathbf{y}^2+\ldots+c_n\mathbf{y}^n=\mathbf{0}$. Consequently, $\mathbf{y}^1,\mathbf{y}^2,\ldots,\mathbf{y}^n$ are linearly dependent. □

We are now in a position to prove Theorem 2.

PROOF OF THEOREM 2. If n linearly independent vectors span $\mathbf{V}$, then, by the definition of dimension, $\dim\mathbf{V}\leqslant n$. By Lemma 2, $n\leqslant\dim\mathbf{V}$. Hence, $\dim\mathbf{V}=n$. □

Example 5. The dimension of R^n is n since $\mathbf{e}^1,\mathbf{e}^2,\ldots,\mathbf{e}^n$ are n linearly independent vectors which span R^n.

Example 6. Let $\mathbf{V}$ be the set of all 3×3 matrices

$$\mathbf{A}=\begin{bmatrix}a_{11}&a_{12}&a_{13}\\a_{21}&a_{22}&a_{23}\\a_{31}&a_{32}&a_{33}\end{bmatrix},$$

and let $\mathbf{E}_{ij}$ denote the matrix with a one in the ith row, jth column and zeros everywhere else. For example,

$$\mathbf{E}_{23}=\begin{bmatrix}0&0&0\\0&0&1\\0&0&0\end{bmatrix}.$$

To determine whether these matrices are linearly dependent or linearly independent, we consider the equation

$$\sum_{i,j=1}^{3} c_{ij}\mathbf{E}_{ij} = \mathbf{0} = \begin{bmatrix} 0 & 0 & 0 \\ 0 & 0 & 0 \\ 0 & 0 & 0 \end{bmatrix}. \tag{7}$$

Now, observe that the left-hand side of (7) is the matrix

$$c_{11}\begin{bmatrix} 1 & 0 & 0 \\ 0 & 0 & 0 \\ 0 & 0 & 0 \end{bmatrix} + c_{12}\begin{bmatrix} 0 & 1 & 0 \\ 0 & 0 & 0 \\ 0 & 0 & 0 \end{bmatrix} + \dots + c_{33}\begin{bmatrix} 0 & 0 & 0 \\ 0 & 0 & 0 \\ 0 & 0 & 1 \end{bmatrix} = \begin{bmatrix} c_{11} & c_{12} & c_{13} \\ c_{21} & c_{22} & c_{23} \\ c_{31} & c_{32} & c_{33} \end{bmatrix}.$$

Equating this matrix to the zero matrix gives $c_{11} = 0, c_{12} = 0, \dots, c_{33} = 0$. Hence the 9 matrices $\mathbf{E}_{ij}$ are linearly independent. Moreover, these 9 matrices also span $\mathbf{V}$ since any matrix

$$\mathbf{A} = \begin{bmatrix} a_{11} & a_{12} & a_{13} \\ a_{21} & a_{22} & a_{23} \\ a_{31} & a_{32} & a_{33} \end{bmatrix}$$

can obviously be written in the form $\mathbf{A} = \sum_{i,j=1}^{3} a_{ij}\mathbf{E}_{ij}$. Hence $\dim \mathbf{V} = 9$.

Definition. If a set of linearly independent vectors span a vector space $\mathbf{V}$, then this set of vectors is said to be a *basis* for $\mathbf{V}$. A basis may also be called a *coordinate system*. For example, the vectors

$$\mathbf{e}^1 = \begin{bmatrix} 1 \\ 0 \\ 0 \\ 0 \end{bmatrix}, \quad \mathbf{e}^2 = \begin{bmatrix} 0 \\ 1 \\ 0 \\ 0 \end{bmatrix}, \quad \mathbf{e}^3 = \begin{bmatrix} 0 \\ 0 \\ 1 \\ 0 \end{bmatrix} \quad \text{and} \quad \mathbf{e}^4 = \begin{bmatrix} 0 \\ 0 \\ 0 \\ 1 \end{bmatrix}$$

are a basis for R^4. If

$$\mathbf{x} = \begin{bmatrix} x_1 \\ x_2 \\ x_3 \\ x_4 \end{bmatrix},$$

then $\mathbf{x} = x_1\mathbf{e}^1 + x_2\mathbf{e}^2 + x_3\mathbf{e}^3 + x_4\mathbf{e}^4$, and relative to this basis the x_i are called "components" or "coordinates."

Corollary. *In a finite dimensional vector space, each basis has the same number of vectors, and this number is the dimension of the space.*

The following theorem is extremely useful in determining whether a set of vectors is a basis for $\mathbf{V}$.

Theorem 3. *Any n linearly independent vectors in an n dimensional space* $\mathbf{V}$ *must also span* $\mathbf{V}$. *That is to say, any n linearly independent vectors in an n dimensional space* $\mathbf{V}$ *are a basis for* $\mathbf{V}$.

PROOF. Let $\mathbf{x}^1,\mathbf{x}^2,\dots,\mathbf{x}^n$ be n linearly independent vectors in an n dimensional space $\mathbf{V}$. To show that they span $\mathbf{V}$, we must show that every $\mathbf{x}$ in $\mathbf{V}$ can be written as a linear combination of $\mathbf{x}^1,\mathbf{x}^2,\dots,\mathbf{x}^n$. To this end, pick any $\mathbf{x}$ in $\mathbf{V}$ and consider the set of vectors $\mathbf{x},\mathbf{x}^1,\mathbf{x}^2,\dots,\mathbf{x}^n$. This is a set of $(n+1)$ vectors in the n dimensional space $\mathbf{V}$; by Lemma 2, they must be linearly dependent. Consequently, there exist constants $c,c_1,c_2,\dots,c_n$, not all zero, such that

$$c\mathbf{x}+c_1\mathbf{x}^1+c_2\mathbf{x}^2+\dots+c_n\mathbf{x}^n=\mathbf{0}. \tag{8}$$

Now $c\neq 0$, for otherwise the set of vectors $\mathbf{x}^1,\mathbf{x}^2,\dots,\mathbf{x}^n$ would be linearly dependent. Therefore, we can divide both sides of (8) by c to obtain that

$$\mathbf{x}=-\frac{c_1}{c}\mathbf{x}^1-\frac{c_2}{c}\mathbf{x}^2-\dots-\frac{c_n}{c}\mathbf{x}^n.$$

Hence, any n linearly independent vectors in an n dimensional space $\mathbf{V}$ must also span $\mathbf{V}$. □

Example 7. Prove that the vectors

$$\mathbf{x}^1=\begin{pmatrix}1\\1\end{pmatrix}\quad\text{and}\quad\mathbf{x}^2=\begin{pmatrix}1\\-1\end{pmatrix}$$

form a basis for R^2.

Solution. To determine whether $\mathbf{x}^1$ and $\mathbf{x}^2$ are linearly dependent or linearly independent, we consider the equation

$$c_1\mathbf{x}^1+c_2\mathbf{x}^2=c_1\begin{pmatrix}1\\1\end{pmatrix}+c_2\begin{pmatrix}1\\-1\end{pmatrix}=\begin{pmatrix}0\\0\end{pmatrix}. \tag{9}$$

Equation (9) implies that $c_1+c_2=0$ and $c_1-c_2=0$. Adding these two equations gives $c_1=0$ while subtracting these two equations gives $c_2=0$. Consequently, $\mathbf{x}^1$ and $\mathbf{x}^2$ are two linearly independent vectors in the two dimensional space R^2. Hence, by Theorem 3, they must also span $\mathbf{V}$.

EXERCISES

In each of Exercises 1–4, determine whether the given set of vectors is linearly dependent or linearly independent.

1. $\begin{pmatrix}1\\1\\1\end{pmatrix}$, $\begin{pmatrix}1\\-1\\1\end{pmatrix}$ and $\begin{pmatrix}-4\\0\\-4\end{pmatrix}$ **2.** $\begin{pmatrix}1\\1\\0\end{pmatrix}$, $\begin{pmatrix}0\\1\\1\end{pmatrix}$ and $\begin{pmatrix}1\\0\\1\end{pmatrix}$

3. $\begin{pmatrix}-1\\1\\1\end{pmatrix}$, $\begin{pmatrix}1\\-1\\1\end{pmatrix}$ and $\begin{pmatrix}1\\1\\-1\end{pmatrix}$

4. $\begin{pmatrix}1\\2\\6\end{pmatrix}$, $\begin{pmatrix}3\\-13\\7\end{pmatrix}$, $\begin{pmatrix}-1\\0\\1\end{pmatrix}$ and $\begin{pmatrix}-1\\-1\\-1\end{pmatrix}$

5. Let **V** be the set of all 2×2 matrices. Determine whether the following sets of matrices are linearly dependent or linearly independent in **V**.

(a) $\begin{pmatrix}1 & 0\\1 & 0\end{pmatrix}$, $\begin{pmatrix}1 & 0\\0 & 1\end{pmatrix}$, $\begin{pmatrix}0 & 1\\1 & 0\end{pmatrix}$ and $\begin{pmatrix}0 & 1\\0 & 1\end{pmatrix}$

(b) $\begin{pmatrix}1 & 0\\1 & 0\end{pmatrix}$, $\begin{pmatrix}1 & 0\\0 & 1\end{pmatrix}$, $\begin{pmatrix}0 & 1\\1 & 0\end{pmatrix}$ and $\begin{pmatrix}2 & -2\\-1 & 1\end{pmatrix}$.

6. Let **V** be the space of all polynomials in t of degree $\leqslant 2$.
(a) Show that $\dim \mathbf{V}=3$.
(b) Let p_1, p_2 and p_3 be the three polynomials whose values at any time t are $(t-1)^2$, $(t-2)^2$, and $(t-1)(t-2)$ respectively. Show that p_1, p_2, and p_3 are linearly independent. Hence, conclude from Theorem 3 that p_1, p_2, and p_3 form a basis for **V**.

7. Let **V** be the set of all solutions of the differential equation $d^2y/dt^2-y=0$.
(a) Show that **V** is a vector space.
(b) Find a basis for **V**.

8. Let **V** be the set of all solutions of the differential equation $(d^3y/dt^3)+y=0$ which satisfy $y(0)=0$. Show that **V** is a vector space and find a basis for it.

9. Let **V** be the set of all polynomials $p(t)=a_0+a_1t+a_2t^2$ which satisfy

$$p(0)+2p'(0)+3p''(0)=0.$$

Show that **V** is a vector space and find a basis for it.

10. Let **V** be the set of all solutions

$$\mathbf{x}=\begin{bmatrix}x_1(t)\\x_2(t)\\x_3(t)\end{bmatrix}$$

of the differential equation

$$\dot{\mathbf{x}}=\begin{pmatrix}0 & 1 & 0\\0 & 0 & 1\\6 & -11 & 6\end{pmatrix}\mathbf{x}.$$

Show that

$$\mathbf{x}^1(t)=\begin{bmatrix}e^t\\e^t\\e^t\end{bmatrix},\qquad \mathbf{x}^2(t)=\begin{bmatrix}e^{2t}\\2e^{2t}\\4e^{2t}\end{bmatrix},\quad\text{and}\quad \mathbf{x}^3(t)=\begin{bmatrix}e^{3t}\\3e^{3t}\\9e^{3t}\end{bmatrix}$$

form a basis for **V**.

11. Let $\mathbf{V}$ be a vector space. We say that $\mathbf{W}$ is a subspace of $\mathbf{V}$ if $\mathbf{W}$ is a subset of $\mathbf{V}$ which is itself a vector space. Let $\mathbf{W}$ be the subset of R^3 which consists of all vectors

$$\mathbf{x}=\begin{pmatrix} x_1\\ x_2\\ x_3\end{pmatrix}$$

which satisfy the equations

$$\begin{aligned} x_1+x_2+2x_3&=0\\ 2x_1-x_2+\ \ x_3&=0\\ 6x_1 \qquad +6x_3&=0.\end{aligned}$$

Show that $\mathbf{W}$ is a subspace of R^3 and find a basis for it.

12. Prove that any n vectors which span an n dimensional vector space $\mathbf{V}$ must be linearly independent. *Hint*: Show that any set of linearly dependent vectors contains a linearly independent subset which also spans $\mathbf{V}$.

13. Let $\mathbf{v}^1,\mathbf{v}^2,\dots,\mathbf{v}^n$ be n vectors in a vector space $\mathbf{V}$. Let $\mathbf{W}$ be the subset of $\mathbf{V}$ which consists of all linear combinations $c_1\mathbf{v}^1+c_2\mathbf{v}^2+\dots+c_n\mathbf{v}^n$ of $\mathbf{v}^1,\mathbf{v}^2,\dots,\mathbf{v}^n$. Show that $\mathbf{W}$ is a subspace of $\mathbf{V}$, and that $\dim\mathbf{W}\leqslant n$.

14. Let $\mathbf{V}$ be the set of all functions $f(t)$ which are analytic for $|t|<1$, that is, $f(t)$ has a power series expansion $f(t)=a_0+a_1t+a_2t^2+\dots$ which converges for $|t|<1$. Show that $\mathbf{V}$ is a vector space, and that its dimension is infinite. *Hint*: $\mathbf{V}$ contains all polynomials.

15. Let $\mathbf{v}^1,\mathbf{v}^2,\dots,\mathbf{v}^m$ be m linearly independent vectors in an n dimensional vector space $\mathbf{V}$, with $n>m$. Show that we can find vectors $\mathbf{v}^{m+1},\dots,\mathbf{v}^n$ so that $\mathbf{v}^1,\mathbf{v}^2,\dots,\mathbf{v}^m,\mathbf{v}^{m+1},\dots,\mathbf{v}^n$ form a basis for $\mathbf{V}$. That is to say, any set of m linearly independent vectors in an $n>m$ dimensional space $\mathbf{V}$ can be completed to form a basis for $\mathbf{V}$.

16. Find a basis for R^3 which includes the vectors

$$\begin{pmatrix}1\\1\\0\end{pmatrix}\quad\text{and}\quad\begin{pmatrix}1\\3\\4\end{pmatrix}.$$

17. (a) Show that

$$\mathbf{v}^1=\begin{pmatrix}1\\0\\0\end{pmatrix},\qquad \mathbf{v}^2=\frac{1}{\sqrt{2}}\begin{pmatrix}0\\1\\1\end{pmatrix},\quad\text{and}\quad \mathbf{v}^3=\frac{1}{\sqrt{2}}\begin{pmatrix}0\\-1\\1\end{pmatrix}$$

are linearly independent in R^3.

(b) Let

$$\mathbf{x}=\begin{pmatrix}x_1\\x_2\\x_3\end{pmatrix}=x_1\mathbf{e}^1+x_2\mathbf{e}^2+x_3\mathbf{e}^3.$$

Since $\mathbf{v}^1, \mathbf{v}^2$, and $\mathbf{v}^3$ are linearly independent they are a basis and $\mathbf{x}=y_1\mathbf{v}^1+y_2\mathbf{v}^2+y_3\mathbf{v}^3$. What is the relationship between the original coordinates x_i and the new coordinates y_j?

(c) Express the relations between coordinates in the form $\mathbf{x}=\mathbf{B}\mathbf{y}$. Show that the columns of $\mathbf{B}$ are $\mathbf{v}^1$, $\mathbf{v}^2$, and $\mathbf{v}^3$.

3.4 Applications of linear algebra to differential equations

Recall that an important tool in solving the second-order linear homogeneous equation $(d^2y/dt^2)+p(t)(dy/dt)+q(t)y=0$ was the existence–uniqueness theorem stated in Section 2.1. In a similar manner, we will make extensive use of Theorem 4 below in solving the homogeneous linear system of differential equations

$$\frac{d\mathbf{x}}{dt}=\mathbf{A}\mathbf{x}, \qquad \mathbf{x}=\begin{bmatrix} x_1 \\ \vdots \\ x_n \end{bmatrix}, \qquad \mathbf{A}=\begin{bmatrix} a_{11} & \dots & a_{1n} \\ \vdots & & \vdots \\ a_{n1} & \dots & a_{nn} \end{bmatrix}. \tag{1}$$

The proof of this theorem will be indicated in Section 4.6.

Theorem 4 (Existence–uniqueness theorem). *There exists one, and only one, solution of the initial-value problem*

$$\frac{d\mathbf{x}}{dt}=\mathbf{A}\mathbf{x}, \qquad \mathbf{x}(t_0)=\mathbf{x}^0=\begin{bmatrix} x_1^0 \\ x_2^0 \\ \vdots \\ x_n^0 \end{bmatrix}. \tag{2}$$

Moreover, this solution exists for $-\infty<t<\infty$.

Theorem 4 is an extremely powerful theorem, and has many implications. In particular, if $\mathbf{x}(t)$ is a nontrivial solution, then $\mathbf{x}(t)\neq\mathbf{0}$ for any t. (If $\mathbf{x}(t^*)=\mathbf{0}$ for some t^*, then $\mathbf{x}(t)$ must be identically zero, since it, and the trivial solution, satisfy the same differential equation and have the same value at $t=t^*$.)

We have already shown (see Example 7, Section 3.2) that the space $\mathbf{V}$ of all solutions of (1) is a vector space. Our next step is to determine the dimension of $\mathbf{V}$.

Theorem 5. *The dimension of the space* $\mathbf{V}$ *of all solutions of the homogeneous linear system of differential equations* (1) *is* n.

PROOF. We will exhibit a basis for $\mathbf{V}$ which contains n elements. To this end, let $\boldsymbol{\phi}^j(t), j=1,\ldots,n$ be the solution of the initial-value problem

$$\frac{d\mathbf{x}}{dt}=\mathbf{A}\mathbf{x}, \qquad \mathbf{x}(0)=\mathbf{e}^j=\begin{bmatrix}0\\ \vdots\\ 0\\ 1\\ 0\\ \vdots\\ 0\end{bmatrix} -j\text{th row}. \tag{3}$$

For example, $\boldsymbol{\phi}^1(t)$ is the solution of the differential equation (1) which satisfies the initial condition

$$\boldsymbol{\phi}^1(0)=\mathbf{e}^1=\begin{bmatrix}1\\ 0\\ \vdots\\ 0\end{bmatrix}.$$

Note from Theorem 4 that $\boldsymbol{\phi}^j(t)$ exists for all t and is unique. To determine whether $\boldsymbol{\phi}^1,\boldsymbol{\phi}^2,\ldots,\boldsymbol{\phi}^n$ are linearly dependent or linearly independent vectors in $\mathbf{V}$, we consider the equation

$$c_1\boldsymbol{\phi}^1+c_2\boldsymbol{\phi}^2+\ldots+c_n\boldsymbol{\phi}^n=\mathbf{0} \tag{4}$$

where the zero on the right-hand side of (4) stands for the zero vector in $\mathbf{V}$ (that is, the vector whose every component is the zero function). We want to show that (4) implies $c_1=c_2=\ldots=c_n=0$. Evaluating both sides of (4) at $t=0$ gives

$$c_1\boldsymbol{\phi}^1(0)+c_2\boldsymbol{\phi}^2(0)+\ldots+c_n\boldsymbol{\phi}^n(0)=\begin{bmatrix}0\\ 0\\ \vdots\\ 0\end{bmatrix}=\mathbf{0}$$

or

$$c_1\mathbf{e}^1+c_2\mathbf{e}^2+\ldots+c_n\mathbf{e}^n=\mathbf{0}.$$

Since we know that $\mathbf{e}^1,\mathbf{e}^2,\ldots,\mathbf{e}^n$ are linearly independent in R^n, $c_1=c_2=\ldots=c_n=0$. We conclude, therefore, that $\boldsymbol{\phi}^1,\boldsymbol{\phi}^2,\ldots,\boldsymbol{\phi}^n$ are linearly independent vectors in $\mathbf{V}$.

Next, we claim that $\boldsymbol{\phi}^1,\boldsymbol{\phi}^2,\ldots,\boldsymbol{\phi}^n$ also span $\mathbf{V}$. To prove this, we must show that any vector $\mathbf{x}$ in $\mathbf{V}$ (that is, any solution $\mathbf{x}(t)$ of (1)) can be written as a linear combination of $\boldsymbol{\phi}^1,\boldsymbol{\phi}^2,\ldots,\boldsymbol{\phi}^n$. To this end, pick any $\mathbf{x}$ in $\mathbf{V}$, and let

$$\mathbf{c}=\begin{bmatrix} c_1 \\ c_2 \\ \vdots \\ c_n \end{bmatrix}$$

be the value of $\mathbf{x}$ at $t=0$ ($\mathbf{x}(0)=\mathbf{c}$). With these constants $c_1,c_2,\ldots,c_n$, construct the vector-valued function

$$\boldsymbol{\phi}(t)=c_1\boldsymbol{\phi}^1(t)+c_2\boldsymbol{\phi}^2(t)+\ldots+c_n\boldsymbol{\phi}^n(t).$$

We know that $\boldsymbol{\phi}(t)$ satisfies (1) since it is a linear combination of solutions. Moreover,

$$\boldsymbol{\phi}(0)=c_1\boldsymbol{\phi}^1(0)+c_2\boldsymbol{\phi}^2(0)+\ldots+c_n\boldsymbol{\phi}^n(0)$$

$$=c_1\begin{bmatrix} 1 \\ 0 \\ \vdots \\ 0 \end{bmatrix}+c_2\begin{bmatrix} 0 \\ 1 \\ \vdots \\ 0 \end{bmatrix}+\ldots+c_n\begin{bmatrix} 0 \\ 0 \\ \vdots \\ 1 \end{bmatrix}=\begin{bmatrix} c_1 \\ c_2 \\ \vdots \\ c_n \end{bmatrix}=\mathbf{x}(0).$$

Now, observe that $\mathbf{x}(t)$ and $\boldsymbol{\phi}(t)$ satisfy the same homogeneous linear system of differential equations, and that $\mathbf{x}(t)$ and $\boldsymbol{\phi}(t)$ have the same value at $t=0$. Consequently, by Theorem 4, $\mathbf{x}(t)$ and $\boldsymbol{\phi}(t)$ must be identical, that is

$$\mathbf{x}(t)\equiv\boldsymbol{\phi}(t)=c_1\boldsymbol{\phi}^1(t)+c_2\boldsymbol{\phi}^2(t)+\ldots+c_n\boldsymbol{\phi}^n(t).$$

Thus, $\boldsymbol{\phi}^1,\boldsymbol{\phi}^2,\ldots,\boldsymbol{\phi}^n$ also span $\mathbf{V}$. Therefore, by Theorem 2 of Section 3.3, $\dim\mathbf{V}=n$. □

Theorem 5 states that the space $\mathbf{V}$ of all solutions of (1) has dimension n. Hence, we need only guess, or by some means find, n linearly independent solutions of (1). Theorem 6 below establishes a test for linear independence of solutions. It reduces the problem of determining whether n solutions $\mathbf{x}^1,\mathbf{x}^2,\ldots,\mathbf{x}^n$ are linearly independent to the much simpler problem of determining whether their values $\mathbf{x}^1(t_0),\mathbf{x}^2(t_0),\ldots,\mathbf{x}^n(t_0)$ at an appropriate time t_0 are linearly independent vectors in R^n.

Theorem 6 (Test for linear independence). *Let $\mathbf{x}^1,\mathbf{x}^2,\ldots,\mathbf{x}^k$ be k solutions of $\dot{\mathbf{x}}=\mathbf{A}\mathbf{x}$. Select a convenient t_0. Then, $\mathbf{x}^1,\ldots,\mathbf{x}^k$ are linear independent solutions if, and only if, $\mathbf{x}^1(t_0),\mathbf{x}^2(t_0),\ldots,\mathbf{x}^k(t_0)$ are linearly independent vectors in R^n.*

Proof. Suppose that $\mathbf{x}^1,\mathbf{x}^2,\ldots,\mathbf{x}^k$ are linearly dependent solutions. Then, there exist constants $c_1,c_2,\ldots,c_k$, not all zero, such that

$$c_1\mathbf{x}^1+c_2\mathbf{x}^2+\ldots+c_k\mathbf{x}^k=\mathbf{0}.$$

Evaluating this equation at $t=t_0$ gives

$$c_1\mathbf{x}^1(t_0)+c_2\mathbf{x}^2(t_0)+\ldots+c_k\mathbf{x}^k(t_0)=\begin{bmatrix}0\\0\\\vdots\\0\end{bmatrix}.$$

Hence $\mathbf{x}^1(t_0),\mathbf{x}^2(t_0),\ldots,\mathbf{x}^k(t_0)$ are linearly dependent vectors in R^n.

Conversely, suppose that the values of $\mathbf{x}^1,\mathbf{x}^2,\ldots,\mathbf{x}^k$ at some time t_0 are linearly dependent vectors in R^n. Then, there exist constants $c_1,c_2,\ldots,c_k$, not all zero, such that

$$c_1\mathbf{x}^1(t_0)+c_2\mathbf{x}^2(t_0)+\ldots+c_k\mathbf{x}^k(t_0)=\begin{bmatrix}0\\0\\\vdots\\0\end{bmatrix}=\mathbf{0}.$$

With this choice of constants $c_1,c_2,\ldots,c_k$, construct the vector-valued function

$$\boldsymbol{\phi}(t)=c_1\mathbf{x}^1(t)+c_2\mathbf{x}^2(t)+\ldots+c_k\mathbf{x}^k(t).$$

This function satisfies (1) since it is a linear combination of solutions. Moreover, $\boldsymbol{\phi}(t_0)=\mathbf{0}$. Hence, by Theorem 4, $\boldsymbol{\phi}(t)=\mathbf{0}$ for all t. This implies that $\mathbf{x}^1,\mathbf{x}^2,\ldots,\mathbf{x}^k$ are linearly dependent solutions. □

Example 1. Consider the system of differential equations

$$\begin{aligned}\frac{dx_1}{dt}&=x_2\\ \frac{dx_2}{dt}&=-x_1-2x_2\end{aligned}\quad\text{or}\quad\frac{d\mathbf{x}}{dt}=\begin{pmatrix}0&1\\-1&-2\end{pmatrix}\mathbf{x},\quad\mathbf{x}=\begin{pmatrix}x_1\\x_2\end{pmatrix}.\tag{5}$$

This system of equations arose from the single second-order equation

$$\frac{d^2y}{dt^2}+2\frac{dy}{dt}+y=0\tag{6}$$

by setting $x_1=y$ and $x_2=dy/dt$. Since $y_1(t)=e^{-t}$ and $y_2(t)=te^{-t}$ are two solutions of (6), we see that

$$\mathbf{x}^1(t)=\begin{pmatrix}e^{-t}\\-e^{-t}\end{pmatrix}$$

and

$$\mathbf{x}^2(t)=\begin{pmatrix}te^{-t}\\(1-t)e^{-t}\end{pmatrix}$$

are two solutions of (5). To determine whether $\mathbf{x}^1$ and $\mathbf{x}^2$ are linearly dependent or linearly independent, we check whether their initial values

$$\mathbf{x}^1(0)=\begin{pmatrix}1\\-1\end{pmatrix}$$

and

$$\mathbf{x}^2(0)=\begin{pmatrix}0\\1\end{pmatrix}$$

are linearly dependent or linearly independent vectors in R^2. Thus, we consider the equation

$$c_1\mathbf{x}^1(0)+c_2\mathbf{x}^2(0)=\begin{pmatrix}c_1\\-c_1+c_2\end{pmatrix}=\begin{pmatrix}0\\0\end{pmatrix}.$$

This equation implies that both c_1 and c_2 are zero. Hence, $\mathbf{x}^1(0)$ and $\mathbf{x}^2(0)$ are linearly independent vectors in R^2. Consequently, by Theorem 6, $\mathbf{x}^1(t)$ and $\mathbf{x}^2(t)$ are linearly independent solutions of (5), and every solution $\mathbf{x}(t)$ of (5) can be written in the form

$$\begin{aligned}\mathbf{x}(t)=\begin{pmatrix}x_1(t)\\x_2(t)\end{pmatrix}&=c_1\begin{pmatrix}e^{-t}\\-e^{-t}\end{pmatrix}+c_2\begin{pmatrix}te^{-t}\\(1-t)e^{-t}\end{pmatrix}\\&=\begin{pmatrix}(c_1+c_2t)e^{-t}\\(c_2-c_1-c_2t)e^{-t}\end{pmatrix}.\end{aligned}\tag{7}$$

Example 2. Solve the initial-value problem

$$\frac{d\mathbf{x}}{dt}=\begin{pmatrix}0&1\\-1&-2\end{pmatrix}\mathbf{x},\qquad \mathbf{x}(0)=\begin{pmatrix}1\\1\end{pmatrix}.$$

Solution. From Example 1, every solution $\mathbf{x}(t)$ must be of the form (7). The constants c_1 and c_2 are determined from the initial conditions

$$\begin{pmatrix}1\\1\end{pmatrix}=\mathbf{x}(0)=\begin{pmatrix}c_1\\c_2-c_1\end{pmatrix}.$$

Therefore, $c_1=1$ and $c_2=1+c_1=2$. Hence

$$\mathbf{x}(t)=\begin{pmatrix}x_1(t)\\x_2(t)\end{pmatrix}=\begin{pmatrix}(1+2t)e^{-t}\\(1-2t)e^{-t}\end{pmatrix}.$$

Up to this point in studying (1) we have found the concepts of linear algebra such as vector space, dependence, dimension, basis, etc., and vector-matrix notation useful, but we might well ask is all this other than simply an appropriate and convenient language. If it were nothing else it would be worth introducing. Good notations are important in expressing mathematical ideas. However, it is more. It is a body of theory with many applications.

In Sections 3.8–3.10 we will reduce the problem of finding all solutions of (1) to the much simpler algebraic problem of solving simultaneous linear equations of the form

$$\begin{aligned} a_{11}x_1 + a_{12}x_2 + \dots + a_{1n}x_n &= b_1 \\ a_{21}x_1 + a_{22}x_2 + \dots + a_{2n}x_n &= b_2 \\ &\vdots \\ a_{n1}x_1 + a_{n2}x_2 + \dots + a_{nn}x_n &= b_n \end{aligned}$$

Therefore, we will now digress to study the theory of simultaneous linear equations. Here too we will see the role played by linear algebra.

EXERCISES

In each of Exercises 1–4 find a basis for the set of solutions of the given differential equation.

1. $\dot{\mathbf{x}} = \begin{pmatrix} 0 & 1 \\ -1 & -1 \end{pmatrix}\mathbf{x}$ (*Hint*: Find a second-order differential equation satisfied by $x_1(t)$.)

2. $\dot{\mathbf{x}} = \begin{pmatrix} 0 & 1 & 0 \\ 0 & 0 & 1 \\ 2 & -1 & 2 \end{pmatrix}\mathbf{x}$ (*Hint*: Find a third-order differential equation satisfied by $x_1(t)$.)

3. $\dot{\mathbf{x}} = \begin{pmatrix} 1 & 0 \\ 2 & 1 \end{pmatrix}\mathbf{x}$

4. $\dot{\mathbf{x}} = \begin{pmatrix} 1 & 0 & 0 \\ 1 & 1 & 0 \\ 1 & 1 & 1 \end{pmatrix}\mathbf{x}$

For each of the differential equations 5–9 determine whether the given solutions are a basis for the set of all solutions.

5. $\dot{\mathbf{x}} = \begin{pmatrix} 0 & -1 \\ -1 & 0 \end{pmatrix}\mathbf{x}$; $\mathbf{x}^1(t) = \begin{pmatrix} e^t \\ -e^t \end{pmatrix}$, $\mathbf{x}^2(t) = \begin{pmatrix} e^{-t} \\ e^{-t} \end{pmatrix}$

6. $\dot{\mathbf{x}} = \begin{pmatrix} 4 & -2 & 2 \\ -1 & 3 & 1 \\ 1 & -1 & 5 \end{pmatrix}\mathbf{x}$; $\mathbf{x}^1(t) = \begin{bmatrix} e^{2t} \\ e^{2t} \\ 0 \end{bmatrix}$, $\mathbf{x}^2(t) = \begin{bmatrix} 0 \\ e^{4t} \\ e^{4t} \end{bmatrix}$, $\mathbf{x}^3(t) = \begin{bmatrix} e^{6t} \\ 0 \\ e^{6t} \end{bmatrix}$

7. $\dot{\mathbf{x}} = \begin{pmatrix} -3 & -2 & 3 \\ 1 & 0 & -3 \\ 1 & -2 & -1 \end{pmatrix}\mathbf{x}$; $\mathbf{x}^1(t) = \begin{bmatrix} e^{-2t} \\ e^{-2t} \\ e^{-2t} \end{bmatrix}$, $\mathbf{x}^2(t) = \begin{bmatrix} e^{2t} \\ -e^{2t} \\ e^{2t} \end{bmatrix}$, $\mathbf{x}^3(t) = \begin{bmatrix} -e^{-4t} \\ e^{-4t} \\ e^{-4t} \end{bmatrix}$

8. $\dot{\mathbf{x}} = \begin{pmatrix} -5 & 2 & -2 \\ 1 & -4 & -1 \\ -1 & 1 & -6 \end{pmatrix}\mathbf{x}$; $\mathbf{x}^1(t) = \begin{bmatrix} e^{-3t} \\ e^{-3t} \\ 0 \end{bmatrix}$, $\mathbf{x}^2(t) = \begin{bmatrix} 0 \\ e^{-5t} \\ e^{-5t} \end{bmatrix}$

9. $\dot{\mathbf{x}}=\begin{pmatrix} -5 & 2 & -2 \\ 1 & -4 & -1 \\ -1 & 1 & -6 \end{pmatrix}\mathbf{x}$; $\mathbf{x}^1(t)=\begin{bmatrix} e^{-3t} \\ e^{-3t} \\ 0 \end{bmatrix}$,

$\mathbf{x}^2(t)=\begin{bmatrix} 0 \\ e^{-5t} \\ e^{-5t} \end{bmatrix}$, $\mathbf{x}^3(t)=\begin{bmatrix} e^{-3t}+e^{-7t} \\ e^{-3t} \\ e^{-7t} \end{bmatrix}$, $\mathbf{x}^4(t)=\begin{bmatrix} 2e^{-7t} \\ e^{-5t} \\ e^{-5t}+2e^{-7t} \end{bmatrix}$

10. Determine the solutions $\boldsymbol{\phi}^1,\boldsymbol{\phi}^2,\dots,\boldsymbol{\phi}^n$ (see proof of Theorem 5) for the system of differential equations in (a) Problem 5; (b) Problem 6; (c) Problem 7.

11. Let $\mathbf{V}$ be the vector space of all continuous functions on $(-\infty,\infty)$ to R^n (the values of $\mathbf{x}(t)$ lie in R^n). Let $\mathbf{x}^1,\mathbf{x}^2,\dots,\mathbf{x}^n$ be functions in $\mathbf{V}$.
(a) Show that $\mathbf{x}^1(t_0),\dots,\mathbf{x}^n(t_0)$ linearly independent vectors in R^n for some t_0 implies $\mathbf{x}^1,\mathbf{x}^2,\dots,\mathbf{x}^n$ are linearly independent functions in $\mathbf{V}$.
(b) Is it true that $\mathbf{x}^1(t_0),\dots,\mathbf{x}^n(t_0)$ linearly dependent in R^n for some t_0 implies $\mathbf{x}^1,\mathbf{x}^2,\dots,\mathbf{x}^n$ are linearly dependent functions in $\mathbf{V}$? Justify your answer.

12. Let $\mathbf{u}$ be a vector in $\mathsf{R}^n(\mathbf{u}\neq 0)$.
(a) Is $\mathbf{x}(t)=t\mathbf{u}$ a solution of a linear homogeneous differential equation $\dot{\mathbf{x}}=\mathbf{A}\mathbf{x}$?
(b) Is $\mathbf{x}(t)=e^{\lambda t}\mathbf{u}$?
(c) Is $\mathbf{x}(t)=(e^t-e^{-t})\mathbf{u}$?
(d) Is $\mathbf{x}(t)=(e^t+e^{-t})\mathbf{u}$?
(e) Is $\mathbf{x}(t)=(e^{\lambda_1 t}+e^{\lambda_2 t})\mathbf{u}$?
(f) For what functions $\phi(t)$ can $\mathbf{x}(t)=\phi(t)\mathbf{u}$ be a solution of some $\dot{\mathbf{x}}=\mathbf{A}\mathbf{x}$?

3.5 The theory of determinants

In this section we will study simultaneous equations of the form

$$\begin{aligned} a_{11}x_1+a_{12}x_2+\dots+a_{1n}x_n&=b_1\\ a_{21}x_1+a_{22}x_2+\dots+a_{2n}x_n&=b_2\\ \vdots\qquad\qquad&\quad\vdots\\ a_{n1}x_1+a_{n2}x_2+\dots+a_{nn}x_n&=b_n. \end{aligned} \tag{1}$$

Our goal is to determine a necessary and sufficient condition for the system of equations (1) to have a unique solution $x_1,x_2,\dots,x_n$.

To gain some insight into this problem, we begin with the simplest case $n=2$. If we multiply the first equation $a_{11}x_1+a_{12}x_2=b_1$ by a_{21}, the second equation $a_{21}x_1+a_{22}x_2=b_2$ by a_{11}, and then subtract the former from the latter, we obtain that

$$(a_{11}a_{22}-a_{12}a_{21})x_2=a_{11}b_2-a_{21}b_1.$$

Similarly, if we multiply the first equation by a_{22}, the second equation by a_{12}, and then subtract the latter from the former, we obtain that

$$(a_{11}a_{22}-a_{12}a_{21})x_1=a_{22}b_1-a_{12}b_2.$$

Consequently, the system of equations (1) has a unique solution

$$x_1=\frac{a_{22}b_1-a_{12}b_2}{a_{11}a_{22}-a_{12}a_{21}},\qquad x_2=\frac{a_{11}b_2-a_{21}b_1}{a_{11}a_{22}-a_{12}a_{21}}$$

if the number $a_{11}a_{22} - a_{12}a_{21}$ is unequal to zero. If this number equals zero, then we may, or may not, have a solution. For example, the system of equations

$$x_1 - x_2 = 1, \qquad 2x_1 - 2x_2 = 4$$

obviously has no solutions, while the system of equations

$$x_1 - x_2 = 0, \qquad 2x_1 - 2x_2 = 0$$

has an infinity of solutions $x_1 = c, x_2 = c$ for any number c. For both these systems of equations,

$$a_{11}a_{22} - a_{12}a_{21} = 1(-2) - (-1)2 = 0.$$

The case of three equations

$$\begin{aligned} a_{11}x_1 + a_{12}x_2 + a_{13}x_3 &= b_1 \\ a_{21}x_1 + a_{22}x_2 + a_{23}x_3 &= b_2 \\ a_{31}x_1 + a_{32}x_2 + a_{33}x_3 &= b_3 \end{aligned} \tag{2}$$

in three unknowns x_1, x_2, x_3 can also be handled quite easily. By eliminating one of the variables from two of the equations (2), and thus reducing ourselves to the case $n = 2$, it is possible to show (see Exercise 1) that the system of equations (2) has a unique solution x_1, x_2, x_3 if, and only if, the number

$$a_{11}a_{22}a_{33} + a_{12}a_{23}a_{31} + a_{13}a_{21}a_{32} - a_{13}a_{22}a_{31} - a_{12}a_{21}a_{33} - a_{11}a_{23}a_{32} \tag{3}$$

is unequal to zero.

We now suspect that the system of equations (1), which we abbreviate in the form

$$\mathbf{Ax} = \mathbf{b}, \quad \mathbf{A} = \begin{pmatrix} a_{11} & \dots & a_{1n} \\ \vdots & & \vdots \\ a_{n1} & \dots & a_{nn} \end{pmatrix}, \quad \mathbf{x} = \begin{pmatrix} x_1 \\ \vdots \\ x_n \end{pmatrix}, \quad \mathbf{b} = \begin{pmatrix} b_1 \\ \vdots \\ b_n \end{pmatrix}, \tag{4}$$

has a unique solution $\mathbf{x}$ if, and only if, a certain number, which depends on the elements a_{ij} of the matrix $\mathbf{A}$, is unequal to zero. We can determine this number for $n = 4$ by eliminating one of the variables from two of the equations (1). However, the algebra is so complex that the resulting number is unintelligible. Instead, we will generalize the number (3) so as to associate with each system of equations $\mathbf{Ax} = \mathbf{b}$ a single number called determinant $\mathbf{A}$ ($\det \mathbf{A}$ for short), which depends on the elements of the matrix $\mathbf{A}$. We will establish several useful properties of this association, and then use these properties to show that the system of equations (4) has a unique solution $\mathbf{x}$ if, and only if, $\det \mathbf{A} \neq 0$.

If we carefully analyze the number (3), we see that it can be described in the following interesting manner. First, we pick an element a_{1j_1} from the first row of the matrix

$$\mathbf{A}=\begin{bmatrix} a_{11} & a_{12} & a_{13} \\ a_{21} & a_{22} & a_{23} \\ a_{31} & a_{32} & a_{33} \end{bmatrix}.$$

This element can be either a_{11}, a_{12}, or a_{13}. Then, we multiply a_{1j_1} by an element a_{2j_2} from the second row of **A**. However, j_2 must not equal j_1. For example, if we choose a_{12} from the first row of **A**, then we must choose either a_{21} or a_{23} from the second row of **A**. Next, we multiply these two numbers by the element in the third row of **A** in the remaining column. We do this for all possible choices of picking one element from each row of **A**, never picking from the same column twice. In this manner, we obtain 6 different products of three elements of **A**, since there are three ways of choosing an element from the first row of **A**, then two ways of picking an element from the second row of **A**, and then only one way of choosing an element from the third row of **A**. Each of these products $a_{1j_1}a_{2j_2}a_{3j_3}$ is multiplied by ± 1, depending on the specific order $j_1 j_2 j_3$. The products $a_{1j_1}a_{2j_2}a_{3j_3}$ with $(j_1 j_2 j_3)=(123)$, (231), and (312) are multiplied by $+1$, while the products $a_{1j_1}a_{2j_2}a_{3j_3}$ with $(j_1 j_2 j_3)=(321)$, (213), and (132) are multiplied by -1. Finally, the resulting numbers are added together.

The six sets of numbers (123), (231), (312), (321), (213), and (132) are called *permutations*, or *scramblings*, of the integers 1, 2, and 3. Observe that each of the three permutations corresponding to the plus terms requires an even number of interchanges of adjacent integers to unscramble the permutation, that is, to bring the integers back to their natural order. Similarly, each of the three permutations corresponding to the minus terms requires an odd number of interchanges of adjacent integers to unscramble the permutation. To verify this, observe that

$$\begin{aligned} &231 \to 213 \to 123 \quad \text{(2 interchanges)} \\ &312 \to 132 \to 123 \quad \text{(2 interchanges)} \\ &321 \to 312 \to 132 \to 123 \quad \text{(3 interchanges)} \\ &213 \to 123 \quad \text{and} \quad 132 \to 123 \quad \text{(1 interchange each).} \end{aligned}$$

This motivates the following definition of the *determinant* of an $n \times n$ matrix **A**.

Definition.

$$\det \mathbf{A} = \sum_{j_1,\dots,j_n} \varepsilon_{j_1 j_2 \dots j_n} a_{1j_1} a_{2j_2} \dots a_{nj_n} \tag{5}$$

where $\varepsilon_{j_1 j_2 \dots j_n} = 1$ if the permutation $(j_1 j_2 \dots j_n)$ is even, that is, if we can bring the integers $j_1 \dots j_n$ back to their natural order by an even number of interchanges of adjacent integers, and $\varepsilon_{j_1 j_2 \dots j_n} = -1$ if the permutation $(j_1 j_2 \dots j_n)$ is odd. In other words, pick an element a_{1j_1} from the first row of the matrix $\mathbf{A}$. Then, multiply it by an element a_{2j_2} from the second row of $\mathbf{A}$, with $j_2 \neq j_1$. Continue this process, going from row to row, and always picking from a new column. Finally, multiply the product $a_{1j_1}a_{2j_2}\dots a_{nj_n}$ by $+1$ if the permutation $(j_1 j_2 \dots j_n)$ is even and by -1 if the permutation is odd. Do this for all possible choices of picking one element from each row of $\mathbf{A}$, never picking from the same column twice. Then, add up all these contributions and denote the resulting number by $\det \mathbf{A}$.

Remark. There are many different ways of bringing a permutation of the integers $1,2,\dots,n$ back to their natural order by successive interchanges of adjacent integers. For example,

$$4312 \to 4132 \to 1432 \to 1423 \to 1243 \to 1234$$

and

$$4312 \to 3412 \to 3142 \to 3124 \to 1324 \to 1234.$$

However, it can be shown that the number of interchanges of adjacent integers necessary to unscramble the permuation $j_1 \dots j_n$ is always odd or always even. Hence $\varepsilon_{j_1\dots j_n}$ is perfectly well defined.

Example 1. Let

$$\mathbf{A} = \begin{pmatrix} a_{11} & a_{12} \\ a_{21} & a_{22} \end{pmatrix}.$$

In this case, there are only two products $a_{11}a_{22}$ and $a_{12}a_{21}$ that enter into the definition of $\det \mathbf{A}$. Since the permutation (12) is even, and the permutation (21) is odd, the term $a_{11}a_{22}$ is multiplied by $+1$ and the term $a_{12}a_{21}$ is multiplied by -1. Hence, $\det \mathbf{A} = a_{11}a_{22} - a_{12}a_{21}$.

Example 2. Compute

$$\det\begin{bmatrix} 1 & 1 & 1 \\ 3 & 2 & 1 \\ -1 & -1 & 2 \end{bmatrix}.$$

Solution. A shorthand method for computing the determinant of a 3×3 matrix is to write the first two columns after the matrix and then take products along the diagonals as shown below.

$$\det\begin{bmatrix} 1 & 1 & 1 \\ 3 & 2 & 1 \\ -1 & -1 & 2 \end{bmatrix} = \begin{array}{ccccc} 1 & 1 & 1 & 1 & 1 \\ 3 & 2 & 1 & 3 & 2 \\ -1 & -1 & 2 & -1 & -1 \end{array}$$

$$= 1\cdot 2\cdot 2 + 1\cdot 1\cdot(-1) + 1\cdot 3(-1) - (-1)\cdot 2\cdot 1 - (-1)\cdot 1\cdot 1 - 2\cdot 3\cdot 1 = -3.$$

If $\mathbf{A}$ is an $n\times n$ matrix, then $\det \mathbf{A}$ will contain, in general, $n!$ products of n elements. The determinant of a 4×4 matrix contains, in general, 24 terms, while the determinant of a 10×10 matrix contains the unacceptably large figure of 3,628,800 terms. Thus, it is practically impossible to compute the determinant of a large matrix $\mathbf{A}$ using the definition (5) alone. The smart way, and the only way, to compute determinants is to (i) find special matrices whose determinants are easy to compute, and (ii) reduce the problem of finding any determinant to the much simpler problem of computing the determinant of one of these special matrices. To this end, observe that there are three special classes of matrices whose determinants are trivial to compute.

1. *Diagonal matrices*: A matrix

$$\mathbf{A}=\begin{bmatrix} a_{11} & 0 & 0 & \dots & 0 \\ 0 & a_{22} & 0 & \dots & 0 \\ \vdots & \vdots & \vdots & & \vdots \\ 0 & 0 & 0 & \dots & a_{nn} \end{bmatrix},$$

whose nondiagonal elements are all zero, is called a diagonal matrix. Its determinant is the product of the diagonal elements $a_{11}, a_{22}, \dots, a_{nn}$. This follows immediately from the observation that the only way we can choose a nonzero element from the first row of $\mathbf{A}$ is to pick a_{11}. Similarly, the only way we can choose a nonzero element from the jth row of $\mathbf{A}$ is to pick a_{jj}. Thus the only nonzero product entering into the definition of $\det \mathbf{A}$ is $a_{11}a_{22}\dots a_{nn}$, and this term is multiplied by $+1$ since the permutation $(12\dots n)$ is even.

2. *Lower diagonal matrices*: A matrix

$$\mathbf{A}=\begin{bmatrix} a_{11} & 0 & \dots & 0 \\ a_{21} & a_{22} & \dots & 0 \\ \vdots & \vdots & & \vdots \\ a_{n1} & a_{n2} & \dots & a_{nn} \end{bmatrix}$$

whose elements above the main diagonal are all zero, is called a lower diagonal matrix, and its determinant too is the product of the diagonal elements $a_{11}, \dots, a_{nn}$. To prove this, observe that the only way we can choose a nonzero element from the first row of $\mathbf{A}$ is to pick a_{11}. The second row of $\mathbf{A}$ has two nonzero elements, but since we have already chosen from the first column, we are forced to pick a_{22} from the second row. Similarly, we are forced to pick a_{jj} from the jth row of $\mathbf{A}$. Thus, $\det \mathbf{A} = a_{11}a_{22}\dots a_{nn}$.

3. *Upper diagonal matrices*: A matrix

$$\mathbf{A}=\begin{bmatrix} a_{11} & a_{12} & \dots & a_{1n} \\ 0 & a_{22} & \dots & a_{2n} \\ \vdots & \vdots & & \vdots \\ 0 & 0 & \dots & a_{nn} \end{bmatrix}$$

whose elements below the main diagonal are all zero is called an upper diagonal matrix and its determinant too is the product of the diagonal elements $a_{11},\dots,a_{nn}$. To prove this, we proceed backwards. The only way we can choose a nonzero element from the last row of $\mathbf{A}$ is to pick a_{nn}. This then forces us to pick $a_{n-1,n-1}$ from the $(n-1)$st row of $\mathbf{A}$. Similarly, we are forced to pick a_{jj} from the jth row of $\mathbf{A}$. Hence $\det\mathbf{A}=a_{nn}\dots a_{22}a_{11}$.

We now derive some simple but extremely useful properties of determinants.

Property 1. If we interchange two adjacent rows of $\mathbf{A}$, then we change the sign of its determinant.

Proof. Let $\mathbf{B}$ be the matrix obtained from $\mathbf{A}$ by interchanging the kth and $(k+1)$st rows. Observe that all of the products entering into the definition of $\det\mathbf{B}$ are exactly the same as the products entering into the definition of $\det\mathbf{A}$. The only difference is that the order in which we choose from the columns of $\mathbf{A}$ and $\mathbf{B}$ is changed. For example, let

$$\mathbf{A}=\begin{bmatrix} 1 & 3 & 4 \\ 2 & -1 & 0 \\ 1 & 2 & -1 \end{bmatrix}$$

and

$$\mathbf{B}=\begin{bmatrix} 1 & 3 & 4 \\ 1 & 2 & -1 \\ 2 & -1 & 0 \end{bmatrix}.$$

The product $4\times2\times2$ appears in $\det\mathbf{A}$ by choosing first from the first row, third column; then from the second row, first column; and finally from the third row, second column. This same product appears in $\det\mathbf{B}$ by choosing first from the first row, third column; then from the second row, second column; and finally from the third row, first column. More generally, the term

$$a_{1j_1}\dots a_{kj_k}a_{k+1,j_{k+1}}\dots a_{nj_n}$$

in $\det\mathbf{A}$ corresponds to the term

$$b_{1j_1}\dots b_{k,j_{k+1}}b_{k+1,j_k}\dots b_{nj_n}$$

in $\det\mathbf{B}$. The sign of the first term is determined by the permutation $(j_1\dots j_k j_{k+1}\dots j_n)$ while the sign of the second term is determined by the permutation $(j_1\dots j_{k+1}j_k\dots j_n)$. Since the second permutation is obtained from the first by interchanging the kth and $(k+1)$st elements, we see that these two terms have opposite signs. Hence $\det\mathbf{B}=-\det\mathbf{A}$. □

Property 2. If we interchange any two rows of $\mathbf{A}$, then we change the sign of its determinant.

PROOF. We will show that the number of interchanges of adjacent rows required to interchange the ith and jth rows of $\mathbf{A}$ is odd. Property 2 will then follow immediately from Property 1. To this end, assume that j is greater than i. We need $j-i$ successive interchanges of adjacent rows to get the jth row into the ith place, and then $j-i-1$ successive interchanges of adjacent rows to get the original ith row into the jth place. Thus the total number of interchanges required is $2(j-i)-1$, and this number is always odd. □

Property 3. If any two rows of $\mathbf{A}$ are equal, then $\det \mathbf{A}=0$.

PROOF. Let the ith and jth rows of $\mathbf{A}$ be equal, and let $\mathbf{B}$ be the matrix obtained from $\mathbf{A}$ by interchanging its ith and jth rows. Obviously, $\mathbf{B}=\mathbf{A}$. But Property 2 states that $\det \mathbf{B}=-\det \mathbf{A}$. Hence, $\det \mathbf{A}=-\det \mathbf{A}$ if two rows of $\mathbf{A}$ are equal, and this is possible only if $\det \mathbf{A}=0$. □

Property 4. $\det c\mathbf{A}=c^n \det \mathbf{A}$.

PROOF. Obvious. □

Property 5. Let $\mathbf{B}$ be the matrix obtained from $\mathbf{A}$ by multiplying its ith row by a constant c. Then, $\det \mathbf{B}=c\det \mathbf{A}$.

PROOF. Obvious. □

Property 6. Let $\mathbf{A}^T$ be the matrix obtained from $\mathbf{A}$ by switching rows and columns. The matrix $\mathbf{A}^T$ is called the transpose of $\mathbf{A}$. For example, if

$$\mathbf{A}=\begin{bmatrix} 1 & 3 & 2\\ 6 & 9 & 4\\ -1 & 2 & 7\end{bmatrix}, \quad \text{then} \quad \mathbf{A}^T=\begin{bmatrix} 1 & 6 & -1\\ 3 & 9 & 2\\ 2 & 4 & 7\end{bmatrix}.$$

A concise way of saying this is $(\mathbf{A}^T)_{ij}=a_{ji}$. Then,

$$\det \mathbf{A}^T=\det \mathbf{A}.$$

PROOF. It is clear that all the products entering into the definition of $\det \mathbf{A}$ and $\det \mathbf{A}^T$ are equal, since we always choose an element from each row and each column. The proof that these products have the same sign, though, is rather difficult, and will not be included here. (Frankly, whenever this author teaches determinants to his students he wishes that he were king, so that he could declare Property 6 true by edict.) □

Remark. It follows immediately from Properties 2, 3, and 6 that we change the sign of the determinant when we interchange two columns of $\mathbf{A}$, and that $\det \mathbf{A}=0$ if two columns of $\mathbf{A}$ are equal.

Property 7. If we add a multiple of one row of $\mathbf{A}$ to another row of $\mathbf{A}$, then we do not change the value of its determinant.

PROOF. We first make the crucial observation that $\det \mathbf{A}$ is a linear function of each row of $\mathbf{A}$ separately. By this we mean the following. Write the matrix $\mathbf{A}$ in the form

$$\mathbf{A}=\begin{bmatrix} a_{11} & a_{12} & \dots & a_{1n} \\ a_{21} & a_{22} & \dots & a_{2n} \\ \vdots & \vdots & & \vdots \\ a_{n1} & a_{n2} & \dots & a_{nn} \end{bmatrix}=\begin{bmatrix} \mathbf{a}^1 \\ \mathbf{a}^2 \\ \vdots \\ \mathbf{a}^n \end{bmatrix}$$

where

$$\begin{aligned} \mathbf{a}^1 &= (a_{11}, a_{12}, \dots, a_{1n}) \\ \mathbf{a}^2 &= (a_{21}, a_{22}, \dots, a_{2n}) \\ &\vdots \\ \mathbf{a}^n &= (a_{n1}, a_{n2}, \dots, a_{nn}). \end{aligned}$$

Then

$$\det\begin{bmatrix} \mathbf{a}^1 \\ \vdots \\ c\mathbf{a}^k \\ \vdots \\ \mathbf{a}^n \end{bmatrix} = c\det\begin{bmatrix} \mathbf{a}^1 \\ \vdots \\ \mathbf{a}^k \\ \vdots \\ \mathbf{a}^n \end{bmatrix} \tag{i}$$

and

$$\det\begin{bmatrix} \mathbf{a}^1 \\ \vdots \\ \mathbf{a}^k+\mathbf{b} \\ \vdots \\ \mathbf{a}^n \end{bmatrix} = \det\begin{bmatrix} \mathbf{a}^1 \\ \vdots \\ \mathbf{a}^k \\ \vdots \\ \mathbf{a}^n \end{bmatrix} + \det\begin{bmatrix} \mathbf{a}^1 \\ \vdots \\ \mathbf{b} \\ \vdots \\ \mathbf{a}^n \end{bmatrix}. \tag{ii}$$

For example,

$$\det\begin{bmatrix} 1 & 5 & 7 \\ 8 & 3 & 7 \\ 6 & -1 & 0 \end{bmatrix} = \det\begin{bmatrix} 1 & 5 & 7 \\ 4 & 1 & 9 \\ 6 & -1 & 0 \end{bmatrix} + \det\begin{bmatrix} 1 & 5 & 7 \\ 4 & 2 & -2 \\ 6 & -1 & 0 \end{bmatrix}$$

since $(4,1,9)+(4,2,-2)=(8,3,7)$. Now Equation (i) is Property 5. To derive Equation (ii), we compute

$$\det\begin{bmatrix}\mathbf{a}^1\\ \vdots\\ \mathbf{a}^k+\mathbf{b}\\ \vdots\\ \mathbf{a}^n\end{bmatrix}=\sum_{j_1,\dots,j_n}\varepsilon_{j_1\dots j_n}a_{1j_1}\dots\left(a_{kj_k}+b_{j_k}\right)\dots a_{nj_n}$$

$$=\sum_{j_1,\dots,j_n}\varepsilon_{j_1\dots j_n}a_{1j_1}\dots a_{kj_k}\dots a_{nj_n}+\sum_{j_1,\dots,j_n}\varepsilon_{j_1\dots j_n}a_{1j_1}\dots b_{j_k}\dots a_{nj_n}$$

$$=\det\begin{bmatrix}\mathbf{a}^1\\ \vdots\\ \mathbf{a}^k\\ \vdots\\ \mathbf{a}^n\end{bmatrix}+\det\begin{bmatrix}\mathbf{a}^1\\ \vdots\\ \mathbf{b}\\ \vdots\\ \mathbf{a}^n\end{bmatrix}.$$

Property 7 now follows immediately from Equation (ii), for if $\mathbf{B}$ is the matrix obtained from $\mathbf{A}$ by adding a multiple c of the kth row of $\mathbf{A}$ to the jth row of $\mathbf{A}$, then

$$\det\mathbf{B}=\det\begin{bmatrix}\mathbf{a}^1\\ \vdots\\ \mathbf{a}^j+c\mathbf{a}^k\\ \vdots\\ \mathbf{a}^k\\ \vdots\\ \mathbf{a}^n\end{bmatrix}=\det\begin{bmatrix}\mathbf{a}^1\\ \vdots\\ \mathbf{a}^j\\ \vdots\\ \mathbf{a}^k\\ \vdots\\ \mathbf{a}^n\end{bmatrix}+\det\begin{bmatrix}\mathbf{a}^1\\ \vdots\\ c\mathbf{a}^k\\ \vdots\\ \mathbf{a}^k\\ \vdots\\ \mathbf{a}^n\end{bmatrix}=\det\mathbf{A}+c\det\begin{bmatrix}\mathbf{a}^1\\ \vdots\\ \mathbf{a}^k\\ \vdots\\ \mathbf{a}^k\\ \vdots\\ \mathbf{a}^n\end{bmatrix}.$$

But

$$\det\begin{bmatrix}\mathbf{a}^1\\ \vdots\\ \mathbf{a}^k\\ \vdots\\ \mathbf{a}^k\\ \vdots\\ \mathbf{a}^n\end{bmatrix}=0$$

since this matrix has two equal rows. Hence, $\det\mathbf{B}=\det\mathbf{A}$. □

Remark 1. Everything we say about rows applies to columns, since $\det \mathbf{A}^T = \det \mathbf{A}$. Thus, we do not change the value of the determinant when we add a multiple of one column of $\mathbf{A}$ to another column of $\mathbf{A}$.

Remark 2. The determinant is a linear function of each row of $\mathbf{A}$ separately. It is not a linear function of $\mathbf{A}$ itself, since, in general,

$$\det c\mathbf{A} \neq c \det \mathbf{A} \quad \text{and} \quad \det(\mathbf{A}+\mathbf{B}) \neq \det \mathbf{A} + \det \mathbf{B}.$$

For example, if

$$\mathbf{A} = \begin{pmatrix} 1 & -2 \\ 0 & 3 \end{pmatrix}$$

and

$$\mathbf{B} = \begin{pmatrix} -1 & 7 \\ 0 & 9 \end{pmatrix},$$

then

$$\det(\mathbf{A}+\mathbf{B}) = \det\begin{pmatrix} 0 & 5 \\ 0 & 12 \end{pmatrix} = 0,$$

while $\det \mathbf{A} + \det \mathbf{B} = 3 - 9 = -6$.

Property 7 is extremely important because it enables us to reduce the problem of computing any determinant to the much simpler problem of computing the determinant of an upper diagonal matrix. To wit, if

$$\mathbf{A} = \begin{bmatrix} a_{11} & a_{12} & \cdots & a_{1n} \\ a_{21} & a_{22} & \cdots & a_{2n} \\ \vdots & \vdots & & \vdots \\ a_{n1} & a_{n2} & \cdots & a_{nn} \end{bmatrix}$$

and $a_{11} \neq 0$, then we can add suitable multiples of the first row of $\mathbf{A}$ to the remaining rows of $\mathbf{A}$ so as to make the resulting values of $a_{21}, \ldots, a_{n1}$ all zero. Similarly, we can add multiples of the resulting second row of $\mathbf{A}$ to the rows beneath it so as to make the resulting values of $a_{32}, \ldots, a_{n2}$ all zero, and so on. We illustrate this method with the following example.

Example 3. Compute

$$\det\begin{bmatrix} 1 & -1 & 2 & 3 \\ 2 & 2 & 0 & 2 \\ 4 & 1 & -1 & -1 \\ 1 & 2 & 3 & 0 \end{bmatrix}.$$

Solution. Subtracting twice the first row from the second row; four times the first row from the third row; and the first row from the last row gives

$$\det\begin{bmatrix} 1 & -1 & 2 & 3 \\ 2 & 2 & 0 & 2 \\ 4 & 1 & -1 & -1 \\ 1 & 2 & 3 & 0 \end{bmatrix} = \det\begin{bmatrix} 1 & -1 & 2 & 3 \\ 0 & 4 & -4 & -4 \\ 0 & 5 & -9 & -13 \\ 0 & 3 & 1 & -3 \end{bmatrix}$$

$$= 4\det\begin{bmatrix} 1 & -1 & 2 & 3 \\ 0 & 1 & -1 & -1 \\ 0 & 5 & -9 & -13 \\ 0 & 3 & 1 & -3 \end{bmatrix}.$$

Next, we subtract five times the second row of this latter matrix from the third row, and three times the second row from the fourth row. Then

$$\det\begin{bmatrix} 1 & -1 & 2 & 3 \\ 2 & 2 & 0 & 2 \\ 4 & 1 & -1 & -1 \\ 1 & 2 & 3 & 0 \end{bmatrix} = 4\det\begin{bmatrix} 1 & -1 & 2 & 3 \\ 0 & 1 & -1 & -1 \\ 0 & 0 & -4 & -8 \\ 0 & 0 & 4 & 0 \end{bmatrix}.$$

Finally, adding the third row of this matrix to the fourth row gives

$$\det\begin{bmatrix} 1 & -1 & 2 & 3 \\ 2 & 2 & 0 & 2 \\ 4 & 1 & -1 & -1 \\ 1 & 2 & 3 & 0 \end{bmatrix} = 4\det\begin{bmatrix} 1 & -1 & 2 & 3 \\ 0 & 1 & -1 & -1 \\ 0 & 0 & -4 & -8 \\ 0 & 0 & 0 & -8 \end{bmatrix}$$
$$= 4(-4)(-8) = 128.$$

(Alternately, we could have interchanged the third and fourth columns of the matrix

$$\begin{bmatrix} 1 & -1 & 2 & 3 \\ 0 & 1 & -1 & -1 \\ 0 & 0 & -4 & -8 \\ 0 & 0 & 4 & 0 \end{bmatrix}$$

to yield the same result.)

Remark 1. If $a_{11}=0$ and $a_{j1}\neq 0$ for some j, then we can interchange the first and jth rows of $\mathbf{A}$ so as to make $a_{11}\neq 0$. (We must remember, of course, to multiply the determinant by -1.) If the entire first column of $\mathbf{A}$ is zero, that is, if $a_{11}=a_{21}=\ldots=a_{n1}=0$ then we need proceed no further for $\det\mathbf{A}=0$.

Remark 2. In exactly the same manner as we reduced the matrix **A** to an upper diagonal matrix, we can reduce the system of equations

$$\begin{aligned} a_{11}x_1+a_{12}x_2+\ldots+a_{1n}x_n&=b_1\\ a_{21}x_1+a_{22}x_2+\ldots+a_{2n}x_n&=b_2\\ &\vdots\\ a_{n1}x_1+a_{n2}x_2+\ldots+a_{nn}x_n&=b_n \end{aligned}$$

to an equivalent system of the form

$$\begin{aligned} c_{11}x_1+c_{12}x_2+\ldots+c_{1n}x_n&=d_1\\ c_{22}x_2+\ldots+c_{2n}x_n&=d_2\\ &\vdots\\ c_{nn}x_n&=d_n. \end{aligned}$$

We can then solve (if $c_{nn}\neq 0$) for x_n from the last equation, for x_{n-1} from the $(n-1)$st equation, and so on.

Example 4. Find all solutions of the system of equations

$$\begin{aligned} x_1+x_2+x_3&=1\\ -x_1+x_2+x_3&=2\\ 2x_1-x_2+x_3&=3. \end{aligned}$$

Solution. Adding the first equation to the second equation and subtracting twice the first equation from the third equation gives

$$\begin{aligned} x_1+x_2+x_3&=1\\ 2x_2+2x_3&=3\\ -3x_2-x_3&=1. \end{aligned}$$

Next, adding $\frac{3}{2}$ the second equation to the third equation gives

$$\begin{aligned} x_1+x_2+x_3&=1\\ 2x_2+2x_3&=3\\ 2x_3&=\tfrac{11}{2}. \end{aligned}$$

Consequently, $x_3=\frac{11}{4}$, $x_2=(3-\frac{11}{2})/2=-\frac{5}{4}$, and $x_1=1+\frac{5}{4}-\frac{11}{4}=-\frac{1}{2}$.

Exercises

1. Show that the system of equations

$$\begin{aligned} a_{11}x_1+a_{12}x_2+a_{13}x_3&=b_1\\ a_{21}x_1+a_{22}x_2+a_{23}x_3&=b_2\\ a_{31}x_1+a_{32}x_2+a_{33}x_3&=b_3 \end{aligned}$$

has a unique solution x_1, x_2, x_3 if, and only if,

$$\det\begin{pmatrix} a_{11} & a_{12} & a_{13} \\ a_{21} & a_{22} & a_{23} \\ a_{31} & a_{32} & a_{33} \end{pmatrix} \neq 0.$$

Hint: Solve for x_1 in terms of x_2 and x_3 from one of these equations.

2. (a) Show that the total number of permutations of the integers $1,2,\ldots,n$ is even.
(b) Prove that exactly half of these permutations are even, and half are odd.

In each of Problems 3–8 compute the determinant of the given matrix.

3. $\begin{pmatrix} 1 & 0 & 2 \\ 1 & 2 & 5 \\ 6 & 8 & 0 \end{pmatrix}$

4. $\begin{bmatrix} 1 & t & t^2 \\ t & t^2 & 1 \\ t^2 & t & 1 \end{bmatrix}$

5. $\begin{bmatrix} 0 & a & b & 0 \\ -a & 0 & c & 0 \\ -b & -c & 0 & 0 \\ 0 & 0 & 0 & 1 \end{bmatrix}$

6. $\begin{bmatrix} 2 & -1 & 6 & 3 \\ 1 & 0 & 1 & -1 \\ 1 & 3 & 0 & 2 \\ 1 & -1 & 1 & 0 \end{bmatrix}$

7. $\begin{bmatrix} 0 & 2 & 3 & -1 \\ 1 & 8 & -1 & 1 \\ 1 & -1 & 0 & -1 \\ 0 & 2 & 6 & 1 \end{bmatrix}$

8. $\begin{bmatrix} 1 & 1 & 1 & 1 & 1 \\ 1 & 0 & 0 & 0 & 2 \\ 0 & 1 & 0 & 0 & 3 \\ 0 & 0 & 1 & 0 & 4 \\ 0 & 0 & 0 & 1 & 5 \end{bmatrix}$

9. Without doing any computations, show that

$$\det\begin{bmatrix} a & b & c \\ d & e & f \\ g & h & k \end{bmatrix} = \det\begin{bmatrix} e & b & h \\ d & a & g \\ f & c & k \end{bmatrix}.$$

In each of Problems 10–15, find all solutions of the given system of equations

10. $$\begin{aligned} x_1+x_2-x_3&=0 \\ 2x_1 \quad + \quad x_3&=14 \\ x_2+x_3&=13 \end{aligned}$$

11. $$\begin{aligned} x_1+x_2+x_3&=6 \\ x_1-x_2-x_3&=-4 \\ x_2+x_3&=-1 \end{aligned}$$

12. $$\begin{aligned} x_1+x_2+x_3&=0 \\ x_1-x_2-x_3&=0 \\ x_2+x_3&=0 \end{aligned}$$

13. $$\begin{aligned} x_1+x_2+x_3-x_4&=1 \\ x_1+2x_2-2x_3+x_4&=1 \\ x_1+3x_2-3x_3-x_4&=1 \\ x_1+4x_2-4x_3-x_4&=1 \end{aligned}$$

14. $$\begin{aligned} x_1+x_2+2x_3-x_4&=1 \\ x_1-x_2+2x_3+x_4&=2 \\ x_1+x_2+2x_3-x_4&=1 \\ -x_1-x_2-2x_3+x_4&=0 \end{aligned}$$

15. $$\begin{aligned} x_1-x_2+x_3+x_4&=0 \\ x_1+2x_2-x_3+3x_4&=0 \\ 3x_1+3x_2-x_3+7x_4&=0 \\ -x_1+2x_2+x_3-x_4&=0 \end{aligned}$$

3.6 Solutions of simultaneous linear equations

In this section we will prove that the system of equations

$$\mathbf{Ax}=\mathbf{b}, \qquad \mathbf{A}=\begin{bmatrix} a_{11} & \dots & a_{1n} \\ \vdots & & \vdots \\ a_{n1} & \dots & a_{nn} \end{bmatrix}, \quad \mathbf{x}=\begin{bmatrix} x_1 \\ \vdots \\ x_n \end{bmatrix}, \quad \mathbf{b}=\begin{bmatrix} b_1 \\ \vdots \\ b_n \end{bmatrix} \tag{1}$$

has a unique solution $\mathbf{x}$ if $\det\mathbf{A}\neq 0$. To this end, we define the product of two $n\times n$ matrices and then derive some additional properties of determinants.

Definition. Let $\mathbf{A}$ and $\mathbf{B}$ be $n\times n$ matrices with elements a_{ij} and b_{ij} respectively. We define their product $\mathbf{AB}$ as the $n\times n$ matrix $\mathbf{C}$ whose ij element c_{ij} is the product of the ith row of $\mathbf{A}$ with the jth column of $\mathbf{B}$. That is to say,

$$c_{ij}=\sum_{k=1}^{n} a_{ik}b_{kj}.$$

Alternately, if we write $\mathbf{B}$ in the form $\mathbf{B}=(\mathbf{b}^1,\mathbf{b}^2,\dots,\mathbf{b}^n)$, where $\mathbf{b}^j$ is the jth column of $\mathbf{B}$, then we can express the product $\mathbf{C}=\mathbf{AB}$ in the form $\mathbf{C}=(\mathbf{Ab}^1,\mathbf{Ab}^2,\dots,\mathbf{Ab}^n)$, since the ith component of the vector $\mathbf{Ab}^j$ is $\sum_{k=1}^{n}a_{ik}b_{kj}$.

Example 1. Let

$$\mathbf{A}=\begin{bmatrix} 3 & 1 & -1 \\ 0 & 2 & 1 \\ 1 & 1 & 1 \end{bmatrix}$$

and

$$\mathbf{B}=\begin{bmatrix} 1 & -1 & 0 \\ 2 & -1 & 1 \\ -1 & 0 & 0 \end{bmatrix}.$$

Compute $\mathbf{AB}$.

Solution.

$$\begin{bmatrix} 3 & 1 & -1 \\ 0 & 2 & 1 \\ 1 & 1 & 1 \end{bmatrix}\begin{bmatrix} 1 & -1 & 0 \\ 2 & -1 & 1 \\ -1 & 0 & 0 \end{bmatrix}=\begin{bmatrix} 3+2+1 & -3-1+0 & 0+1+0 \\ 0+4-1 & 0-2+0 & 0+2+0 \\ 1+2-1 & -1-1+0 & 0+1+0 \end{bmatrix}$$

$$=\begin{bmatrix} 6 & -4 & 1 \\ 3 & -2 & 2 \\ 2 & -2 & 1 \end{bmatrix}$$

Example 2. Let **A** and **B** be the matrices in Example 1. Compute **BA**.
Solution.

$$\begin{bmatrix} 1 & -1 & 0 \\ 2 & -1 & 1 \\ -1 & 0 & 0 \end{bmatrix}\begin{bmatrix} 3 & 1 & -1 \\ 0 & 2 & 1 \\ 1 & 1 & 1 \end{bmatrix}=\begin{bmatrix} 3+0+0 & 1-2+0 & -1-1+0 \\ 6+0+1 & 2-2+1 & -2-1+1 \\ -3+0+0 & -1+0+0 & 1+0+0 \end{bmatrix}$$

$$=\begin{bmatrix} 3 & -1 & -2 \\ 7 & 1 & -2 \\ -3 & -1 & 1 \end{bmatrix}$$

Remark 1. As Examples 1 and 2 indicate, it is generally not true that $\mathbf{AB}=\mathbf{BA}$. It can be shown, though, that

$$\mathbf{A}(\mathbf{BC})=(\mathbf{AB})\mathbf{C} \tag{2}$$

for any three $n\times n$ matrices **A**, **B**, and **C**. We will give an extremely simple proof of (2) in the next section.

Remark 2. Let **I** denote the diagonal matrix

$$\begin{bmatrix} 1 & 0 & \dots & 0 \\ 0 & 1 & \dots & 0 \\ \vdots & \vdots & & \vdots \\ 0 & 0 & \dots & 1 \end{bmatrix}.$$

I is called the identity matrix since $\mathbf{IA}=\mathbf{AI}=\mathbf{A}$ (see Exercise 5) for any $n\times n$ matrix **A**.

The following two properties of determinants are extremely useful in many applications.

Property 8.

$$\det \mathbf{AB}=\det \mathbf{A}\times\det \mathbf{B}.$$

That is to say, the determinant of the product is the product of the determinants.

Property 9. Let $\mathbf{A}(i|j)$ denote the $(n-1)\times(n-1)$ matrix obtained from **A** by deleting the ith row and jth column of **A**. For example, if

$$\mathbf{A}=\begin{bmatrix} 1 & 2 & 6 \\ 0 & -1 & -3 \\ -4 & -5 & 0 \end{bmatrix}, \quad \text{then} \quad \mathbf{A}(2|3)=\begin{pmatrix} 1 & 2 \\ -4 & -5 \end{pmatrix}.$$

Let $c_{ij}=(-1)^{i+j}\det\mathbf{A}(i|j)$. Then

$$\det\mathbf{A}=\sum_{i=1}^{n} a_{ij}c_{ij}$$

for any choice of j between 1 and n. This process of computing determinants is known as "expansion by the elements of columns," and Property 9 states that it does not matter which column we choose to expand about. For example, let

$$\mathbf{A}=\begin{bmatrix} 1 & 3 & 2 & 6 \\ 9 & -1 & 0 & 7 \\ 2 & 1 & 3 & 4 \\ -1 & 6 & 3 & 5 \end{bmatrix}.$$

Expanding about the first, second, third, and fourth columns of $\mathbf{A}$, respectively, gives

$$\begin{aligned}
\det\mathbf{A} &= \det\begin{bmatrix} -1 & 0 & 7 \\ 1 & 3 & 4 \\ 6 & 3 & 5 \end{bmatrix} - 9\det\begin{bmatrix} 3 & 2 & 6 \\ 1 & 3 & 4 \\ 6 & 3 & 5 \end{bmatrix} \\
&\quad + 2\det\begin{bmatrix} 3 & 2 & 6 \\ -1 & 0 & 7 \\ 6 & 3 & 5 \end{bmatrix} + \det\begin{bmatrix} 3 & 2 & 6 \\ -1 & 0 & 7 \\ 1 & 3 & 4 \end{bmatrix} \\
&= -3\det\begin{bmatrix} 9 & 0 & 7 \\ 2 & 3 & 4 \\ -1 & 3 & 5 \end{bmatrix} - \det\begin{bmatrix} 1 & 2 & 6 \\ 2 & 3 & 4 \\ -1 & 3 & 5 \end{bmatrix} \\
&\quad - \det\begin{bmatrix} 1 & 2 & 6 \\ 9 & 0 & 7 \\ -1 & 3 & 5 \end{bmatrix} + 6\det\begin{bmatrix} 1 & 2 & 6 \\ 9 & 0 & 7 \\ 2 & 3 & 4 \end{bmatrix} \\
&= 2\det\begin{bmatrix} 9 & -1 & 7 \\ 2 & 1 & 4 \\ -1 & 6 & 5 \end{bmatrix} + 3\det\begin{bmatrix} 1 & 3 & 6 \\ 9 & -1 & 7 \\ -1 & 6 & 5 \end{bmatrix} \\
&\quad - 3\det\begin{bmatrix} 1 & 3 & 6 \\ 9 & -1 & 7 \\ 2 & 1 & 4 \end{bmatrix} \\
&= -6\det\begin{bmatrix} 9 & -1 & 0 \\ 2 & 1 & 3 \\ -1 & 6 & 3 \end{bmatrix} + 7\det\begin{bmatrix} 1 & 3 & 2 \\ 2 & 1 & 3 \\ -1 & 6 & 3 \end{bmatrix} \\
&\quad - 4\det\begin{bmatrix} 1 & 3 & 2 \\ 9 & -1 & 0 \\ -1 & 6 & 3 \end{bmatrix} + 5\det\begin{bmatrix} 1 & 3 & 2 \\ 9 & -1 & 0 \\ 2 & 1 & 3 \end{bmatrix}.
\end{aligned}$$

We will derive Properties 8 and 9 with the aid of the following lemma.

Lemma 1. Let $D=D(\mathbf{A})$ be a function that assigns to each $n\times n$ matrix $\mathbf{A}$ a number $D(\mathbf{A})$. Suppose, moreover, that D is a linear function of each column (row) of $\mathbf{A}$ separately, i.e.

$$D(\mathbf{a}^1,\ldots,\mathbf{a}^j+c\mathbf{b}^j,\ldots,\mathbf{a}^n)=D(\mathbf{a}^1,\ldots,\mathbf{a}^j,\ldots,\mathbf{a}^n)+cD(\mathbf{a}^1,\ldots,\mathbf{b}^j,\ldots,\mathbf{a}^n),$$

and $D(\mathbf{B})=-D(\mathbf{A})$ if $\mathbf{B}$ is obtained from $\mathbf{A}$ by interchanging two columns (rows) of $\mathbf{A}$. Then

$$D(\mathbf{A})=\det\mathbf{A}\times D(\mathbf{I}).$$

A function D that assigns to each $n\times n$ matrix a number is called alternating if $D(\mathbf{B})=-D(\mathbf{A})$ whenever $\mathbf{B}$ is obtained from $\mathbf{A}$ by interchanging two columns (rows) of $\mathbf{A}$. Lemma 1 shows that the properties of being alternating and linear in the columns (rows) of $\mathbf{A}$ serve almost completely to characterize the determinant function $\det\mathbf{A}$. More precisely, any function $D(\mathbf{A})$ which is alternating and linear in the columns (rows) of $\mathbf{A}$ must be a constant multiple of $\det\mathbf{A}$. If, in addition, $D(\mathbf{I})=1$, then $D(\mathbf{A})=\det\mathbf{A}$ for all $n\times n$ matrices $\mathbf{A}$. It also follows immediately from Lemma 1 that if $D(\mathbf{A})$ is alternating and linear in the columns of $\mathbf{A}$, then it is also alternating and linear in the rows of $\mathbf{A}$.

Proof of Lemma 1. We first write $\mathbf{A}$ in the form $\mathbf{A}=(\mathbf{a}^1,\mathbf{a}^2,\ldots,\mathbf{a}^n)$ where

$$\mathbf{a}^1=\begin{pmatrix}a_{11}\\a_{21}\\\vdots\\a_{n1}\end{pmatrix},\qquad \mathbf{a}^2=\begin{pmatrix}a_{12}\\a_{22}\\\vdots\\a_{n2}\end{pmatrix},\ldots,\mathbf{a}^n=\begin{pmatrix}a_{1n}\\a_{2n}\\\vdots\\a_{nn}\end{pmatrix}.$$

Then, writing $\mathbf{a}^1$ in the form $\mathbf{a}^1=a_{11}\mathbf{e}^1+\ldots+a_{n1}\mathbf{e}^n$ we see that

$$\begin{aligned}D(\mathbf{A})&=D\left(a_{11}\mathbf{e}^1+\ldots+a_{n1}\mathbf{e}^n,\mathbf{a}^2,\ldots,\mathbf{a}^n\right)\\&=a_{11}D(\mathbf{e}^1,\mathbf{a}^2,\ldots,\mathbf{a}^n)+\ldots+a_{n1}D(\mathbf{e}^n,\mathbf{a}^2,\ldots,\mathbf{a}^n)\\&=\sum_{j_1}a_{1j_1}D(\mathbf{e}^{j_1},\mathbf{a}^2,\ldots,\mathbf{a}^n).\end{aligned}$$

Similarly, writing $\mathbf{a}^2$ in the form $\mathbf{a}^2=a_{12}\mathbf{e}^1+\ldots+a_{n2}\mathbf{e}^n$ we see that

$$D(\mathbf{A})=\sum_{j_1,j_2}a_{1j_1}a_{2j_2}D(\mathbf{e}^{j_1},\mathbf{e}^{j_2},\mathbf{a}^3,\ldots,\mathbf{a}^n).$$

Proceeding inductively, we see that

$$D(\mathbf{A})=\sum_{j_1,\ldots,j_n}a_{1j_1}a_{2j_2}\cdots a_{nj_n}D(\mathbf{e}^{j_1},\mathbf{e}^{j_2},\ldots,\mathbf{e}^{j_n}).$$

Now, we need only sum over those integers $j_1,j_2,\ldots,j_n$ with $j_i\neq j_k$ since $D(\mathbf{A})$ is zero if $\mathbf{A}$ has two equal columns. Moreover,

$$D(\mathbf{e}^{j_1},\mathbf{e}^{j_2},\ldots,\mathbf{e}^{j_n})=\varepsilon_{j_1j_2\ldots j_n}D(\mathbf{e}^1,\mathbf{e}^2,\ldots,\mathbf{e}^n)=\varepsilon_{j_1\ldots j_n}D(\mathbf{I}).$$

Consequently,

$$D(\mathbf{A}) = \sum_{j_1,\ldots,j_n} \varepsilon_{j_1\ldots j_n} a_{1j_1}\cdots a_{nj_n} D(\mathbf{I}) = \det\mathbf{A}\times D(\mathbf{I}). \qquad \square$$

We are now in a position to derive Properties 8 and 9.

PROOF OF PROPERTY 8. Let $\mathbf{A}$ be a fixed $n\times n$ matrix, and define the function $D(\mathbf{B})$ by the formula

$$D(\mathbf{B}) = \det\mathbf{AB}.$$

Observe that $D(\mathbf{B})$ is alternating and linear in the columns $\mathbf{b}^1,\ldots,\mathbf{b}^n$ of $\mathbf{B}$. This follows immediately from the fact that the columns of $\mathbf{AB}$ are $\mathbf{Ab}^1,\ldots,\mathbf{Ab}^n$. Hence, by Lemma 1.

$$D(\mathbf{B}) = \det\mathbf{B}\times D(\mathbf{I}) = \det\mathbf{B}\times\det\mathbf{AI} = \det\mathbf{A}\times\det\mathbf{B}. \qquad \square$$

PROOF OF PROPERTY 9. Pick any integer j between 1 and n and let

$$D(\mathbf{A}) = \sum_{i=1}^{n} a_{ij}c_{ij},$$

where $c_{ij} = (-1)^{i+j}\det\mathbf{A}(i|j)$. It is trivial to verify that D is alternating and linear in the columns of $\mathbf{A}$. Hence, by Lemma 1,

$$D(\mathbf{A}) = \det\mathbf{A}\times D(\mathbf{I}) = \det\mathbf{A}. \qquad \square$$

The key to solving the system of equations (1) is the crucial observation that

$$\sum_{i=1}^{n} a_{ik}c_{ij} = 0 \quad \text{for} \quad k\neq j \tag{3}$$

where $c_{ij} = (-1)^{i+j}\det\mathbf{A}(i|j)$. The proof of (3) is very simple: Let $\mathbf{B}$ denote the matrix obtained from $\mathbf{A}$ by replacing the jth column of $\mathbf{A}$ by its kth column, leaving everything else unchanged. For example, if

$$\mathbf{A} = \begin{bmatrix} 3 & 5 & 6 \\ 4 & 2 & 1 \\ -1 & 0 & -1 \end{bmatrix}, \qquad j=2, \quad \text{and} \quad k=3,$$

then

$$\mathbf{B} = \begin{bmatrix} 3 & 6 & 6 \\ 4 & 1 & 1 \\ -1 & -1 & -1 \end{bmatrix}.$$

Now, the determinant of $\mathbf{B}$ is zero, since $\mathbf{B}$ has two equal columns. On the other hand, expanding about the jth column of $\mathbf{B}$ gives

$$\det\mathbf{B} = \sum_{i=1}^{n} b_{ij}\hat{c}_{ij} = \sum_{i=1}^{n} a_{ik}\hat{c}_{ij}$$

where

$$\hat{c}_{ij} = (-1)^{i+j} \det \mathbf{B}(i|j) = (-1)^{i+j} \det \mathbf{A}(i|j) = c_{ij}.$$

Hence, $\sum_{i=1}^{n} a_{ik} c_{ij} = 0$ if k is unequal to j.

Now, whenever we see a sum from 1 to n involving the product of terms with two fixed indices j and k (as in (3)), we try and write it as the jk element of the product of two matrices. If we let $\mathbf{C}$ denote the matrix whose ij element is c_{ij} and set

$$\operatorname{adj} \mathbf{A} \equiv \mathbf{C}^T,$$

then

$$\sum_{i=1}^{n} a_{ik} c_{ij} = \sum_{i=1}^{n} (\operatorname{adj} \mathbf{A})_{ji} a_{ik} = (\operatorname{adj} \mathbf{A} \times \mathbf{A})_{jk}.$$

Hence, from (3),

$$(\operatorname{adj} \mathbf{A} \times \mathbf{A})_{jk} = 0 \quad \text{for} \quad j \neq k.$$

Combining this result with the identity

$$\det \mathbf{A} = \sum_{i=1}^{n} a_{ij} c_{ij} = (\operatorname{adj} \mathbf{A} \times \mathbf{A})_{jj},$$

we see that

$$\operatorname{adj} \mathbf{A} \times \mathbf{A} = \begin{bmatrix} \det \mathbf{A} & 0 & \dots & 0 \\ 0 & \det \mathbf{A} & \dots & 0 \\ \vdots & \vdots & & \vdots \\ 0 & 0 & \dots & \det \mathbf{A} \end{bmatrix} = (\det \mathbf{A})\mathbf{I}. \tag{4}$$

Similarly, by working with $\mathbf{A}^T$ instead of $\mathbf{A}$ (see Exercise 8) we see that

$$\mathbf{A} \times \operatorname{adj} \mathbf{A} = \begin{bmatrix} \det \mathbf{A} & 0 & \dots & 0 \\ 0 & \det \mathbf{A} & \dots & 0 \\ \vdots & \vdots & & \vdots \\ 0 & 0 & \dots & \det \mathbf{A} \end{bmatrix} = (\det \mathbf{A})\mathbf{I}. \tag{5}$$

Example 3. Let

$$\mathbf{A} = \begin{bmatrix} 1 & 0 & -1 \\ 1 & 1 & -1 \\ 1 & 2 & 1 \end{bmatrix}.$$

Compute $\operatorname{adj} \mathbf{A}$ and verify directly the identities (4) and (5).

Solution.

$$\operatorname{adj} \mathbf{A} = \mathbf{C}^T = \begin{bmatrix} 3 & -2 & 1 \\ -2 & 2 & -2 \\ 1 & 0 & 1 \end{bmatrix}^T = \begin{bmatrix} 3 & -2 & 1 \\ -2 & 2 & 0 \\ 1 & -2 & 1 \end{bmatrix}$$

so that

$$\operatorname{adj}\mathbf{A}\times\mathbf{A}=\begin{bmatrix}3&-2&1\\-2&2&0\\1&-2&1\end{bmatrix}\begin{bmatrix}1&0&-1\\1&1&-1\\1&2&1\end{bmatrix}=\begin{bmatrix}2&0&0\\0&2&0\\0&0&2\end{bmatrix}=2\mathbf{I}$$

and

$$\mathbf{A}\times\operatorname{adj}\mathbf{A}=\begin{bmatrix}1&0&-1\\1&1&-1\\1&2&1\end{bmatrix}\begin{bmatrix}3&-2&1\\-2&2&0\\1&-2&1\end{bmatrix}=\begin{bmatrix}2&0&0\\0&2&0\\0&0&2\end{bmatrix}=2\mathbf{I}.$$

Now, $\det\mathbf{A}=1-2+1+2=2$. Hence,

$$\operatorname{adj}\mathbf{A}\times\mathbf{A}=\mathbf{A}\times\operatorname{adj}\mathbf{A}=(\det\mathbf{A})\mathbf{I}.$$

We return now to the system of equations

$$\mathbf{A}\mathbf{x}=\mathbf{b},\qquad \mathbf{A}=\begin{bmatrix}a_{11}&\dots&a_{1n}\\\vdots&&\vdots\\a_{n1}&\dots&a_{nn}\end{bmatrix},\quad \mathbf{x}=\begin{bmatrix}x_1\\\vdots\\x_n\end{bmatrix},\quad \mathbf{b}=\begin{bmatrix}b_1\\\vdots\\b_n\end{bmatrix}.\tag{6}$$

If $\mathbf{A}$ were a nonzero number instead of a matrix, we would divide both sides of (6) by $\mathbf{A}$ to obtain that $\mathbf{x}=\mathbf{b}/\mathbf{A}$. This expression, of course, does not make sense if $\mathbf{A}$ is a matrix. However, there is a way of deriving the solution $\mathbf{x}=\mathbf{b}/\mathbf{A}$ which *does* generalize to the case where $\mathbf{A}$ is an $n\times n$ matrix. To wit, if the number $\mathbf{A}$ were unequal to zero, then we can multiply both sides of (6) by the number $\mathbf{A}^{-1}$ to obtain that

$$\mathbf{A}^{-1}\mathbf{A}\mathbf{x}=\mathbf{x}=\mathbf{A}^{-1}\mathbf{b}.$$

Now, if we can define $\mathbf{A}^{-1}$ as an $n\times n$ matrix when $\mathbf{A}$ is an $n\times n$ matrix, then the expression $\mathbf{A}^{-1}\mathbf{b}$ would make perfectly good sense. This leads us to ask the following two questions.

Question 1: Given an $n\times n$ matrix $\mathbf{A}$, does there exist another $n\times n$ matrix, which we will call $\mathbf{A}^{-1}$, with the property that

$$\mathbf{A}^{-1}\mathbf{A}=\mathbf{A}\mathbf{A}^{-1}=\mathbf{I}?$$

Question 2: If $\mathbf{A}^{-1}$ exists, is it unique? That is to say, can there exist two *distinct* matrices $\mathbf{B}$ and $\mathbf{C}$ with the property that

$$\mathbf{B}\mathbf{A}=\mathbf{A}\mathbf{B}=\mathbf{I}\quad\text{and}\quad\mathbf{C}\mathbf{A}=\mathbf{A}\mathbf{C}=\mathbf{I}?$$

The answers to these two questions are supplied in Theorems 7 and 8.

Theorem 7. *An $n\times n$ matrix $\mathbf{A}$ has at most one inverse.*

PROOF. Suppose that $\mathbf{A}$ has two distinct inverses. Then, there exist two distinct matrices $\mathbf{B}$ and $\mathbf{C}$ with the property that

$$\mathbf{A}\mathbf{B}=\mathbf{B}\mathbf{A}=\mathbf{I}\quad\text{and}\quad\mathbf{A}\mathbf{C}=\mathbf{C}\mathbf{A}=\mathbf{I}.$$

If we multiply both sides of the equation $\mathbf{AC}=\mathbf{I}$ by $\mathbf{B}$ we obtain that

$$\mathbf{BI}=\mathbf{B}(\mathbf{AC})=(\mathbf{BA})\mathbf{C}=\mathbf{IC}.$$

Hence, $\mathbf{B}=\mathbf{C}$, which is a contradiction. □

Theorem 8. $\mathbf{A}^{-1}$ *exists if, and only if,* $\det\mathbf{A}\neq 0$, *and in this case*

$$\mathbf{A}^{-1}=\frac{1}{\det\mathbf{A}}\operatorname{adj}\mathbf{A}. \tag{7}$$

PROOF. Suppose that $\det\mathbf{A}\neq 0$. Then, we can divide both sides of the identities (4) and (5) by $\det\mathbf{A}$ to obtain that

$$\frac{\operatorname{adj}\mathbf{A}}{\det\mathbf{A}}\times\mathbf{A}=\mathbf{I}=\frac{1}{\det\mathbf{A}}\mathbf{A}\times\operatorname{adj}\mathbf{A}=\mathbf{A}\times\frac{\operatorname{adj}\mathbf{A}}{\det\mathbf{A}}.$$

Hence, $\mathbf{A}^{-1}=\operatorname{adj}\mathbf{A}/\det\mathbf{A}$.

Conversely, suppose that $\mathbf{A}^{-1}$ exists. Taking determinants of both sides of the equation $\mathbf{A}^{-1}\mathbf{A}=\mathbf{I}$, and using Property 8 gives

$$(\det\mathbf{A}^{-1})\det\mathbf{A}=\det\mathbf{I}=1.$$

But this equation implies that $\det\mathbf{A}$ cannot equal zero. □

Finally, suppose that $\det\mathbf{A}\neq 0$. Then, $\mathbf{A}^{-1}$ exists, and multiplying both sides of (6) by this matrix gives

$$\mathbf{A}^{-1}\mathbf{A}\mathbf{x}=\mathbf{I}\mathbf{x}=\mathbf{x}=\mathbf{A}^{-1}\mathbf{b}.$$

Hence, if a solution exists, it must be $\mathbf{A}^{-1}\mathbf{b}$. Moreover, this vector is a solution of (6) since

$$\mathbf{A}(\mathbf{A}^{-1}\mathbf{b})=\mathbf{A}\mathbf{A}^{-1}\mathbf{b}=\mathbf{I}\mathbf{b}=\mathbf{b}.$$

Thus, Equation (6) has a unique solution $\mathbf{x}=\mathbf{A}^{-1}\mathbf{b}$ if $\det\mathbf{A}\neq 0$.

Example 4. Find all solutions of the equation

$$\begin{bmatrix}1&1&1\\2&-2&2\\3&3&-3\end{bmatrix}\mathbf{x}=\begin{bmatrix}0\\0\\0\end{bmatrix},\qquad \mathbf{x}=\begin{bmatrix}x_1\\x_2\\x_3\end{bmatrix}. \tag{8}$$

Solution.

$$\det\begin{bmatrix}1&1&1\\2&-2&2\\3&3&-3\end{bmatrix}=\det\begin{bmatrix}1&1&1\\0&-4&0\\0&0&-6\end{bmatrix}=24.$$

Hence, Equation (8) has a *unique* solution $\mathbf{x}$. But

$$\mathbf{x}=\begin{bmatrix}0\\0\\0\end{bmatrix}$$

is obviously one solution. Therefore,

$$\mathbf{x}=\begin{bmatrix}0\\0\\0\end{bmatrix}$$

is the unique solution of (8).

Remark. It is often quite cumbersome and time consuming to compute the inverse of an $n\times n$ matrix $\mathbf{A}$ from (7). This is especially true for $n\geqslant 4$. An alternate, and much more efficient way of computing $\mathbf{A}^{-1}$, is by means of "elementary row operations."

Definition. An elementary row operation on a matrix $\mathbf{A}$ is either
(i) an interchange of two rows,
(ii) the multiplication of one row by a nonzero number,
or
(iii) the addition of a multiple of one row to another row.

It can be shown that every matrix $\mathbf{A}$, with $\det\mathbf{A}\neq 0$, can be transformed into the identity $\mathbf{I}$ by a systematic sequence of these operations. Moreover, if the same sequence of operations is then performed upon $\mathbf{I}$, it is transformed into $\mathbf{A}^{-1}$. We illustrate this method with the following example.

Example 5. Find the inverse of the matrix

$$\mathbf{A}=\begin{bmatrix}1&0&-1\\1&1&-1\\1&2&1\end{bmatrix}.$$

Solution. The matrix $\mathbf{A}$ can be transformed into $\mathbf{I}$ by the following sequence of elementary row operations. The result of each step appears below the operation performed.

(a) We obtain zeros in the off-diagonal positions in the first column by subtracting the first row from both the second and third rows.

$$\begin{bmatrix}1&0&-1\\0&1&0\\0&2&2\end{bmatrix}$$

(b) We obtain zeros in the off-diagonal positions in the second column by adding (-2) times the second row to the third row.

$$\begin{bmatrix}1&0&-1\\0&1&0\\0&0&2\end{bmatrix}$$

(c) We obtain a one in the diagonal position in the third column by multiplying the third row by $\frac{1}{2}$.

$$\begin{bmatrix} 1 & 0 & -1 \\ 0 & 1 & 0 \\ 0 & 0 & 1 \end{bmatrix}$$

(d) Finally, we obtain zeros in the off-diagonal positions in the third column by adding the third row to the first row.

$$\begin{bmatrix} 1 & 0 & 0 \\ 0 & 1 & 0 \\ 0 & 0 & 1 \end{bmatrix}$$

If we perform the same sequence of elementary row operations upon **I**, we obtain the following sequence of matrices:

$$\begin{bmatrix} 1 & 0 & 0 \\ 0 & 1 & 0 \\ 0 & 0 & 1 \end{bmatrix}, \begin{bmatrix} 1 & 0 & 0 \\ -1 & 1 & 0 \\ -1 & 0 & 1 \end{bmatrix}, \begin{bmatrix} 1 & 0 & 0 \\ -1 & 1 & 0 \\ 1 & -2 & 1 \end{bmatrix},$$

$$\begin{bmatrix} 1 & 0 & 0 \\ -1 & 1 & 0 \\ \frac{1}{2} & -1 & \frac{1}{2} \end{bmatrix}, \begin{bmatrix} \frac{3}{2} & -1 & \frac{1}{2} \\ -1 & 1 & 0 \\ \frac{1}{2} & -1 & \frac{1}{2} \end{bmatrix}.$$

The last of these matrices is $\mathbf{A}^{-1}$.

Exercises

In each of Problems 1–4, compute **AB** and **BA** for the given matrices **A** and **B**.

1. $\mathbf{A}=\begin{pmatrix} 3 & 2 & 0 \\ 1 & 1 & 1 \\ 2 & 6 & 0 \end{pmatrix}$, $\mathbf{B}=\begin{pmatrix} -1 & 0 & 9 \\ 2 & 1 & -1 \\ 0 & 6 & 2 \end{pmatrix}$ **2.** $\mathbf{A}=\begin{pmatrix} 1 & 2 \\ 3 & 4 \end{pmatrix}$, $\mathbf{B}=\begin{pmatrix} 4 & 3 \\ 2 & 1 \end{pmatrix}$

3. $\mathbf{A}=\begin{pmatrix} 1 & 0 & 1 \\ 1 & 1 & 0 \\ 0 & 1 & 1 \end{pmatrix}$, $\mathbf{B}=\begin{pmatrix} 0 & 1 & 1 \\ 1 & 1 & 0 \\ 1 & 0 & 1 \end{pmatrix}$

4. $\mathbf{A}=\begin{bmatrix} 1 & 1 & 1 & 1 \\ 1 & 1 & 1 & 1 \\ 1 & 1 & 1 & 1 \\ 1 & 1 & 1 & 1 \end{bmatrix}$, $\mathbf{B}=\begin{bmatrix} 1 & 1 & 1 & 1 \\ 2 & 2 & 2 & 2 \\ 3 & 3 & 3 & 3 \\ 4 & 4 & 4 & 4 \end{bmatrix}$

5. Show that $\mathbf{IA}=\mathbf{AI}=\mathbf{A}$ for all matrices **A**.

6. Show that any two diagonal matrices **A** and **B** *commute*, that is $\mathbf{AB}=\mathbf{BA}$, if **A** and **B** are diagonal matrices.

7. Suppose that $\mathbf{AD}=\mathbf{DA}$ for all matrices $\mathbf{A}$. Prove that $\mathbf{D}$ is a multiple of the identity matrix.

8. Prove that $\mathbf{A}\times\operatorname{adj}\mathbf{A}=\det\mathbf{A}\times\mathbf{I}$.

In each of Problems 9–14, find the inverse, if it exists, of the given matrix.

9. $\begin{pmatrix} -2 & 3 & 2 \\ 6 & 0 & 3 \\ 4 & 1 & -1 \end{pmatrix}$ **10.** $\begin{pmatrix} \cos\theta & 0 & -\sin\theta \\ 0 & 1 & 0 \\ \sin\theta & 0 & \cos\theta \end{pmatrix}$

11. $\begin{pmatrix} 1 & 1 & 1 \\ -1 & i & -i \\ 2 & 1 & 1 \end{pmatrix}$ **12.** $\begin{pmatrix} 1 & 2 & -1 \\ 3 & 1 & 2 \\ 2 & 3 & -1 \end{pmatrix}$

13. $\begin{pmatrix} 1 & 1+i & 1-i \\ 1 & 0 & 0 \\ 0 & 1 & 1 \end{pmatrix}$ **14.** $\begin{pmatrix} 0 & 1 & 1 \\ -1 & 0 & 1 \\ -1 & -1 & 0 \end{pmatrix}$

15. Let

$$\mathbf{A}=\begin{pmatrix} a_{11} & a_{12} \\ a_{21} & a_{22} \end{pmatrix}.$$

Show that

$$\mathbf{A}^{-1}=\frac{1}{\det\mathbf{A}}\begin{pmatrix} a_{22} & -a_{12} \\ -a_{21} & a_{11} \end{pmatrix}$$

if $\det\mathbf{A}\neq 0$.

16. Show that $(\mathbf{AB})^{-1}=\mathbf{B}^{-1}\mathbf{A}^{-1}$ if $\det\mathbf{A}\times\det\mathbf{B}\neq 0$.

In each of Problems 17–20 show that $\mathbf{x}=\mathbf{0}$ is the unique solution of the given system of equations.

17. $\begin{aligned} x_1-x_2-x_3&=0\\ 3x_1-x_2+2x_3&=0\\ 2x_1+2x_2+3x_3&=0 \end{aligned}$

18. $\begin{aligned} x_1+2x_2+4x_3&=0\\ x_2+x_3&=0\\ x_1+x_2+x_3&=0 \end{aligned}$

19. $\begin{aligned} x_1+2x_2-x_3&=0\\ 2x_1+3x_2+x_3-x_4&=0\\ -x_1+2x_3+2x_4&=0\\ 3x_1-x_2+x_3+3x_4&=0 \end{aligned}$

20. $\begin{aligned} x_1+2x_2-x_3+3x_4&=0\\ 2x_1+3x_2-x_4&=0\\ -x_1+x_2+2x_3+x_4&=0\\ -x_2+2x_3+3x_4&=0 \end{aligned}$

3.7 Linear transformations

In the previous section we approached the problem of solving the equation

$$\mathbf{Ax}=\mathbf{b},\quad \mathbf{A}=\begin{bmatrix} a_{11} & \dots & a_{1n} \\ \vdots & & \vdots \\ a_{n1} & \dots & a_{nn} \end{bmatrix},\quad \mathbf{x}=\begin{bmatrix} x_1 \\ \vdots \\ x_n \end{bmatrix},\quad \mathbf{b}=\begin{bmatrix} b_1 \\ \vdots \\ b_n \end{bmatrix} \tag{1}$$

by asking whether the matrix $\mathbf{A}^{-1}$ exists. This approach led us to the conclusion that Equation (1) has a unique solution $\mathbf{x}=\mathbf{A}^{-1}\mathbf{b}$ if $\det\mathbf{A}\neq 0$. In order to determine what happens when $\det \mathbf{A}=0$, we will approach the problem of solving (1) in an entirely different manner. To wit, we will look at the set $\mathbf{V}$ of all vectors obtained by multiplying every vector in R^n by $\mathbf{A}$ and see if $\mathbf{b}$ is in this set. Obviously, Equation (1) has at least one solution $\mathbf{x}$ if, and only if, $\mathbf{b}$ is in $\mathbf{V}$. We begin with the following simple but extremely useful lemma.

Lemma 1. *Let* $\mathbf{A}$ *be an* $n\times n$ *matrix with elements* a_{ij}, *and let* $\mathbf{x}$ *be a vector with components* $x_1,x_2,\ldots,x_n$. *Let*

$$\mathbf{a}^j=\begin{bmatrix} a_{1j} \\ \vdots \\ a_{nj} \end{bmatrix}$$

denote the jth *column of* $\mathbf{A}$. *Then*

$$\mathbf{Ax}=x_1\mathbf{a}^1+x_2\mathbf{a}^2+\ldots+x_n\mathbf{a}^n.$$

PROOF. We will show that the vectors $\mathbf{Ax}$ and $x_1\mathbf{a}^1+\ldots+x_n\mathbf{a}^n$ have the same components. To this end, observe that $(\mathbf{Ax})_j$, the jth component of $\mathbf{Ax}$, is $a_{j1}x_1+\ldots+a_{jn}x_n$, while the jth component of the vector $x_1\mathbf{a}^1+\ldots+x_n\mathbf{a}^n$ is

$$x_1\mathbf{a}_j^1+\ldots+x_n\mathbf{a}_j^n=x_1a_{j1}+\ldots+x_na_{jn}=(\mathbf{Ax})_j.$$

Hence, $\mathbf{A}x=x_1\mathbf{a}^1+x_2\mathbf{a}^2+\ldots+x_n\mathbf{a}^n$. □

Now, let $\mathbf{V}$ be the set of vectors obtained by multiplying every vector $\mathbf{x}$ in R^n by the matrix $\mathbf{A}$. It follows immediately from Lemma 1 that $\mathbf{V}$ is the set of all linear combinations of the vectors $\mathbf{a}^1,\mathbf{a}^2,\ldots,\mathbf{a}^n$, that is, $\mathbf{V}$ is spanned by the columns of $\mathbf{A}$. Hence the equation $\mathbf{Ax}=\mathbf{b}$ has a solution if, and only if, $\mathbf{b}$ is a linear combination of the columns of $\mathbf{A}$. With the aid of this observation, we can now prove the following theorem.

Theorem 9. (a) *The equation* $\mathbf{Ax}=\mathbf{b}$ *has a unique solution if the columns of* $\mathbf{A}$ *are linearly independent.*

(b) *The equation* $\mathbf{Ax}=\mathbf{b}$ *has either no solution, or infinitely many solutions, if the columns of* $\mathbf{A}$ *are linearly dependent.*

PROOF. (a) Suppose that the columns of $\mathbf{A}$ are linearly independent. Then, $\mathbf{a}^1,\mathbf{a}^2,\ldots,\mathbf{a}^n$ form a basis for R^n. In particular, every vector $\mathbf{b}$ can be written as a linear combination of $\mathbf{a}^1,\mathbf{a}^2,\ldots,\mathbf{a}^n$. Consequently, Equation (1) has at least one solution. To prove that (1) has exactly one solution, we show that any two solutions

$$\mathbf{x}=\begin{bmatrix} x_1 \\ \vdots \\ x_n \end{bmatrix}$$

and

$$\mathbf{y}=\begin{bmatrix} y_1 \\ \vdots \\ y_n \end{bmatrix}$$

must be equal. To this end, observe that if $\mathbf{x}$ and $\mathbf{y}$ are two solutions of (1), then

$$\mathbf{A}(\mathbf{x}-\mathbf{y})=\mathbf{A}\mathbf{x}-\mathbf{A}\mathbf{y}=\mathbf{b}-\mathbf{b}=\mathbf{0}.$$

By Lemma 1, therefore,

$$(x_1-y_1)\mathbf{a}^1+(x_2-y_2)\mathbf{a}^2+\ldots+(x_n-y_n)\mathbf{a}^n=\mathbf{0}. \tag{2}$$

But this implies that $x_1=y_1$, $x_2=y_2,\ldots,$ and $x_n=y_n$, since $\mathbf{a}^1,\mathbf{a}^2,\ldots,\mathbf{a}^n$ are linearly independent. Consequently, $\mathbf{x}=\mathbf{y}$.

(b) If the columns of $\mathbf{A}$ are linearly dependent, then we can extract from $\mathbf{a}^1,\mathbf{a}^2,\ldots,\mathbf{a}^n$ a smaller set of independent vectors which also span $\mathbf{V}$ (see Exercise 12, Section 3.3). Consequently, the space $\mathbf{V}$ has dimension at most $n-1$. In other words, the space $\mathbf{V}$ is distinctly smaller than R^n. Hence, there are vectors in R^n which are not in $\mathbf{V}$. If $\mathbf{b}$ is one of these vectors, then the equation $\mathbf{A}\mathbf{x}=\mathbf{b}$ obviously has no solutions. On the other hand, if $\mathbf{b}$ is in $\mathbf{V}$, that is, there exists at least one vector $\mathbf{x}^*$ such that $\mathbf{A}\mathbf{x}^*=\mathbf{b}$, then Equation (1) has infinitely many solutions. To prove this, observe first that $\mathbf{x}=\mathbf{x}^*$ is certainly one solution of (1). Second, observe that if $\mathbf{A}\boldsymbol{\xi}=\mathbf{0}$ for some vector $\boldsymbol{\xi}$ in R^n, then $\mathbf{x}=\mathbf{x}^*+\boldsymbol{\xi}$ is also a solution of (1), since

$$\mathbf{A}(\mathbf{x}^*+\boldsymbol{\xi})=\mathbf{A}\mathbf{x}^*+\mathbf{A}\boldsymbol{\xi}=\mathbf{b}+\mathbf{0}=\mathbf{b}.$$

Finally, observe that there exist numbers $c_1,c_2,\ldots,c_n$ not all zero, such that $c_1\mathbf{a}^1+c_2\mathbf{a}^2+\ldots+c_n\mathbf{a}^n=\mathbf{0}$. By Lemma 1, therefore, $\mathbf{A}\boldsymbol{\xi}=\mathbf{0}$, where

$$\boldsymbol{\xi}=\begin{bmatrix} c_1 \\ \vdots \\ c_n \end{bmatrix}.$$

But if $\mathbf{A}\boldsymbol{\xi}$ equals zero, then $\mathbf{A}(\alpha\boldsymbol{\xi})$ also equals zero, for any constant α. Thus, there are infinitely many vectors $\boldsymbol{\xi}$ with the property that $\mathbf{A}\boldsymbol{\xi}=\mathbf{0}$. Consequently, the equation $\mathbf{A}\mathbf{x}=\mathbf{b}$ has infinitely many solutions. □

Example 1. (a) For which vectors

$$\mathbf{b}=\begin{bmatrix} b_1 \\ b_2 \\ b_3 \end{bmatrix}$$

can we solve the equation

$$\mathbf{A}\mathbf{x}=\begin{bmatrix} 1 & 1 & 2 \\ 1 & 0 & 1 \\ 2 & 1 & 3 \end{bmatrix}\mathbf{x}=\mathbf{b}?$$

(b) Find all solutions of the equation

$$\begin{bmatrix} 1 & 1 & 2 \\ 1 & 0 & 1 \\ 2 & 1 & 3 \end{bmatrix} \mathbf{x} = \begin{bmatrix} 0 \\ 0 \\ 0 \end{bmatrix}.$$

(c) Find all solutions of the equation

$$\begin{bmatrix} 1 & 1 & 2 \\ 1 & 0 & 1 \\ 2 & 1 & 3 \end{bmatrix} \mathbf{x} = \begin{bmatrix} 1 \\ 0 \\ 1 \end{bmatrix}.$$

Solution. (a) The columns of the matrix $\mathbf{A}$ are

$$\mathbf{a}^1 = \begin{bmatrix} 1 \\ 1 \\ 2 \end{bmatrix}, \quad \mathbf{a}^2 = \begin{bmatrix} 1 \\ 0 \\ 1 \end{bmatrix} \quad \text{and} \quad \mathbf{a}^3 = \begin{bmatrix} 2 \\ 1 \\ 3 \end{bmatrix}.$$

Notice that $\mathbf{a}^1$ and $\mathbf{a}^2$ are linearly independent while $\mathbf{a}^3$ is the sum of $\mathbf{a}^1$ and $\mathbf{a}^2$. Hence, we can solve the equation $\mathbf{Ax}=\mathbf{b}$ if, and only if,

$$\mathbf{b} = c_1 \begin{bmatrix} 1 \\ 1 \\ 2 \end{bmatrix} + c_2 \begin{bmatrix} 1 \\ 0 \\ 1 \end{bmatrix} = \begin{bmatrix} c_1 + c_2 \\ c_1 \\ 2c_1 + c_2 \end{bmatrix}$$

for some constants c_1 and c_2. Equivalently, (see Exercise 25) $b_3 = b_1 + b_2$.

(b) Consider the three equations

$$\begin{aligned} x_1 + x_2 + 2x_3 &= 0 \\ x_1 \qquad + \ x_3 &= 0 \\ 2x_1 + x_2 + 3x_3 &= 0. \end{aligned}$$

Notice that the third equation is the sum of the first two equations. Hence, we need only consider the first two equations. The second equation says that $x_1 = -x_3$. Substituting this value of x_1 into the first equation gives $x_2 = -x_3$. Hence, all solutions of the equation $\mathbf{Ax}=\mathbf{0}$ are of the form

$$\mathbf{x} = c \begin{bmatrix} -1 \\ -1 \\ 1 \end{bmatrix}.$$

(c) Observe first that

$$\mathbf{x}^1 = \begin{bmatrix} 0 \\ 1 \\ 0 \end{bmatrix}$$

is clearly one solution of this equation. Next, let $\mathbf{x}^2$ be any other solution of this equation. It follows immediately that $\mathbf{x}^2 = \mathbf{x}^1 + \boldsymbol{\xi}$, where $\boldsymbol{\xi}$ is a solution of the homogeneous equation $\mathbf{Ax}=\mathbf{0}$. Moreover, the sum of any solution of the nonhomogeneous equation

$$\mathbf{A}\mathbf{x}=\begin{bmatrix}1\\0\\1\end{bmatrix}$$

with a solution of the homogeneous equation is again a solution of the nonhomogeneous equation. Hence, any solution of the equation

$$\mathbf{A}\mathbf{x}=\begin{bmatrix}1\\0\\1\end{bmatrix}$$

is of the form

$$\mathbf{x}=\begin{bmatrix}0\\1\\0\end{bmatrix}+c\begin{bmatrix}-1\\-1\\1\end{bmatrix}.$$

Theorem 9 is an extremely useful theorem since it establishes necessary and sufficient conditions for the equation $\mathbf{A}\mathbf{x}=\mathbf{b}$ to have a unique solution. However, it is often very difficult to apply Theorem 9 since it is usually quite difficult to determine whether n vectors are linearly dependent or linearly independent. Fortunately, we can relate the question of whether the columns of $\mathbf{A}$ are linearly dependent or linearly independent to the much simpler problem of determining whether the determinant of $\mathbf{A}$ is zero or nonzero. There are several different ways of accomplishing this. In this section, we will present a very elegant method which utilizes the important concept of a linear transformation.

Definition. A linear transformation $\mathcal{Q}$ taking R^n into R^n is a function which assigns to each vector $\mathbf{x}$ in R^n a new vector which we call $\mathcal{Q}(\mathbf{x})=\mathcal{Q}(x_1,\ldots,x_n)$. Moreover, this association obeys the following rules.

$$\mathcal{Q}(c\mathbf{x})=c\mathcal{Q}(\mathbf{x}) \tag{i}$$

and

$$\mathcal{Q}(\mathbf{x}+\mathbf{y})=\mathcal{Q}(\mathbf{x})+\mathcal{Q}(\mathbf{y}). \tag{ii}$$

Example 2. The transformation

$$\mathbf{x}=\begin{bmatrix}x_1\\\vdots\\x_n\end{bmatrix}\rightarrow\mathcal{Q}(\mathbf{x})=\begin{bmatrix}x_1\\\vdots\\x_n\end{bmatrix}$$

is obviously a linear transformation of R^n into R^n since

$$\mathcal{Q}(c\mathbf{x})=c\mathbf{x}=c\mathcal{Q}(\mathbf{x})$$

and

$$\mathcal{Q}(\mathbf{x}+\mathbf{y})=\mathbf{x}+\mathbf{y}=\mathcal{Q}(\mathbf{x})+\mathcal{Q}(\mathbf{y}).$$

Example 3. The transformation

$$\mathbf{x}=\begin{bmatrix} x_1 \\ x_2 \\ x_3 \end{bmatrix} \rightarrow \mathcal{Q}(\mathbf{x})=\begin{bmatrix} x_1+x_2+x_3 \\ x_1+x_2-x_3 \\ x_1 \end{bmatrix}$$

is a linear transformation taking R^3 into R^3 since

$$\mathcal{Q}(c\mathbf{x})=\begin{bmatrix} cx_1+cx_2+cx_3 \\ cx_1+cx_2-cx_3 \\ cx_1 \end{bmatrix}=c\begin{bmatrix} x_1+x_2+x_3 \\ x_1+x_2-x_3 \\ x_1 \end{bmatrix}=c\mathcal{Q}(\mathbf{x})$$

and

$$\mathcal{Q}(\mathbf{x}+\mathbf{y})=\begin{bmatrix} (x_1+y_1)+(x_2+y_2)+(x_3+y_3) \\ (x_1+y_1)+(x_2+y_2)-(x_3+y_3) \\ x_1+y_1 \end{bmatrix}$$

$$=\begin{bmatrix} x_1+x_2+x_3 \\ x_1+x_2-x_3 \\ x_1 \end{bmatrix}+\begin{bmatrix} y_1+y_2+y_3 \\ y_1+y_2-y_3 \\ y_1 \end{bmatrix}=\mathcal{Q}(\mathbf{x})+\mathcal{Q}(\mathbf{y}).$$

Example 4. Let $\mathcal{Q}(x_1,x_2)$ be the point obtained from $\mathbf{x}=(x_1,x_2)$ by rotating $\mathbf{x}$ 30° in a counterclockwise direction. It is intuitively obvious that any rotation is a linear transformation. If the reader is not convinced of this, though, he can compute

$$\mathcal{Q}(x_1,x_2)=\begin{bmatrix} \frac{1}{2}\sqrt{3}\,x_1-\frac{1}{2}x_2 \\ \frac{1}{2}x_1+\frac{1}{2}\sqrt{3}\,x_2 \end{bmatrix}$$

and then verify directly that $\mathcal{Q}$ is a linear transformation taking R^2 into R^2.

Example 5. The transformation

$$\mathbf{x}=\begin{pmatrix} x_1 \\ x_2 \end{pmatrix} \rightarrow \begin{pmatrix} 1 \\ x_1^2+x_2^2 \end{pmatrix}$$

takes R^2 into R^2 but is not linear, since

$$\mathcal{Q}(2\mathbf{x})=\mathcal{Q}(2x_1,2x_2)=\begin{pmatrix} 1 \\ 4x_1^2+4x_2^2 \end{pmatrix}\neq 2\mathcal{Q}(\mathbf{x}).$$

Now, every $n\times n$ matrix $\mathbf{A}$ defines, in a very natural manner, a linear transformation $\mathcal{Q}$ taking $\mathsf{R}^n\rightarrow\mathsf{R}^n$. To wit, consider the transformation of $\mathsf{R}^n\rightarrow\mathsf{R}^n$ defined by

$$\mathbf{x}\rightarrow\mathcal{Q}(\mathbf{x})=\mathbf{A}\mathbf{x}.$$

In Section 3.1, we showed that $\mathbf{A}(c\mathbf{x})=c\mathbf{A}\mathbf{x}$ and $\mathbf{A}(\mathbf{x}+\mathbf{y})=\mathbf{A}\mathbf{x}+\mathbf{A}\mathbf{y}$. Hence, the association $\mathbf{x}\to\mathcal{A}(\mathbf{x})=\mathbf{A}\mathbf{x}$ is linear. Conversely, any linear transformation $\mathcal{A}$ taking $\mathsf{R}^n\to\mathsf{R}^n$ must be of the form $\mathcal{A}(\mathbf{x})=\mathbf{A}\mathbf{x}$ for some matrix $\mathbf{A}$. This is the content of the following theorem.

Theorem 10. *Any linear transformation $\mathbf{x}\to\mathcal{A}(\mathbf{x})$ taking R^n into R^n must be of the form $\mathcal{A}(\mathbf{x})=\mathbf{A}\mathbf{x}$. In other words, given any linear transformation $\mathcal{A}$ taking R^n into R^n, we can find an $n\times n$ matrix $\mathbf{A}$ such that*

$$\mathcal{A}(\mathbf{x})=\mathbf{A}\mathbf{x}$$

for all $\mathbf{x}$.

PROOF. Let $\mathbf{e}^j$ denote the vector whose jth component is one, and whose remaining components are zero, and let

$$\mathbf{a}^j=\mathcal{A}(\mathbf{e}^j).$$

We claim that

$$\mathcal{A}(\mathbf{x})=\mathbf{A}\mathbf{x},\quad \mathbf{A}=(\mathbf{a}^1,\mathbf{a}^2,\ldots,\mathbf{a}^n) \tag{3}$$

for every

$$\mathbf{x}=\begin{bmatrix} x_1\\ \vdots\\ x_n\end{bmatrix}$$

in R^n. To prove this, observe that any vector $\mathbf{x}$ can be written in the form $\mathbf{x}=x_1\mathbf{e}^1+\ldots+x_n\mathbf{e}^n$. Hence, by the linearity of $\mathcal{A}$,

$$\begin{aligned}\mathcal{A}(\mathbf{x})&=\mathcal{A}\left(x_1\mathbf{e}^1+\ldots+x_n\mathbf{e}^n\right)=x_1\mathcal{A}(\mathbf{e}^1)+\ldots+x_n\mathcal{A}(\mathbf{e}^n)\\&=x_1\mathbf{a}^1+\ldots+x_n\mathbf{a}^n=\mathbf{A}\mathbf{x}.\end{aligned}$$ □

Remark 1. The simplest way of evaluating a linear transformation $\mathcal{A}$ is to compute $\mathbf{a}^1=\mathcal{A}(\mathbf{e}^1),\ldots,\mathbf{a}^n=\mathcal{A}(\mathbf{e}^n)$ and then observe from Theorem 10 and Lemma 1 that $\mathcal{A}(\mathbf{x})=\mathbf{A}\mathbf{x}$, where $\mathbf{A}=(\mathbf{a}^1,\mathbf{a}^2,\ldots,\mathbf{a}^n)$. Thus, to evaluate the linear transformation in Example 4 above, we observe that under a rotation of 30° in a counterclockwise direction, the point $(1,0)$ goes into the point $(\frac{1}{2}\sqrt{3}\,,\frac{1}{2})$ and the point $(0,1)$ goes into the point $(-\frac{1}{2},\frac{1}{2}\sqrt{3}\,)$. Hence, any point $\mathbf{x}=(x_1,x_2)$ goes into the point

$$\begin{bmatrix}\frac{1}{2}\sqrt{3} & -\frac{1}{2}\\ \frac{1}{2} & \frac{1}{2}\sqrt{3}\end{bmatrix}\begin{bmatrix}x_1\\ x_2\end{bmatrix}=\begin{bmatrix}\frac{1}{2}\sqrt{3}\,x_1-\frac{1}{2}x_2\\ \frac{1}{2}x_1+\frac{1}{2}\sqrt{3}\,x_2\end{bmatrix}.$$

Remark 2. If $\mathcal{A}$ and $\mathcal{B}$ are linear transformations taking R^n into R^n, then the composition transformation $\mathcal{A}\circ\mathcal{B}$ defined by the relation

$$\mathcal{A}\circ\mathcal{B}(\mathbf{x})=\mathcal{A}(\mathcal{B}(\mathbf{x}))$$

is again a linear transformation taking $\mathsf{R}^n \to \mathsf{R}^n$. To prove this, observe that

$$\begin{aligned}\mathcal{A} \circ \mathcal{B}(c\mathbf{x}) &= \mathcal{A}(\mathcal{B}(c\mathbf{x})) = \mathcal{A}(c\mathcal{B}(\mathbf{x})) = c\mathcal{A}(\mathcal{B}(\mathbf{x}))\\ &= c\mathcal{A} \circ \mathcal{B}(\mathbf{x})\end{aligned}$$

and

$$\begin{aligned}\mathcal{A} \circ \mathcal{B}(\mathbf{x}+\mathbf{y}) &= \mathcal{A}(\mathcal{B}(\mathbf{x}+\mathbf{y})) = \mathcal{A}(\mathcal{B}(\mathbf{x})+\mathcal{B}(\mathbf{y}))\\ &= \mathcal{A}(\mathcal{B}(\mathbf{x}))+\mathcal{A}(\mathcal{B}(\mathbf{y})) = \mathcal{A} \circ \mathcal{B}(\mathbf{x}) + \mathcal{A} \circ \mathcal{B}(\mathbf{y}).\end{aligned}$$

Moreover, it is a simple matter to show (see Exercise 15) that if $\mathcal{A}(\mathbf{x}) = \mathbf{A}\mathbf{x}$ and $\mathcal{B}(\mathbf{x}) = \mathbf{B}\mathbf{x}$, then

$$\mathcal{A} \circ \mathcal{B}(\mathbf{x}) = \mathbf{AB}\mathbf{x}. \tag{4}$$

Similarly, if $\mathcal{A}$, $\mathcal{B}$, and $\mathcal{C}$ are 3 linear transformations taking R^n into R^n, with $\mathcal{A}(\mathbf{x}) = \mathbf{A}\mathbf{x}$, $\mathcal{B}(\mathbf{x}) = \mathbf{B}\mathbf{x}$, and $\mathcal{C}(\mathbf{x}) = \mathbf{C}\mathbf{x}$ then

$$(\mathcal{A} \circ \mathcal{B}) \circ \mathcal{C}(\mathbf{x}) = (\mathbf{AB})\mathbf{C}\mathbf{x} \tag{5}$$

and

$$\mathcal{A} \circ (\mathcal{B} \circ \mathcal{C})(\mathbf{x}) = \mathbf{A}(\mathbf{BC})\mathbf{x}. \tag{6}$$

Now, clearly, $(\mathcal{A} \circ \mathcal{B}) \circ \mathcal{C}(\mathbf{x}) = \mathcal{A} \circ (\mathcal{B} \circ \mathcal{C})(\mathbf{x})$. Hence,

$$(\mathbf{AB})\mathbf{C}\mathbf{x} = \mathbf{A}(\mathbf{BC})\mathbf{x}$$

for all vectors $\mathbf{x}$ in R^n. This implies (see Exercise 14) that

$$(\mathbf{AB})\mathbf{C} = \mathbf{A}(\mathbf{BC})$$

for any three $n \times n$ matrices $\mathbf{A}$, $\mathbf{B}$, and $\mathbf{C}$.

In most applications, it is usually desirable, and often absolutely essential, that the inverse of a linear transformation exists. Heuristically, the inverse of a transformation $\mathcal{A}$ undoes the effect of $\mathcal{A}$. That is to say, if $\mathcal{A}(\mathbf{x}) = \mathbf{y}$, then the inverse transformation applied to $\mathbf{y}$ must yield $\mathbf{x}$. More precisely, we define $\mathcal{A}^{-1}(\mathbf{y})$ as the unique element $\mathbf{x}$ in R^n for which $\mathcal{A}(\mathbf{x}) = \mathbf{y}$. Of course, the transformation $\mathcal{A}^{-1}$ may not exist. There may be some vectors $\mathbf{y}$ with the property that $\mathbf{y} \neq \mathcal{A}(\mathbf{x})$ for all $\mathbf{x}$ in R^n. Or, there may be some vectors $\mathbf{y}$ which come from more than one $\mathbf{x}$, that is, $\mathcal{A}(\mathbf{x}^1) = \mathbf{y}$ and $\mathcal{A}(\mathbf{x}^2) = \mathbf{y}$. In both these cases, the transformation $\mathcal{A}$ does not possess an inverse. In fact, it is clear that $\mathcal{A}$ possesses an inverse, which we will call $\mathcal{A}^{-1}$ if, and only if, the equation $\mathcal{A}(\mathbf{x}) = \mathbf{y}$ has a unique solution $\mathbf{x}$ for every $\mathbf{y}$ in R^n. In addition, it is clear that if $\mathcal{A}$ is a linear transformation and $\mathcal{A}^{-1}$ exists, then $\mathcal{A}^{-1}$ must also be linear. To prove this, observe first that $\mathcal{A}^{-1}(c\mathbf{y}) = c\mathcal{A}^{-1}(\mathbf{y})$ since $\mathcal{A}(c\mathbf{x}) = c\mathbf{y}$ if $\mathcal{A}(\mathbf{x}) = \mathbf{y}$. Second, observe that $\mathcal{A}^{-1}(\mathbf{y}^1 + \mathbf{y}^2) = \mathcal{A}^{-1}(\mathbf{y}^1) + \mathcal{A}^{-1}(\mathbf{y}^2)$ since $\mathcal{A}(\mathbf{x}^1 + \mathbf{x}^2) = \mathbf{y}^1 + \mathbf{y}^2$ if $\mathcal{A}(\mathbf{x}^1) = \mathbf{y}^1$ and $\mathcal{A}(\mathbf{x}^2) = \mathbf{y}^2$. Thus $\mathcal{A}^{-1}$, if it exists, must be linear.

At this point the reader should feel that there is an intimate connection between the linear transformation $\mathcal{A}^{-1}$ and the matrix $\mathbf{A}^{-1}$. This is the content of the following lemma.

Lemma 2. *Let* $\mathbf{A}$ *be an* $n\times n$ *matrix and let* $\mathcal{A}$ *be the linear transformation defined by the equation* $\mathcal{A}(\mathbf{x})=\mathbf{A}\mathbf{x}$. *Then,* $\mathcal{A}$ *has an inverse if, and only if, the matrix* $\mathbf{A}$ *has an inverse. Moreover, if* $\mathbf{A}^{-1}$ *exists, then* $\mathcal{A}^{-1}(\mathbf{x})=\mathbf{A}^{-1}\mathbf{x}$.

PROOF. Suppose that $\mathcal{A}^{-1}$ exists. Clearly,

$$\mathcal{A}\circ\mathcal{A}^{-1}(\mathbf{x})=\mathcal{A}^{-1}\circ\mathcal{A}(\mathbf{x})=\mathbf{x} \tag{7}$$

and $\mathcal{A}^{-1}$ is linear. Moreover, there exists a matrix $\mathbf{B}$ with the property that $\mathcal{A}^{-1}(\mathbf{x})=\mathbf{B}\mathbf{x}$. Therefore, from (4) and (7)

$$\mathbf{A}\mathbf{B}\mathbf{x}=\mathbf{B}\mathbf{A}\mathbf{x}=\mathbf{x}$$

for all $\mathbf{x}$ in R^n. But this immediately implies (see Exercise 14) that

$$\mathbf{A}\mathbf{B}=\mathbf{B}\mathbf{A}=\mathbf{I}.$$

Hence $\mathbf{B}=\mathbf{A}^{-1}$.

Conversely, suppose that $\mathbf{A}^{-1}$ exists. Then, the equation

$$\mathcal{A}(\mathbf{x})=\mathbf{A}\mathbf{x}=\mathbf{y}$$

has a unique solution $\mathbf{x}=\mathbf{A}^{-1}\mathbf{y}$ for all $\mathbf{y}$ in R^n. Thus, $\mathcal{A}^{-1}$ also exists, and

$$\mathcal{A}^{-1}(\mathbf{x})=\mathbf{A}^{-1}\mathbf{x}.$$ □

We are now ready to relate the problem of determining whether the columns of an $n\times n$ matrix $\mathbf{A}$ are linearly dependent or linearly independent to the much simpler problem of determining whether the determinant of $\mathbf{A}$ is zero or nonzero.

Lemma 3. *The columns of an* $n\times n$ *matrix* $\mathbf{A}$ *are linearly independent if, and only if,* $\det\mathbf{A}\neq 0$.

PROOF. We prove Lemma 3 by the following complex, but very clever argument.

(1) The columns of $\mathbf{A}$ are linearly independent if, and only if, the equation $\mathbf{A}\mathbf{x}=\mathbf{b}$ has a unique solution $\mathbf{x}$ for every $\mathbf{b}$ in R^n. This statement is just a reformulation of Theorem 9.
(2) From the remarks preceding Lemma 2, we conclude that the equation $\mathbf{A}\mathbf{x}=\mathbf{b}$ has a unique solution $\mathbf{x}$ for every $\mathbf{b}$ if, and only if, the linear transformation $\mathcal{A}(\mathbf{x})=\mathbf{A}\mathbf{x}$ has an inverse.
(3) From Lemma 2, the linear transformation $\mathcal{A}$ has an inverse if, and only if, the matrix $\mathbf{A}^{-1}$ exists.
(4) Finally, the matrix $\mathbf{A}^{-1}$ exists if, and only if, $\det\mathbf{A}\neq 0$. This is the content of Theorem 8, Section 3.6. Therefore, we conclude that the columns of $\mathbf{A}$ are linearly independent if, and only if, $\det\mathbf{A}\neq 0$. □

We summarize the results of this section by the following theorem.

Theorem 11. *The equation* $\mathbf{Ax}=\mathbf{b}$ *has a unique solution* $\mathbf{x}=\mathbf{A}^{-1}\mathbf{b}$ *if* $\det\mathbf{A}\neq 0$. *The equation* $\mathbf{Ax}=\mathbf{b}$ *has either no solutions, or infinitely many solutions if* $\det\mathbf{A}=0$.

PROOF. Theorem 11 follows immediately from Theorem 9 and Lemma 3. □

Corollary. *The equation* $\mathbf{Ax}=\mathbf{0}$ *has a nontrivial solution (that is, a solution*

$$\mathbf{x}=\begin{bmatrix} x_1 \\ \vdots \\ x_n \end{bmatrix}$$

with not all the x_i *equal to zero) if, and only if,* $\det\mathbf{A}=0$.

PROOF. Observe that

$$\mathbf{x}=\begin{bmatrix} 0 \\ \vdots \\ 0 \end{bmatrix}$$

is always one solution of the equation $\mathbf{Ax}=\mathbf{0}$. Hence, it is the only solution if $\det\mathbf{A}\neq 0$. On the other hand, there exist infinitely many solutions if $\det\mathbf{A}=0$, and all but one of these are nontrivial. □

Example 6. For which values of λ does the equation

$$\begin{bmatrix} 1 & \lambda & \lambda \\ 1 & 0 & 1 \\ 1 & 1 & 1 \end{bmatrix}\mathbf{x}=\mathbf{0}$$

have a nontrivial solution?

Solution.

$$\det\begin{bmatrix} 1 & \lambda & \lambda \\ 1 & 0 & 1 \\ 1 & 1 & 1 \end{bmatrix}=\lambda+\lambda-1-\lambda=\lambda-1.$$

Hence, the equation

$$\begin{bmatrix} 1 & \lambda & \lambda \\ 1 & 0 & 1 \\ 1 & 1 & 1 \end{bmatrix}\mathbf{x}=\mathbf{0}$$

has a nontrivial solution if, and only if, $\lambda=1$.

Remark 1. Everything we've said about the equation $\mathbf{Ax}=\mathbf{b}$ applies equally well when the elements of $\mathbf{A}$ and the components of $\mathbf{x}$ and $\mathbf{b}$ are complex numbers. In this case, we interpret $\mathbf{x}$ and $\mathbf{b}$ as vectors in $\mathbf{C}^n$, and the matrix $\mathbf{A}$ as inducing a linear transformation of $\mathbf{C}^n$ into itself.

Remark 2. Suppose that we seek to determine n numbers $x_1, x_2, \ldots, x_n$. Our intuitive feeling is that we must be given n equations which are satisfied by these unknowns. This is certainly the case if we are given n linear equations of the form

$$a_{j1}x_1+a_{j2}x_2+\dots+a_{jn}x_n=b_j,\quad j=1,2,\dots,n \tag{9}$$

and $\det \mathbf{A}\neq 0$, where

$$\mathbf{A}=\begin{bmatrix} a_{11} & \dots & a_{1n}\\ \vdots & & \vdots \\ a_{n1} & \dots & a_{nn}\end{bmatrix}.$$

On the other hand, our intuition would seem to be wrong when $\det \mathbf{A}=0$. This is not the case, though. To wit, if $\det \mathbf{A}=0$, then the columns of $\mathbf{A}$ are linearly dependent. But then the columns of $\mathbf{A}^T$, which are the rows of $\mathbf{A}$, are also linearly dependent, since $\det \mathbf{A}^T=\det \mathbf{A}$. Consequently, one of the rows of $\mathbf{A}$ is a linear combination of the other rows. Now, this implies that the left-hand side of one of the equations (9), say the kth equation, is a linear combination of the other left-hand sides. Obviously, the equation $\mathbf{Ax}=\mathbf{b}$ has no solution if b_k is not the exact same linear combination of $b_1,\dots,b_{k-1},b_{k+1},\dots,b_n$. For example, the system of equations

$$\begin{aligned} x_1-x_2+\ x_3&=1\\ x_1+x_2+\ x_3&=1\\ 2x_1\qquad +2x_3&=3\end{aligned}$$

obviously has no solution. On the other hand, if b_k is the same linear combination of $b_1,\dots,b_{k-1},b_{k+1},\dots,b_n$, then the kth equation is redundant. In this case, therefore, we really have only $n-1$ equations for the n unknowns $x_1,x_2,\dots,x_n$.

Remark 3. Once we introduce the concept of a linear transformation we no longer need to view an $n\times n$ matrix as just a square array of numbers. Rather, we can now view an $n\times n$ matrix $\mathbf{A}$ as inducing a linear transformation $\mathcal{A}(\mathbf{x})=\mathbf{Ax}$ on R^n. The benefit of this approach is that we can derive properties of $\mathbf{A}$ by deriving the equivalent properties of the linear transformation $\mathcal{A}$. For example, we showed that $(\mathbf{AB})\mathbf{C}=\mathbf{A}(\mathbf{BC})$ for any 3 $n\times n$ matrices $\mathbf{A}$, $\mathbf{B}$, and $\mathbf{C}$ by showing that the induced linear transformations $\mathcal{A}$, $\mathcal{B}$, and $\mathcal{C}$ satisfy the relation $(\mathcal{A}\circ\mathcal{B})\circ\mathcal{C}=\mathcal{A}\circ(\mathcal{B}\circ\mathcal{C})$. Now, this result can be proven directly, but it requires a great deal more work (see Exercise 24).

Exercises

In each of Problems 1–3 find all vectors $\mathbf{b}$ for which the given system of equations has a solution.

1. $\begin{pmatrix} 3 & -1 & 1\\ 1 & 1 & 1\\ 5 & 1 & 3\end{pmatrix}\mathbf{x}=\mathbf{b}$

2. $\begin{bmatrix} 1 & 2 & 2 & 3\\ 2 & 4 & 4 & 6\\ 3 & 6 & 6 & 9\\ 4 & 8 & 8 & 12\end{bmatrix}\mathbf{x}=\mathbf{b}$

3. $\begin{pmatrix} 11 & 4 & 8 \\ 1 & -1 & 1 \\ 3 & 2 & 2 \end{pmatrix}\mathbf{x}=\mathbf{b}$

In each of Problems 4–9, find all solutions of the given system of equations.

4. $\begin{pmatrix} 1 & 2 & -3 \\ 2 & 3 & -1 \\ 1 & 3 & 10 \end{pmatrix}\mathbf{x}=\mathbf{0}$

5. $\begin{pmatrix} 1 & 2 & 1 \\ -3 & -2 & 1 \\ 6 & 8 & 2 \end{pmatrix}\mathbf{x}=\begin{pmatrix} 3 \\ 5 \\ 4 \end{pmatrix}$

6. $\begin{bmatrix} 1 & 1 & 1 & 1 \\ 1 & -1 & 1 & 1 \\ 1 & 1 & -1 & 1 \\ 1 & 1 & 1 & -1 \end{bmatrix}\mathbf{x}=\mathbf{0}$

7. $\begin{pmatrix} 2 & -1 & -1 \\ -5 & 3 & 1 \\ -1 & 1 & -1 \end{pmatrix}\mathbf{x}=\begin{pmatrix} -1 \\ 2 \\ 3 \end{pmatrix}$

8. $\begin{bmatrix} 1 & 2 & 3 & 4 \\ 0 & 1 & -1 & 6 \\ 2 & 0 & 0 & 1 \\ 2 & 4 & 6 & 8 \end{bmatrix}\mathbf{x}=\mathbf{0}$

9. $\begin{pmatrix} 1 & 1 & 1 \\ 1 & -1 & 1 \\ 1 & 1 & -1 \end{pmatrix}\mathbf{x}=\begin{pmatrix} 1 \\ 1 \\ 1 \end{pmatrix}$

In each of Problems 10–12, determine all values of λ for which the given system of equations has a nontrivial solution.

10. $\begin{pmatrix} \lambda & 1 & 3 \\ 1 & \lambda & 3 \\ -1 & 3 & \lambda \end{pmatrix}\mathbf{x}=\mathbf{0}$

11. $\begin{bmatrix} 1 & 1 & \lambda & 1 \\ 1 & -1 & \lambda & \lambda \\ 0 & 1 & 0 & 1 \\ -1 & 0 & -1 & 1 \end{bmatrix}\mathbf{x}=\mathbf{0}$

12. $\begin{pmatrix} \lambda & -1 & -1 \\ 2 & -1 & 0 \\ 3 & \lambda & 5 \end{pmatrix}\mathbf{x}=\mathbf{0}.$

13. (a) For which value of λ does the system of equations

$$\begin{pmatrix} -2 & -4 & -5 \\ 1 & -1 & -1 \\ 4 & 2 & 3 \end{pmatrix}\mathbf{x}=\begin{pmatrix} \lambda \\ 1 \\ 3 \end{pmatrix}$$

have a solution?

(b) Find all solutions for this value of λ.

14. Suppose that $\mathbf{Ax}=\mathbf{Bx}$ for all vectors

$$\mathbf{x}=\begin{bmatrix} x_1 \\ \vdots \\ x_n \end{bmatrix}.$$

Prove that $\mathbf{A}=\mathbf{B}$.

15. Let $\mathcal{A}$ and $\mathcal{B}$ be two linear transformations taking R^n into R^n. Then, there exist $n\times n$ matrices $\mathbf{A}$ and $\mathbf{B}$ such that $\mathcal{A}(\mathbf{x})=\mathbf{Ax}$ and $\mathcal{B}(\mathbf{x})=\mathbf{Bx}$. Show that $\mathcal{A}\circ\mathcal{B}(\mathbf{x}) =\mathbf{ABx}$. *Hint*: $\mathcal{A}\circ\mathcal{B}$ is a linear transformation taking R^n into R^n. Hence, there exists an $n\times n$ matrix $\mathbf{C}$ such that $\mathcal{A}\circ\mathcal{B}(\mathbf{x})=\mathbf{Cx}$. The jth column of $\mathbf{C}$ is $\mathcal{A}\circ\mathcal{B}(\mathbf{e}^j)$. Thus show that $\mathcal{A}\circ\mathcal{B}(\mathbf{e}^j)$ is the jth column of the matrix $\mathbf{AB}$.

16. Let $\mathcal{A}$ be a linear transformation taking $\mathsf{R}^n \to \mathsf{R}^n$. Show that $\mathcal{A}(\mathbf{0}) = \mathbf{0}$.

17. Let $\mathcal{R}(\theta)$ be the linear transformation which rotates each point in the plane by an angle θ in the counterclockwise direction. Show that

$$\mathcal{R}(\theta)(\mathbf{x}) = \begin{pmatrix} \cos\theta & -\sin\theta \\ \sin\theta & \cos\theta \end{pmatrix}\begin{pmatrix} x_1 \\ x_2 \end{pmatrix}.$$

18. Let $\mathcal{R}_1$ and $\mathcal{R}_2$ be the linear transformations which rotate each point in the plane by angles θ_1 and θ_2 respectively. Then the linear transformation $\mathcal{R}_3 = \mathcal{R}_1 \circ \mathcal{R}_2$ rotates each point in the plane by an angle $\theta_1 + \theta_2$ (in the counterclockwise direction). Using Exercise (15), show that

$$\begin{pmatrix} \cos(\theta_1+\theta_2) & -\sin(\theta_1+\theta_2) \\ \sin(\theta_1+\theta_2) & \cos(\theta_1+\theta_2) \end{pmatrix} = \begin{pmatrix} \cos\theta_1 & -\sin\theta_1 \\ \sin\theta_1 & \cos\theta_1 \end{pmatrix}\begin{pmatrix} \cos\theta_2 & -\sin\theta_2 \\ \sin\theta_2 & \cos\theta_2 \end{pmatrix}.$$

Thus, derive the trigonometric identities

$$\sin(\theta_1+\theta_2) = \sin\theta_1\cos\theta_2 + \cos\theta_1\sin\theta_2$$
$$\cos(\theta_1+\theta_2) = \cos\theta_1\cos\theta_2 - \sin\theta_1\sin\theta_2.$$

19. Let

$$\mathcal{A}(x_1,x_2) = \begin{pmatrix} x_1 + x_2 \\ x_1 - x_2 \end{pmatrix}.$$

(a) Verify that $\mathcal{A}$ is linear.
(b) Show that every point (x_1,x_2) on the unit circle $x_1^2 + x_2^2 = 1$ goes into a point on the circle, $x_1^2 + x_2^2 = 2$.

20. Let $\mathbf{V}$ be the space of all polynomials $p(t)$ of degree less than or equal to 3 and let $(Dp)(t) = dp(t)/dt$.
(a) Show that D is a linear transformation taking $\mathbf{V}$ into $\mathbf{V}$.
(b) Show that D does not possess an inverse.

21. Let $\mathbf{V}$ be the space of all continuous functions $f(t)$, $-\infty < t < \infty$ and let $(Kf)(t) = \int_0^t f(s)\,ds$.
(a) Show that K is a linear transformation taking $\mathbf{V}$ into $\mathbf{V}$.
(b) Show that $(DK)f = f$ where $Df = f'$.
(c) Let $f(t)$ be differentiable. Show that

$$[(KD)f](t) = f(t) - f(0).$$

22. A linear transformation $\mathcal{A}$ is said to be $1-1$ if $\mathcal{A}(\mathbf{x}) \neq \mathcal{A}(\mathbf{y})$ whenever $\mathbf{x} \neq \mathbf{y}$. In other words, no two vectors go into the same vector under $\mathcal{A}$. Show that $\mathcal{A}$ is $1-1$ if, and only if, $\mathcal{A}(\mathbf{x}) = \mathbf{0}$ implies that $\mathbf{x} = \mathbf{0}$.

23. A linear transformation $\mathcal{A}$ is said to be *onto* if the equation $\mathcal{A}(\mathbf{x}) = \mathbf{y}$ has at least one solution for every $\mathbf{y}$ in R^n. Prove that $\mathcal{A}$ is onto if, and only if, $\mathcal{A}$ is $1-1$. *Hint*: Show first that $\mathcal{A}$ is onto if, and only if, the vectors $\mathcal{A}(\mathbf{e}^1), \ldots, \mathcal{A}(\mathbf{e}^n)$ are linearly independent. Then, use Lemma 1 to show that we can find a nonzero solution of the equation $\mathcal{A}(\mathbf{x}) = \mathbf{0}$ if $\mathcal{A}(\mathbf{e}^1), \ldots, \mathcal{A}(\mathbf{e}^n)$ are linearly dependent. Finally, show that $\mathcal{A}(\mathbf{e}^1), \ldots, \mathcal{A}(\mathbf{e}^n)$ are linearly dependent if the equation $\mathcal{A}(\mathbf{x}) = \mathbf{0}$ has a nonzero solution.

24. Prove directly that $(\mathbf{AB})\mathbf{C}=\mathbf{A}(\mathbf{BC})$. *Hint*: Show that these matrices have the same elements.

25. Show that

$$\mathbf{b}=\begin{bmatrix} b_1 \\ b_2 \\ b_3 \end{bmatrix}=c_1\begin{pmatrix} 1 \\ 1 \\ 2 \end{pmatrix}+c_2\begin{pmatrix} 1 \\ 0 \\ 1 \end{pmatrix}$$

if, and only if, $b_3=b_1+b_2$.

3.8 The eigenvalue–eigenvector method of finding solutions

We return now to the first-order linear homogeneous differential equation

$$\dot{\mathbf{x}}=\mathbf{A}\mathbf{x},\quad \mathbf{x}=\begin{bmatrix} x_1 \\ \vdots \\ x_n \end{bmatrix},\quad \mathbf{A}=\begin{bmatrix} a_{11} & \dots & a_{1n} \\ \vdots & & \vdots \\ a_{n1} & \dots & a_{nn} \end{bmatrix}. \tag{1}$$

Our goal is to find n linearly independent solutions $\mathbf{x}^1(t),\dots,\mathbf{x}^n(t)$. Now, recall that both the first-order and second-order linear homogeneous scalar equations have exponential functions as solutions. This suggests that we try $\mathbf{x}(t)=e^{\lambda t}\mathbf{v}$, where $\mathbf{v}$ is a constant vector, as a solution of (1). To this end, observe that

$$\frac{d}{dt}e^{\lambda t}\mathbf{v}=\lambda e^{\lambda t}\mathbf{v}$$

and

$$\mathbf{A}(e^{\lambda t}\mathbf{v})=e^{\lambda t}\mathbf{A}\mathbf{v}.$$

Hence, $\mathbf{x}(t)=e^{\lambda t}\mathbf{v}$ is a solution of (1) if, and only if, $\lambda e^{\lambda t}\mathbf{v}=e^{\lambda t}\mathbf{A}\mathbf{v}$. Dividing both sides of this equation by $e^{\lambda t}$ gives

$$\mathbf{A}\mathbf{v}=\lambda\mathbf{v}. \tag{2}$$

Thus, $\mathbf{x}(t)=e^{\lambda t}\mathbf{v}$ is a solution of (1) if, and only if, λ and $\mathbf{v}$ satisfy (2).

Definition. A nonzero vector $\mathbf{v}$ satisfying (2) is called an *eigenvector* of $\mathbf{A}$ with *eigenvalue* λ.

Remark. The vector $\mathbf{v}=\mathbf{0}$ is excluded because it is uninteresting. Obviously, $\mathbf{A0}=\lambda\cdot\mathbf{0}$ for any number λ.

An eigenvector of a matrix $\mathbf{A}$ is a rather special vector: under the linear transformation $\mathbf{x}\to\mathbf{A}\mathbf{x}$, it goes into a multiple λ of itself. Vectors which are transformed into multiples of themselves play an important role in many applications. To find such vectors, we rewrite Equation (2) in the form

$$\mathbf{0}=\mathbf{A}\mathbf{v}-\lambda\mathbf{v}=(\mathbf{A}-\lambda\mathbf{I})\mathbf{v}. \tag{3}$$

But, Equation (3) has a nonzero solution $\mathbf{v}$ only if $\det(\mathbf{A}-\lambda\mathbf{I})=0$. Hence the eigenvalues λ of $\mathbf{A}$ are the roots of the equation

$$0=\det(\mathbf{A}-\lambda\mathbf{I})=\det\begin{bmatrix} a_{11}-\lambda & a_{12} & \dots & a_{1n} \\ a_{21} & a_{22}-\lambda & \dots & a_{2n} \\ \vdots & \vdots & & \vdots \\ a_{n1} & a_{n2} & \dots & a_{nn}-\lambda \end{bmatrix},$$

and the eigenvectors of $\mathbf{A}$ are then the nonzero solutions of the equations $(\mathbf{A}-\lambda\mathbf{I})\mathbf{v}=\mathbf{0}$, for these values of λ.

The determinant of the matrix $\mathbf{A}-\lambda\mathbf{I}$ is clearly a polynomial in λ of degree n, with leading term $(-1)^n\lambda^n$. It is customary to call this polynomial the characteristic polynomial of $\mathbf{A}$ and to denote it by $p(\lambda)$. For each root λ_j of $p(\lambda)$, that is, for each number λ_j such that $p(\lambda_j)=0$, there exists at least one nonzero vector $\mathbf{v}^j$ such that $\mathbf{A}\mathbf{v}^j=\lambda_j\mathbf{v}^j$. Now, every polynomial of degree $n\geqslant 1$ has at least one (possibly complex) root. Therefore, every matrix has at least one eigenvalue, and consequently, at least one eigenvector. On the other hand, $p(\lambda)$ has at most n distinct roots. Therefore, every $n\times n$ matrix has at most n eigenvalues. Finally, observe that every $n\times n$ matrix has at most n linearly independent eigenvectors, since the space of all vectors

$$\mathbf{v}=\begin{bmatrix} v_1 \\ \vdots \\ v_n \end{bmatrix}$$

has dimension n.

Remark. Let $\mathbf{v}$ be an eigenvector of $\mathbf{A}$ with eigenvalue λ. Observe that

$$\mathbf{A}(c\mathbf{v})=c\mathbf{A}\mathbf{v}=c\lambda\mathbf{v}=\lambda(c\mathbf{v})$$

for any constant c. Hence, any constant multiple $(c\neq 0)$ of an eigenvector of $\mathbf{A}$ is again an eigenvector of $\mathbf{A}$, with the same eigenvalue.

For each eigenvector $\mathbf{v}^j$ of $\mathbf{A}$ with eigenvalue λ_j, we have a solution $\mathbf{x}^j(t)=e^{\lambda_j t}\mathbf{v}^j$ of (1). If $\mathbf{A}$ has n linearly independent eigenvectors $\mathbf{v}^1,\dots,\mathbf{v}^n$ with eigenvalues $\lambda_1,\dots,\lambda_n$ respectively ($\lambda_1,\dots,\lambda_n$ need not be distinct), then $\mathbf{x}^j(t)=e^{\lambda_j t}\mathbf{v}^j$, $j=1,\dots,n$ are n linearly independent solutions of (1). This follows immediately from Theorem 6 of Section 3.4 and the fact that $\mathbf{x}^j(0)=\mathbf{v}^j$. In this case, then, every solution $\mathbf{x}(t)$ of (1) is of the form

$$\mathbf{x}(t)=c_1e^{\lambda_1 t}\mathbf{v}^1+c_2e^{\lambda_2 t}\mathbf{v}^2+\dots+c_ne^{\lambda_n t}\mathbf{v}^n. \tag{4}$$

This is sometimes called the "general solution" of (1).

The situation is simplest when $\mathbf{A}$ has n distinct real eigenvalues $\lambda_1,\lambda_2,\dots,\lambda_n$ with eigenvectors $\mathbf{v}^1,\mathbf{v}^2,\dots,\mathbf{v}^n$ respectively, for in this case we are guaranteed that $\mathbf{v}^1, \mathbf{v}^2, \ldots, \mathbf{v}^n$ are linearly independent. This is the content of Theorem 12.

Theorem 12. *Any k eigenvectors $\mathbf{v}^1, \ldots, \mathbf{v}^k$ of $\mathbf{A}$ with distinct eigenvalues $\lambda_1, \ldots, \lambda_k$ respectively, are linearly independent.*

PROOF. We will prove Theorem 12 by induction on k, the number of eigenvectors. Observe that this theorem is certainly true for $k=1$. Next, we assume that Theorem 12 is true for $k=j$. That is to say, we assume that any set of j eigenvectors of $\mathbf{A}$ with distinct eigenvalues is linearly independent. We must show that any set of $j+1$ eigenvectors of $\mathbf{A}$ with distinct eigenvalues is also linearly independent. To this end, let $\mathbf{v}^1, \ldots, \mathbf{v}^{j+1}$ be $j+1$ eigenvectors of $\mathbf{A}$ with distinct eigenvalues $\lambda_1, \ldots, \lambda_{j+1}$ respectively. To determine whether these vectors are linearly dependent or linearly independent, we consider the equation

$$c_1\mathbf{v}^1 + c_2\mathbf{v}^2 + \ldots + c_{j+1}\mathbf{v}^{j+1} = \mathbf{0}. \tag{5}$$

Applying $\mathbf{A}$ to both sides of (5) gives

$$\lambda_1 c_1\mathbf{v}^1 + \lambda_2 c_2\mathbf{v}^2 + \ldots + \lambda_{j+1} c_{j+1}\mathbf{v}^{j+1} = \mathbf{0}. \tag{6}$$

Thus, if we multiply both sides of (5) by λ_1 and subtract the resulting equation from (6), we obtain that

$$(\lambda_2 - \lambda_1)c_2\mathbf{v}^2 + \ldots + (\lambda_{j+1} - \lambda_1)c_{j+1}\mathbf{v}^{j+1} = \mathbf{0}. \tag{7}$$

But $\mathbf{v}^2, \ldots, \mathbf{v}^{j+1}$ are j eigenvectors of $\mathbf{A}$ with distinct eigenvalues $\lambda_2, \ldots, \lambda_{j+1}$ respectively. By the induction hypothesis, they are linearly independent. Consequently,

$$(\lambda_2 - \lambda_1)c_2 = 0, \qquad (\lambda_3 - \lambda_1)c_3 = 0, \ldots, \qquad \text{and} \quad (\lambda_{j+1} - \lambda_1)c_{j+1} = 0.$$

Since $\lambda_1, \lambda_2, \ldots, \lambda_{j+1}$ are distinct, we conclude that $c_2, c_3, \ldots, c_{j+1}$ are all zero. Equation (5) now forces c_1 to be zero. Hence, $\mathbf{v}^1, \mathbf{v}^2, \ldots, \mathbf{v}^{j+1}$ are linearly independent. By induction, therefore, every set of k eigenvectors of $\mathbf{A}$ with distinct eigenvalues is linearly independent. □

Example 1. Find all solutions of the equation

$$\dot{\mathbf{x}} = \begin{pmatrix} 1 & -1 & 4 \\ 3 & 2 & -1 \\ 2 & 1 & -1 \end{pmatrix} \mathbf{x}.$$

Solution. The characteristic polynomial of the matrix

$$\mathbf{A} = \begin{pmatrix} 1 & -1 & 4 \\ 3 & 2 & -1 \\ 2 & 1 & -1 \end{pmatrix}$$

is

$$p(\lambda)=\det(\mathbf{A}-\lambda\mathbf{I})=\det\begin{bmatrix}1-\lambda & -1 & 4\\ 3 & 2-\lambda & -1\\ 2 & 1 & -1-\lambda\end{bmatrix}$$
$$=-(1+\lambda)(1-\lambda)(2-\lambda)+2+12-8(2-\lambda)+(1-\lambda)-3(1+\lambda)$$
$$=(1-\lambda)(\lambda-3)(\lambda+2).$$

Thus the eigenvalues of $\mathbf{A}$ are $\lambda_1=1$, $\lambda_2=3$, and $\lambda_3=-2$.

(i) $\lambda_1=1$: We seek a nonzero vector $\mathbf{v}$ such that

$$(\mathbf{A}-\mathbf{I})\mathbf{v}=\begin{bmatrix}0 & -1 & 4\\ 3 & 1 & -1\\ 2 & 1 & -2\end{bmatrix}\begin{bmatrix}v_1\\ v_2\\ v_3\end{bmatrix}=\begin{bmatrix}0\\ 0\\ 0\end{bmatrix}.$$

This implies that

$$-v_2+4v_3=0,\qquad 3v_1+v_2-v_3=0,\quad\text{and}\quad 2v_1+v_2-2v_3=0.$$

Solving for v_1 and v_2 in terms of v_3 from the first two equations gives $v_1=-v_3$ and $v_2=4v_3$. Hence, each vector

$$\mathbf{v}=c\begin{bmatrix}-1\\ 4\\ 1\end{bmatrix}$$

is an eigenvector of $\mathbf{A}$ with eigenvalue one. Consequently,

$$ce^{t}\begin{bmatrix}-1\\ 4\\ 1\end{bmatrix}$$

is a solution of the differential equation for any constant c. For simplicity, we take

$$\mathbf{x}^1(t)=e^{t}\begin{bmatrix}-1\\ 4\\ 1\end{bmatrix}.$$

(ii) $\lambda_2=3$: We seek a nonzero vector $\mathbf{v}$ such that

$$(\mathbf{A}-3\mathbf{I})\mathbf{v}=\begin{bmatrix}-2 & -1 & 4\\ 3 & -1 & -1\\ 2 & 1 & -4\end{bmatrix}\begin{bmatrix}v_1\\ v_2\\ v_3\end{bmatrix}=\begin{bmatrix}0\\ 0\\ 0\end{bmatrix}.$$

This implies that

$$-2v_1-v_2+4v_3=0,\qquad 3v_1-v_2-v_3=0,\quad\text{and}\quad 2v_1+v_2-4v_3=0.$$

Solving for v_1 and v_2 in terms of v_3 from the first two equations gives $v_1=v_3$ and $v_2=2v_3$. Consequently, each vector

$$\mathbf{v}=c\begin{bmatrix}1\\ 2\\ 1\end{bmatrix}$$

is an eigenvector of $\mathbf{A}$ with eigenvalue 3. Therefore,

$$\mathbf{x}^2(t)=e^{3t}\begin{bmatrix}1\\2\\1\end{bmatrix}$$

is a second solution of the differential equation.

(iii) $\lambda_3=-2$: We seek a nonzero vector $\mathbf{v}$ such that

$$(\mathbf{A}+2\mathbf{I})\mathbf{v}=\begin{bmatrix}3&-1&4\\3&4&-1\\2&1&1\end{bmatrix}\begin{bmatrix}v_1\\v_2\\v_3\end{bmatrix}=\begin{bmatrix}0\\0\\0\end{bmatrix}.$$

This implies that

$$3v_1-v_2+4v_3=0,\qquad 3v_1+4v_2-v_3=0\quad\text{and}\quad 2v_1+v_2+v_3=0.$$

Solving for v_1 and v_2 in terms of v_3 gives $v_1=-v_3$ and $v_2=v_3$. Hence, each vector

$$\mathbf{v}=c\begin{bmatrix}-1\\1\\1\end{bmatrix}$$

is an eigenvector of $\mathbf{A}$ with eigenvalue -2. Consequently,

$$\mathbf{x}^3(t)=e^{-2t}\begin{bmatrix}-1\\1\\1\end{bmatrix}$$

is a third solution of the differential equation. These solutions must be linearly independent, since $\mathbf{A}$ has distinct eigenvalues. Therefore, every solution $\mathbf{x}(t)$ must be of the form

$$\mathbf{x}(t)=c_1e^t\begin{bmatrix}-1\\4\\1\end{bmatrix}+c_2e^{3t}\begin{bmatrix}1\\2\\1\end{bmatrix}+c_3e^{-2t}\begin{bmatrix}-1\\1\\1\end{bmatrix}$$

$$=\begin{bmatrix}-c_1e^t+c_2e^{3t}-c_3e^{-2t}\\4c_1e^t+2c_2e^{3t}+c_3e^{-2t}\\c_1e^t+c_2e^{3t}+c_3e^{-2t}\end{bmatrix}.$$

Remark. If λ is an eigenvalue of $\mathbf{A}$, then the n equations

$$a_{j1}v_1+\ldots+(a_{jj}-\lambda)v_j+\ldots+a_{jn}v_n=0,\qquad j=1,\ldots,n$$

are not independent; at least one of them is a linear combination of the others. Consequently, we have at most $n-1$ independent equations for the n unknowns $v_1,\ldots,v_n$. This implies that at least one of the unknowns $v_1,\ldots,v_n$ can be chosen arbitrarily.

Example 2. Solve the initial-value problem

$$\dot{\mathbf{x}}=\begin{pmatrix}1 & 12\\ 3 & 1\end{pmatrix}\mathbf{x}, \qquad \mathbf{x}(0)=\begin{pmatrix}0\\ 1\end{pmatrix}.$$

Solution. The characteristic polynomial of the matrix

$$\mathbf{A}=\begin{pmatrix}1 & 12\\ 3 & 1\end{pmatrix}$$

is

$$p(\lambda)=\det\begin{pmatrix}1-\lambda & 12\\ 3 & 1-\lambda\end{pmatrix}=(1-\lambda)^2-36=(\lambda-7)(\lambda+5).$$

Thus, the eigenvalues of A are $\lambda_1=7$ and $\lambda_2=-5$.

(i) $\lambda_1=7$: We seek a nonzero vector $\mathbf{v}$ such that

$$(\mathbf{A}-7\mathbf{I})\mathbf{v}=\begin{pmatrix}-6 & 12\\ 3 & -6\end{pmatrix}\begin{pmatrix}v_1\\ v_2\end{pmatrix}=\begin{pmatrix}0\\ 0\end{pmatrix}.$$

This implies that $v_1=2v_2$. Consequently, every vector

$$\mathbf{v}=c\begin{pmatrix}2\\ 1\end{pmatrix}$$

is an eigenvector of $\mathbf{A}$ with eigenvalue 7. Therefore,

$$\mathbf{x}^1(t)=e^{7t}\begin{pmatrix}2\\ 1\end{pmatrix}$$

is a solution of the differential equation.

(ii) $\lambda_2=-5$: We seek a nonzero vector $\mathbf{v}$ such that

$$(\mathbf{A}+5\mathbf{I})\mathbf{v}=\begin{pmatrix}6 & 12\\ 3 & 6\end{pmatrix}\begin{pmatrix}v_1\\ v_2\end{pmatrix}=\begin{pmatrix}0\\ 0\end{pmatrix}.$$

This implies that $v_1=-2v_2$. Consequently,

$$\begin{pmatrix}-2\\ 1\end{pmatrix}$$

is an eigenvector of $\mathbf{A}$ with eigenvalue -5, and

$$\mathbf{x}^2(t)=e^{-5t}\begin{pmatrix}-2\\ 1\end{pmatrix}$$

is a second solution of the differential equation. These solutions are linearly independent since $\mathbf{A}$ has distinct eigenvalues. Hence, $\mathbf{x}(t)=c_1\mathbf{x}^1(t)+c_2\mathbf{x}^2(t)$. The constants c_1 and c_2 are determined from the initial condition

$$\begin{pmatrix}0\\ 1\end{pmatrix}=c_1\mathbf{x}^1(0)+c_2\mathbf{x}^2(0)=\begin{pmatrix}2c_1\\ c_1\end{pmatrix}+\begin{pmatrix}-2c_2\\ c_2\end{pmatrix}.$$

Thus, $2c_1 - 2c_2 = 0$ and $c_1 + c_2 = 1$. The solution of these two equations is $c_1 = \frac{1}{2}$ and $c_2 = \frac{1}{2}$. Consequently,

$$\mathbf{x}(t) = \frac{1}{2}e^{7t}\begin{pmatrix}2\\1\end{pmatrix} + \frac{1}{2}e^{-5t}\begin{pmatrix}-2\\1\end{pmatrix} = \begin{bmatrix} e^{7t} - e^{-5t} \\ \frac{1}{2}e^{7t} + \frac{1}{2}e^{-5t} \end{bmatrix}.$$

Example 3. Find all solutions of the equation

$$\dot{\mathbf{x}} = \mathbf{A}\mathbf{x} = \begin{bmatrix} 1 & 1 & 1 & 1 & 1 \\ 2 & 2 & 2 & 2 & 2 \\ 3 & 3 & 3 & 3 & 3 \\ 4 & 4 & 4 & 4 & 4 \\ 5 & 5 & 5 & 5 & 5 \end{bmatrix}\mathbf{x}.$$

Solution. It is not necessary to compute the characteristic polynomial of $\mathbf{A}$ in order to find the eigenvalues and eigenvectors of $\mathbf{A}$. To wit, observe that

$$\mathbf{A}\mathbf{x} = \mathbf{A}\begin{bmatrix} x_1 \\ x_2 \\ x_3 \\ x_4 \\ x_5 \end{bmatrix} = (x_1 + x_2 + x_3 + x_4 + x_5)\begin{bmatrix} 1 \\ 2 \\ 3 \\ 4 \\ 5 \end{bmatrix}.$$

Hence, any vector $\mathbf{x}$ whose components add up to zero is an eigenvector of $\mathbf{A}$ with eigenvalue 0. In particular

$$\mathbf{v}^1 = \begin{bmatrix} 1 \\ 0 \\ 0 \\ 0 \\ -1 \end{bmatrix}, \quad \mathbf{v}^2 = \begin{bmatrix} 0 \\ 1 \\ 0 \\ 0 \\ -1 \end{bmatrix}, \quad \mathbf{v}^3 = \begin{bmatrix} 0 \\ 0 \\ 1 \\ 0 \\ -1 \end{bmatrix}, \quad \text{and} \quad \mathbf{v}^4 = \begin{bmatrix} 0 \\ 0 \\ 0 \\ 1 \\ -1 \end{bmatrix}$$

are four independent eigenvectors of $\mathbf{A}$ with eigenvalue zero. Moreover, observe that

$$\mathbf{v}^5 = \begin{bmatrix} 1 \\ 2 \\ 3 \\ 4 \\ 5 \end{bmatrix}$$

is an eigenvector of $\mathbf{A}$ with eigenvalue 15 since

$$\mathbf{A}\begin{bmatrix} 1 \\ 2 \\ 3 \\ 4 \\ 5 \end{bmatrix} = (1+2+3+4+5)\begin{bmatrix} 1 \\ 2 \\ 3 \\ 4 \\ 5 \end{bmatrix} = 15\begin{bmatrix} 1 \\ 2 \\ 3 \\ 4 \\ 5 \end{bmatrix}.$$

The five vectors $\mathbf{v}^1$, $\mathbf{v}^2$, $\mathbf{v}^3$, $\mathbf{v}^4$, and $\mathbf{v}^5$ are easily seen to be linearly independent. Hence, every solution $\mathbf{x}(t)$ is of the form

$$\mathbf{x}(t)=c_1\begin{bmatrix}1\\0\\0\\0\\-1\end{bmatrix}+c_2\begin{bmatrix}0\\1\\0\\0\\-1\end{bmatrix}+c_3\begin{bmatrix}0\\0\\1\\0\\-1\end{bmatrix}+c_4\begin{bmatrix}0\\0\\0\\1\\-1\end{bmatrix}+c_5e^{15t}\begin{bmatrix}1\\2\\3\\4\\5\end{bmatrix}.$$

Exercises

In each of Problems 1–6 find all solutions of the given differential equation.

1. $\dot{\mathbf{x}}=\begin{pmatrix}6 & -3\\ 2 & 1\end{pmatrix}\mathbf{x}$

2. $\dot{\mathbf{x}}=\begin{pmatrix}-2 & 1\\ -4 & 3\end{pmatrix}\mathbf{x}$

3. $\dot{\mathbf{x}}=\begin{pmatrix}3 & 2 & 4\\ 2 & 0 & 2\\ 4 & 2 & 3\end{pmatrix}\mathbf{x}$

4. $\dot{\mathbf{x}}=\begin{pmatrix}7 & -1 & 6\\ -10 & 4 & -12\\ -2 & 1 & -1\end{pmatrix}\mathbf{x}$

5. $\dot{\mathbf{x}}=\begin{pmatrix}-7 & 0 & 6\\ 0 & 5 & 0\\ 6 & 0 & 2\end{pmatrix}\mathbf{x}$

6. $\dot{\mathbf{x}}=\begin{bmatrix}1 & 2 & 3 & 6\\ 3 & 6 & 9 & 18\\ 5 & 10 & 15 & 30\\ 7 & 14 & 21 & 42\end{bmatrix}\mathbf{x}$

In each of Problems 7–12, solve the given initial-value problem.

7. $\dot{\mathbf{x}}=\begin{pmatrix}1 & 1\\ 4 & 1\end{pmatrix}\mathbf{x},\quad \mathbf{x}(0)=\begin{pmatrix}2\\ 3\end{pmatrix}$

8. $\dot{\mathbf{x}}=\begin{pmatrix}1 & -3\\ -2 & 2\end{pmatrix}\mathbf{x},\quad \mathbf{x}(0)=\begin{pmatrix}0\\ 5\end{pmatrix}$

9. $\dot{\mathbf{x}}=\begin{pmatrix}3 & 1 & -1\\ 1 & 3 & -1\\ 3 & 3 & -1\end{pmatrix}\mathbf{x},\quad \mathbf{x}(0)=\begin{pmatrix}1\\ -2\\ -1\end{pmatrix}$

10. $\dot{\mathbf{x}}=\begin{pmatrix}1 & -1 & 0\\ 1 & 2 & 1\\ 1 & 10 & 2\end{pmatrix}\mathbf{x},\quad \mathbf{x}(0)=\begin{pmatrix}-1\\ -4\\ 13\end{pmatrix}$

11. $\dot{\mathbf{x}}=\begin{pmatrix}1 & -3 & 2\\ 0 & -1 & 0\\ 0 & -1 & -2\end{pmatrix}\mathbf{x},\quad \mathbf{x}(0)=\begin{pmatrix}-2\\ 0\\ 3\end{pmatrix}$

12. $\dot{\mathbf{x}}=\begin{pmatrix}3 & 1 & -2\\ -1 & 2 & 1\\ 4 & 1 & -3\end{pmatrix}\mathbf{x},\quad \mathbf{x}(0)=\begin{pmatrix}1\\ 4\\ -7\end{pmatrix}$

13. (a) Show that $e^{\lambda(t-t_0)}\mathbf{v}, t_0$ constant, is a solution of $\dot{\mathbf{x}}=\mathbf{A}\mathbf{x}$ if $\mathbf{A}\mathbf{v}=\lambda\mathbf{v}$.
(b) Solve the initial-value problem

$$\dot{\mathbf{x}}=\begin{pmatrix}3 & 1 & -2\\ -1 & 2 & 1\\ 4 & 1 & -3\end{pmatrix}\mathbf{x},\quad \mathbf{x}(1)=\begin{pmatrix}1\\ 4\\ -7\end{pmatrix}$$

(see Exercise 12).

14. Three solutions of the equation $\dot{\mathbf{x}}=\mathbf{A}\mathbf{x}$ are

$$\begin{bmatrix} e^{t}+e^{2t} \\ e^{2t} \\ 0 \end{bmatrix}, \quad \begin{bmatrix} e^{t}+e^{3t} \\ e^{3t} \\ e^{3t} \end{bmatrix} \quad \text{and} \quad \begin{bmatrix} e^{t}-e^{3t} \\ -e^{3t} \\ -e^{3t} \end{bmatrix}.$$

Find the eigenvalues and eigenvectors of $\mathbf{A}$.

15. Show that the eigenvalues of $\mathbf{A}^{-1}$ are the reciprocals of the eigenvalues of $\mathbf{A}$.

16. Show that the eigenvalues of $\mathbf{A}^{n}$ are the nth power of the eigenvalues of $\mathbf{A}$.

17. Show that $\lambda=0$ is an eigenvalue of $\mathbf{A}$ if $\det\mathbf{A}=0$.

18. Show, by example, that the eigenvalues of $\mathbf{A}+\mathbf{B}$ are not necessarily the sum of an eigenvalue of $\mathbf{A}$ and an eigenvalue of $\mathbf{B}$.

19. Show, by example, that the eigenvalues of $\mathbf{AB}$ are not necessarily the product of an eigenvalue of $\mathbf{A}$ with an eigenvalue of $\mathbf{B}$.

20. Show that the matrices $\mathbf{A}$ and $\mathbf{T}^{-1}\mathbf{AT}$ have the same characteristic polynomial.

21. Suppose that either $\mathbf{B}^{-1}$ or $\mathbf{A}^{-1}$ exists. Prove that $\mathbf{AB}$ and $\mathbf{BA}$ have the same eigenvalues. *Hint*: Use Exercise 20. (This result is true even if neither $\mathbf{B}^{-1}$ or $\mathbf{A}^{-1}$ exist; however, it is more difficult to prove then.)

3.9 Complex roots

If $\lambda=\alpha+i\beta$ is a complex eigenvalue of $\mathbf{A}$ with eigenvector $\mathbf{v}=\mathbf{v}^1+i\mathbf{v}^2$, then $\mathbf{x}(t)=e^{\lambda t}\mathbf{v}$ is a complex-valued solution of the differential equation

$$\dot{\mathbf{x}}=\mathbf{A}\mathbf{x}. \tag{1}$$

This complex-valued solution gives rise to *two* real-valued solutions, as we now show.

Lemma 1. *Let* $\mathbf{x}(t)=\mathbf{y}(t)+i\mathbf{z}(t)$ *be a complex-valued solution of* (1). *Then, both* $\mathbf{y}(t)$ *and* $\mathbf{z}(t)$ *are real-valued solutions of* (1).

PROOF. If $\mathbf{x}(t)=\mathbf{y}(t)+i\mathbf{z}(t)$ is a complex-valued solution of (1), then

$$\dot{\mathbf{y}}(t)+i\dot{\mathbf{z}}(t)=\mathbf{A}(\mathbf{y}(t)+i\mathbf{z}(t))=\mathbf{A}\mathbf{y}(t)+i\mathbf{A}\mathbf{z}(t). \tag{2}$$

Equating real and imaginary parts of (2) gives $\dot{\mathbf{y}}(t)=\mathbf{A}\mathbf{y}(t)$ and $\dot{\mathbf{z}}(t)=\mathbf{A}\mathbf{z}(t)$. Consequently, both $\mathbf{y}(t)=\mathrm{Re}\{\mathbf{x}(t)\}$ and $\mathbf{z}(t)=\mathrm{Im}\{\mathbf{x}(t)\}$ are real-valued solutions of (1). □

The complex-valued function $\mathbf{x}(t)=e^{(\alpha+i\beta)t}(\mathbf{v}^1+i\mathbf{v}^2)$ can be written in the form

$$\begin{aligned}\mathbf{x}(t)&=e^{\alpha t}(\cos\beta t+i\sin\beta t)(\mathbf{v}^1+i\mathbf{v}^2)\\&=e^{\alpha t}\left[(\mathbf{v}^1\cos\beta t-\mathbf{v}^2\sin\beta t)+i(\mathbf{v}^1\sin\beta t+\mathbf{v}^2\cos\beta t)\right].\end{aligned}$$

Hence, if $\lambda=\alpha+i\beta$ is an eigenvalue of $\mathbf{A}$ with eigenvector $\mathbf{v}=\mathbf{v}^1+i\mathbf{v}^2$, then

$$\mathbf{y}(t)=e^{\alpha t}(\mathbf{v}^1\cos\beta t-\mathbf{v}^2\sin\beta t)$$

and

$$\mathbf{z}(t)=e^{\alpha t}(\mathbf{v}^1\sin\beta t+\mathbf{v}^2\cos\beta t)$$

are two real-valued solutions of (1). Moreover, these two solutions must be linearly independent (see Exercise 10).

Example 1. Solve the initial-value problem

$$\dot{\mathbf{x}}=\begin{pmatrix}1&0&0\\0&1&-1\\0&1&1\end{pmatrix}\mathbf{x},\qquad \mathbf{x}(0)=\begin{pmatrix}1\\1\\1\end{pmatrix}.$$

Solution. The characteristic polynomial of the matrix

$$\mathbf{A}=\begin{pmatrix}1&0&0\\0&1&-1\\0&1&1\end{pmatrix}$$

is

$$p(\lambda)=\det(\mathbf{A}-\lambda\mathbf{I})=\det\begin{pmatrix}1-\lambda&0&0\\0&1-\lambda&-1\\0&1&1-\lambda\end{pmatrix}$$
$$=(1-\lambda)^3+(1-\lambda)=(1-\lambda)(\lambda^2-2\lambda+2).$$

Hence the eigenvalues of $\mathbf{A}$ are

$$\lambda=1\quad\text{and}\quad\lambda=\frac{2\pm\sqrt{4-8}}{2}=1\pm i.$$

(i) $\lambda=1$: Clearly,

$$\begin{pmatrix}1\\0\\0\end{pmatrix}$$

is an eigenvector of $\mathbf{A}$ with eigenvalue 1. Hence

$$\mathbf{x}^1(t)=e^t\begin{pmatrix}1\\0\\0\end{pmatrix}$$

is one solution of the differential equation $\dot{\mathbf{x}}=\mathbf{A}\mathbf{x}$.

(ii) $\lambda=1+i$: We seek a nonzero vector $\mathbf{v}$ such that

$$\left[\mathbf{A}-(1+i)\mathbf{I}\right]\mathbf{v}=\begin{pmatrix}-i&0&0\\0&-i&-1\\0&1&-i\end{pmatrix}\begin{pmatrix}v_1\\v_2\\v_3\end{pmatrix}=\begin{pmatrix}0\\0\\0\end{pmatrix}.$$

This implies that $-iv_1=0$, $-iv_2-v_3=0$, and $v_2-iv_3=0$. The first equation says that $v_1=0$ and the second and third equations both say that $v_2=iv_3$. Consequently, each vector

$$\mathbf{v}=c\begin{bmatrix}0\\ i\\ 1\end{bmatrix}$$

is an eigenvector of $\mathbf{A}$ with eigenvalue $1+i$. Thus,

$$\mathbf{x}(t)=e^{(1+i)t}\begin{bmatrix}0\\ i\\ 1\end{bmatrix}$$

is a complex-valued solution of the differential equation $\dot{\mathbf{x}}=\mathbf{A}\mathbf{x}$. Now,

$$\begin{aligned}
e^{(1+i)t}\begin{bmatrix}0\\ i\\ 1\end{bmatrix} &= e^{t}(\cos t+i\sin t)\left[\begin{bmatrix}0\\ 0\\ 1\end{bmatrix}+i\begin{bmatrix}0\\ 1\\ 0\end{bmatrix}\right]\\
&= e^{t}\left[\cos t\begin{bmatrix}0\\ 0\\ 1\end{bmatrix}-\sin t\begin{bmatrix}0\\ 1\\ 0\end{bmatrix}+i\sin t\begin{bmatrix}0\\ 0\\ 1\end{bmatrix}+i\cos t\begin{bmatrix}0\\ 1\\ 0\end{bmatrix}\right]\\
&= e^{t}\begin{bmatrix}0\\ -\sin t\\ \cos t\end{bmatrix}+ie^{t}\begin{bmatrix}0\\ \cos t\\ \sin t\end{bmatrix}.
\end{aligned}$$

Consequently, by Lemma 1,

$$\mathbf{x}^2(t)=e^{t}\begin{bmatrix}0\\ -\sin t\\ \cos t\end{bmatrix}\quad\text{and}\quad \mathbf{x}^3(t)=e^{t}\begin{bmatrix}0\\ \cos t\\ \sin t\end{bmatrix}$$

are real-valued solutions. The three solutions $\mathbf{x}^1(t)$, $\mathbf{x}^2(t)$, and $\mathbf{x}^3(t)$ are linearly independent since their initial values

$$\mathbf{x}^1(0)=\begin{bmatrix}1\\ 0\\ 0\end{bmatrix},\qquad \mathbf{x}^2(0)=\begin{bmatrix}0\\ 1\\ 0\end{bmatrix},\quad\text{and}\quad \mathbf{x}^3(0)=\begin{bmatrix}0\\ 0\\ 1\end{bmatrix}$$

are linearly independent vectors in R^3. Therefore, the solution $\mathbf{x}(t)$ of our initial-value problem must have the form

$$\mathbf{x}(t)=c_1e^{t}\begin{bmatrix}1\\ 0\\ 0\end{bmatrix}+c_2e^{t}\begin{bmatrix}0\\ -\sin t\\ \cos t\end{bmatrix}+c_3e^{t}\begin{bmatrix}0\\ \cos t\\ \sin t\end{bmatrix}.$$

Setting $t=0$, we see that

$$\begin{bmatrix}1\\ 1\\ 1\end{bmatrix}=c_1\begin{bmatrix}1\\ 0\\ 0\end{bmatrix}+c_2\begin{bmatrix}0\\ 0\\ 1\end{bmatrix}+c_3\begin{bmatrix}0\\ 1\\ 0\end{bmatrix}=\begin{bmatrix}c_1\\ c_3\\ c_2\end{bmatrix}.$$

Consequently $c_1=c_2=c_3=1$ and

$$\mathbf{x}(t)=e^t\begin{bmatrix}1\\0\\0\end{bmatrix}+e^t\begin{bmatrix}0\\-\sin t\\\cos t\end{bmatrix}+e^t\begin{bmatrix}0\\\cos t\\\sin t\end{bmatrix}$$

$$=e^t\begin{bmatrix}1\\\cos t-\sin t\\\cos t+\sin t\end{bmatrix}.$$

Remark. If $\mathbf{v}$ is an eigenvector of $\mathbf{A}$ with eigenvalue λ, then $\bar{\mathbf{v}}$, the complex conjugate of $\mathbf{v}$, is an eigenvector of $\mathbf{A}$ with eigenvalue $\bar{\lambda}$. (Each component of $\bar{\mathbf{v}}$ is the complex conjugate of the corresponding component of $\mathbf{v}$.) To prove this, we take complex conjugates of both sides of the equation $\mathbf{A}\mathbf{v}=\lambda\mathbf{v}$ and observe that the complex conjugate of the vector $\mathbf{A}\mathbf{v}$ is $\mathbf{A}\bar{\mathbf{v}}$ if $\mathbf{A}$ is real. Hence, $\mathbf{A}\bar{\mathbf{v}}=\bar{\lambda}\bar{\mathbf{v}}$, which shows that $\bar{\mathbf{v}}$ is an eigenvector of $\mathbf{A}$ with eigenvalue $\bar{\lambda}$.

Exercises

In each of Problems 1–4 find the general solution of the given system of differential equations.

1. $\dot{\mathbf{x}}=\begin{pmatrix}-3 & 2\\-1 & -1\end{pmatrix}\mathbf{x}$

2. $\dot{\mathbf{x}}=\begin{pmatrix}1 & -5 & 0\\1 & -3 & 0\\0 & 0 & 1\end{pmatrix}\mathbf{x}$

3. $\dot{\mathbf{x}}=\begin{pmatrix}1 & 0 & 0\\3 & 1 & -2\\2 & 2 & 1\end{pmatrix}\mathbf{x}$

4. $\dot{\mathbf{x}}=\begin{pmatrix}1 & 0 & 1\\0 & 1 & -1\\-2 & 0 & -1\end{pmatrix}\mathbf{x}$

In each of Problems 5–8, solve the given initial-value problem.

5. $\dot{\mathbf{x}}=\begin{pmatrix}1 & -1\\5 & -3\end{pmatrix}\mathbf{x},\quad \mathbf{x}(0)=\begin{pmatrix}1\\2\end{pmatrix}$

6. $\dot{\mathbf{x}}=\begin{pmatrix}3 & -2\\4 & -1\end{pmatrix}\mathbf{x},\quad \mathbf{x}(0)=\begin{pmatrix}1\\5\end{pmatrix}$

7. $\dot{\mathbf{x}}=\begin{pmatrix}-3 & 0 & 2\\1 & -1 & 0\\-2 & -1 & 0\end{pmatrix}\mathbf{x},\quad \mathbf{x}(0)=\begin{pmatrix}0\\-1\\-2\end{pmatrix}$

8. $\dot{\mathbf{x}}=\begin{bmatrix}0 & 2 & 0 & 0\\-2 & 0 & 0 & 0\\0 & 0 & 0 & -3\\0 & 0 & 3 & 0\end{bmatrix}\mathbf{x},\quad \mathbf{x}(0)=\begin{bmatrix}1\\1\\1\\0\end{bmatrix}$

9. Determine all vectors $\mathbf{x}^0$ such that the solution of the initial-value problem

$$\dot{\mathbf{x}}=\begin{pmatrix}1 & 0 & -2\\0 & 1 & 0\\1 & -1 & -1\end{pmatrix}\mathbf{x},\qquad \mathbf{x}(0)=\mathbf{x}^0$$

is a periodic function of time.

10. Let $\mathbf{x}(t)=e^{\lambda t}\mathbf{v}$ be a solution of $\dot{\mathbf{x}}=\mathbf{A}\mathbf{x}$. Prove that $\mathbf{y}(t)=\text{Re}\{\mathbf{x}(t)\}$ and $\mathbf{z}(t)=\text{Im}\{\mathbf{x}(t)\}$ are linearly independent. *Hint*: Observe that $\mathbf{v}$ and $\bar{\mathbf{v}}$ are linearly independent in $\mathbf{C}^n$ since they are eigenvectors of $\mathbf{A}$ with distinct eigenvalues.

3.10 Equal roots

If the characteristic polynomial of $\mathbf{A}$ does not have n distinct roots, then $\mathbf{A}$ may not have n linearly independent eigenvectors. For example, the matrix

$$\mathbf{A}=\begin{pmatrix} 1 & 1 & 0 \\ 0 & 1 & 0 \\ 0 & 0 & 2 \end{pmatrix}$$

has only two distinct eigenvalues $\lambda_1=1$ and $\lambda_2=2$ and two linearly independent eigenvectors, which we take to be

$$\begin{pmatrix} 1 \\ 0 \\ 0 \end{pmatrix} \quad \text{and} \quad \begin{pmatrix} 0 \\ 0 \\ 1 \end{pmatrix}.$$

Consequently, the differential equation $\dot{\mathbf{x}}=\mathbf{A}\mathbf{x}$ has only two linearly independent solutions

$$e^{t}\begin{pmatrix} 1 \\ 0 \\ 0 \end{pmatrix} \quad \text{and} \quad e^{2t}\begin{pmatrix} 0 \\ 0 \\ 1 \end{pmatrix}$$

of the form $e^{\lambda t}\mathbf{v}$. Our problem, in this case, is to find a third linearly independent solution. More generally, suppose that the $n\times n$ matrix $\mathbf{A}$ has only $k<n$ linearly independent eigenvectors. Then, the differential equation $\dot{\mathbf{x}}=\mathbf{A}\mathbf{x}$ has only k linearly independent solutions of the form $e^{\lambda t}\mathbf{v}$. Our problem is to find an additional $n-k$ linearly independent solutions.

We approach this problem in the following ingenious manner. Recall that $x(t)=e^{at}c$ is a solution of the scalar differential equation $\dot{x}=ax$, for every constant c. Analogously, we would like to say that $\mathbf{x}(t)=e^{\mathbf{A}t}\mathbf{v}$ is a solution of the vector differential equation

$$\dot{\mathbf{x}}=\mathbf{A}\mathbf{x} \tag{1}$$

for every constant vector $\mathbf{v}$. However, $e^{\mathbf{A}t}$ is not defined if $\mathbf{A}$ is an $n\times n$ matrix. This is not a serious difficulty, though. There is a very natural way of defining $e^{\mathbf{A}t}$ so that it resembles the scalar exponential e^{at}; simply set

$$e^{\mathbf{A}t}\equiv\mathbf{I}+\mathbf{A}t+\frac{\mathbf{A}^2t^2}{2!}+\ldots+\frac{\mathbf{A}^nt^n}{n!}+\ldots. \tag{2}$$

It can be shown that the infinite series (2) converges for all t, and can be differentiated term by term. In particular

$$\frac{d}{dt}e^{\mathbf{A}t} = \mathbf{A} + \mathbf{A}^2 t + \ldots + \frac{\mathbf{A}^{n+1}}{n!}t^n + \ldots$$
$$= \mathbf{A}\left[\mathbf{I} + \mathbf{A}t + \ldots + \frac{\mathbf{A}^n t^n}{n!} + \ldots\right] = \mathbf{A}e^{\mathbf{A}t}.$$

This implies that $e^{\mathbf{A}t}\mathbf{v}$ is a solution of (1) for every constant vector $\mathbf{v}$, since

$$\frac{d}{dt}e^{\mathbf{A}t}\mathbf{v} = \mathbf{A}e^{\mathbf{A}t}\mathbf{v} = \mathbf{A}(e^{\mathbf{A}t}\mathbf{v}).$$

Remark. The matrix exponential $e^{\mathbf{A}t}$ and the scalar exponential e^{at} satisfy many similar properties. For example,

$$\left(e^{\mathbf{A}t}\right)^{-1} = e^{-\mathbf{A}t} \quad \text{and} \quad e^{\mathbf{A}(t+s)} = e^{\mathbf{A}t}e^{\mathbf{A}s}. \tag{3}$$

Indeed, the same proofs which show that $(e^{at})^{-1} = e^{-at}$ and $e^{a(t+s)} = e^{at}e^{as}$ can be used to establish the identities (3): we need only replace every a by $\mathbf{A}$ and every 1 by $\mathbf{I}$. However, $e^{\mathbf{A}t+\mathbf{B}t}$ equals $e^{\mathbf{A}t}e^{\mathbf{B}t}$ only if $\mathbf{AB} = \mathbf{BA}$ (see Exercise 15, Section 3.11).

There are several classes of matrices $\mathbf{A}$ (see Problems 9–11) for which the infinite series (2) can be summed exactly. In general, though, it does not seem possible to express $e^{\mathbf{A}t}$ in closed form. Yet, the remarkable fact is that we can always find n linearly independent vectors $\mathbf{v}$ for which the infinite series $e^{\mathbf{A}t}\mathbf{v}$ can be summed exactly. Moreover, once we know n linearly independent solutions of (1), we can even compute $e^{\mathbf{A}t}$ exactly. (This latter property will be proven in the next section.)

We now show how to find n linearly independent vectors $\mathbf{v}$ for which the infinite series $e^{\mathbf{A}t}\mathbf{v}$ can be summed exactly. Observe that

$$e^{\mathbf{A}t}\mathbf{v} = e^{(\mathbf{A}-\lambda\mathbf{I})t}e^{\lambda\mathbf{I}t}\mathbf{v}$$

for any constant λ, since $(\mathbf{A}-\lambda\mathbf{I})(\lambda\mathbf{I}) = (\lambda\mathbf{I})(\mathbf{A}-\lambda\mathbf{I})$. Moreover,

$$e^{\lambda\mathbf{I}t}\mathbf{v} = \left[\mathbf{I} + \lambda\mathbf{I}t + \frac{\lambda^2\mathbf{I}^2t^2}{2!} + \ldots\right]\mathbf{v}$$
$$= \left[1 + \lambda t + \frac{\lambda^2 t^2}{2!} + \ldots\right]\mathbf{v} = e^{\lambda t}\mathbf{v}.$$

Hence, $e^{\mathbf{A}t}\mathbf{v} = e^{\lambda t}e^{(\mathbf{A}-\lambda\mathbf{I})t}\mathbf{v}$.

Next, we make the crucial observation that if $\mathbf{v}$ satisfies $(\mathbf{A}-\lambda\mathbf{I})^m\mathbf{v} = \mathbf{0}$ for some integer m, then the infinite series $e^{(\mathbf{A}-\lambda\mathbf{I})t}\mathbf{v}$ terminates after m terms. If $(\mathbf{A}-\lambda\mathbf{I})^m\mathbf{v} = \mathbf{0}$, then $(\mathbf{A}-\lambda\mathbf{I})^{m+l}\mathbf{v}$ is also zero, for every positive integer l, since

$$(\mathbf{A}-\lambda\mathbf{I})^{m+l}\mathbf{v} = (\mathbf{A}-\lambda\mathbf{I})^l\left[(\mathbf{A}-\lambda\mathbf{I})^m\mathbf{v}\right] = \mathbf{0}.$$

Consequently,

$$e^{(\mathbf{A}-\lambda\mathbf{I})t}\mathbf{v}=\mathbf{v}+t(\mathbf{A}-\lambda\mathbf{I})\mathbf{v}+\ldots+\frac{t^{m-1}}{(m-1)!}(\mathbf{A}-\lambda\mathbf{I})^{m-1}\mathbf{v}$$

and

$$\begin{aligned}e^{\mathbf{A}t}\mathbf{v}&=e^{\lambda t}e^{(\mathbf{A}-\lambda\mathbf{I})t}\mathbf{v}\\&=e^{\lambda t}\left[\mathbf{v}+t(\mathbf{A}-\lambda\mathbf{I})\mathbf{v}+\ldots+\frac{t^{m-1}}{(m-1)!}(\mathbf{A}-\lambda\mathbf{I})^{m-1}\mathbf{v}\right].\end{aligned}$$

This suggests the following algorithm for finding n linearly independent solutions of (1).

(1) Find all the eigenvalues and eigenvectors of $\mathbf{A}$. If $\mathbf{A}$ has n linearly independent eigenvectors, then the differential equation $\dot{\mathbf{x}}=\mathbf{A}\mathbf{x}$ has n linearly independent solutions of the form $e^{\lambda t}\mathbf{v}$. (Observe that the infinite series $e^{(\mathbf{A}-\lambda\mathbf{I})t}\mathbf{v}$ terminates after one term if $\mathbf{v}$ is an eigenvector of $\mathbf{A}$ with eigenvalue λ.)

(2) Suppose that $\mathbf{A}$ has only $k<n$ linearly independent eigenvectors. Then, we have only k linearly independent solutions of the form $e^{\lambda t}\mathbf{v}$. To find additional solutions we pick an eigenvalue λ of $\mathbf{A}$ and find all vectors $\mathbf{v}$ for which $(\mathbf{A}-\lambda\mathbf{I})^2\mathbf{v}=\mathbf{0}$, but $(\mathbf{A}-\lambda\mathbf{I})\mathbf{v}\neq\mathbf{0}$. For each such vector $\mathbf{v}$

$$e^{\mathbf{A}t}\mathbf{v}=e^{\lambda t}e^{(\mathbf{A}-\lambda\mathbf{I})t}\mathbf{v}=e^{\lambda t}\left[\mathbf{v}+t(\mathbf{A}-\lambda\mathbf{I})\mathbf{v}\right]$$

is an additional solution of $\dot{\mathbf{x}}=\mathbf{A}\mathbf{x}$. We do this for all the eigenvalues λ of $\mathbf{A}$.

(3) If we still do not have enough solutions, then we find all vectors $\mathbf{v}$ for which $(\mathbf{A}-\lambda\mathbf{I})^3\mathbf{v}=\mathbf{0}$, but $(\mathbf{A}-\lambda\mathbf{I})^2\mathbf{v}\neq\mathbf{0}$. For each such vector $\mathbf{v}$,

$$e^{\mathbf{A}t}\mathbf{v}=e^{\lambda t}\left[\mathbf{v}+t(\mathbf{A}-\lambda\mathbf{I})\mathbf{v}+\frac{t^2}{2!}(\mathbf{A}-\lambda\mathbf{I})^2\mathbf{v}\right]$$

is an additional solution of $\dot{\mathbf{x}}=\mathbf{A}\mathbf{x}$.

(4) We keep proceeding in this manner until, hopefully, we obtain n linearly independent solutions.

The following lemma from linear algebra, which we accept without proof, guarantees that this algorithm always works. Moreover, it puts an upper bound on the number of steps we have to perform in this algorithm.

Lemma 1. *Let the characteristic polynomial of* $\mathbf{A}$ *have* k *distinct roots* $\lambda_1,\ldots,\lambda_k$ *with multiplicities* $n_1,\ldots,n_k$ *respectively. (This means that* $p(\lambda)$ *can be factored into the form* $(\lambda_1-\lambda)^{n_1}\ldots(\lambda_k-\lambda)^{n_k}$.) *Suppose that* $\mathbf{A}$ *has only* $\nu_j<n_j$ *linearly independent eigenvectors with eigenvalue* λ_j. *Then the equation* $(\mathbf{A}-\lambda_j\mathbf{I})^2\mathbf{v}=\mathbf{0}$ *has at least* ν_j+1 *independent solutions. More generally, if the equation* $(\mathbf{A}-\lambda_j\mathbf{I})^{m}\mathbf{v}=\mathbf{0}$ *has only* $m_j<n_j$ *independent solutions, then the equation* $(\mathbf{A}-\lambda_j\mathbf{I})^{m+1}\mathbf{v}=\mathbf{0}$ *has at least* m_j+1 *independent solutions.*

Lemma 1 clearly implies that there exists an integer d_j with $d_j \leq n_j$, such that the equation $(\mathbf{A}-\lambda_j\mathbf{I})^{d_j}\mathbf{v}=\mathbf{0}$ has at least n_j linearly independent solutions. Thus, for each eigenvalue λ_j of $\mathbf{A}$, we can compute n_j linearly independent solutions of $\dot{\mathbf{x}}=\mathbf{A}\mathbf{x}$. All these solutions have the form

$$\mathbf{x}(t)=e^{\lambda_j t}\left[\mathbf{v}+t(\mathbf{A}-\lambda_j\mathbf{I})\mathbf{v}+\ldots+\frac{t^{d_j-1}}{(d_j-1)!}(\mathbf{A}-\lambda_j\mathbf{I})^{d_j-1}\mathbf{v}\right].$$

In addition, it can be shown that the set of $n_1+\ldots+n_k=n$ solutions thus obtained must be linearly independent.

Example 1. Find three linearly independent solutions of the differential equation

$$\dot{\mathbf{x}}=\begin{pmatrix}1 & 1 & 0\\ 0 & 1 & 0\\ 0 & 0 & 2\end{pmatrix}\mathbf{x}.$$

Solution. The characteristic polynomial of the matrix

$$\mathbf{A}=\begin{pmatrix}1 & 1 & 0\\ 0 & 1 & 0\\ 0 & 0 & 2\end{pmatrix}$$

is $(1-\lambda)^2(2-\lambda)$. Hence $\lambda=1$ is an eigenvalue of $\mathbf{A}$ with multiplicity two, and $\lambda=2$ is an eigenvalue of $\mathbf{A}$ with multiplicity one.

(i) $\lambda=1$: We seek all nonzero vectors $\mathbf{v}$ such that

$$(\mathbf{A}-\mathbf{I})\mathbf{v}=\begin{pmatrix}0 & 1 & 0\\ 0 & 0 & 0\\ 0 & 0 & 1\end{pmatrix}\begin{pmatrix}v_1\\ v_2\\ v_3\end{pmatrix}=\begin{pmatrix}0\\ 0\\ 0\end{pmatrix}.$$

This implies that $v_2=v_3=0$, and v_1 is arbitrary. Consequently,

$$\mathbf{x}^1(t)=e^t\begin{pmatrix}1\\ 0\\ 0\end{pmatrix}$$

is one solution of $\dot{\mathbf{x}}=\mathbf{A}\mathbf{x}$. Since $\mathbf{A}$ has only one linearly independent eigenvector with eigenvalue 1, we look for all solutions of the equation

$$(\mathbf{A}-\mathbf{I})^2\mathbf{v}=\begin{pmatrix}0 & 1 & 0\\ 0 & 0 & 0\\ 0 & 0 & 1\end{pmatrix}\begin{pmatrix}0 & 1 & 0\\ 0 & 0 & 0\\ 0 & 0 & 1\end{pmatrix}\mathbf{v}=\begin{pmatrix}0 & 0 & 0\\ 0 & 0 & 0\\ 0 & 0 & 1\end{pmatrix}\begin{pmatrix}v_1\\ v_2\\ v_3\end{pmatrix}=\begin{pmatrix}0\\ 0\\ 0\end{pmatrix}$$

This implies that $v_3=0$ and both v_1 and v_2 are arbitrary. Now, the vector

$$\mathbf{v}=\begin{pmatrix}0\\ 1\\ 0\end{pmatrix}$$

satisfies $(\mathbf{A}-\mathbf{I})^2\mathbf{v}=\mathbf{0}$, but $(\mathbf{A}-\mathbf{I})\mathbf{v}\neq\mathbf{0}$. (We could just as well choose any

$$\mathbf{v}=\begin{bmatrix} v_1\\ v_2\\ 0\end{bmatrix}$$

for which $v_2\neq 0$.) Hence,

$$\mathbf{x}^2(t)=e^{\mathbf{A}t}\begin{bmatrix}0\\1\\0\end{bmatrix}=e^te^{(\mathbf{A}-\mathbf{I})t}\begin{bmatrix}0\\1\\0\end{bmatrix}$$

$$=e^t\left[\mathbf{I}+t(\mathbf{A}-\mathbf{I})\right]\begin{bmatrix}0\\1\\0\end{bmatrix}=e^t\left[\begin{bmatrix}0\\1\\0\end{bmatrix}+t\begin{bmatrix}0&1&0\\0&0&0\\0&0&1\end{bmatrix}\begin{bmatrix}0\\1\\0\end{bmatrix}\right]$$

$$=e^t\begin{bmatrix}0\\1\\0\end{bmatrix}+te^t\begin{bmatrix}1\\0\\0\end{bmatrix}=e^t\begin{bmatrix}t\\1\\0\end{bmatrix}$$

is a second linearly independent solution.

(ii) $\lambda=2$: We seek a nonzero vector $\mathbf{v}$ such that

$$(\mathbf{A}-2\mathbf{I})\mathbf{v}=\begin{bmatrix}-1&1&0\\0&-1&0\\0&0&0\end{bmatrix}\begin{bmatrix}v_1\\v_2\\v_3\end{bmatrix}=\begin{bmatrix}0\\0\\0\end{bmatrix}.$$

This implies that $v_1=v_2=0$ and v_3 is arbitrary. Hence

$$\mathbf{x}^3(t)=e^{2t}\begin{bmatrix}0\\0\\1\end{bmatrix}$$

is a third linearly independent solution.

Example 2. Solve the initial-value problem

$$\dot{\mathbf{x}}=\begin{bmatrix}2&1&3\\0&2&-1\\0&0&2\end{bmatrix}\mathbf{x},\qquad \mathbf{x}(0)=\begin{bmatrix}1\\2\\1\end{bmatrix}.$$

Solution. The characteristic polynomial of the matrix

$$\mathbf{A}=\begin{bmatrix}2&1&3\\0&2&-1\\0&0&2\end{bmatrix}$$

is $(2-\lambda)^3$. Hence $\lambda=2$ is an eigenvalue of $\mathbf{A}$ with multiplicity three. The eigenvectors of $\mathbf{A}$ satisfy the equation

$$(\mathbf{A}-2\mathbf{I})\mathbf{v}=\begin{bmatrix}0&1&3\\0&0&-1\\0&0&0\end{bmatrix}\begin{bmatrix}v_1\\v_2\\v_3\end{bmatrix}=\begin{bmatrix}0\\0\\0\end{bmatrix}.$$

This implies that $v_2=v_3=0$ and v_1 is arbitrary. Hence

$$\mathbf{x}^1(t)=e^{2t}\begin{bmatrix}1\\0\\0\end{bmatrix}$$

is one solution of $\dot{\mathbf{x}}=\mathbf{A}\mathbf{x}$.

Since $\mathbf{A}$ has only one linearly independent eigenvector we look for all solutions of the equation

$$(\mathbf{A}-2\mathbf{I})^2\mathbf{v}=\begin{bmatrix}0&1&3\\0&0&-1\\0&0&0\end{bmatrix}\begin{bmatrix}0&1&3\\0&0&-1\\0&0&0\end{bmatrix}\mathbf{v}=\begin{bmatrix}0&0&-1\\0&0&0\\0&0&0\end{bmatrix}\begin{bmatrix}v_1\\v_2\\v_3\end{bmatrix}=\begin{bmatrix}0\\0\\0\end{bmatrix}.$$

This implies that $v_3=0$ and both v_1 and v_2 are arbitrary. Now, the vector

$$\mathbf{v}=\begin{bmatrix}0\\1\\0\end{bmatrix}$$

satisfies $(\mathbf{A}-2\mathbf{I})^2\mathbf{v}=\mathbf{0}$, but $(\mathbf{A}-2\mathbf{I})\mathbf{v}\neq\mathbf{0}$. Hence

$$\begin{aligned}\mathbf{x}^2(t)&=e^{\mathbf{A}t}\begin{bmatrix}0\\1\\0\end{bmatrix}=e^{2t}e^{(\mathbf{A}-2\mathbf{I})t}\begin{bmatrix}0\\1\\0\end{bmatrix}\\&=e^{2t}\left[\mathbf{I}+t(\mathbf{A}-2\mathbf{I})\right]\begin{bmatrix}0\\1\\0\end{bmatrix}=e^{2t}\left[\mathbf{I}+t\begin{bmatrix}0&1&3\\0&0&-1\\0&0&0\end{bmatrix}\right]\begin{bmatrix}0\\1\\0\end{bmatrix}\\&=e^{2t}\left[\begin{bmatrix}0\\1\\0\end{bmatrix}+t\begin{bmatrix}1\\0\\0\end{bmatrix}\right]=e^{2t}\begin{bmatrix}t\\1\\0\end{bmatrix}\end{aligned}$$

is a second solution of $\dot{\mathbf{x}}=\mathbf{A}\mathbf{x}$.

Since the equation $(\mathbf{A}-2\mathbf{I})^2\mathbf{v}=\mathbf{0}$ has only two linearly independent solutions, we look for all solutions of the equation

$$(\mathbf{A}-2\mathbf{I})^3\mathbf{v}=\begin{bmatrix}0&1&3\\0&0&-1\\0&0&0\end{bmatrix}^3\mathbf{v}=\begin{bmatrix}0&0&0\\0&0&0\\0&0&0\end{bmatrix}\begin{bmatrix}v_1\\v_2\\v_3\end{bmatrix}=\begin{bmatrix}0\\0\\0\end{bmatrix}.$$

Obviously, every vector $\mathbf{v}$ is a solution of this equation. The vector

$$\mathbf{v}=\begin{bmatrix}0\\0\\1\end{bmatrix}$$

does not satisfy $(\mathbf{A}-2\mathbf{I})^2\mathbf{v}=\mathbf{0}$. Hence

$$\mathbf{x}^3(t)=e^{\mathbf{A}t}\begin{bmatrix}0\\0\\1\end{bmatrix}=e^{2t}e^{(\mathbf{A}-2\mathbf{I})t}\begin{bmatrix}0\\0\\1\end{bmatrix}$$

$$=e^{2t}\left[\mathbf{I}+t(\mathbf{A}-2\mathbf{I})+\frac{t^2}{2}(\mathbf{A}-2\mathbf{I})^2\right]\begin{bmatrix}0\\0\\1\end{bmatrix}$$

$$=e^{2t}\left[\begin{bmatrix}0\\0\\1\end{bmatrix}+t\begin{bmatrix}3\\-1\\0\end{bmatrix}+\frac{t^2}{2}\begin{bmatrix}-1\\0\\0\end{bmatrix}\right]=e^{2t}\begin{bmatrix}3t-\frac{1}{2}t^2\\-t\\1\end{bmatrix}$$

is a third linearly independent solution. Therefore,

$$\mathbf{x}(t)=e^{2t}\left[c_1\begin{bmatrix}1\\0\\0\end{bmatrix}+c_2\begin{bmatrix}t\\1\\0\end{bmatrix}+c_3\begin{bmatrix}3t-\frac{1}{2}t^2\\-t\\1\end{bmatrix}\right].$$

The constants c_1, c_2, and c_3 are determined from the initial conditions

$$\begin{bmatrix}1\\2\\1\end{bmatrix}=c_1\begin{bmatrix}1\\0\\0\end{bmatrix}+c_2\begin{bmatrix}0\\1\\0\end{bmatrix}+c_3\begin{bmatrix}0\\0\\1\end{bmatrix}.$$

This implies that $c_1=1$, $c_2=2$, and $c_3=1$. Hence

$$\mathbf{x}(t)=e^{2t}\begin{bmatrix}1+5t-\frac{1}{2}t^2\\2-t\\1\end{bmatrix}.$$

For the matrix $\mathbf{A}$ in Example 2, $p(\lambda)=(2-\lambda)^3$ and $(2\mathbf{I}-\mathbf{A})^3=\mathbf{0}$. This is not an accident. Every matrix $\mathbf{A}$ satisfies its own characteristic equation. This is the content of the following theorem.

Theorem 13 (Cayley–Hamilton). *Let $p(\lambda)=p_0+p_1\lambda+\ldots+(-1)^n\lambda^n$ be the characteristic polynomial of $\mathbf{A}$. Then,*

$$p(\mathbf{A})\equiv p_0\mathbf{I}+p_1\mathbf{A}+\ldots+(-1)^n\mathbf{A}^n=\mathbf{0}.$$

FAKE PROOF. Setting $\lambda=\mathbf{A}$ in the equation $p(\lambda)=\det(\mathbf{A}-\lambda\mathbf{I})$ gives $p(\mathbf{A})=\det(\mathbf{A}-\mathbf{A}\mathbf{I})=\det\mathbf{0}=0$. The fallacy in this proof is that we cannot set $\lambda=\mathbf{A}$ in the expression $\det(\mathbf{A}-\lambda\mathbf{I})$ since we cannot subtract a matrix from the diagonal elements of $\mathbf{A}$. However, there is a very clever way to make this proof kosher. Let $\mathbf{C}(\lambda)$ be the classical adjoint (see Section 3.6) of the matrix $(\mathbf{A}-\lambda\mathbf{I})$. Then,

$$(\mathbf{A}-\lambda\mathbf{I})\mathbf{C}(\lambda)=p(\lambda)\mathbf{I}. \tag{4}$$

Each element of the matrix $\mathbf{C}(\lambda)$ is a polynomial in λ of degree at most $(n-1)$. Therefore, we can write $\mathbf{C}(\lambda)$ in the form

$$\mathbf{C}(\lambda)=\mathbf{C}_0+\mathbf{C}_1\lambda+\ldots+\mathbf{C}_{n-1}\lambda^{n-1}$$

where $\mathbf{C}_0,\ldots,\mathbf{C}_{n-1}$ are $n\times n$ matrices. For example,

$$\begin{pmatrix}\lambda+\lambda^2 & 2\lambda\\ \lambda^2 & 1-\lambda\end{pmatrix}=\begin{pmatrix}0&0\\0&1\end{pmatrix}+\begin{pmatrix}\lambda & 2\lambda\\ 0 & -\lambda\end{pmatrix}+\begin{pmatrix}\lambda^2 & 0\\ \lambda^2 & 0\end{pmatrix}$$
$$=\begin{pmatrix}0&0\\0&1\end{pmatrix}+\lambda\begin{pmatrix}1&2\\0&-1\end{pmatrix}+\lambda^2\begin{pmatrix}1&0\\1&0\end{pmatrix}.$$

Thus, Equation (4) can be written in the form

$$(\mathbf{A}-\lambda\mathbf{I})\left[\mathbf{C}_0+\mathbf{C}_1\lambda+\ldots+\mathbf{C}_{n-1}\lambda^{n-1}\right]=p_0\mathbf{I}+p_1\lambda\mathbf{I}+\ldots+(-1)^n\lambda^n\mathbf{I}. \quad (5)$$

Observe that both sides of (5) are polynomials in λ, whose coefficients are $n\times n$ matrices. Since these two polynomials are equal for all values of λ, their coefficients must agree. But if the coefficients of like powers of λ agree, then we can put in anything we want for λ and still have equality. In particular, set $\lambda=\mathbf{A}$. Then,

$$p(\mathbf{A})=p_0\mathbf{I}+p_1\mathbf{A}+\ldots+(-1)^n\mathbf{A}^n$$
$$=(\mathbf{A}-\mathbf{A}\mathbf{I})\left[\mathbf{C}_0+\mathbf{C}_1\mathbf{A}+\ldots+\mathbf{C}_{n-1}\mathbf{A}^{n-1}\right]=\mathbf{0}. \qquad \square$$

Exercises

In each of Problems 1–4 find the general solution of the given system of differential equations.

1. $\dot{\mathbf{x}}=\begin{pmatrix}0&-1&1\\2&-3&1\\1&-1&-1\end{pmatrix}\mathbf{x}$

2. $\dot{\mathbf{x}}=\begin{pmatrix}1&1&1\\2&1&-1\\-3&2&4\end{pmatrix}\mathbf{x}$

3. $\dot{\mathbf{x}}=\begin{pmatrix}-1&-1&0\\0&-1&0\\0&0&-2\end{pmatrix}\mathbf{x}$ *Hint*: Look at Example 1 of text.

4. $\dot{\mathbf{x}}=\begin{bmatrix}2&0&-1&0\\0&2&1&0\\0&0&2&0\\0&0&-1&2\end{bmatrix}\mathbf{x}$

In each of Problems 5–8, solve the given initial-value problem

5. $\dot{\mathbf{x}}=\begin{pmatrix}-1&1&2\\-1&1&1\\-2&1&3\end{pmatrix}\mathbf{x},\quad \mathbf{x}(0)=\begin{pmatrix}1\\0\\1\end{pmatrix}$

6. $\dot{\mathbf{x}}=\begin{pmatrix}-4&-4&0\\10&9&1\\-4&-3&1\end{pmatrix}\mathbf{x},\quad \mathbf{x}(0)=\begin{pmatrix}2\\1\\-1\end{pmatrix}$

7. $\dot{\mathbf{x}}=\begin{pmatrix}1&2&-3\\1&1&2\\1&-1&4\end{pmatrix}\mathbf{x},\quad \mathbf{x}(0)=\begin{pmatrix}1\\0\\0\end{pmatrix}$

8. $\dot{\mathbf{x}}=\begin{bmatrix}3&0&0&0\\1&3&0&0\\0&0&3&0\\0&0&2&3\end{bmatrix}\mathbf{x},\quad \mathbf{x}(0)=\begin{bmatrix}1\\1\\1\\1\end{bmatrix}$

9. Let

$$\mathbf{A}=\begin{bmatrix} \lambda_1 & 0 & \dots & 0 \\ 0 & \lambda_2 & \dots & 0 \\ \vdots & \vdots & & \vdots \\ 0 & 0 & \dots & \lambda_n \end{bmatrix}.$$

Show that

$$e^{\mathbf{A}t}=\begin{bmatrix} e^{\lambda_1 t} & 0 & 0 \\ 0 & e^{\lambda_2 t} & 0 \\ \vdots & \vdots & \vdots \\ 0 & 0 & e^{\lambda_n t} \end{bmatrix}.$$

10. Let

$$\mathbf{A}=\begin{pmatrix} \lambda & 1 & 0 \\ 0 & \lambda & 1 \\ 0 & 0 & \lambda \end{pmatrix}.$$

Prove that

$$e^{\mathbf{A}t}=e^{\lambda t}\begin{bmatrix} 1 & t & \frac{1}{2}t^2 \\ 0 & 1 & t \\ 0 & 0 & 1 \end{bmatrix}.$$

Hint: Write **A** in the form

$$\mathbf{A}=\lambda\mathbf{I}+\begin{pmatrix} 0 & 1 & 0 \\ 0 & 0 & 1 \\ 0 & 0 & 0 \end{pmatrix}$$

and observe that

$$e^{\mathbf{A}t}=e^{\lambda t}\exp\left[\begin{pmatrix} 0 & 1 & 0 \\ 0 & 0 & 1 \\ 0 & 0 & 0 \end{pmatrix}t\right].$$

11. Let **A** be the $n\times n$ matrix

$$\begin{bmatrix} \lambda & 1 & 0 & \dots & 0 \\ 0 & \lambda & 1 & \dots & 0 \\ \vdots & \vdots & \vdots & & \vdots \\ 0 & 0 & 0 & \dots & 1 \\ 0 & 0 & 0 & \dots & \lambda \end{bmatrix},$$

and let **P** be the $n\times n$ matrix

$$\begin{bmatrix} 0 & 1 & 0 & \dots & 0 \\ 0 & 0 & 1 & \dots & 0 \\ \vdots & \vdots & \vdots & & \vdots \\ 0 & 0 & 0 & \dots & 1 \\ 0 & 0 & 0 & \dots & 0 \end{bmatrix}.$$

(a) Show that $\mathbf{P}^n=\mathbf{0}$. (b) Show that $(\lambda\mathbf{I})\mathbf{P}=\mathbf{P}(\lambda\mathbf{I})$.
(c) Show that

$$e^{\mathbf{A}t}=e^{\lambda t}\left[\mathbf{I}+t\mathbf{P}+\frac{t^2\mathbf{P}^2}{2!}+\ldots+\frac{t^{n-1}}{(n-1)!}\mathbf{P}^{n-1}\right].$$

12. Compute $e^{\mathbf{A}t}$ if

(a) $\mathbf{A}=\begin{bmatrix}2&1&0&0\\0&2&1&0\\0&0&2&1\\0&0&0&2\end{bmatrix}$,

(b) $\mathbf{A}=\begin{bmatrix}2&1&0&0\\0&2&1&0\\0&0&2&0\\0&0&0&2\end{bmatrix}$, (c) $\mathbf{A}=\begin{bmatrix}2&1&0&0\\0&2&0&0\\0&0&2&0\\0&0&0&2\end{bmatrix}$.

13. (a) Show that $e^{\mathbf{T}^{-1}\mathbf{A}\mathbf{T}}=\mathbf{T}^{-1}e^{\mathbf{A}}\mathbf{T}$.
(b) Given that

$$\mathbf{T}^{-1}\mathbf{A}\mathbf{T}=\begin{pmatrix}1&0&0\\0&2&0\\0&0&-1\end{pmatrix}$$

with

$$\mathbf{T}=\begin{pmatrix}1&-1&-1\\3&-1&2\\2&2&3\end{pmatrix},$$

compute $e^{\mathbf{A}t}$.

14. Suppose that $p(\lambda)=\det(\mathbf{A}-\lambda\mathbf{I})$ has n distinct roots $\lambda_1,\ldots,\lambda_n$. Prove directly that $p(\mathbf{A})\equiv(-1)^n(\mathbf{A}-\lambda_1\mathbf{I})\ldots(\mathbf{A}-\lambda_n\mathbf{I})=\mathbf{0}$. *Hint*: Write any vector $\mathbf{x}$ in the form $\mathbf{x}=x_1\mathbf{v}^1+\ldots+x_n\mathbf{v}^n$ where $\mathbf{v}^1,\ldots,\mathbf{v}^n$ are n independent eigenvectors of $\mathbf{A}$ with eigenvalues $\lambda_1,\ldots,\lambda_n$ respectively, and conclude that $p(\mathbf{A})\mathbf{x}=\mathbf{0}$ for all vectors $\mathbf{x}$.

15. Suppose that $\mathbf{A}^2=\alpha\mathbf{A}$. Find $e^{\mathbf{A}t}$.

16. Let

$$\mathbf{A}=\begin{bmatrix}1&1&1&1&1\\1&1&1&1&1\\1&1&1&1&1\\1&1&1&1&1\\1&1&1&1&1\end{bmatrix}.$$

(a) Show that $\mathbf{A}(\mathbf{A}-5\mathbf{I})=\mathbf{0}$.
(b) Find $e^{\mathbf{A}t}$.

17. Let

$$\mathbf{A}=\begin{pmatrix}0&1\\-1&0\end{pmatrix}.$$

(a) Show that $\mathbf{A}^2=-\mathbf{I}$.

(b) Show that

$$e^{\mathbf{A}t}=\begin{pmatrix} \cos t & \sin t \\ -\sin t & \cos t \end{pmatrix}.$$

In each of Problems 18–20 verify directly the Cayley–Hamilton Theorem for the given matrix **A**.

18. $\mathbf{A}=\begin{pmatrix} 1 & -1 & -1 \\ 2 & 1 & -1 \\ -1 & -1 & 1 \end{pmatrix}$ **19.** $\mathbf{A}=\begin{pmatrix} 2 & 3 & 1 \\ 2 & 3 & 1 \\ 2 & 3 & 1 \end{pmatrix}$ **20.** $\mathbf{A}=\begin{pmatrix} -1 & 0 & 1 \\ -1 & 3 & 0 \\ 2 & 4 & 6 \end{pmatrix}$

3.11 Fundamental matrix solutions; $e^{\mathbf{A}t}$

If $\mathbf{x}^1(t),\dots,\mathbf{x}^n(t)$ are n linearly independent solutions of the differential equation

$$\dot{\mathbf{x}}=\mathbf{A}\mathbf{x}, \tag{1}$$

then every solution $\mathbf{x}(t)$ can be written in the form

$$\mathbf{x}(t)=c_1\mathbf{x}^1(t)+c_2\mathbf{x}^2(t)+\dots+c_n\mathbf{x}^n(t). \tag{2}$$

Let $\mathbf{X}(t)$ be the matrix whose columns are $\mathbf{x}^1(t),\dots,\mathbf{x}^n(t)$. Then, Equation (2) can be written in the concise form $\mathbf{x}(t)=\mathbf{X}(t)\mathbf{c}$, where

$$\mathbf{c}=\begin{bmatrix} c_1 \\ \vdots \\ c_n \end{bmatrix}.$$

Definition. A matrix $\mathbf{X}(t)$ is called a *fundamental matrix solution* of (1) if its columns form a set of n linearly independent solutions of (1).

Example 1. Find a fundamental matrix solution of the system of differential equations

$$\dot{\mathbf{x}}=\begin{bmatrix} 1 & -1 & 4 \\ 3 & 2 & -1 \\ 2 & 1 & -1 \end{bmatrix}\mathbf{x}. \tag{3}$$

Solution. We showed in Section 3.8 (see Example 1) that

$$e^{t}\begin{bmatrix} -1 \\ 4 \\ 1 \end{bmatrix},\quad e^{3t}\begin{bmatrix} 1 \\ 2 \\ 1 \end{bmatrix}\quad\text{and}\quad e^{-2t}\begin{bmatrix} -1 \\ 1 \\ 1 \end{bmatrix}$$

are three linearly independent solutions of (3). Hence

$$\mathbf{X}(t)=\begin{bmatrix} -e^{t} & e^{3t} & -e^{-2t} \\ 4e^{t} & 2e^{3t} & e^{-2t} \\ e^{t} & e^{3t} & e^{-2t} \end{bmatrix}$$

is a fundamental matrix solution of (3).

In this section we will show that the matrix $e^{\mathbf{A}t}$ can be computed directly from any fundamental matrix solution of (1). This is rather remarkable since it does not appear possible to sum the infinite series $[\mathbf{I}+\mathbf{A}t+(\mathbf{A}t)^2/2!+\dots]$ exactly, for an arbitrary matrix $\mathbf{A}$. Specifically, we have the following theorem.

Theorem 14. *Let* $\mathbf{X}(t)$ *be a fundamental matrix solution of the differential equation* $\dot{\mathbf{x}}=\mathbf{A}\mathbf{x}$. *Then,*

$$e^{\mathbf{A}t}=\mathbf{X}(t)\mathbf{X}^{-1}(0). \tag{4}$$

In other words, the product of any fundamental matrix solution of (1) *with its inverse at* $t=0$ *must yield* $e^{\mathbf{A}t}$.

We prove Theorem 14 in three steps. First, we establish a simple test to determine whether a matrix-valued function is a fundamental matrix solution of (1). Then, we use this test to show that $e^{\mathbf{A}t}$ is a fundamental matrix solution of (1). Finally, we establish a connection between any two fundamental matrix solutions of (1).

Lemma 1. *A matrix* $\mathbf{X}(t)$ *is a fundamental matrix solution of* (1) *if, and only if,* $\dot{\mathbf{X}}(t)=\mathbf{A}\mathbf{X}(t)$ *and* $\det\mathbf{X}(0)\neq 0$. (*The derivative of a matrix-valued function* $\mathbf{X}(t)$ *is the matrix whose components are the derivatives of the corresponding components of* $\mathbf{X}(t)$.)

PROOF. Let $\mathbf{x}^1(t),\dots,\mathbf{x}^n(t)$ denote the n columns of $\mathbf{X}(t)$. Observe that

$$\dot{\mathbf{X}}(t)=\left(\dot{\mathbf{x}}^1(t),\dots,\dot{\mathbf{x}}^n(t)\right)$$

and

$$\mathbf{A}\mathbf{X}(t)=\left(\mathbf{A}\mathbf{x}^1(t),\dots,\mathbf{A}\mathbf{x}^n(t)\right).$$

Hence, the n vector equations $\dot{\mathbf{x}}^1(t)=\mathbf{A}\mathbf{x}^1(t),\dots,\dot{\mathbf{x}}^n(t)=\mathbf{A}\mathbf{x}^n(t)$ are equivalent to the single matrix equation $\dot{\mathbf{X}}(t)=\mathbf{A}\mathbf{X}(t)$. Moreover, n solutions $\mathbf{x}^1(t),\dots,\mathbf{x}^n(t)$ of (1) are linearly independent if, and only if, $\mathbf{x}^1(0),\dots,\mathbf{x}^n(0)$ are linearly independent vectors of R^n. These vectors, in turn, are linearly independent if, and only if, $\det\mathbf{X}(0)\neq 0$. Consequently, $\mathbf{X}(t)$ is a fundamental matrix solution of (1) if, and only if, $\dot{\mathbf{X}}(t)=\mathbf{A}\mathbf{X}(t)$ and $\det\mathbf{X}(0)\neq 0$. □

Lemma 2. *The matrix-valued function* $e^{\mathbf{A}t}\equiv\mathbf{I}+\mathbf{A}t+\mathbf{A}^2t^2/2!+\dots$ *is a fundamental matrix solution of* (1).

PROOF. We showed in Section 3.10 that $(d/dt)e^{\mathbf{A}t}=\mathbf{A}e^{\mathbf{A}t}$. Hence $e^{\mathbf{A}t}$ is a solution of the matrix differential equation $\dot{\mathbf{X}}(t)=\mathbf{A}\mathbf{X}(t)$. Moreover, its determinant, evaluated at $t=0$, is one since $e^{\mathbf{A}0}=\mathbf{I}$. Therefore, by Lemma 1, $e^{\mathbf{A}t}$ is a fundamental matrix solution of (1). □

Lemma 3. *Let* $\mathbf{X}(t)$ *and* $\mathbf{Y}(t)$ *be two fundamental matrix solutions of* (1). *Then, there exists a constant matrix* $\mathbf{C}$ *such that* $\mathbf{Y}(t)=\mathbf{X}(t)\mathbf{C}$.

PROOF. By definition, the columns $\mathbf{x}^1(t),\dots,\mathbf{x}^n(t)$ of $\mathbf{X}(t)$ and $\mathbf{y}^1(t),\dots,$ $\mathbf{y}^n(t)$ of $\mathbf{Y}(t)$ are linearly independent sets of solutions of (1). In particular, therefore, each column of $\mathbf{Y}(t)$ can be written as a linear combination of the columns of $\mathbf{X}(t)$; i.e., there exist constants $c_1^j,\dots,c_n^j$ such that

$$\mathbf{y}^j(t)=c_1^j\mathbf{x}^1(t)+c_2^j\mathbf{x}^2(t)+\dots+c_n^j\mathbf{x}^n(t),\qquad j=1,\dots,n. \tag{5}$$

Let $\mathbf{C}$ be the matrix $(\mathbf{c}^1,\mathbf{c}^2,\dots,\mathbf{c}^n)$ where

$$\mathbf{c}^j=\begin{bmatrix} c_1^j\\ \vdots\\ c_n^j\end{bmatrix}.$$

Then, the n equations (5) are equivalent to the single matrix equation $\mathbf{Y}(t)$ $=\mathbf{X}(t)\mathbf{C}$. □

We are now in a position to prove Theorem 14.

PROOF OF THEOREM 14. Let $\mathbf{X}(t)$ be a fundamental matrix solution of (1). Then, by Lemmas 2 and 3 there exists a constant matrix $\mathbf{C}$ such that

$$e^{\mathbf{A}t}=\mathbf{X}(t)\mathbf{C}. \tag{6}$$

Setting $t=0$ in (6) gives $\mathbf{I}=\mathbf{X}(0)\mathbf{C}$, which implies that $\mathbf{C}=\mathbf{X}^{-1}(0)$. Hence, $e^{\mathbf{A}t}=\mathbf{X}(t)\mathbf{X}^{-1}(0)$. □

Example 1. Find $e^{\mathbf{A}t}$ if

$$\mathbf{A}=\begin{bmatrix}1&1&1\\0&3&2\\0&0&5\end{bmatrix}.$$

Solution. Our first step is to find 3 linearly independent solutions of the differential equation

$$\dot{\mathbf{x}}=\begin{bmatrix}1&1&1\\0&3&2\\0&0&5\end{bmatrix}\mathbf{x}.$$

To this end we compute

$$p(\lambda)=\det(\mathbf{A}-\lambda\mathbf{I})=\det\begin{bmatrix}1-\lambda&1&1\\0&3-\lambda&2\\0&0&5-\lambda\end{bmatrix}=(1-\lambda)(3-\lambda)(5-\lambda).$$

Thus, $\mathbf{A}$ has 3 distinct eigenvalues $\lambda=1$, $\lambda=3$, and $\lambda=5$.

(i) $\lambda=1$: Clearly,

$$\mathbf{v}^1=\begin{bmatrix}1\\0\\0\end{bmatrix}$$

is an eigenvector of $\mathbf{A}$ with eigenvalue one. Hence

$$\mathbf{x}^1(t)=e^t\begin{bmatrix}1\\0\\0\end{bmatrix}$$

is one solution of $\dot{\mathbf{x}}=\mathbf{A}\mathbf{x}$.

(ii) $\lambda=3$: We seek a nonzero solution of the equation

$$(\mathbf{A}-3\mathbf{I})\mathbf{v}=\begin{bmatrix}-2&1&1\\0&0&2\\0&0&2\end{bmatrix}\begin{bmatrix}v_1\\v_2\\v_3\end{bmatrix}=\begin{bmatrix}0\\0\\0\end{bmatrix}.$$

This implies that $v_3=0$ and $v_2=2v_1$. Hence,

$$\mathbf{v}^2=\begin{bmatrix}1\\2\\0\end{bmatrix}$$

is an eigenvector of $\mathbf{A}$ with eigenvalue 3. Consequently,

$$\mathbf{x}^2(t)=e^{3t}\begin{bmatrix}1\\2\\0\end{bmatrix}$$

is a second solution of $\dot{\mathbf{x}}=\mathbf{A}\mathbf{x}$.

(iii) $\lambda=5$: We seek a nonzero solution of the equation

$$(\mathbf{A}-5\mathbf{I})\mathbf{v}=\begin{bmatrix}-4&1&1\\0&-2&2\\0&0&0\end{bmatrix}\begin{bmatrix}v_1\\v_2\\v_3\end{bmatrix}=\begin{bmatrix}0\\0\\0\end{bmatrix}.$$

This implies that $v_2=v_3$ and $v_1=v_3/2$. Hence,

$$\mathbf{v}^3=\begin{bmatrix}1\\2\\2\end{bmatrix}$$

is an eigenvector of $\mathbf{A}$ with eigenvalue 5. Consequently,

$$\mathbf{x}^3(t)=e^{5t}\begin{bmatrix}1\\2\\2\end{bmatrix}$$

is a third solution of $\dot{\mathbf{x}}=\mathbf{A}\mathbf{x}$. These solutions are clearly linearly independent. Therefore,

$$\mathbf{X}(t)=\begin{bmatrix}e^t&e^{3t}&e^{5t}\\0&2e^{3t}&2e^{5t}\\0&0&2e^{5t}\end{bmatrix}$$

is a fundamental matrix solution. Using the methods of Section 3.6, we compute

$$\mathbf{X}^{-1}(0)=\begin{bmatrix}1&1&1\\0&2&2\\0&0&2\end{bmatrix}^{-1}=\begin{bmatrix}1&-\frac{1}{2}&0\\0&\frac{1}{2}&-\frac{1}{2}\\0&0&\frac{1}{2}\end{bmatrix}.$$

Therefore,

$$\exp\left[\begin{pmatrix}1&1&1\\0&3&2\\0&0&5\end{pmatrix}t\right]=\begin{bmatrix}e^{t}&e^{3t}&e^{5t}\\0&2e^{3t}&2e^{5t}\\0&0&2e^{5t}\end{bmatrix}\begin{bmatrix}1&-\frac{1}{2}&0\\0&\frac{1}{2}&-\frac{1}{2}\\0&0&\frac{1}{2}\end{bmatrix}$$

$$=\begin{bmatrix}e^{t}&-\frac{1}{2}e^{t}+\frac{1}{2}e^{3t}&-\frac{1}{2}e^{3t}+\frac{1}{2}e^{5t}\\0&e^{3t}&-e^{3t}+e^{5t}\\0&0&e^{5t}\end{bmatrix}.$$

Exercises

Compute $e^{\mathbf{A}t}$ for $\mathbf{A}$ equal

1. $\begin{pmatrix}1&-1&-1\\1&3&1\\-3&1&-1\end{pmatrix}$ **2.** $\begin{pmatrix}1&1&-1\\2&3&-4\\4&1&-4\end{pmatrix}$ **3.** $\begin{pmatrix}2&0&1\\1&0&1\\1&-2&0\end{pmatrix}$

4. $\begin{pmatrix}1&0&0\\2&1&-2\\3&2&1\end{pmatrix}$ **5.** $\begin{pmatrix}0&2&1\\-1&-3&-1\\1&1&-1\end{pmatrix}$ **6.** $\begin{pmatrix}1&1&1\\2&1&-1\\0&-1&1\end{pmatrix}$

7. Find $\mathbf{A}$ if

$$e^{\mathbf{A}t}=\begin{bmatrix}2e^{2t}-e^{t}&e^{2t}-e^{t}&e^{t}-e^{2t}\\e^{2t}-e^{t}&2e^{2t}-e^{t}&e^{t}-e^{2t}\\3e^{2t}-3e^{t}&3e^{2t}-3e^{t}&3e^{t}-2e^{2t}\end{bmatrix}.$$

In each of Problems 8–11, determine whether the given matrix is a fundamental matrix solution of $\dot{\mathbf{x}}=\mathbf{A}\mathbf{x}$, for some $\mathbf{A}$; if yes, find $\mathbf{A}$.

8. $\begin{bmatrix}e^{t}&e^{-t}&e^{t}+2e^{-t}\\e^{t}&-e^{-t}&e^{t}-2e^{-t}\\2e^{t}&e^{-t}&2(e^{t}+e^{-t})\end{bmatrix}$

9. $\begin{bmatrix}-5\cos 2t&-5\sin 2t&3e^{2t}\\-2(\cos 2t+\sin 2t)&2(\cos 2t-\sin 2t)&0\\\cos 2t&\sin 2t&e^{2t}\end{bmatrix}$

10. $e^t\begin{bmatrix} 1 & t+1 & t^2+1 \\ 1 & 2(t+1) & 4t^2 \\ 1 & t+2 & 3 \end{bmatrix}$ **11.** $\begin{bmatrix} e^{2t} & 2e^{-t} & e^{3t} \\ 2e^t & 2e^{-t} & e^{3t} \\ 3e^t & e^{-t} & 2e^{3t} \end{bmatrix}$

12. Let $\boldsymbol{\phi}^j(t)$ be the solution of the initial-value problem $\dot{\mathbf{x}}=\mathbf{A}\mathbf{x}$, $\mathbf{x}(0)=\mathbf{e}^j$. Show that $e^{\mathbf{A}t}=(\boldsymbol{\phi}^1,\boldsymbol{\phi}^2,\ldots,\boldsymbol{\phi}^n)$.

13. Suppose that $\mathbf{Y}(t)=\mathbf{X}(t)\mathbf{C}$, where $\mathbf{X}(t)$ and $\mathbf{Y}(t)$ are fundamental matrix solutions of $\dot{\mathbf{x}}=\mathbf{A}\mathbf{x}$, and $\mathbf{C}$ is a constant matrix. Prove that $\det\mathbf{C}\neq 0$.

14. Let $\mathbf{X}(t)$ be a fundamental matrix solution of (1), and $\mathbf{C}$ a constant matrix with $\det\mathbf{C}\neq 0$. Show that $\mathbf{Y}(t)=\mathbf{X}(t)\mathbf{C}$ is again a fundamental matrix solution of (1).

15. Let $\mathbf{X}(t)$ be a fundamental matrix solution of $\dot{\mathbf{x}}=\mathbf{A}\mathbf{x}$. Prove that the solution $\mathbf{x}(t)$ of the initial-value problem $\dot{\mathbf{x}}=\mathbf{A}\mathbf{x}$, $\mathbf{x}(t_0)=\mathbf{x}^0$ is $\mathbf{x}(t)=\mathbf{X}(t)\mathbf{X}^{-1}(t_0)\mathbf{x}^0$.

16. Let $\mathbf{X}(t)$ be a fundamental matrix solution of $\dot{\mathbf{x}}=\mathbf{A}\mathbf{x}$. Prove that $\mathbf{X}(t)\mathbf{X}^{-1}(t_0)=e^{\mathbf{A}(t-t_0)}$.

17. Here is an elegant proof of the identity $e^{\mathbf{A}t+\mathbf{B}t}=e^{\mathbf{A}t}e^{\mathbf{B}t}$ if $\mathbf{AB}=\mathbf{BA}$.

(a) Show that $\mathbf{X}(t)=e^{\mathbf{A}t+\mathbf{B}t}$ satisfies the initial-value problem $\dot{\mathbf{X}}=(\mathbf{A}+\mathbf{B})\mathbf{X}$, $\mathbf{X}(0)=\mathbf{I}$.

(b) Show that $e^{\mathbf{A}t}\mathbf{B}=\mathbf{B}e^{\mathbf{A}t}$ if $\mathbf{AB}=\mathbf{BA}$. (*Hint*: $\mathbf{A}^j\mathbf{B}=\mathbf{B}\mathbf{A}^j$ if $\mathbf{AB}=\mathbf{BA}$). Then, conclude that $(d/dt)e^{\mathbf{A}t}e^{\mathbf{B}t}=(\mathbf{A}+\mathbf{B})e^{\mathbf{A}t}e^{\mathbf{B}t}$.

(c) It follows immediately from Theorem 4, Section 3.4 that the solution $\mathbf{X}(t)$ of the initial-value problem $\dot{\mathbf{X}}=(\mathbf{A}+\mathbf{B})\mathbf{X}$, $\mathbf{X}(0)=\mathbf{I}$, is unique. Conclude, therefore, that $e^{\mathbf{A}t+\mathbf{B}t}=e^{\mathbf{A}t}e^{\mathbf{B}t}$.

3.12 The nonhomogeneous equation; variation of parameters

Consider now the nonhomogeneous equation $\dot{\mathbf{x}}=\mathbf{A}\mathbf{x}+\mathbf{f}(t)$. In this case, we can use our knowledge of the solutions of the homogeneous equation

$$\dot{\mathbf{x}}=\mathbf{A}\mathbf{x} \tag{1}$$

to help us find the solution of the initial-value problem

$$\dot{\mathbf{x}}=\mathbf{A}\mathbf{x}+\mathbf{f}(t), \qquad \mathbf{x}(t_0)=\mathbf{x}^0. \tag{2}$$

Let $\mathbf{x}^1(t),\ldots,\mathbf{x}^n(t)$ be n linearly independent solutions of the homogeneous equation (1). Since the general solution of (1) is $c_1\mathbf{x}^1(t)+\ldots+c_n\mathbf{x}^n(t)$, it is natural to seek a solution of (2) of the form

$$\mathbf{x}(t)=u_1(t)\mathbf{x}^1(t)+u_2(t)\mathbf{x}^2(t)+\ldots+u_n(t)\mathbf{x}^n(t). \tag{3}$$

This equation can be written concisely in the form $\mathbf{x}(t)=\mathbf{X}(t)\mathbf{u}(t)$ where

$\mathbf{X}(t)=(\mathbf{x}^1(t),\ldots,\mathbf{x}^n(t))$ and

$$\mathbf{u}(t)=\begin{bmatrix} u_1(t) \\ \vdots \\ u_n(t) \end{bmatrix}.$$

Plugging this expression into the differential equation $\dot{\mathbf{x}}=\mathbf{A}\mathbf{x}+\mathbf{f}(t)$ gives

$$\dot{\mathbf{X}}(t)\mathbf{u}(t)+\mathbf{X}(t)\dot{\mathbf{u}}(t)=\mathbf{A}\mathbf{X}(t)\mathbf{u}(t)+\mathbf{f}(t). \tag{4}$$

The matrix $\mathbf{X}(t)$ is a fundamental matrix solution of (1). Hence, $\dot{\mathbf{X}}(t)=\mathbf{A}\mathbf{X}(t)$, and Equation (4) reduces to

$$\mathbf{X}(t)\dot{\mathbf{u}}(t)=\mathbf{f}(t). \tag{5}$$

Recall that the columns of $\mathbf{X}(t)$ are linearly independent vectors of R^n at every time t. Hence $\mathbf{X}^{-1}(t)$ exists, and

$$\dot{\mathbf{u}}(t)=\mathbf{X}^{-1}(t)\mathbf{f}(t). \tag{6}$$

Integrating this expression between t_0 and t gives

$$\begin{aligned}\mathbf{u}(t)&=\mathbf{u}(t_0)+\int_{t_0}^{t}\mathbf{X}^{-1}(s)\mathbf{f}(s)\,ds\\ &=\mathbf{X}^{-1}(t_0)\mathbf{x}^0+\int_{t_0}^{t}\mathbf{X}^{-1}(s)\mathbf{f}(s)\,ds.\end{aligned}$$

Consequently,

$$\mathbf{x}(t)=\mathbf{X}(t)\mathbf{X}^{-1}(t_0)\mathbf{x}^0+\mathbf{X}(t)\int_{t_0}^{t}\mathbf{X}^{-1}(s)\mathbf{f}(s)\,ds. \tag{7}$$

If $\mathbf{X}(t)$ is the fundamental matrix solution $e^{\mathbf{A}t}$, then Equation (7) simplifies considerably. To wit, if $\mathbf{X}(t)=e^{\mathbf{A}t}$, then $\mathbf{X}^{-1}(s)=e^{-\mathbf{A}s}$. Hence

$$\begin{aligned}\mathbf{x}(t)&=e^{\mathbf{A}t}e^{-\mathbf{A}t_0}\mathbf{x}^0+e^{\mathbf{A}t}\int_{t_0}^{t}e^{-\mathbf{A}s}\mathbf{f}(s)\,ds\\ &=e^{\mathbf{A}(t-t_0)}\mathbf{x}^0+\int_{t_0}^{t}e^{\mathbf{A}(t-s)}\mathbf{f}(s)\,ds.\end{aligned} \tag{8}$$

Example 1. Solve the initial-value problem

$$\dot{\mathbf{x}}=\begin{bmatrix} 1 & 0 & 0 \\ 2 & 1 & -2 \\ 3 & 2 & 1 \end{bmatrix}\mathbf{x}+\begin{bmatrix} 0 \\ 0 \\ e^{t}\cos 2t \end{bmatrix}, \qquad \mathbf{x}(0)=\begin{bmatrix} 0 \\ 1 \\ 1 \end{bmatrix}.$$

Solution. We first find $e^{\mathbf{A}t}$, where

$$\mathbf{A}=\begin{bmatrix} 1 & 0 & 0 \\ 2 & 1 & -2 \\ 3 & 2 & 1 \end{bmatrix}.$$

To this end compute

$$\det(\mathbf{A}-\lambda\mathbf{I})=\det\begin{bmatrix} 1-\lambda & 0 & 0 \\ 2 & 1-\lambda & -2 \\ 3 & 2 & 1-\lambda \end{bmatrix}=(1-\lambda)(\lambda^2-2\lambda+5).$$

Thus the eigenvalues of $\mathbf{A}$ are

$$\lambda=1 \quad \text{and} \quad \lambda=\frac{2\pm\sqrt{4-20}}{2}=1\pm 2i.$$

(i) $\lambda=1$: We seek a nonzero vector $\mathbf{v}$ such that

$$(\mathbf{A}-\mathbf{I})\mathbf{v}=\begin{bmatrix} 0 & 0 & 0 \\ 2 & 0 & -2 \\ 3 & 2 & 0 \end{bmatrix}\begin{bmatrix} v_1 \\ v_2 \\ v_3 \end{bmatrix}=\begin{bmatrix} 0 \\ 0 \\ 0 \end{bmatrix}.$$

This implies that $v_1=v_3$ and $v_2=-3v_1/2$. Hence

$$\mathbf{v}=\begin{bmatrix} 2 \\ -3 \\ 2 \end{bmatrix}$$

is an eigenvector of $\mathbf{A}$ with eigenvalue 1. Consequently,

$$\mathbf{x}^1(t)=e^t\begin{bmatrix} 2 \\ -3 \\ 2 \end{bmatrix}$$

is a solution of the homogeneous equation $\dot{\mathbf{x}}=\mathbf{A}\mathbf{x}$.

(ii) $\lambda=1+2i$: We seek a nonzero vector $\mathbf{v}$ such that

$$\left[\mathbf{A}-(1+2i)\mathbf{I}\right]\mathbf{v}=\begin{bmatrix} -2i & 0 & 0 \\ 2 & -2i & -2 \\ 3 & 2 & -2i \end{bmatrix}\begin{bmatrix} v_1 \\ v_2 \\ v_3 \end{bmatrix}=\begin{bmatrix} 0 \\ 0 \\ 0 \end{bmatrix}.$$

This implies that $v_1=0$ and $v_3=-iv_2$. Hence,

$$\mathbf{v}=\begin{bmatrix} 0 \\ 1 \\ -i \end{bmatrix}$$

is an eigenvector of $\mathbf{A}$ with eigenvalue $1+2i$. Therefore,

$$\mathbf{x}(t)=\begin{bmatrix} 0 \\ 1 \\ -i \end{bmatrix}e^{(1+2i)t}$$

is a complex-valued solution of $\dot{\mathbf{x}}=\mathbf{A}\mathbf{x}$. Now,

$$\begin{bmatrix} 0 \\ 1 \\ -i \end{bmatrix} e^{(1+2i)t} = e^{t}(\cos 2t + i\sin 2t)\left[\begin{pmatrix} 0 \\ 1 \\ 0 \end{pmatrix} - i\begin{pmatrix} 0 \\ 0 \\ 1 \end{pmatrix}\right]$$

$$= e^{t}\left[\cos 2t\begin{pmatrix} 0 \\ 1 \\ 0 \end{pmatrix} + \sin 2t\begin{pmatrix} 0 \\ 0 \\ 1 \end{pmatrix}\right]$$

$$+ ie^{t}\left[\sin 2t\begin{pmatrix} 0 \\ 1 \\ 0 \end{pmatrix} - \cos 2t\begin{pmatrix} 0 \\ 0 \\ 1 \end{pmatrix}\right].$$

Consequently,

$$\mathbf{x}^2(t) = e^{t}\begin{bmatrix} 0 \\ \cos 2t \\ \sin 2t \end{bmatrix} \quad \text{and} \quad \mathbf{x}^3(t) = e^{t}\begin{bmatrix} 0 \\ \sin 2t \\ -\cos 2t \end{bmatrix}$$

are real-valued solutions of $\dot{\mathbf{x}}=\mathbf{A}\mathbf{x}$. The solutions $\mathbf{x}^1$, $\mathbf{x}^2$, and $\mathbf{x}^3$ are linearly independent since their values at $t=0$ are clearly linearly independent vectors of R^3. Therefore,

$$\mathbf{X}(t) = \begin{bmatrix} 2e^{t} & 0 & 0 \\ -3e^{t} & e^{t}\cos 2t & e^{t}\sin 2t \\ 2e^{t} & e^{t}\sin 2t & -e^{t}\cos 2t \end{bmatrix}$$

is a fundamental matrix solution of $\dot{\mathbf{x}}=\mathbf{A}\mathbf{x}$. Computing

$$\mathbf{X}^{-1}(0) = \begin{bmatrix} 2 & 0 & 0 \\ -3 & 1 & 0 \\ 2 & 0 & -1 \end{bmatrix}^{-1} = \begin{bmatrix} \frac{1}{2} & 0 & 0 \\ \frac{3}{2} & 1 & 0 \\ 1 & 0 & -1 \end{bmatrix}$$

we see that

$$e^{\mathbf{A}t} = \begin{bmatrix} 2e^{t} & 0 & 0 \\ -3e^{t} & e^{t}\cos 2t & e^{t}\sin 2t \\ 2e^{t} & e^{t}\sin 2t & -e^{t}\cos 2t \end{bmatrix}\begin{bmatrix} \frac{1}{2} & 0 & 0 \\ \frac{3}{2} & 1 & 0 \\ 1 & 0 & -1 \end{bmatrix}$$

$$= e^{t}\begin{bmatrix} 1 & 0 & 0 \\ -\frac{3}{2}+\frac{3}{2}\cos 2t + \sin 2t & \cos 2t & -\sin 2t \\ 1+\frac{3}{2}\sin 2t - \cos 2t & \sin 2t & \cos 2t \end{bmatrix}.$$

Consequently,

$$\mathbf{x}(t)=e^{\mathbf{A}t}\begin{bmatrix}0\\1\\1\end{bmatrix}$$

$$+e^{\mathbf{A}t}\int_0^t e^{-s}\begin{bmatrix}1 & 0 & 0\\ -\frac{3}{2}+\frac{3}{2}\cos 2s-\sin 2s & \cos 2s & \sin 2s\\ 1-\frac{3}{2}\sin 2s-\cos 2s & -\sin 2s & \cos 2s\end{bmatrix}\begin{bmatrix}0\\0\\e^{s}\cos 2s\end{bmatrix}ds$$

$$=e^{t}\begin{bmatrix}0\\ \cos 2t-\sin 2t\\ \cos 2t+\sin 2t\end{bmatrix}+e^{\mathbf{A}t}\int_0^t\begin{bmatrix}0\\ \sin 2s\cos 2s\\ \cos^2 2s\end{bmatrix}ds$$

$$=e^{t}\begin{bmatrix}0\\ \cos 2t-\sin 2t\\ \cos 2t+\sin 2t\end{bmatrix}+e^{\mathbf{A}t}\begin{bmatrix}0\\ (1-\cos 4t)/8\\ t/2+(\sin 4t)/8\end{bmatrix}$$

$$=e^{t}\begin{bmatrix}0\\ \cos 2t-\sin 2t\\ \cos 2t+\sin 2t\end{bmatrix}$$

$$+e^{t}\begin{bmatrix}0\\ -\dfrac{t\sin 2t}{2}+\dfrac{\cos 2t-\cos 4t\cos 2t-\sin 4t\sin 2t}{8}\\ \dfrac{t\cos 2t}{2}+\dfrac{\sin 4t\cos 2t-\sin 2t\cos 4t+\sin 2t}{8}\end{bmatrix}$$

$$=e^{t}\begin{bmatrix}0\\ \cos 2t-\left(1+\frac{1}{2}t\right)\sin 2t\\ \left(1+\frac{1}{2}t\right)\cos 2t+\frac{5}{4}\sin 2t\end{bmatrix}.$$

As Example 1 indicates, the method of variation of parameters is often quite tedious and laborious. One way of avoiding many of these calculations is to "guess" a particular solution $\psi(t)$ of the nonhomogeneous equation and then to observe (see Exercise 9) that every solution $\mathbf{x}(t)$ of the nonhomogeneous equation must be of the form $\phi(t)+\psi(t)$ where $\phi(t)$ is a solution of the homogeneous equation.

Example 2. Find all solutions of the differential equation

$$\dot{\mathbf{x}}=\begin{bmatrix}1 & 0 & 0\\ 2 & 1 & -2\\ 3 & 2 & 1\end{bmatrix}\mathbf{x}+\begin{bmatrix}1\\0\\0\end{bmatrix}e^{ct},\qquad c\neq 1. \tag{9}$$

Solution. Let

$$\mathbf{A}=\begin{bmatrix}1&0&0\\2&1&-2\\3&2&1\end{bmatrix}.$$

We "guess" a particular solution $\psi(t)$ of the form $\psi(t)=\mathbf{b}e^{ct}$. Plugging this expression into (9) gives

$$c\mathbf{b}e^{ct}=\mathbf{A}\mathbf{b}e^{ct}+\begin{bmatrix}1\\0\\0\end{bmatrix}e^{ct},$$

or

$$(\mathbf{A}-c\mathbf{I})\mathbf{b}=\begin{bmatrix}-1\\0\\0\end{bmatrix}.$$

This implies that

$$\mathbf{b}=\frac{-1}{1-c}\begin{bmatrix}1\\ \dfrac{2(c-4)}{4+(1-c)^2}\\ \dfrac{1+3c}{4+(1-c)^2}\end{bmatrix}$$

Hence, every solution $\mathbf{x}(t)$ of (9) is of the form

$$\mathbf{x}(t)=e^{t}\left[c_1\begin{bmatrix}2\\-3\\2\end{bmatrix}+c_2\begin{bmatrix}0\\\cos 2t\\\sin 2t\end{bmatrix}+c_3\begin{bmatrix}0\\-\sin 2t\\\cos 2t\end{bmatrix}\right]$$

$$-\frac{e^{ct}}{1-c}\begin{bmatrix}1\\ \dfrac{2(c-4)}{4+(1-c)^2}\\ \dfrac{1+3c}{4+(1-c)^2}\end{bmatrix}$$

Remark. We run into trouble when $c=1$ because one is an eigenvalue of the matrix

$$\begin{bmatrix} 1 & 0 & 0 \\ 2 & 1 & -2 \\ 3 & 2 & 1 \end{bmatrix}.$$

More generally, the differential equation $\dot{\mathbf{x}}=\mathbf{A}\mathbf{x}+\mathbf{v}e^{ct}$ may not have a solution of the form $\mathbf{b}e^{ct}$ if c is an eigenvalue of $\mathbf{A}$. In this case we have to guess a particular solution of the form

$$\psi(t)=e^{ct}\left[\mathbf{b}_0+\mathbf{b}_1t+\ldots+\mathbf{b}_{k-1}t^{k-1}\right]$$

for some appropriate integer k. (See Exercises 10–18).

Exercises

In each of Problems 1–6 use the method of variation of parameters to solve the given initial-value problem.

1. $\dot{\mathbf{x}}=\begin{pmatrix} 4 & 5 \\ -2 & -2 \end{pmatrix}\mathbf{x}+\begin{pmatrix} 4e^t\cos t \\ 0 \end{pmatrix}, \quad \mathbf{x}(0)=\begin{pmatrix} 0 \\ 0 \end{pmatrix}$

2. $\dot{\mathbf{x}}=\begin{pmatrix} 3 & -4 \\ 1 & -1 \end{pmatrix}\mathbf{x}+\begin{pmatrix} 1 \\ 1 \end{pmatrix}e^t, \quad \mathbf{x}(0)=\begin{pmatrix} 1 \\ 1 \end{pmatrix}$

3. $\dot{\mathbf{x}}=\begin{pmatrix} 2 & -5 \\ 1 & -2 \end{pmatrix}\mathbf{x}+\begin{pmatrix} \sin t \\ \tan t \end{pmatrix}, \quad \mathbf{x}(0)=\begin{pmatrix} 0 \\ 0 \end{pmatrix}$

4. $\dot{\mathbf{x}}=\begin{pmatrix} 0 & 1 \\ -1 & 0 \end{pmatrix}\mathbf{x}+\begin{pmatrix} f_1(t) \\ f_2(t) \end{pmatrix}, \quad \mathbf{x}(0)=\begin{pmatrix} 0 \\ 0 \end{pmatrix}$

5. $\dot{\mathbf{x}}=\begin{pmatrix} 2 & 0 & 1 \\ 0 & 2 & 0 \\ 0 & 1 & 3 \end{pmatrix}\mathbf{x}+\begin{pmatrix} 1 \\ 0 \\ 1 \end{pmatrix}e^{2t}, \quad \mathbf{x}(0)=\begin{pmatrix} 1 \\ 1 \\ 1 \end{pmatrix}$

6. $\dot{\mathbf{x}}=\begin{pmatrix} -1 & -1 & -2 \\ 1 & 1 & 1 \\ 2 & 1 & 3 \end{pmatrix}\mathbf{x}+\begin{pmatrix} 1 \\ 0 \\ 0 \end{pmatrix}e^t, \quad \mathbf{x}(0)=\begin{pmatrix} 0 \\ 0 \\ 0 \end{pmatrix}$

7. Consider the nth-order scalar differential equation

$$L[y]=\frac{d^ny}{dt^n}+a_1\frac{d^{n-1}y}{dt^{n-1}}+\ldots+a_ny=f(t). \tag{*}$$

Let $v(t)$ be the solution of $L[y]=0$ which satisfies the initial conditions $y(0)=\ldots=y^{(n-2)}(0)=0$, $y^{(n-1)}(0)=1$. Show that

$$y(t)=\int_0^t v(t-s)f(s)\,ds$$

is the solution of (*) which satisfies the initial conditions $y(0)=\ldots=y^{(n-1)}(0)=0$. *Hint*: Convert (*) to a system of n first-order equations of the form $\dot{\mathbf{x}}=\mathbf{A}\mathbf{x}$, and show that

$$\begin{bmatrix} v(t) \\ v'(t) \\ \vdots \\ v^{(n-1)}(t) \end{bmatrix}$$

is the nth column of $e^{\mathbf{A}t}$.

8. Find the solution of the initial-value problem

$$\frac{d^3y}{dt^3}+\frac{dy}{dt}=\sec t\tan t,\qquad y(0)=y'(0)=y''(0)=0.$$

9. (a) Let $\psi(t)$ be a solution of the nonhomogeneous equation $\dot{\mathbf{x}}=\mathbf{A}\mathbf{x}+\mathbf{f}(t)$ and let $\phi(t)$ be a solution of the homogeneous equation $\dot{\mathbf{x}}=\mathbf{A}\mathbf{x}$. Show that $\phi(t)+\psi(t)$ is a solution of the nonhomogeneous equation.

(b) Let $\psi_1(t)$ and $\psi_2(t)$ be two solutions of the nonhomogeneous equation. Show that $\psi_1(t)-\psi_2(t)$ is a solution of the homogeneous equation.

(c) Let $\psi(t)$ be a particular solution of the nonhomogeneous equation. Show that any other solution $\mathbf{y}(t)$ must be of the form $\mathbf{y}(t)=\phi(t)+\psi(t)$ where $\phi(t)$ is a solution of the homogeneous equation.

In each of Problems 10–14 use the method of judicious guessing to find a particular solution of the given differential equation.

10. $\dot{\mathbf{x}}=\begin{pmatrix} 2 & 1 \\ 3 & -2 \end{pmatrix}\mathbf{x}+\begin{pmatrix} 1 \\ 1 \end{pmatrix}e^{3t}$

11. $\dot{\mathbf{x}}=\begin{pmatrix} 1 & -1 \\ 1 & 3 \end{pmatrix}\mathbf{x}+\begin{pmatrix} -t^2 \\ 2t \end{pmatrix}$

12. $\dot{\mathbf{x}}=\begin{pmatrix} 1 & 3 & 2 \\ -1 & 2 & 1 \\ 4 & -1 & -1 \end{pmatrix}\mathbf{x}+\begin{pmatrix} \sin t \\ 0 \\ 0 \end{pmatrix}$

13. $\dot{\mathbf{x}}=\begin{pmatrix} 1 & 2 & -3 \\ 1 & 1 & 2 \\ 1 & -1 & 4 \end{pmatrix}\mathbf{x}+\begin{pmatrix} 1 \\ 0 \\ -1 \end{pmatrix}e^{t}$

14. $\dot{\mathbf{x}}=\begin{pmatrix} -1 & -1 & 0 \\ 0 & -4 & -1 \\ 0 & 5 & 0 \end{pmatrix}\mathbf{x}+\begin{pmatrix} 1 \\ t \\ e^{t} \end{pmatrix}$

15. Consider the differential equation

$$\dot{\mathbf{x}}=\mathbf{A}\mathbf{x}+\mathbf{v}e^{\lambda t} \tag{*}$$

where $\mathbf{v}$ is an eigenvector of $\mathbf{A}$ with eigenvalue λ. Suppose moreover, that $\mathbf{A}$ has n linearly independent eigenvectors $\mathbf{v}^1,\mathbf{v}^2,\ldots,\mathbf{v}^n$, with distinct eigenvalues $\lambda_1,\ldots,\lambda_n$ respectively.

(a) Show that (*) has no solution $\psi(t)$ of the form $\psi(t)=\mathbf{a}e^{\lambda t}$. *Hint*: Write $\mathbf{a}=a_1\mathbf{v}^1+\ldots+a_n\mathbf{v}^n$.

(b) Show that (*) has a solution $\psi(t)$ of the form

$$\psi(t)=\mathbf{a}e^{\lambda t}+\mathbf{b}te^{\lambda t}.$$

Hint: Show that $\mathbf{b}$ is an eigenvector of $\mathbf{A}$ with eigenvalue λ and choose it so that we can solve the equation

$$(\mathbf{A}-\lambda\mathbf{I})\mathbf{a}=\mathbf{b}-\mathbf{v}.$$

In each of Problems 16–18, find a particular solution $\psi(t)$ of the given differential equation of the form $\psi(t)=e^{\lambda t}(\mathbf{a}+\mathbf{b}t)$.

16. $\dot{\mathbf{x}}=\begin{pmatrix} 1 & 1 & -1 \\ 2 & 3 & -4 \\ 4 & 1 & -4 \end{pmatrix}\mathbf{x}+\begin{pmatrix} 1 \\ 2 \\ 1 \end{pmatrix}e^{2t}$

17. $\dot{\mathbf{x}}=\begin{pmatrix} 1 & -1 & -1 \\ 1 & 3 & 1 \\ -3 & 1 & -1 \end{pmatrix}\mathbf{x}+\begin{pmatrix} 1 \\ -1 \\ -1 \end{pmatrix}e^{3t}$

18. $\dot{\mathbf{x}}=\begin{pmatrix} 3 & 2 & 4 \\ 2 & 0 & 2 \\ 4 & 2 & 3 \end{pmatrix}\mathbf{x}+\begin{pmatrix} 2 \\ 1 \\ 2 \end{pmatrix}e^{8t}$

3.13 Solving systems by Laplace transforms

The method of Laplace transforms introduced in Chapter 2 can also be used to solve the initial-value problem

$$\dot{\mathbf{x}}=\mathbf{A}\mathbf{x}+\mathbf{f}(t), \qquad \mathbf{x}(0)=\mathbf{x}^0. \tag{1}$$

Let

$$\mathbf{X}(s)=\begin{bmatrix} X_1(s) \\ \vdots \\ X_n(s) \end{bmatrix}=\mathcal{L}\{\mathbf{x}(t)\}=\begin{bmatrix} \int_0^\infty e^{-st}x_1(t)\,dt \\ \vdots \\ \int_0^\infty e^{-st}x_n(t)\,dt \end{bmatrix}$$

and

$$\mathbf{F}(s)=\begin{bmatrix} F_1(s) \\ \vdots \\ F_n(s) \end{bmatrix}=\mathcal{L}\{\mathbf{f}(t)\}=\begin{bmatrix} \int_0^\infty e^{-st}f_1(t)\,dt \\ \vdots \\ \int_0^\infty e^{-st}f_n(t)\,dt \end{bmatrix}.$$

Taking Laplace transforms of both sides of (1) gives

$$\begin{aligned}\mathcal{L}\{\dot{\mathbf{x}}(t)\}&=\mathcal{L}\{\mathbf{A}\mathbf{x}(t)+\mathbf{f}(t)\}=\mathbf{A}\mathcal{L}\{\mathbf{x}(t)\}+\mathcal{L}\{\mathbf{f}(t)\}\\&=\mathbf{A}\mathbf{X}(s)+\mathbf{F}(s),\end{aligned}$$

and from Lemma 3 of Section 2.9,

$$\mathcal{L}\{\dot{\mathbf{x}}(t)\}=\begin{bmatrix}\mathcal{L}\{\dot{x}_1(t)\}\\ \vdots\\ \mathcal{L}\{\dot{x}_n(t)\}\end{bmatrix}=\begin{bmatrix}sX_1(s)-x_1(0)\\ \vdots\\ sX_n(s)-x_n(0)\end{bmatrix}$$

$$=s\mathbf{X}(s)-\mathbf{x}^0.$$

Hence,

$$s\mathbf{X}(s)-\mathbf{x}^0=\mathbf{A}\mathbf{X}(s)+\mathbf{F}(s)$$

or

$$(s\mathbf{I}-\mathbf{A})\mathbf{X}(s)=\mathbf{x}^0+\mathbf{F}(s),\qquad \mathbf{I}=\begin{bmatrix}1&0&\dots&0\\ 0&1&\dots&0\\ \vdots&\vdots&&\vdots\\ 0&0&\dots&1\end{bmatrix}. \tag{2}$$

Equation (2) is a system of n simultaneous equations for $X_1(s),\dots,X_n(s)$, and it can be solved in a variety of ways. (One way, in particular, is to multiply both sides of (2) by $(s\mathbf{I}-\mathbf{A})^{-1}$.) Once we know $X_1(s),\dots,X_n(s)$ we can find $x_1(t),\dots,x_n(t)$ by inverting these Laplace transforms.

Example 1. Solve the initial-value problem

$$\dot{\mathbf{x}}=\begin{pmatrix}1&4\\1&1\end{pmatrix}\mathbf{x}+\begin{pmatrix}1\\1\end{pmatrix}e^t,\qquad \mathbf{x}(0)=\begin{pmatrix}2\\1\end{pmatrix}. \tag{3}$$

Solution. Taking Laplace transforms of both sides of the differential equation gives

$$s\mathbf{X}(s)-\begin{pmatrix}2\\1\end{pmatrix}=\begin{pmatrix}1&4\\1&1\end{pmatrix}\mathbf{X}(s)+\frac{1}{s-1}\begin{pmatrix}1\\1\end{pmatrix}$$

or

$$(s-1)X_1(s)-4X_2(s)=2+\frac{1}{s-1}-$$

$$X_1(s)+(s-1)X_2(s)=1+\frac{1}{s-1}.$$

The solution of these equations is

$$X_1(s)=\frac{2}{s-3}+\frac{1}{s^2-1},\qquad X_2(s)=\frac{1}{s-3}+\frac{s}{(s-1)(s+1)(s-3)}.$$

Now,

$$\frac{2}{s-3}=\mathcal{L}\{2e^{3t}\},\quad\text{and}\quad \frac{1}{s^2-1}=\mathcal{L}\{\sinh t\}=\mathcal{L}\left\{\frac{e^t-e^{-t}}{2}\right\}.$$

Hence,

$$x_1(t) = 2e^{3t} + \frac{e^t - e^{-t}}{2}.$$

To invert $X_2(s)$, we use partial fractions. Let

$$\frac{s}{(s-1)(s+1)(s-3)} = \frac{A}{s-1} + \frac{B}{s+1} + \frac{C}{s-3}.$$

This implies that

$$A(s+1)(s-3) + B(s-1)(s-3) + C(s-1)(s+1) = s. \tag{4}$$

Setting $s=1$, -1, and 3 respectively in (4) gives $A = -\frac{1}{4}$, $B = -\frac{1}{8}$, and $C = \frac{3}{8}$. Consequently,

$$\begin{aligned} x_2(t) &= \mathcal{L}^{-1}\left\{ -\frac{1}{4}\frac{1}{s-1} - \frac{1}{8}\frac{1}{s+1} + \frac{11}{8}\frac{1}{s-3} \right\} \\ &= -\frac{1}{8}e^{-t} - \frac{1}{4}e^{t} + \frac{11}{8}e^{3t}. \end{aligned}$$

Exercises

Find the solution of each of the following initial-value problems.

1. $\dot{\mathbf{x}} = \begin{pmatrix} 1 & -3 \\ -2 & 2 \end{pmatrix}\mathbf{x}, \quad \mathbf{x}(0) = \begin{pmatrix} 0 \\ 5 \end{pmatrix}$

2. $\dot{\mathbf{x}} = \begin{pmatrix} 1 & -1 \\ 5 & -3 \end{pmatrix}\mathbf{x}, \quad \mathbf{x}(0) = \begin{pmatrix} 1 \\ 2 \end{pmatrix}$

3. $\dot{\mathbf{x}} = \begin{pmatrix} 3 & -2 \\ 2 & -2 \end{pmatrix}\mathbf{x} + \begin{pmatrix} t \\ 3e^t \end{pmatrix}, \quad \mathbf{x}(0) = \begin{pmatrix} 2 \\ 1 \end{pmatrix}$

4. $\dot{\mathbf{x}} = \begin{pmatrix} 1 & 1 \\ 4 & 1 \end{pmatrix}\mathbf{x} + \begin{pmatrix} 2 \\ -1 \end{pmatrix}e^t, \quad \mathbf{x}(0) = \begin{pmatrix} 0 \\ 0 \end{pmatrix}$

5. $\dot{\mathbf{x}} = \begin{pmatrix} 3 & -4 \\ 1 & -1 \end{pmatrix}\mathbf{x} + \begin{pmatrix} 1 \\ 1 \end{pmatrix}e^t, \quad \mathbf{x}(0) = \begin{pmatrix} 1 \\ 1 \end{pmatrix}$

6. $\dot{\mathbf{x}} = \begin{pmatrix} 2 & -5 \\ 1 & -2 \end{pmatrix}\mathbf{x} + \begin{pmatrix} \sin t \\ \tan t \end{pmatrix}, \quad \mathbf{x}(0) = \begin{pmatrix} -1 \\ 0 \end{pmatrix}$

7. $\dot{\mathbf{x}} = \begin{pmatrix} 4 & 5 \\ -2 & -2 \end{pmatrix}\mathbf{x} + \begin{pmatrix} 4e^t\cos t \\ 0 \end{pmatrix}, \quad \mathbf{x}(0) = \begin{pmatrix} 1 \\ 1 \end{pmatrix}$

8. $\dot{\mathbf{x}} = \begin{pmatrix} 0 & 1 \\ -1 & 0 \end{pmatrix}\mathbf{x} + \begin{pmatrix} f_1(t) \\ f_2(t) \end{pmatrix}, \quad \mathbf{x}(0) = \begin{pmatrix} 0 \\ 0 \end{pmatrix}$

9. $\dot{\mathbf{x}} = \begin{pmatrix} 2 & -2 \\ 4 & -2 \end{pmatrix}\mathbf{x} + \begin{pmatrix} 0 \\ \delta(t-\pi) \end{pmatrix}, \quad \mathbf{x}(0) = \begin{pmatrix} 1 \\ 0 \end{pmatrix}$

10. $\dot{\mathbf{x}} = \begin{pmatrix} 3 & -2 \\ 2 & -2 \end{pmatrix}\mathbf{x} + \begin{pmatrix} 1 - H_\pi(t) \\ 0 \end{pmatrix}, \quad \mathbf{x}(0) = \begin{pmatrix} 0 \\ 0 \end{pmatrix}$

11. $\dot{\mathbf{x}}=\begin{pmatrix}1 & 2 & -3\\ 1 & 1 & 2\\ 1 & -1 & 4\end{pmatrix}\mathbf{x},\quad \mathbf{x}(0)=\begin{pmatrix}1\\0\\0\end{pmatrix}$

12. $\dot{\mathbf{x}}=\begin{pmatrix}2 & 0 & 1\\ 0 & 2 & 0\\ 0 & 0 & 3\end{pmatrix}\mathbf{x}+\begin{pmatrix}1\\0\\1\end{pmatrix}e^{2t},\quad \mathbf{x}(0)=\begin{pmatrix}1\\1\\1\end{pmatrix}$

13. $\dot{\mathbf{x}}=\begin{pmatrix}-1 & -1 & 2\\ 1 & 1 & 1\\ 2 & 1 & 3\end{pmatrix}\mathbf{x}+\begin{pmatrix}1\\0\\0\end{pmatrix}e^{t},\quad \mathbf{x}(0)=\begin{pmatrix}0\\0\\0\end{pmatrix}$

14. $\dot{\mathbf{x}}=\begin{pmatrix}1 & 0 & 0\\ 2 & 1 & -2\\ 3 & 2 & 1\end{pmatrix}\mathbf{x}+\begin{pmatrix}0\\0\\e^{t}\cos 2t\end{pmatrix},\quad \mathbf{x}(0)=\begin{pmatrix}0\\1\\1\end{pmatrix}$

15. $\dot{\mathbf{x}}=\begin{bmatrix}3 & 0 & 0 & 0\\ 1 & 3 & 0 & 0\\ 0 & 0 & 3 & 0\\ 0 & 0 & 2 & 3\end{bmatrix}\mathbf{x},\quad \mathbf{x}(0)=\begin{bmatrix}1\\1\\1\\1\end{bmatrix}$

4 Qualitative theory of differential equations

4.1 Introduction

In this chapter we consider the differential equation

$$\dot{\mathbf{x}} = \mathbf{f}(t, \mathbf{x}) \tag{1}$$

where

$$\mathbf{x} = \begin{bmatrix} x_1(t) \\ \vdots \\ x_n(t) \end{bmatrix},$$

and

$$\mathbf{f}(t,\mathbf{x}) = \begin{bmatrix} f_1(t, x_1, \ldots, x_n) \\ \vdots \\ f_n(t, x_1, \ldots, x_n) \end{bmatrix}$$

is a nonlinear function of $x_1, \ldots, x_n$. Unfortunately, there are no known methods of solving Equation (1). This, of course, is very disappointing. However, it is not necessary, in most applications, to find the solutions of (1) explicitly. For example, let $x_1(t)$ and $x_2(t)$ denote the populations, at time t, of two species competing amongst themselves for the limited food and living space in their microcosm. Suppose, moreover, that the rates of growth of $x_1(t)$ and $x_2(t)$ are governed by the differential equation (1). In this case, we are not really interested in the values of $x_1(t)$ and $x_2(t)$ at every time t. Rather, we are interested in the qualitative properties of $x_1(t)$ and $x_2(t)$. Specically, we wish to answer the following questions.

1. Do there exist values ξ_1 and ξ_2 at which the two species coexist together in a steady state? That is to say, are there numbers ξ_1, ξ_2 such that $x_1(t) \equiv \xi_1$, $x_2(t) \equiv \xi_2$ is a solution of (1)? Such values ξ_1, ξ_2, if they exist, are called *equilibrium points* of (1).

2. Suppose that the two species are coexisting in equilibrium. Suddenly, we add a few members of species 1 to the microcosm. Will $x_1(t)$ and $x_2(t)$ remain close to their equilibrium values for all future time? Or perhaps the extra few members give species 1 a large advantage and it will proceed to annihilate species 2.

3. Suppose that x_1 and x_2 have arbitrary values at $t=0$. What happens as t approaches infinity? Will one species ultimately emerge victorious, or will the struggle for existence end in a draw?

More generally, we are interested in determining the following properties of solutions of (1).

1. Do there exist equilibrium values

$$\mathbf{x}^0 = \begin{bmatrix} x_1^0 \\ \vdots \\ x_n^0 \end{bmatrix}$$

for which $\mathbf{x}(t) \equiv \mathbf{x}^0$ is a solution of (1)?

2. Let $\boldsymbol{\phi}(t)$ be a solution of (1). Suppose that $\boldsymbol{\psi}(t)$ is a second solution with $\boldsymbol{\psi}(0)$ very close to $\boldsymbol{\phi}(0)$; that is, $\psi_j(0)$ is very close to $\phi_j(0)$, $j=1,\ldots,n$. Will $\boldsymbol{\psi}(t)$ remain close to $\boldsymbol{\phi}(t)$ for all future time, or will $\boldsymbol{\psi}(t)$ diverge from $\boldsymbol{\phi}(t)$ as t approaches infinity? This question is often referred to as the problem of *stability*. It is the most fundamental problem in the qualitative theory of differential equations, and has occupied the attention of many mathematicians for the past hundred years.

3. What happens to solutions $\mathbf{x}(t)$ of (1) as t approaches infinity? Do all solutions approach equilibrium values? If they don't approach equilibrium values, do they at least approach a periodic solution?

This chapter is devoted to answering these three questions. Remarkably, we can often give satisfactory answers to these questions, even though we cannot solve Equation (1) explicitly. Indeed, the first question can be answered immediately. Observe that $\dot{\mathbf{x}}(t)$ is identically zero if $\mathbf{x}(t) \equiv \mathbf{x}^0$. Hence, $\mathbf{x}^0$ is an equilibrium value of (1), if, and only if,

$$\mathbf{f}(t, \mathbf{x}^0) \equiv \mathbf{0}. \tag{2}$$

Example 1. Find all equilibrium values of the system of differential equations

$$\frac{dx_1}{dt} = 1 - x_2, \qquad \frac{dx_2}{dt} = x_1^3 + x_2.$$

Solution.

$$\mathbf{x}^0=\begin{pmatrix} x_1^0 \\ x_2^0 \end{pmatrix}$$

is an equilibrium value if, and only if, $1-x_2^0=0$ and $(x_1^0)^3+x_2^0=0$. This implies that $x_2^0=1$ and $x_1^0=-1$. Hence $\begin{pmatrix} -1 \\ 1 \end{pmatrix}$ is the only equilibrium value of this system.

Example 2. Find all equilibrium solutions of the system

$$\frac{dx}{dt}=(x-1)(y-1), \qquad \frac{dy}{dt}=(x+1)(y+1).$$

Solution.

$$\begin{pmatrix} x_0 \\ y_0 \end{pmatrix}$$

is an equilibrium value of this system if, and only if, $(x_0-1)(y_0-1)=0$ and $(x_0+1)(y_0+1)=0$. The first equation is satisfied if either x_0 or y_0 is 1, while the second equation is satisfied if either x_0 or y_0 is -1. Hence, $x=1$, $y=-1$ and $x=-1$, $y=1$ are the equilibrium solutions of this system.

The question of stability is of paramount importance in all physical applications, since we can never measure initial conditions exactly. For example, consider the case of a particle of mass one kgm attached to an elastic spring of force constant 1 N/m which is moving in a frictionless medium. In addition, an external force $F(t)=\cos 2t$ N is acting on the particle. Let $y(t)$ denote the position of the particle relative to its equilibrium position. Then $(d^2y/dt^2)+y=\cos 2t$. We convert this second-order equation into a system of two first-order equations by setting $x_1=y$, $x_2=y'$. Then,

$$\frac{dx_1}{dt}=x_2, \qquad \frac{dx_2}{dt}=-x_1+\cos 2t. \tag{3}$$

The functions $y_1(t)=\sin t$ and $y_2(t)=\cos t$ are two independent solutions of the homogeneous equation $y''+y=0$. Moreover, $y=-\frac{1}{3}\cos 2t$ is a particular solution of the nonhomogeneous equation. Therefore, every solution

$$\mathbf{x}(t)=\begin{pmatrix} x_1(t) \\ x_2(t) \end{pmatrix}$$

of (3) is of the form

$$\mathbf{x}(t)=c_1\begin{pmatrix} \sin t \\ \cos t \end{pmatrix}+c_2\begin{pmatrix} \cos t \\ -\sin t \end{pmatrix}+\begin{bmatrix} -\frac{1}{3}\cos 2t \\ \frac{2}{3}\sin 2t \end{bmatrix}. \tag{4}$$

At time $t=0$ we measure the position and velocity of the particle and obtain $y(0)=1, y'(0)=0$. This implies that $c_1=0$ and $c_2=\frac{4}{3}$. Consequently, the position and velocity of the particle for all future time are given by the equation

$$\begin{pmatrix} y(t) \\ y'(t) \end{pmatrix} = \begin{pmatrix} x_1(t) \\ x_2(t) \end{pmatrix} = \begin{bmatrix} \frac{4}{3}\cos t - \frac{1}{3}\cos 2t \\ -\frac{4}{3}\sin t + \frac{2}{3}\sin 2t \end{bmatrix}. \tag{5}$$

However, suppose that our measurements permit an error of magnitude 10^{-4}. Will the position and velocity of the particle remain close to the values predicted by (5)? The answer to this question had better be yes, for otherwise, Newtonian mechanics would be of no practical value to us. Fortunately, it is quite easy to show, in this case, that the position and velocity of the particle remain very close to the values predicted by (5). Let $\hat{y}(t)$ and $\hat{y}'(t)$ denote the true values of $y(t)$ and $y'(t)$ respectively. Clearly,

$$y(t)-\hat{y}(t) = \left(\tfrac{4}{3}-c_2\right)\cos t - c_1 \sin t$$

$$y'(t)-\hat{y}'(t) = -c_1\cos t - \left(\tfrac{4}{3}-c_2\right)\sin t$$

where c_1 and c_2 are two constants satisfying

$$-10^{-4} \leqslant c_1 \leqslant 10^{-4}, \qquad \tfrac{4}{3}-10^{-4} \leqslant c_2 \leqslant \tfrac{4}{3}+10^{-4}.$$

We can rewrite these equations in the form

$$y(t)-\hat{y}(t) = \left[c_1^2+\left(\tfrac{4}{3}-c_2\right)^2\right]^{1/2}\cos(t-\delta_1), \qquad \tan\delta_1 = \frac{c_1}{c_2-\frac{4}{3}}$$

$$y'(t)-\hat{y}'(t) = \left[c_1^2+\left(\tfrac{4}{3}-c_2\right)^2\right]^{1/2}\cos(t-\delta_2), \qquad \tan\delta_2 = \frac{\frac{4}{3}-c_2}{c_1}.$$

Hence, both $y(t)-\hat{y}(t)$ and $y'(t)-\hat{y}'(t)$ are bounded in absolute value by $[c_1^2+(\frac{4}{3}-c_2)^2]^{1/2}$. This quantity is at most $\sqrt{2}\,10^{-4}$. Therefore, the true values of $y(t)$ and $y'(t)$ are indeed close to the values predicted by (5).

As a second example of the concept of stability, consider the case of a particle of mass m which is supported by a wire, or inelastic string, of length l and of negligible mass. The wire is always straight, and the system is free to vibrate in a vertical plane. This configuration is usually referred to as a simple pendulum. The equation of motion of the pendulum is

$$\frac{d^2y}{dt^2} + \frac{g}{l}\sin y = 0$$

where y is the angle which the wire makes with the vertical line A0 (see

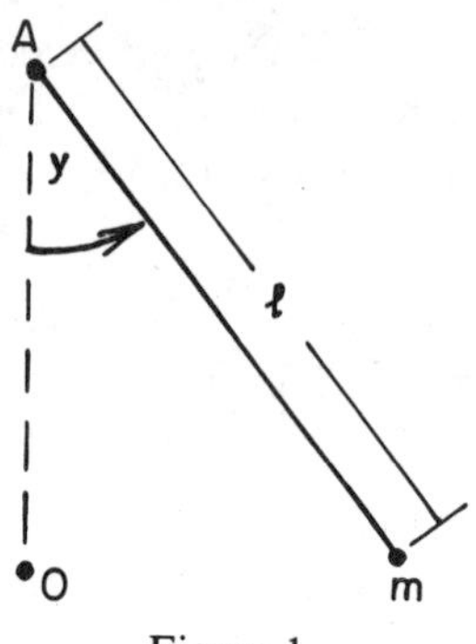

Figure 1

Figure 1). Setting $x_1 = y$ and $x_2 = dy/dt$ we see that

$$\frac{dx_1}{dt} = x_2, \qquad \frac{dx_2}{dt} = -\frac{g}{l}\sin x_1. \tag{6}$$

The system of equations (6) has equilibrium solutions $x_1 = 0$, $x_2 = 0$, and $x_1 = \pi$, $x_2 = 0$. (If the pendulum is suspended in the upright position $y = \pi$ with zero velocity, then it will remain in this upright position for all future time.) These two equilibrium solutions have very different properties. If we disturb the pendulum slightly from the equilibrium position $x_1 = 0$, $x_2 = 0$, by either displacing it slightly, or giving it a small velocity, then it will execute small oscillations about $x_1 = 0$. On the other hand, if we disturb the pendulum slightly from the equilibrium position $x_1 = \pi$, $x_2 = 0$, then it will either execute very large oscillations about $x_1 = 0$, or it will rotate around and around ad infinitum. Thus, the slightest disturbance causes the pendulum to deviate drastically from its equilibrium position $x_1 = \pi$, $x_2 = 0$. Intuitively, we would say that the equilibrium value $x_1 = 0$, $x_2 = 0$ of (6) is stable, while the equilibrium value $x_1 = \pi$, $x_2 = 0$ of (6) is unstable. This concept will be made precise in Section 4.2.

The question of stability is usually very difficult to resolve, because we cannot solve (1) explicitly. The only case which is manageable is when $\mathbf{f}(t, \mathbf{x})$ does not depend explicitly on t; that is, $\mathbf{f}$ is a function of $\mathbf{x}$ alone. Such differential equations are called *autonomous*. And even for autonomous differential equations, there are only two instances, generally, where we can completely resolve the stability question. The first case is when $\mathbf{f}(\mathbf{x}) = \mathbf{A}\mathbf{x}$ and it will be treated in the next section. The second case is when we are only interested in the stability of an equilibrium solution of $\dot{\mathbf{x}} = \mathbf{f}(\mathbf{x})$. This case will be treated in Section 4.3.

Question 3 is extremely important in many applications since an answer to this question is a prediction concerning the long time evolution of the system under consideration. We answer this question, when possible, in Sections 4.6–4.8 and apply our results to some extremely important applications in Sections 4.9–4.12.

EXERCISES

In each of Problems 1–8, find all equilibrium values of the given system of differential equations.

1. $\dfrac{dx}{dt}=x-x^2-2xy$

$\dfrac{dy}{dt}=2y-2y^2-3xy$

2. $\dfrac{dx}{dt}=-\beta xy+\mu$

$\dfrac{dy}{dt}=\beta xy-\gamma y$

3. $\dfrac{dx}{dt}=ax-bxy$

$\dfrac{dy}{dt}=-cy+dxy$

$\dfrac{dz}{dt}=z+x^2+y^2$

4. $\dfrac{dx}{dt}=-x-xy^2$

$\dfrac{dy}{dt}=-y-yx^2$

$\dfrac{dz}{dt}=1-z+x^2$

5. $\dfrac{dx}{dt}=xy^2-x$

$\dfrac{dy}{dt}=x\sin\pi y$

6. $\dfrac{dx}{dt}=\cos y$

$\dfrac{dy}{dt}=\sin x-1$

7. $\dfrac{dx}{dt}=-1-y-e^x$

$\dfrac{dy}{dt}=x^2+y(e^x-1)$

$\dfrac{dz}{dt}=x+\sin z$

8. $\dfrac{dx}{dt}=x-y^2$

$\dfrac{dy}{dt}=x^2-y$

$\dfrac{dz}{dt}=e^z-x$

9. Consider the system of differential equations

$$\frac{dx}{dt}=ax+by,\qquad \frac{dy}{dt}=cx+dy. \tag{*}$$

(i) Show that $x=0, y=0$ is the only equilibrium point of (*) if $ad-bc\neq 0$.
(ii) Show that (*) has a line of equilibrium points if $ad-bc=0$.

10. Let $x=x(t), y=y(t)$ be the solution of the initial-value problem

$$\frac{dx}{dt}=-x-y,\qquad \frac{dy}{dt}=2x-y,\qquad x(0)=y(0)=1.$$

Suppose that we make an error of magnitude 10^{-4} in measuring $x(0)$ and $y(0)$. What is the largest error we make in evaluating $x(t), y(t)$ for $0\leqslant t<\infty$?

11. (a) Verify that

$$\mathbf{x}=\begin{pmatrix}1\\0\\0\end{pmatrix}e^{-t}$$

is the solution of the initial-value problem

$$\dot{\mathbf{x}}=\begin{pmatrix}1&1&1\\2&1&-1\\-3&2&4\end{pmatrix}\mathbf{x}-\begin{pmatrix}2\\2\\-3\end{pmatrix}e^{-t},\qquad \mathbf{x}(0)=\begin{pmatrix}1\\0\\0\end{pmatrix}.$$

(b) Let $\mathbf{x}=\boldsymbol{\psi}(t)$ be the solution of the above differential equation which satisfies the initial condition

$$\mathbf{x}(0)=\mathbf{x}^0\neq\begin{pmatrix}1\\0\\0\end{pmatrix}.$$

Show that each component of $\boldsymbol{\psi}(t)$ approaches infinity, in absolute value, as $t\to\infty$.

4.2 Stability of linear systems

In this section we consider the stability question for solutions of autonomous differential equations. Let $\mathbf{x}=\boldsymbol{\phi}(t)$ be a solution of the differential equation

$$\dot{\mathbf{x}}=\mathbf{f}(\mathbf{x}). \tag{1}$$

We are interested in determining whether $\boldsymbol{\phi}(t)$ is stable or unstable. That is to say, we seek to determine whether every solution $\boldsymbol{\psi}(t)$ of (1) which starts sufficiently close to $\boldsymbol{\phi}(t)$ at $t=0$ must remain close to $\boldsymbol{\phi}(t)$ for all future time $t\geqslant 0$. We begin with the following formal definition of stability.

Definition. The solution $\mathbf{x}=\boldsymbol{\phi}(t)$ of (1) is stable if every solution $\boldsymbol{\psi}(t)$ of (1) which starts sufficiently close to $\boldsymbol{\phi}(t)$ at $t=0$ must remain close to $\boldsymbol{\phi}(t)$ for all future time t. The solution $\boldsymbol{\phi}(t)$ is unstable if there exists at least one solution $\boldsymbol{\psi}(t)$ of (1) which starts near $\boldsymbol{\phi}(t)$ at $t=0$ but which does not remain close to $\boldsymbol{\phi}(t)$ for all future time. More precisely, the solution $\boldsymbol{\phi}(t)$ is stable if for every $\varepsilon>0$ there exists $\delta=\delta(\varepsilon)$ such that

$$|\psi_j(t)-\phi_j(t)|<\varepsilon \quad \text{if} \quad |\psi_j(0)-\phi_j(0)|<\delta(\varepsilon), \qquad j=1,\ldots,n$$

for every solution $\boldsymbol{\psi}(t)$ of (1).

The stability question can be completely resolved for each solution of the linear differential equation

$$\dot{\mathbf{x}}=\mathbf{A}\mathbf{x}. \tag{2}$$

This is not surprising, of course, since we can solve Equation (2) exactly. We have the following important theorem.

Theorem 1. (a) *Every solution* $\mathbf{x}=\boldsymbol{\phi}(t)$ *of* (2) *is stable if all the eigenvalues of* $\mathbf{A}$ *have negative real part.*

(b) *Every solution* $\mathbf{x}=\boldsymbol{\phi}(t)$ *of* (2) *is unstable if at least one eigenvalue of* $\mathbf{A}$ *has positive real part.*

(c) *Suppose that all the eigenvalues of* $\mathbf{A}$ *have real part* $\leqslant 0$ *and* $\lambda_1=i\sigma_1,\ldots,\lambda_l=i\sigma_l$ *have zero real part. Let* $\lambda_j=i\sigma_j$ *have multiplicity* k_j. *This means that the characteristic polynomial of* $\mathbf{A}$ *can be factored into the form*

$$p(\lambda)=(\lambda-i\sigma_1)^{k_1}\ldots(\lambda-i\sigma_l)^{k_l}q(\lambda)$$

where all the roots of $q(\lambda)$ have negative real part. Then, every solution $\mathbf{x}=\boldsymbol{\phi}(t)$ of (1) *is stable if* **A** *has k_j linearly independent eigenvectors for each eigenvalue $\lambda_j = i\sigma_j$. Otherwise, every solution $\boldsymbol{\phi}(t)$ is unstable.*

Our first step in proving Theorem 1 is to show that every solution $\boldsymbol{\phi}(t)$ is stable if the equilibrium solution $\mathbf{x}(t)\equiv\mathbf{0}$ is stable, and every solution $\boldsymbol{\phi}(t)$ is unstable if $\mathbf{x}(t)\equiv\mathbf{0}$ is unstable. To this end, let $\boldsymbol{\psi}(t)$ be any solution of (2). Observe that $\mathbf{z}(t)=\boldsymbol{\phi}(t)-\boldsymbol{\psi}(t)$ is again a solution of (2). Therefore, if the equilibrium solution $\mathbf{x}(t)\equiv\mathbf{0}$ is stable, then $\mathbf{z}(t)=\boldsymbol{\phi}(t)-\boldsymbol{\psi}(t)$ will always remain small if $\mathbf{z}(0)=\boldsymbol{\phi}(0)-\boldsymbol{\psi}(0)$ is sufficiently small. Consequently, every solution $\boldsymbol{\phi}(t)$ of (2) is stable. On the other hand suppose that $\mathbf{x}(t)\equiv\mathbf{0}$ is unstable. Then, there exists a solution $\mathbf{x}=\mathbf{h}(t)$ which is very small initially, but which becomes large as t approaches infinity. The function $\boldsymbol{\psi}(t)=\boldsymbol{\phi}(t)+\mathbf{h}(t)$ is clearly a solution of (2). Moreover, $\boldsymbol{\psi}(t)$ is close to $\boldsymbol{\phi}(t)$ initially, but diverges from $\boldsymbol{\phi}(t)$ as t increases. Therefore, every solution $\mathbf{x}=\boldsymbol{\phi}(t)$ of (2) is unstable.

Our next step in proving Theorem 1 is to reduce the problem of showing that n quantities $\psi_j(t), j=1,\ldots,n$ are small to the much simpler problem of showing that only one quantity is small. This is accomplished by introducing the concept of length, or magnitude, of a vector.

Definition. Let

$$\mathbf{x}=\begin{pmatrix} x_1 \\ \vdots \\ x_n \end{pmatrix}$$

be a vector with n components. The numbers $x_1,\ldots,x_n$ may be real or complex. We define the length of $\mathbf{x}$, denoted by $\|\mathbf{x}\|$ as

$$\|\mathbf{x}\|\equiv\max\{|x_1|,|x_2|,\ldots,|x_n|\}.$$

For example, if

$$\mathbf{x}=\begin{pmatrix} 1 \\ 2 \\ -3 \end{pmatrix},$$

then $\|\mathbf{x}\|=3$ and if

$$\mathbf{x}=\begin{pmatrix} 1+2i \\ 2 \\ -1 \end{pmatrix}$$

then $\|\mathbf{x}\|=\sqrt{5}$.

The concept of the length, or magnitude of a vector corresponds to the concept of the length, or magnitude of a number. Observe that $\|\mathbf{x}\|\geqslant 0$ for any vector $\mathbf{x}$ and $\|\mathbf{x}\|=0$ only if $\mathbf{x}=\mathbf{0}$. Second, observe that

$$\|\lambda\mathbf{x}\|=\max\{|\lambda x_1|,\ldots,|\lambda x_n|\}=|\lambda|\max\{|x_1|,\ldots,|x_n|\}=|\lambda|\|\mathbf{x}\|.$$

Finally, observe that

$$\begin{aligned}\|\mathbf{x}+\mathbf{y}\|&=\max\{|x_1+y_1|,\ldots,|x_n+y_n|\}\leqslant\max\{|x_1|+|y_1|,\ldots,|x_n|+|y_n|\}\\&\leqslant\max\{|x_1|,\ldots,|x_n|\}+\max\{|y_1|,\ldots,|y_n|\}=\|\mathbf{x}\|+\|\mathbf{y}\|.\end{aligned}$$

Thus, our definition really captures the meaning of length.

In Section 4.7 we give a simple geometric proof of Theorem 1 for the case $n=2$. The following proof is valid for arbitrary n.

PROOF OF THEOREM 1. (a) Every solution $\mathbf{x}=\boldsymbol{\psi}(t)$ of $\dot{\mathbf{x}}=\mathbf{A}\mathbf{x}$ is of the form $\boldsymbol{\psi}(t)=e^{\mathbf{A}t}\boldsymbol{\psi}(0)$. Let $\phi_{ij}(t)$ be the ij element of the matrix $e^{\mathbf{A}t}$, and let $\psi_1^0,\ldots,\psi_n^0$ be the components of $\boldsymbol{\psi}(0)$. Then, the ith component of $\boldsymbol{\psi}(t)$ is

$$\psi_i(t)=\phi_{i1}(t)\psi_1^0+\ldots+\phi_{in}(t)\psi_n^0\equiv\sum_{j=1}^{n}\phi_{ij}(t)\psi_j^0.$$

Suppose that all the eigenvalues of $\mathbf{A}$ have negative real part. Let $-\alpha_1$ be the largest of the real parts of the eigenvalues of $\mathbf{A}$. It is a simple matter to show (see Exercise 17) that for every number $-\alpha$, with $-\alpha_1<-\alpha<0$, we can find a number K such that $|\phi_{ij}(t)|\leqslant Ke^{-\alpha t}$, $t\geqslant 0$. Consequently,

$$|\psi_i(t)|\leqslant\sum_{j=1}^{n}Ke^{-\alpha t}|\psi_j^0|=Ke^{-\alpha t}\sum_{j=1}^{n}|\psi_j^0|$$

for some positive constants K and α. Now, $|\psi_j^0|\leqslant\|\boldsymbol{\psi}(0)\|$. Hence,

$$\|\boldsymbol{\psi}(t)\|=\max\{|\psi_1(t)|,\ldots,|\psi_n(t)|\}\leqslant nKe^{-\alpha t}\|\boldsymbol{\psi}(0)\|.$$

Let $\varepsilon>0$ be given. Choose $\delta(\varepsilon)=\varepsilon/nK$. Then, $\|\boldsymbol{\psi}(t)\|<\varepsilon$ if $\|\boldsymbol{\psi}(0)\|<\delta(\varepsilon)$ and $t\geqslant 0$, since

$$\|\boldsymbol{\psi}(t)\|\leqslant nKe^{-\alpha t}\|\boldsymbol{\psi}(0)\|<nK\varepsilon/nK=\varepsilon.$$

Consequently, the equilibrium solution $\mathbf{x}(t)\equiv\mathbf{0}$ is stable.

(b) Let λ be an eigenvalue of $\mathbf{A}$ with positive real part and let $\mathbf{v}$ be an eigenvector of $\mathbf{A}$ with eigenvalue λ. Then, $\boldsymbol{\psi}(t)=ce^{\lambda t}\mathbf{v}$ is a solution of $\dot{\mathbf{x}}=\mathbf{A}\mathbf{x}$ for any constant c. If λ is real then $\mathbf{v}$ is also real and $\|\boldsymbol{\psi}(t)\|=|c|e^{\lambda t}\|\mathbf{v}\|$. Clearly, $\|\boldsymbol{\psi}(t)\|$ approaches infinity as t approaches infinity, for any choice of $c\neq 0$, no matter how small. Therefore, $\mathbf{x}(t)\equiv\mathbf{0}$ is unstable. If $\lambda=\alpha+i\beta$ is complex, then $\mathbf{v}=\mathbf{v}^1+i\mathbf{v}^2$ is also complex. In this case

$$\begin{aligned}e^{(\alpha+i\beta)t}(\mathbf{v}^1+i\mathbf{v}^2)&=e^{\alpha t}(\cos\beta t+i\sin\beta t)(\mathbf{v}^1+i\mathbf{v}^2)\\&=e^{\alpha t}\left[(\mathbf{v}^1\cos\beta t-\mathbf{v}^2\sin\beta t)+i(\mathbf{v}^1\sin\beta t+\mathbf{v}^2\cos\beta t)\right]\end{aligned}$$

is a complex-valued solution of (2). Therefore

$$\boldsymbol{\psi}^1(t)=ce^{\alpha t}(\mathbf{v}^1\cos\beta t-\mathbf{v}^2\sin\beta t)$$

is a real-valued solution of (2), for any choice of constant c. Clearly,

$\|\boldsymbol{\psi}^1(t)\|$ is unbounded as t approaches infinity if c and either $\mathbf{v}^1$ or $\mathbf{v}^2$ is nonzero. Thus, $\mathbf{x}(t)\equiv\mathbf{0}$ is unstable.

(c) If $\mathbf{A}$ has k_j linearly independent eigenvectors for each eigenvalue $\lambda_j = i\sigma_j$ of multiplicity k_j, then we can find a constant K such that $|(e^{\mathbf{A}t})_{ij}| \leqslant K$ (see Exercise 18). There, $\|\boldsymbol{\psi}(t)\| \leqslant nK\|\boldsymbol{\psi}(0)\|$ for every solution $\boldsymbol{\psi}(t)$ of (2). It now follows immediately from the proof of (a) that $\mathbf{x}(t)\equiv\mathbf{0}$ is stable.

On the other hand, if $\mathbf{A}$ has fewer than k_j linearly independent eigenvectors with eigenvalue $\lambda_j = i\sigma_j$, then $\dot{\mathbf{x}} = \mathbf{A}\mathbf{x}$ has solutions $\boldsymbol{\psi}(t)$ of the form

$$\boldsymbol{\psi}(t) = ce^{i\sigma_j t}\left[\mathbf{v} + t(\mathbf{A} - i\sigma_j\mathbf{I})\mathbf{v}\right]$$

where $(\mathbf{A} - i\sigma_j\mathbf{I})\mathbf{v} \neq \mathbf{0}$. If $\sigma_j = 0$, then $\boldsymbol{\psi}(t) = c(\mathbf{v} + t\mathbf{A}\mathbf{v})$ is real-valued. Moreover, $\|\boldsymbol{\psi}(t)\|$ is unbounded as t approaches infinity for any choice of $c \neq 0$. Similarly, both the real and imaginary parts of $\boldsymbol{\psi}(t)$ are unbounded in magnitude for arbitrarily small $\boldsymbol{\psi}(0) \neq \mathbf{0}$, if $\sigma_j \neq 0$. Therefore, the equilibrium solution $\mathbf{x}(t)\equiv\mathbf{0}$ is unstable. □

If all the eigenvalues of $\mathbf{A}$ have negative real part, then every solution $\mathbf{x}(t)$ of $\dot{\mathbf{x}} = \mathbf{A}\mathbf{x}$ approaches zero as t approaches infinity. This follows immediately from the estimate $\|\mathbf{x}(t)\| \leqslant Ke^{-\alpha t}\|\mathbf{x}(0)\|$ which we derived in the proof of part (a) of Theorem 1. Thus, not only is the equilibrium solution $\mathbf{x}(t)\equiv\mathbf{0}$ stable, but every solution $\boldsymbol{\psi}(t)$ of (2) approaches it as t approaches infinity. This very strong type of stability is known as *asymptotic stability*.

Definition. A solution $\mathbf{x} = \boldsymbol{\phi}(t)$ of (1) is asymptotically stable if it is stable, and if every solution $\boldsymbol{\psi}(t)$ which starts sufficiently close to $\boldsymbol{\phi}(t)$ must approach $\boldsymbol{\phi}(t)$ as t approaches infinity. In particular, an equilibrium solution $\mathbf{x}(t) = \mathbf{x}^0$ of (1) is asymptotically stable if every solution $\mathbf{x} = \boldsymbol{\psi}(t)$ of (1) which starts sufficiently close to $\mathbf{x}^0$ at time $t = 0$ not only remains close to $\mathbf{x}^0$ for all future time, but ultimately approaches $\mathbf{x}^0$ as t approaches infinity.

Remark. The asymptotic stability of any solution $\mathbf{x} = \boldsymbol{\phi}(t)$ of (2) is clearly equivalent to the asymptotic stability of the equilibrium solution $\mathbf{x}(t)\equiv\mathbf{0}$.

Example 1. Determine whether each solution $\mathbf{x}(t)$ of the differential equation

$$\dot{\mathbf{x}} = \begin{pmatrix} -1 & 0 & 0 \\ -2 & -1 & 2 \\ -3 & -2 & -1 \end{pmatrix}\mathbf{x}$$

is stable, asymptotically stable, or unstable.

Solution. The characteristic polynomial of the matrix

$$\mathbf{A} = \begin{pmatrix} -1 & 0 & 0 \\ -2 & -1 & 2 \\ -3 & -2 & -1 \end{pmatrix}$$

is

$$p(\lambda)=(\det \mathbf{A}-\lambda\mathbf{I})=\det\begin{bmatrix} -1-\lambda & 0 & 0 \\ -2 & -1-\lambda & 2 \\ -3 & -2 & -1-\lambda \end{bmatrix}$$

$$=-(1+\lambda)^3-4(1+\lambda)=-(1+\lambda)(\lambda^2+2\lambda+5).$$

Hence, $\lambda=-1$ and $\lambda=-1\pm 2i$ are the eigenvalues of $\mathbf{A}$. Since all three eigenvalues have negative real part, we conclude that every solution of the differential equation $\dot{\mathbf{x}}=\mathbf{A}\mathbf{x}$ is asymptotically stable.

Example 2. Prove that every solution of the differential equation

$$\dot{\mathbf{x}}=\begin{pmatrix} 1 & 5 \\ 5 & 1 \end{pmatrix}\mathbf{x}$$

is unstable.

Solution. The characteristic polynomial of the matrix

$$\mathbf{A}=\begin{pmatrix} 1 & 5 \\ 5 & 1 \end{pmatrix}$$

is

$$p(\lambda)=\det(\mathbf{A}-\lambda\mathbf{I})=\det\begin{pmatrix} 1-\lambda & 5 \\ 5 & 1-\lambda \end{pmatrix}=(1-\lambda)^2-25.$$

Hence $\lambda=6$ and $\lambda=-4$ are the eigenvalues of $\mathbf{A}$. Since one eigenvalue of $\mathbf{A}$ is positive, we conclude that every solution $\mathbf{x}=\boldsymbol{\phi}(t)$ of $\dot{\mathbf{x}}=\mathbf{A}\mathbf{x}$ is unstable.

Example 3. Show that every solution of the differential equation

$$\dot{\mathbf{x}}=\begin{pmatrix} 0 & -3 \\ 2 & 0 \end{pmatrix}\mathbf{x}$$

is stable, but not asymptotically stable.

Solution. The characteristic polynomial of the matrix

$$\mathbf{A}=\begin{pmatrix} 0 & -3 \\ 2 & 0 \end{pmatrix}$$

is

$$p(\lambda)=\det(\mathbf{A}-\lambda\mathbf{I})=\det\begin{pmatrix} -\lambda & -3 \\ 2 & -\lambda \end{pmatrix}=\lambda^2+6.$$

Thus, the eigenvalues of $\mathbf{A}$ are $\lambda=\pm\sqrt{6}\,i$. Therefore, by part (c) of Theorem 1, every solution $\mathbf{x}=\boldsymbol{\phi}(t)$ of $\dot{\mathbf{x}}=\mathbf{A}\mathbf{x}$ is stable. However, no solution is asymptotically stable. This follows immediately from the fact that the general solution of $\dot{\mathbf{x}}=\mathbf{A}\mathbf{x}$ is

$$\mathbf{x}(t)=c_1\begin{pmatrix} -\sqrt{6}\,\sin\sqrt{6}\,t \\ 2\cos\sqrt{6}\,t \end{pmatrix}+c_2\begin{pmatrix} \sqrt{6}\,\cos\sqrt{6}\,t \\ 2\sin\sqrt{6}\,t \end{pmatrix}.$$

Hence, every solution $\mathbf{x}(t)$ is periodic, with period $2\pi/\sqrt{6}$, and no solution $\mathbf{x}(t)$ (except $\mathbf{x}(t)\equiv\mathbf{0}$) approaches 0 as t approaches infinity.

Example 4. Show that every solution of the differential equation

$$\dot{\mathbf{x}}=\begin{pmatrix} 2 & -3 & 0 \\ 0 & -6 & -2 \\ -6 & 0 & -3 \end{pmatrix}\mathbf{x}$$

is unstable.

Solution. The characteristic polynomial of the matrix

$$\mathbf{A}=\begin{pmatrix} 2 & -3 & 0 \\ 0 & -6 & -2 \\ -6 & 0 & -3 \end{pmatrix}$$

is

$$p(\lambda)=\det(\mathbf{A}-\lambda\mathbf{I})=\det\begin{pmatrix} 2-\lambda & -3 & 0 \\ 0 & -6-\lambda & -2 \\ -6 & 0 & -3-\lambda \end{pmatrix}=-\lambda^2(\lambda+7).$$

Hence, the eigenvalues of $\mathbf{A}$ are $\lambda=-7$ and $\lambda=0$. Every eigenvector $\mathbf{v}$ of $\mathbf{A}$ with eigenvalue 0 must satisfy the equation

$$\mathbf{A}\mathbf{v}=\begin{pmatrix} 2 & -3 & 0 \\ 0 & -6 & -2 \\ -6 & 0 & -3 \end{pmatrix}\begin{pmatrix} v_1 \\ v_2 \\ v_3 \end{pmatrix}=\begin{pmatrix} 0 \\ 0 \\ 0 \end{pmatrix}.$$

This implies that $v_1=3v_2/2$ and $v_3=-3v_2$, so that every eigenvector $\mathbf{v}$ of $\mathbf{A}$ with eigenvalue 0 must be of the form

$$\mathbf{v}=c\begin{pmatrix} 3 \\ 2 \\ -6 \end{pmatrix}.$$

Consequently, every solution $\mathbf{x}=\boldsymbol{\phi}(t)$ of $\dot{\mathbf{x}}=\mathbf{A}\mathbf{x}$ is unstable, since $\lambda=0$ is an eigenvalue of multiplicity two and $\mathbf{A}$ has only one linearly independent eigenvector with eigenvalue 0.

Exercises

Determine the stability or instability of all solutions of the following systems of differential equations.

1. $\dot{\mathbf{x}}=\begin{pmatrix} 1 & 1 \\ -2 & -2 \end{pmatrix}\mathbf{x}$

2. $\dot{\mathbf{x}}=\begin{pmatrix} -3 & -4 \\ 2 & 1 \end{pmatrix}\mathbf{x}$

3. $\dot{\mathbf{x}}=\begin{pmatrix} -5 & 3 \\ -1 & 1 \end{pmatrix}\mathbf{x}$

4. $\dot{\mathbf{x}}=\begin{pmatrix} 1 & -4 \\ 4 & -7 \end{pmatrix}\mathbf{x}$

5. $\dot{\mathbf{x}}=\begin{pmatrix} -7 & 1 & -6 \\ 10 & -4 & 12 \\ 2 & -1 & 1 \end{pmatrix}\mathbf{x}$

6. $\dot{\mathbf{x}}=\begin{pmatrix} 3 & 2 & 4 \\ 2 & 0 & 2 \\ 4 & 2 & 3 \end{pmatrix}\mathbf{x}$

7. $\dot{\mathbf{x}}=\begin{pmatrix} 0 & 2 & 1 \\ -1 & -3 & -1 \\ 1 & 1 & -1 \end{pmatrix}\mathbf{x}$ 8. $\dot{\mathbf{x}}=\begin{pmatrix} -2 & 1 & 1 \\ -3 & 2 & 3 \\ 1 & -1 & -2 \end{pmatrix}\mathbf{x}$

9. $\dot{\mathbf{x}}=\begin{bmatrix} 0 & 2 & 0 & 0 \\ -2 & 0 & 0 & 0 \\ 0 & 0 & 0 & 2 \\ 0 & 0 & -2 & 0 \end{bmatrix}\mathbf{x}$ 10. $\dot{\mathbf{x}}=\begin{bmatrix} 0 & 2 & 1 & 0 \\ -2 & 0 & 0 & 1 \\ 0 & 0 & 0 & 2 \\ 0 & 0 & -2 & 0 \end{bmatrix}\mathbf{x}$

11. Determine whether the solutions $x(t)\equiv 0$ and $x(t)\equiv 1$ of the single scalar equation $\dot{x}=x(1-x)$ are stable or unstable.

12. Determine whether the solutions $x(t)\equiv 0$ and $x(t)\equiv 1$ of the single scalar equation $\dot{x}=-x(1-x)$ are stable or unstable.

13. Consider the differential equation $\dot{x}=x^2$. Show that all solutions $x(t)$ with $x(0)\geqslant 0$ are unstable while all solutions $x(t)$ with $x(0)<0$ are asymptotically stable.

14. Consider the system of differential equations

$$\dot{x}_1=x_2$$
$$\dot{x}_2=-\tfrac{1}{2}\left[x_1^2+\left(x_1^2+4x_2^2\right)^{1/2}\right]x_1. \qquad (*)$$

(a) Show that

$$\mathbf{x}(t)=\begin{pmatrix} x_1(t) \\ x_2(t) \end{pmatrix}=\begin{pmatrix} c\sin(ct+d) \\ c^2\cos(ct+d) \end{pmatrix}$$

is a solution of (*) for any choice of constants c and d.

(b) Assume that a solution $\mathbf{x}(t)$ of (*) is uniquely determined once $x_1(0)$ and $x_2(0)$ are prescribed. Prove that (a) represents the general solution of (*).

(c) Show that the solution $\mathbf{x}=\mathbf{0}$ of (*) is stable, but not asymptotically stable.

(d) Show that every solution $\mathbf{x}(t)\not\equiv\mathbf{0}$ of (*) is unstable.

15. Show that the stability of any solution $\mathbf{x}(t)$ of the nonhomogeneous equation $\dot{\mathbf{x}}=\mathbf{A}\mathbf{x}+\mathbf{f}(t)$ is equivalent to the stability of the equilibrium solution $\mathbf{x}\equiv\mathbf{0}$ of the homogeneous equation $\dot{\mathbf{x}}=\mathbf{A}\mathbf{x}$.

16. Determine the stability or instability of all solutions $\mathbf{x}(t)$ of the differential equation

$$\dot{\mathbf{x}}=\begin{pmatrix} -1 & -1 \\ 2 & -1 \end{pmatrix}\mathbf{x}+\begin{pmatrix} 1 \\ 5 \end{pmatrix}.$$

17. (a) Let $f(t)=t^a e^{-bt}$, for some positive constants a and b, and let c be a positive number smaller than b. Show that we can find a positive constant K such that $|f(t)|\leqslant Ke^{-ct}$, $0\leqslant t<\infty$. *Hint*: Show that $f(t)/e^{-ct}$ approaches zero as t approaches infinity.

(b) Suppose that all the eigenvalues of $\mathbf{A}$ have negative real part. Show that we can find positive constants K and α such that $|(e^{\mathbf{A}t})_{ij}|\leqslant Ke^{-\alpha t}$ for $1\leqslant i$, $j\leqslant n$. *Hint*: Each component of $e^{\mathbf{A}t}$ is a finite linear combination of functions of the form $q(t)e^{\lambda t}$, where $q(t)$ is a polynomial in t (of degree $\leqslant n-1$) and λ is an eigenvalue of $\mathbf{A}$.

18. (a) Let $\mathbf{x}(t)=e^{i\sigma t}\mathbf{v}$, σ real, be a complex-valued solution of $\dot{\mathbf{x}}=\mathbf{A}\mathbf{x}$. Show that both the real and imaginary parts of $\mathbf{x}(t)$ are bounded solutions of $\dot{\mathbf{x}}=\mathbf{A}\mathbf{x}$.
(b) Suppose that all the eigenvalues of $\mathbf{A}$ have real part $\leqslant 0$ and $\lambda_1=i\sigma_1,\ldots,\lambda_l=i\sigma_l$ have zero real part. Let $\lambda_j=i\sigma_j$ have multiplicity k_j, and suppose that $\mathbf{A}$ has k_j linearly independent eigenvectors for each eigenvalue λ_j, $j=1,\ldots,l$. Prove that we can find a constant K such that $|(e^{\mathbf{A}t})_{ij}|\leqslant K$.

19. Let

$$\mathbf{x}=\begin{pmatrix} x_1 \\ \vdots \\ x_n \end{pmatrix}$$

and define $\|x\|_1=|x_1|+\ldots+|x_n|$. Show that
(i) $\|\mathbf{x}\|_1\geqslant 0$ and $\|\mathbf{x}\|_1=0$ only if $\mathbf{x}=\mathbf{0}$
(ii) $\|\lambda\mathbf{x}\|_1=|\lambda|\|\mathbf{x}\|_1$
(iii) $\|\mathbf{x}+\mathbf{y}\|_1\leqslant\|\mathbf{x}\|_1+\|\mathbf{y}\|_1$

20. Let

$$\mathbf{x}=\begin{pmatrix} x_1 \\ \vdots \\ x_n \end{pmatrix}$$

and define $\|\mathbf{x}\|_2=[|x_1|^2+\ldots+|x_n|^2]^{1/2}$. Show that
(i) $\|\mathbf{x}\|_2\geqslant 0$ and $\|\mathbf{x}\|_2=0$ only if $\mathbf{x}=\mathbf{0}$
(ii) $\|\lambda\mathbf{x}\|_2=|\lambda|\|\mathbf{x}\|_2$
(iii) $\|\mathbf{x}+\mathbf{y}\|_2\leqslant\|\mathbf{x}\|_2+\|\mathbf{y}\|_2$

21. Show that there exist constants M and N such that

$$M\|\mathbf{x}\|_1\leqslant\|\mathbf{x}\|_2\leqslant N\|\mathbf{x}\|_1.$$

4.3 Stability of equilibrium solutions

In Section 4.2 we treated the simple equation $\dot{\mathbf{x}}=\mathbf{A}\mathbf{x}$. The next simplest equation is

$$\dot{\mathbf{x}}=\mathbf{A}\mathbf{x}+\mathbf{g}(\mathbf{x}) \tag{1}$$

where

$$\mathbf{g}(\mathbf{x})=\begin{pmatrix} g_1(\mathbf{x}) \\ \vdots \\ g_n(\mathbf{x}) \end{pmatrix}$$

is very small compared to $\mathbf{x}$. Specifically we assume that

$$\frac{g_1(\mathbf{x})}{\max\{|x_1|,\ldots,|x_n|\}},\ldots,\frac{g_n(\mathbf{x})}{\max\{|x_1|,\ldots,|x_n|\}}$$

are continuous functions of $x_1,\ldots,x_n$ which vanish for $x_1=\ldots=x_n=0$. This is always the case if each component of $\mathbf{g}(\mathbf{x})$ is a polynomial in $x_1,\ldots,x_n$ which begins with terms of order 2 or higher. For example, if

$$\mathbf{g}(\mathbf{x})=\begin{pmatrix} x_1x_2^2 \\ x_1x_2 \end{pmatrix},$$

then both $x_1x_2^2/\max\{|x_1|,|x_2|\}$ and $x_1x_2/\max\{|x_1|,|x_2|\}$ are continuous functions of x_1,x_2 which vanish for $x_1=x_2=0$.

If $\mathbf{g}(\mathbf{0})=\mathbf{0}$ then $\mathbf{x}(t)\equiv\mathbf{0}$ is an equilibrium solution of (1). We would like to determine whether it is stable or unstable. At first glance this would seem impossible to do, since we cannot solve Equation (1) explicitly. However, if $\mathbf{x}$ is very small, then $\mathbf{g}(\mathbf{x})$ is very small compared to $\mathbf{Ax}$. Therefore, it seems plausible that the stability of the equilibrium solution $\mathbf{x}(t)\equiv\mathbf{0}$ of (1) should be determined by the stability of the "approximate" equation $\dot{\mathbf{x}}=\mathbf{Ax}$. This is almost the case as the following theorem indicates.

Theorem 2. *Suppose that the vector-valued function*

$$\mathbf{g}(\mathbf{x})/\|\mathbf{x}\|\equiv\mathbf{g}(\mathbf{x})/\max\{|x_1|,\ldots,|x_n|\}$$

is a continuous function of $x_1,\ldots,x_n$ which vanishes for $\mathbf{x}=\mathbf{0}$. Then,
(a) The equilibrium solution $\mathbf{x}(t)\equiv\mathbf{0}$ of (1) is asymptotically stable if the equilibrium solution $\mathbf{x}(t)\equiv\mathbf{0}$ of the "linearized" equation $\dot{\mathbf{x}}=\mathbf{Ax}$ is asymptotically stable. Equivalently, the solution $\mathbf{x}(t)\equiv\mathbf{0}$ of (1) is asymptotically stable if all the eigenvalues of $\mathbf{A}$ have negative real part.
(b) The equilibrium solution $\mathbf{x}(t)\equiv\mathbf{0}$ of (1) is unstable if at least one eigenvalue of $\mathbf{A}$ has positive real part.
(c) The stability of the equilibrium solution $\mathbf{x}(t)\equiv\mathbf{0}$ of (1) cannot be determined from the stability of the equilibrium solution $\mathbf{x}(t)\equiv\mathbf{0}$ of $\dot{\mathbf{x}}=\mathbf{Ax}$ if all the eigenvalues of $\mathbf{A}$ have real part $\leqslant 0$ but at least one eigenvalue of $\mathbf{A}$ has zero real part.

PROOF. (a) The key step in many stability proofs is to use the variation of parameters formula of Section 3.12. This formula implies that any solution $\mathbf{x}(t)$ of (1) can be written in the form

$$\mathbf{x}(t)=e^{\mathbf{A}t}\mathbf{x}(0)+\int_0^t e^{\mathbf{A}(t-s)}\mathbf{g}(\mathbf{x}(s))\,ds. \tag{2}$$

We wish to show that $\|\mathbf{x}(t)\|$ approaches zero as t approaches infinity. To this end recall that if all the eigenvalues of $\mathbf{A}$ have negative real part, then we can find positive constants K and α such that (see Exercise 17, Section 4.2).

$$\|e^{\mathbf{A}t}\mathbf{x}(0)\|\leqslant Ke^{-\alpha t}\|\mathbf{x}(0)\|$$

and

$$\|e^{\mathbf{A}(t-s)}\mathbf{g}(\mathbf{x}(s))\|\leqslant Ke^{-\alpha(t-s)}\|\mathbf{g}(\mathbf{x}(s))\|.$$

Moreover, we can find a positive constant σ such that

$$\|\mathbf{g}(\mathbf{x})\| \leqslant \frac{\alpha}{2K}\|\mathbf{x}\| \quad \text{if} \quad \|\mathbf{x}\| \leqslant \sigma.$$

This follows immediately from our assumption that $g(\mathbf{x})/\|\mathbf{x}\|$ is continuous and vanishes at $\mathbf{x}=\mathbf{0}$. Consequently, Equation (2) implies that

$$\begin{aligned}\|\mathbf{x}(t)\| &\leqslant \|e^{\mathbf{A}t}\mathbf{x}(0)\| + \int_0^t \|e^{\mathbf{A}(t-s)}\mathbf{g}(\mathbf{x}(s))\|\,ds \\ &\leqslant Ke^{-\alpha t}\|\mathbf{x}(0)\| + \frac{\alpha}{2}\int_0^t e^{-\alpha(t-s)}\|\mathbf{x}(s)\|\,ds\end{aligned}$$

as long as $\|\mathbf{x}(s)\| \leqslant \sigma$, $0 \leqslant s \leqslant t$. Multiplying both sides of this inequality by $e^{\alpha t}$ gives

$$e^{\alpha t}\|\mathbf{x}(t)\| \leqslant K\|\mathbf{x}(0)\| + \frac{\alpha}{2}\int_0^t e^{\alpha s}\|\mathbf{x}(s)\|\,ds. \tag{3}$$

The inequality (3) can be simplified by setting $z(t)=e^{\alpha t}\|\mathbf{x}(t)\|$, for then

$$z(t) \leqslant K\|\mathbf{x}(0)\| + \frac{\alpha}{2}\int_0^t z(s)\,ds. \tag{4}$$

We would like to differentiate both sides of (4) with respect to t. However, we cannot, in general, differentiate both sides of an inequality and still preserve the sense of the inequality. We circumvent this difficulty by the clever trick of setting

$$U(t) = \frac{\alpha}{2}\int_0^t z(s)\,ds.$$

Then

$$\frac{dU(t)}{dt} = \frac{\alpha}{2}z(t) \leqslant \frac{\alpha}{2}K\|\mathbf{x}(0)\| + \frac{\alpha}{2}U(t)$$

or

$$\frac{dU(t)}{dt} - \frac{\alpha}{2}U(t) \leqslant \frac{\alpha K}{2}\|\mathbf{x}(0)\|.$$

Multiplying both sides of this inequality by the integrating factor $e^{-\alpha t/2}$ gives

$$\frac{d}{dt}e^{-\alpha t/2}U \leqslant \frac{\alpha K}{2}\|\mathbf{x}(0)\|e^{-\alpha t/2},$$

or

$$\frac{d}{dt}e^{-\alpha t/2}\left[U(t) + K\|\mathbf{x}(0)\|\right] \leqslant 0.$$

Consequently,

$$e^{-\alpha t/2}\left[U(t) + K\|\mathbf{x}(0)\|\right] \leqslant U(0) + K\|\mathbf{x}(0)\| = K\|\mathbf{x}(0)\|,$$

so that $U(t) \leqslant -K\|\mathbf{x}(0)\| + K\|\mathbf{x}(0)\|e^{\alpha t/2}$. Returning to the inequality (4), we see that

$$\|\mathbf{x}(t)\| = e^{-\alpha t}z(t) \leqslant e^{-\alpha t}\left[K\|\mathbf{x}(0)\| + U(t)\right]$$
$$\leqslant K\|\mathbf{x}(0)\|e^{-\alpha t/2} \tag{5}$$

as long as $\|\mathbf{x}(s)\| \leqslant \sigma$, $0 \leqslant s \leqslant t$. Now, if $\|\mathbf{x}(0)\| \leqslant \sigma/K$, then the inequality (5) guarantees that $\|\mathbf{x}(t)\| \leqslant \sigma$ for all future time t. Consequently, the inequality (5) is true for all $t \geqslant 0$ if $\|\mathbf{x}(0)\| \leqslant \sigma/K$. Finally, observe from (5) that $\|\mathbf{x}(t)\| \leqslant K\|\mathbf{x}(0)\|$ and $\|\mathbf{x}(t)\|$ approaches zero as t approaches infinity. Therefore, the equilibrium solution $\mathbf{x}(t) \equiv \mathbf{0}$ of (1) is asymptotically stable.
(b) The proof of (b) is too difficult to present here.
(c) We will present two differential equations of the form (1) where the nonlinear term $\mathbf{g}(\mathbf{x})$ determines the stability of the equilibrium solution $\mathbf{x}(t) \equiv \mathbf{0}$. Consider first the system of differential equations

$$\frac{dx_1}{dt} = x_2 - x_1(x_1^2 + x_2^2), \qquad \frac{dx_2}{dt} = -x_1 - x_2(x_1^2 + x_2^2). \tag{6}$$

The linearized equation is

$$\frac{d}{dt}\begin{pmatrix} x_1 \\ x_2 \end{pmatrix} = \begin{pmatrix} 0 & 1 \\ -1 & 0 \end{pmatrix}\begin{pmatrix} x_1 \\ x_2 \end{pmatrix},$$

and the eigenvalues of the matrix

$$\begin{pmatrix} 0 & 1 \\ -1 & 0 \end{pmatrix}$$

are $\pm i$. To analyze the behavior of the nonlinear system (6) we multiply the first equation by x_1, the second equation by x_2 and add; this gives

$$x_1\frac{dx_1}{dt} + x_2\frac{dx_2}{dt} = -x_1^2(x_1^2 + x_2^2) - x_2^2(x_1^2 + x_2^2)$$
$$= -(x_1^2 + x_2^2)^2.$$

But

$$x_1\frac{dx_1}{dt} + x_2\frac{dx_2}{dt} = \frac{1}{2}\frac{d}{dt}(x_1^2 + x_2^2).$$

Hence,

$$\frac{d}{dt}(x_1^2 + x_2^2) = -2(x_1^2 + x_2^2)^2.$$

This implies that

$$x_1^2(t) + x_2^2(t) = \frac{c}{1 + 2ct},$$

where

$$c = x_1^2(0) + x_2^2(0).$$

Thus, $x_1^2(t)+x_2^2(t)$ approaches zero as t approaches infinity for any solution $x_1(t)$, $x_2(t)$ of (6). Moreover, the value of $x_1^2+x_2^2$ at any time t is always less than its value at $t=0$. We conclude, therefore, that $x_1(t)\equiv 0$, $x_2(t)\equiv 0$ is asymptotically stable.

On the other hand, consider the system of equations

$$\frac{dx_1}{dt}=x_2+x_1(x_1^2+x_2^2), \qquad \frac{dx_2}{dt}=-x_1-x_2(x_1^2+x_2^2). \tag{7}$$

Here too, the linearized system is

$$\dot{\mathbf{x}}=\begin{pmatrix} 0 & 1 \\ -1 & 0 \end{pmatrix}\mathbf{x}.$$

In this case, though, $(d/dt)(x_1^2+x_2^2)=2(x_1^2+x_2^2)^2$. This implies that

$$x_1^2(t)+x_2^2(t)=\frac{c}{1-2ct}, \qquad c=x_1^2(0)+x_2^2(0).$$

Notice that every solution $x_1(t)$, $x_2(t)$ of (7) with $x_1^2(0)+x_2^2(0)\neq 0$ approaches infinity in finite time. We conclude, therefore, that the equilibrium solution $x_1(t)\equiv 0$, $x_2(t)\equiv 0$ is unstable. □

Example 1. Consider the system of differential equations

$$\begin{aligned}
\frac{dx_1}{dt}&=-2x_1+x_2+3x_3+9x_2^3\\
\frac{dx_2}{dt}&=-6x_2-5x_3+7x_3^5\\
\frac{dx_3}{dt}&=-x_3+x_1^2+x_2^2.
\end{aligned}$$

Determine, if possible, whether the equilibrium solution $x_1(t)\equiv 0$, $x_2(t)\equiv 0$, $x_3(t)\equiv 0$ is stable or unstable.

Solution. We rewrite this system in the form $\dot{\mathbf{x}}=\mathbf{A}\mathbf{x}+\mathbf{g}(\mathbf{x})$ where

$$\mathbf{x}=\begin{bmatrix} x_1\\ x_2\\ x_3\end{bmatrix}, \quad \mathbf{A}=\begin{bmatrix} -2 & 1 & 3\\ 0 & -6 & -5\\ 0 & 0 & -1\end{bmatrix} \quad\text{and}\quad \mathbf{g}(\mathbf{x})=\begin{bmatrix} 9x_2^3\\ 7x_3^5\\ x_1^2+x_2^2\end{bmatrix}.$$

The function $\mathbf{g}(\mathbf{x})$ satisfies the hypotheses of Theorem 2, and the eigenvalues of $\mathbf{A}$ are -2, -6 and -1. Hence, the equilibrium solution $\mathbf{x}(t)\equiv\mathbf{0}$ is asymptotically stable.

Theorem 2 can also be used to determine the stability of equilibrium solutions of arbitrary autonomous differential equations. Let $\mathbf{x}^0$ be an equilibrium value of the differential equation

$$\dot{\mathbf{x}}=\mathbf{f}(\mathbf{x}) \tag{8}$$

and set $\mathbf{z}(t)=\mathbf{x}(t)-\mathbf{x}^0$. Then

$$\dot{\mathbf{z}}=\dot{\mathbf{x}}=\mathbf{f}(\mathbf{x}^0+\mathbf{z}). \tag{9}$$

Clearly, $\mathbf{z}(t)\equiv\mathbf{0}$ is an equilibrium solution of (9) and the stability of $\mathbf{x}(t)\equiv\mathbf{x}^0$ is equivalent to the stability of $\mathbf{z}(t)\equiv\mathbf{0}$.

Next, we show that $\mathbf{f}(\mathbf{x}^0+\mathbf{z})$ can be written in the form $\mathbf{f}(\mathbf{x}^0+\mathbf{z})=\mathbf{A}\mathbf{z}+\mathbf{g}(\mathbf{z})$ where $\mathbf{g}(\mathbf{z})$ is small compared to $\mathbf{z}$.

Lemma 1. *Let $\mathbf{f}(\mathbf{x})$ have two continuous partial derivatives with respect to each of its variables $x_1,\ldots,x_n$. Then, $\mathbf{f}(\mathbf{x}^0+\mathbf{z})$ can be written in the form*

$$\mathbf{f}(\mathbf{x}^0+\mathbf{z})=\mathbf{f}(\mathbf{x}^0)+\mathbf{A}\mathbf{z}+\mathbf{g}(\mathbf{z}) \tag{10}$$

where $\mathbf{g}(\mathbf{z})/\max\{|z_1|,\ldots,|z_n|\}$ is a continuous function of $\mathbf{z}$ which vanishes for $\mathbf{z}=\mathbf{0}$.

PROOF #1. Equation (10) is an immediate consequence of Taylor's Theorem which states that each component $f_j(\mathbf{x}^0+\mathbf{z})$ of $\mathbf{f}(\mathbf{x}^0+\mathbf{z})$ can be written in the form

$$f_j(\mathbf{x}^0+\mathbf{z})=f_j(\mathbf{x}^0)+\frac{\partial f_j(\mathbf{x}^0)}{\partial x_1}z_1+\ldots+\frac{\partial f_j(\mathbf{x}^0)}{\partial x_n}z_n+g_j(\mathbf{z})$$

where $g_j(\mathbf{z})/\max\{|z_1|,\ldots,|z_n|\}$ is a continuous function of $\mathbf{z}$ which vanishes for $\mathbf{z}=\mathbf{0}$. Hence,

$$\mathbf{f}(\mathbf{x}^0+\mathbf{z})=\mathbf{f}(\mathbf{x}^0)+\mathbf{A}\mathbf{z}+\mathbf{g}(\mathbf{z})$$

where

$$\mathbf{A}=\begin{bmatrix} \dfrac{\partial f_1(\mathbf{x}^0)}{\partial x_1} & \cdots & \dfrac{\partial f_1(\mathbf{x}^0)}{\partial x_n} \\ \vdots & & \vdots \\ \dfrac{\partial f_n(\mathbf{x}^0)}{\partial x_1} & \cdots & \dfrac{\partial f_n(\mathbf{x}^0)}{\partial x_n} \end{bmatrix}. \qquad \square$$

PROOF #2. If each component of $\mathbf{f}(\mathbf{x})$ is a polynomial (possibly infinite) in $x_1,\ldots,x_n$, then each component of $\mathbf{f}(\mathbf{x}^0+\mathbf{z})$ is a polynomial in $z_1,\ldots,z_n$. Thus,

$$f_j(\mathbf{x}^0+\mathbf{z})=a_{j0}+a_{j1}z_1+\ldots+a_{jn}z_n+g_j(\mathbf{z}) \tag{11}$$

where $g_j(\mathbf{z})$ is a polynomial in $z_1,\ldots,z_n$ beginning with terms of order two. Setting $\mathbf{z}=\mathbf{0}$ in (11) gives $f_j(\mathbf{x}^0)=a_{j0}$. Hence,

$$\mathbf{f}(\mathbf{x}^0+\mathbf{z})=\mathbf{f}(\mathbf{x}^0)+\mathbf{A}\mathbf{z}+\mathbf{g}(\mathbf{z}), \qquad \mathbf{A}=\begin{bmatrix} a_{11} & \cdots & a_{1n} \\ \vdots & & \vdots \\ a_{n1} & \cdots & a_{nn} \end{bmatrix}$$

and each component of $\mathbf{g}(\mathbf{z})$ is a polynomial in $z_1,\dots,z_n$ beginning with terms of order two. □

Theorem 2 and Lemma 1 provide us with the following algorithm for determining whether an equilibrium solution $\mathbf{x}(t)\equiv\mathbf{x}^0$ of $\dot{\mathbf{x}}=\mathbf{f}(\mathbf{x})$ is stable or unstable:

1. Set $\mathbf{z}=\mathbf{x}-\mathbf{x}^0$.
2. Write $\mathbf{f}(\mathbf{x}^0+\mathbf{z})$ in the form $\mathbf{A}\mathbf{z}+\mathbf{g}(\mathbf{z})$ where $\mathbf{g}(\mathbf{z})$ is a vector-valued polynomial in $z_1,\dots,z_n$ beginning with terms of order two or more.
3. Compute the eigenvalues of $\mathbf{A}$. If all the eigenvalues of $\mathbf{A}$ have negative real part, then $\mathbf{x}(t)\equiv\mathbf{x}^0$ is asymptotically stable. If one eigenvalue of $\mathbf{A}$ has positive real part, then $\mathbf{x}(t)\equiv\mathbf{x}^0$ is unstable.

Example 2. Find all equilibrium solutions of the system of differential equations

$$\frac{dx}{dt}=1-xy,\qquad \frac{dy}{dt}=x-y^3 \tag{12}$$

and determine (if possible) whether they are stable or unstable.

Solution. The equations $1-xy=0$ and $x-y^3=0$ imply that $x=1, y=1$ or $x=-1, y=-1$. Hence, $x(t)\equiv 1, y(t)\equiv 1$, and $x(t)\equiv -1, y(t)\equiv -1$ are the only equilibrium solutions of (12).

(i) $x(t)=1, y(t)=1$: Set $u=x-1,\ v=y-1$. Then,

$$\frac{du}{dt}=\frac{dx}{dt}=1-(1+u)(1+v)=-u-v-uv$$

$$\frac{dv}{dt}=\frac{dy}{dt}=(1+u)-(1+v)^3=u-3v-3v^2-v^3.$$

We rewrite this system in the form

$$\frac{d}{dt}\begin{pmatrix}u\\v\end{pmatrix}=\begin{pmatrix}-1&-1\\1&-3\end{pmatrix}\begin{pmatrix}u\\v\end{pmatrix}-\begin{pmatrix}uv\\3v^2+v^3\end{pmatrix}.$$

The matrix

$$\begin{pmatrix}-1&-1\\1&-3\end{pmatrix}$$

has a single eigenvalue $\lambda=-2$ since

$$\det\begin{pmatrix}-1-\lambda&-1\\1&-3-\lambda\end{pmatrix}=(1+\lambda)(3+\lambda)+1=(\lambda+2)^2.$$

Hence, the equilibrium solution $x(t)\equiv 1, y(t)\equiv 1$ of (12) is asymptotically stable.

(ii) $x(t)\equiv -1, y(t)\equiv -1$: Set $u=x+1,\ v=y+1$. Then,

$$\frac{du}{dt}=\frac{dx}{dt}=1-(u-1)(v-1)=u+v-uv$$

$$\frac{dv}{dt}=\frac{dy}{dt}=(u-1)-(v-1)^3=u-3v+3v^2-v^3.$$

We rewrite this system in the form

$$\frac{d}{dt}\begin{pmatrix} u \\ v \end{pmatrix}=\begin{pmatrix} 1 & 1 \\ 1 & -3 \end{pmatrix}\begin{pmatrix} u \\ v \end{pmatrix}+\begin{pmatrix} -uv \\ 3v^2-v^3 \end{pmatrix}.$$

The eigenvalues of the matrix

$$\begin{pmatrix} 1 & 1 \\ 1 & -3 \end{pmatrix}$$

are $\lambda_1=-1-\sqrt{5}$, which is negative, and $\lambda_2=-1+\sqrt{5}$, which is positive. Therefore, the equilibrium solution $x(t)\equiv -1$, $y(t)\equiv -1$ of (12) is unstable.

Example 3. Find all equilibrium solutions of the system of differential equations

$$\frac{dx}{dt}=\sin(x+y), \qquad \frac{dy}{dt}=e^x-1 \tag{13}$$

and determine whether they are stable or unstable.

Solution. The equilibrium points of (13) are determined by the two equations $\sin(x+y)=0$ and $e^x-1=0$. The second equation implies that $x=0$, while the first equation implies that $x+y=n\pi$, n an integer. Consequently, $x(t)\equiv 0$, $y(t)\equiv n\pi$, $n=0,\pm 1,\pm 2,\ldots$, are the equilibrium solutions of (13). Setting $u=x$, $v=y-n\pi$, gives

$$\frac{du}{dt}=\sin(u+v+n\pi), \qquad \frac{dv}{dt}=e^u-1.$$

Now, $\sin(u+v+n\pi)=\cos n\pi\sin(u+v)=(-1)^n\sin(u+v)$. Therefore,

$$\frac{du}{dt}=(-1)^n\sin(u+v), \qquad \frac{dv}{dt}=e^u-1.$$

Next, observe that

$$\sin(u+v)=u+v-\frac{(u+v)^3}{3!}+\ldots, \qquad e^u-1=u+\frac{u^2}{2!}+\ldots.$$

Hence,

$$\frac{du}{dt}=(-1)^n\left[(u+v)-\frac{(u+v)^3}{3!}+\ldots\right], \qquad \frac{dv}{dt}=u+\frac{u^2}{2!}+\ldots.$$

We rewrite this system in the form

$$\frac{d}{dt}\begin{pmatrix} u \\ v \end{pmatrix}=\begin{pmatrix} (-1)^n & (-1)^n \\ 1 & 0 \end{pmatrix}\begin{pmatrix} u \\ v \end{pmatrix}+\text{terms of order 2 or higher in } u \text{ and } v.$$

The eigenvalues of the matrix

$$\begin{pmatrix} (-1)^n & (-1)^n \\ 1 & 0 \end{pmatrix}$$

are

$$\lambda_1=\frac{(-1)^n-\sqrt{1+4(-1)^n}}{2}, \qquad \lambda_2=\frac{(-1)^n+\sqrt{1+4(-1)^n}}{2}.$$

When n is even, $\lambda_1=(1-\sqrt{5}\,)/2$ is negative and $\lambda_2=(1+\sqrt{5}\,)/2$ is positive. Hence, $x(t)\equiv 0$, $y(t)\equiv n\pi$ is unstable if n is even. When n is odd, both $\lambda_1=(-1-\sqrt{3}\,i)/2$ and $\lambda_2=(-1+\sqrt{3}\,i)/2$ have negative real part. Therefore, the equilibrium solution $x(t)\equiv 0$, $y(t)\equiv n\pi$ is asymptotically stable if n is odd.

Exercises

Find all equilibrium solutions of each of the following systems of equations and determine, if possible, whether they are stable or unstable.

1. $\dot{x}=x-x^3-xy^2$
$\dot{y}=2y-y^5-yx^4$

2. $\dot{x}=x^2+y^2-1$
$\dot{y}=x^2-y^2$

3. $\dot{x}=x^2+y^2-1$
$\dot{y}=2xy$

4. $\dot{x}=6x-6x^2-2xy$
$\dot{y}=4y-4y^2-2xy$

5. $\dot{x}=\tan(x+y)$
$\dot{y}=x+x^3$

6. $\dot{x}=e^y-x$
$\dot{y}=e^x+y$

Verify that the origin is an equilibrium point of each of the following systems of equations and determine, if possible, whether it is stable or unstable.

7. $\dot{x}=y+3x^2$
$\dot{y}=x-3y^2$

8. $\dot{x}=y+\cos y-1$
$\dot{y}=-\sin x+x^3$

9. $\dot{x}=e^{x+y}-1$
$\dot{y}=\sin(x+y)$

10. $\dot{x}=\ln(1+x+y^2)$
$\dot{y}=-y+x^3$

11. $\dot{x}=\cos y-\sin x-1$
$\dot{y}=x-y-y^2$

12. $\dot{x}=8x-3y+e^y-1$
$\dot{y}=\sin x^2-\ln(1-x-y)$

13. $\dot{x}=-x-y-(x^2+y^2)^{3/2}$
$\dot{y}=x-y+(x^2+y^2)^{3/2}$

14. $\dot{x}=x-y+z^2$
$\dot{y}=y+z-x^2$
$\dot{z}=z-x+y^2$

15. $\dot{x}=e^{x+y+z}-1$
$\dot{y}=\sin(x+y+z)$
$\dot{z}=x-y-z^2$

16. $\dot{x}=\ln(1-z)$
$\dot{y}=\ln(1-x)$
$\dot{z}=\ln(1-y)$

17. $\dot{x}=x-\cos y-z+1$
$\dot{y}=y-\cos z-x+1$
$\dot{z}=z-\cos x-y+1$

18 (a) Find all equilibrium solutions of the system of differential equations

$$\frac{dx}{dt}=gz-hx, \qquad \frac{dy}{dt}=\frac{c}{a+bx}-ky, \qquad \frac{dz}{dt}=ey-fz.$$

(This system is a model for the control of protein synthesis.)

(b) Determine the stability or instability of these solutions if either g, e, or c is zero.

4.4 The phase-plane

In this section we begin our study of the "geometric" theory of differential equations. For simplicity, we will restrict ourselves, for the most part, to the case $n=2$. Our aim is to obtain as complete a description as possible of all solutions of the system of differential equations

$$\frac{dx}{dt}=f(x,y), \qquad \frac{dy}{dt}=g(x,y). \tag{1}$$

To this end, observe that every solution $x=x(t)$, $y=y(t)$ of (1) defines a curve in the three-dimensional space t,x,y. That is to say, the set of all points $(t,x(t),y(t))$ describe a curve in the three-dimensional space t,x,y. For example, the solution $x=\cos t$, $y=\sin t$ of the system of differential equations

$$\frac{dx}{dt}=-y, \qquad \frac{dy}{dt}=x$$

describes a helix (see Figure 1) in (t,x,y) space.

The geometric theory of differential equations begins with the important observation that every solution $x=x(t)$, $y=y(t)$, $t_0 \leqslant t \leqslant t_1$, of (1) also defines a curve in the $x-y$ plane. To wit, as t runs from t_0 to t_1, the set of points $(x(t),y(t))$ trace out a curve C in the $x-y$ plane. This curve is called the *orbit*, or *trajectory*, of the solution $x=x(t)$, $y=y(t)$, and the $x-y$ plane is called the *phase-plane* of the solutions of (1). Equivalently, we can think of the orbit of $x(t)$, $y(t)$ as the path that the solution traverses in the $x-y$ plane.

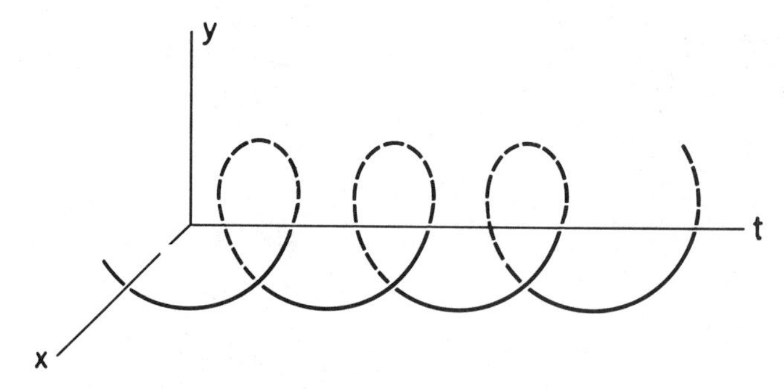

Figure 1. Graph of the solution $x=\cos t$, $y=\sin t$

Example 1. It is easily verified that $x=\cos t$, $y=\sin t$ is a solution of the system of differential equations $\dot{x}=-y$, $\dot{y}=x$. As t runs from 0 to 2π, the set of points $(\cos t,\sin t)$ trace out the unit circle $x^2+y^2=1$ in the $x-y$ plane. Hence, the unit circle $x^2+y^2=1$ is the orbit of the solution $x=\cos t$, $y=\sin t$, $0\leqslant t\leqslant 2\pi$. As t runs from 0 to ∞, the set of points $(\cos t,\sin t)$ trace out this circle infinitely often.

Example 2. It is easily verified that $x=e^{-t}\cos t$, $y=e^{-t}\sin t$, $-\infty<t<\infty$, is a solution of the system of differential equations $dx/dt=-x-y$, $dy/dt=x-y$. As t runs from $-\infty$ to ∞, the set of points $(e^{-t}\cos t, e^{-t}\sin t)$ trace out a spiral in the $x-y$ plane. Hence, the orbit of the solution $x=e^{-t}\cos t$, $y=e^{-t}\sin t$ is the spiral shown in Figure 2.

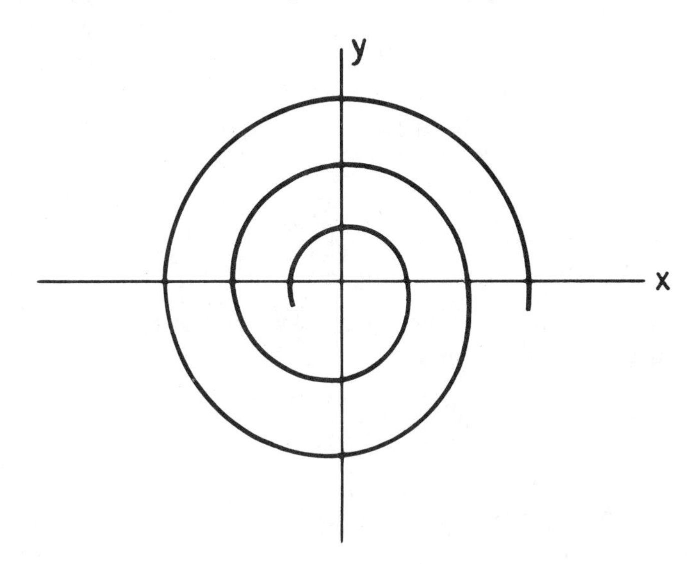

Figure 2. Orbit of $x=e^{-t}\cos t$, $y=e^{-t}\sin t$

Example 3. It is easily verified that $x=3t+2$, $y=5t+7$, $-\infty<t<\infty$ is a solution of the system of differential equations $dx/dt=3$, $dy/dt=5$. As t runs from $-\infty$ to ∞, the set of points $(3t+2, 5t+7)$ trace out the straight line through the point $(2,7)$ with slope $\frac{5}{3}$. Hence, the orbit of the solution $x=3t+2$, $y=5t+7$ is the straight line $y=\frac{5}{3}(x-2)+7$, $-\infty<x<\infty$.

Example 4. It is easily verified that $x=3t^2+2$, $y=5t^2+7$, $0\leqslant t<\infty$ is a solution of the system of differential equations

$$\frac{dx}{dt}=6\left[(y-7)/5\right]^{1/2}, \qquad \frac{dy}{dt}=10\left[(x-2)/3\right]^{1/2}.$$

All of the points $(3t^2+2, 5t^2+7)$ lie on the line through $(2,7)$ with slope $\frac{5}{3}$. However, x is always greater than or equal to 2, and y is always greater than or equal to 7. Hence, the orbit of the solution $x=3t^2+2$, $y=5t^2+7$, $0\leqslant t<\infty$, is the straight line $y=\frac{5}{3}(x-2)+7$, $2\leqslant x<\infty$.

Example 5. It is easily verified that $x=3t+2$, $y=5t^2+7$, $-\infty<t<\infty$, is a solution of the system of differential equations

$$\frac{dx}{dt}=y-\frac{5}{9}(x-2)^2-4, \qquad \frac{dy}{dt}=\frac{10}{3}(x-2).$$

The orbit of this solution is the set of all points $(x,y)=(3t+2,5t^2+7)$. Solving for $t=\frac{1}{3}(x-2)$, we see that $y=\frac{5}{9}(x-2)^2+7$. Hence, the orbit of the solution $x=3t+2$, $y=5t^2+7$ is the parabola $y=\frac{5}{9}(x-2)^2+7$, $|x|<\infty$.

One of the advantages of considering the orbit of the solution rather than the solution itself is that is is often possible to obtain the orbit of a solution without prior knowledge of the solution. Let $x=x(t)$, $y=y(t)$ be a solution of (1). If $x'(t)$ is unequal to zero at $t=t_1$, then we can solve for $t=t(x)$ in a neighborhood of the point $x_1=x(t_1)$ (see Exercise 4). Thus, for t near t_1. the orbit of the solution $x(t)$, $y(t)$ is the curve $y=y(t(x))$. Next, observe that

$$\frac{dy}{dx}=\frac{dy}{dt}\frac{dt}{dx}=\frac{dy/dt}{dx/dt}=\frac{g(x,y)}{f(x,y)}.$$

Thus, the orbits of the solutions $x=x(t)$, $y=y(t)$ of (1) are the solution curves of the first-order scalar equation

$$\frac{dy}{dx}=\frac{g(x,y)}{f(x,y)}. \tag{2}$$

Therefore, it is not necessary to find a solution $x(t)$, $y(t)$ of (1) in order to compute its orbit; we need only solve the single first-order scalar differential equation (2).

Remark. From now on, we will use the phrase "the orbits of (1)" to denote the totality of orbits of solutions of (1).

Example 6. The orbits of the system of differential equations

$$\frac{dx}{dt}=y^2, \qquad \frac{dy}{dt}=x^2 \tag{3}$$

are the solution curves of the scalar equation $dy/dx=x^2/y^2$. This equation is separable, and it is easily seen that every solution is of the form $y(x)=(x^3-c)^{1/3}$, c constant. Thus, the orbits of (3) are the set of all curves $y=(x^3-c)^{1/3}$.

Example 7. The orbits of the system of differential equations

$$\frac{dx}{dt}=y(1+x^2+y^2), \qquad \frac{dy}{dt}=-2x(1+x^2+y^2) \tag{4}$$

are the solution curves of the scalar equation

$$\frac{dy}{dx}=-\frac{2x(1+x^2+y^2)}{y(1+x^2+y^2)}=-\frac{2x}{y}.$$

This equation is separable, and all solutions are of the form $\frac{1}{2}y^2+x^2=c^2$. Hence, the orbits of (4) are the families of ellipses $\frac{1}{2}y^2+x^2=c^2$.

Warning. A solution curve of (2) is an orbit of (1) only if dx/dt and dy/dt are not zero simultaneously along the solution. If a solution curve of (2) passes through an equilibrium point of (1), then the entire solution curve is not an orbit. Rather, it is the union of several distinct orbits. For example, consider the system of differential equations

$$\frac{dx}{dt}=y(1-x^2-y^2), \qquad \frac{dy}{dt}=-x(1-x^2-y^2). \tag{5}$$

The solution curves of the scalar equation

$$\frac{dy}{dx}=\frac{dy/dt}{dx/dt}=-\frac{x}{y}$$

are the family of concentric circles $x^2+y^2=c^2$. Observe, however, that every point on the unit circle $x^2+y^2=1$ is an equilibrium point of (5). Thus, the orbits of this system are the circles $x^2+y^2=c^2$, for $c\neq 1$, and all points on the unit circle $x^2+y^2=1$. Similarly, the orbits of (3) are the curves $y=(x^3-c)^{1/3}$, $c\neq 0$; the half-lines $y=x$, $x>0$, and $y=x$, $x<0$; and the point $(0,0)$.

It is not possible, in general, to explicitly solve Equation (2). Hence, we cannot, in general, find the orbits of (1). Nevertheless, it is still possible to obtain an accurate description of all orbits of (1). This is because the system of differential equations (1) sets up a *direction field* in the $x-y$ plane. That is to say, the system of differential equations (1) tells us how fast a solution moves along its orbit, and in what direction it is moving. More precisely, let $x=x(t)$, $y=y(t)$ be a solution of (1). As t increases, the point $(x(t),y(t))$ moves along the orbit of this solution. Its velocity in the x-direction is dx/dt; its velocity in the y-direction is dy/dt; and the magnitude of its velocity is $[(dx(t)/dt)^2+(dy(t)/dt)^2]^{1/2}$. But $dx(t)/dt=f(x(t),y(t))$, and $dy(t)/dt=g(x(t),y(t))$. Hence, at each point (x,y) in the phase plane of (1) we know (i), the tangent to the orbit through (x,y) (the line through (x,y) with direction numbers $f(x,y)$, $g(x,y)$ respectively) and (ii), the speed $[f^2(x,y)+g^2(x,y)]^{1/2}$ with which the solution is traversing its orbit. As we shall see in Sections 4.8–13, this information can often be used to deduce important properties of the orbits of (1).

The notion of orbit can easily be extended to the case $n>2$. Let $\mathbf{x}=\mathbf{x}(t)$ be a solution of the vector differential equation

$$\dot{\mathbf{x}}=\mathbf{f}(\mathbf{x}), \qquad \mathbf{x}=\begin{bmatrix} x_1\\ \vdots \\ x_n\end{bmatrix}, \qquad \mathbf{f}(\mathbf{x})=\begin{bmatrix} f_1(x_1,\dots,x_n)\\ \vdots \\ f_n(x_1,\dots,x_n)\end{bmatrix} \tag{6}$$

on the interval $t_0\leqslant t\leqslant t_1$. As t runs from t_0 to t_1, the set of points $(x_1(t),\dots,x_n(t))$ trace out a curve C in the n-dimensional space $x_1,x_2,\dots,x_n$. This curve is called the orbit of the solution $\mathbf{x}=\mathbf{x}(t)$, for $t_0\leqslant t\leqslant t_1$, and the n-dimensional space $x_1,\dots,x_n$ is called the phase-space of the solutions of (6).

EXERCISES

In each of Problems 1–3, verify that $x(t), y(t)$ is a solution of the given system of equations, and find its orbit.

1. $\dot{x}=1, \quad \dot{y}=2(1-x)\sin(1-x)^2$
$x(t)=1+t, \quad y(t)=\cos t^2$

2. $\dot{x}=e^{-x}, \quad \dot{y}=e^{e^x-1}$
$x(t)=\ln(1+t), \quad y(t)=e^t$

3. $\dot{x}=1+x^2, \quad \dot{y}=(1+x^2)\sec^2 x$
$x(t)=\tan t, \quad y(t)=\tan(\tan t)$

4. Suppose that $x'(t_1)\neq 0$. Show that we can solve the equation $x=x(t)$ for $t=t(x)$ in a neighborhood of the point $x_1=x(t_1)$. *Hint*: If $x'(t_1)\neq 0$, then $x(t)$ is a strictly monotonic function of t in a neighborhood of $t=t_1$.

Find the orbits of each of the following systems.

5. $\dot{x}=y,$
$\dot{y}=-x$

6. $\dot{x}=y(1+x^2+y^2),$
$\dot{y}=-x(1+x^2+y^2)$

7. $\dot{x}=y(1+x+y),$
$\dot{y}=-x(1+x+y)$

8. $\dot{x}=y+x^2y,$
$\dot{y}=3x+xy^2$

9. $\dot{x}=xye^{-3x},$
$\dot{y}=-2xy^2$

10. $\dot{x}=4y,$
$\dot{y}=x+xy^2$

11. $\dot{x}=ax-bxy,$
$\dot{y}=cx-dxy$
$(a,b,c,d$ positive)

12. $\dot{x}=x^2+\cos y,$
$\dot{y}=-2xy$

13. $\dot{x}=2xy,$
$\dot{y}=x^2-y^2$

14. $\dot{x}=y+\sin x,$
$\dot{y}=x-y\cos x$

4.5 Mathematical theories of war

4.5.1. L. F. Richardson's theory of conflict

In this section we construct a mathematical model which describes the relation between two nations, each determined to defend itself against a possible attack by the other. Each nation considers the possibility of attack quite real, and reasonably enough, bases its apprehensions on the readiness of the other to wage war. Our model is based on the work of Lewis Fry Richardson. It is not an attempt to make scientific statements about foreign politics or to predict the date at which the next war will break out. This, of course, is clearly impossible. Rather, it is a description of what people would do if they did not stop to think. As Richardson writes: "Why are so many nations reluctantly but steadily increasing their armaments as if they were mechanically compelled to do so? Because, I say, they follow their traditions which are fixtures and their instincts which are mechanical;

and because they have not yet made a sufficiently strenuous intellectual and moral effort to control the situation. The process described by the ensuing equations is not to be thought of as inevitable. It is what would occur if instinct and tradition were allowed to act uncontrolled."

Let $x = x(t)$ denote the war potential, or armaments, of the first nation, which we will call Jedesland, and let $y(t)$ denote the war potential of the second nation, which we will call Andersland. The rate of change of $x(t)$ depends, obviously, on the war readiness $y(t)$ of Andersland, and on the grievances that Jedesland feels towards Andersland. In the most simplistic model we represent these terms by ky and g respectively, where k and g are positive constants. These two terms cause x to increase. On the other hand, the cost of armaments has a restraining effect on dx/dt. We represent this term by $-\alpha x$, where α is a positive constant. A similar analysis holds for dy/dt. Consequently, $x = x(t), y = y(t)$ is a solution of the linear system of differential equations

$$\frac{dx}{dt} = ky - \alpha x + g, \qquad \frac{dy}{dt} = lx - \beta y + h. \tag{1}$$

Remark. The model (1) is not limited to two nations; it can also represent the relation between two alliances. For example, Andersland and Jedesland can represent the alliances of France with Russia, and Germany with Austria–Hungary during the years immediately prior to World War I.

Throughout history, there has been a constant debate on the cause of war. Over two thousand years ago, Thucydides claimed that armaments cause war. In his account of the Peloponnesian war he writes: "The real though unavowed cause I believe to have been the growth of Athenian power, which terrified the Lacedaemonians and forced them into war." Sir Edward Grey, the British Foreign Secretary during World War I agrees. He writes: "The increase of armaments that is intended in each nation to produce consciousness of strength, and a sense of security, does not produce these effects. On the contrary, it produces a consciousness of the strength of other nations and a sense of fear. The enormous growth of armaments in Europe, the sense of insecurity and fear caused by them—it was these that made war inevitable. This is the real and final account of the origin of the Great War."

On the other hand, L. S. Amery, a member of Britain's parliament during the 1930's vehemently disagrees. When the opinion of Sir Edward Grey was quoted in the House of Commons, Amery replied: "With all due respect to the memory of an eminent statesman, I believe that statement to be entirely mistaken. The armaments were only the symptoms of the conflict of ambitions and ideals, of those nationalist forces which created the War. The War was brought about because Serbia, Italy and Rumania passionately desired the incorporation in their states of territories which at that time belonged to the Austrian Empire and which the Austrian government was not prepared to abandon without a struggle. France was prepared, if the opportunity ever came, to make an effort to recover Alsace–Lorraine. It was in those facts, in those insoluble conflicts of ambitions, and not in the armaments themselves, that the cause of the War lay."

The system of equations (1) takes both conflicting theories into account. Thucydides and Sir Edward Grey would take g and h small compared to k and l, while Mr. Amery would take k and l small compared to g and h.

The system of equations (1) has several important implications. Suppose that g and h are both zero. Then, $x(t)\equiv 0$, $y(t)\equiv 0$ is an equilibrium solution of (1). That is, if x, y, g, and h are all made zero simultaneously, then $x(t)$ and $y(t)$ will always remain zero. This ideal condition is permanent peace by disarmament and satisfaction. It has existed since 1817 on the border between Canada and the United States, and since 1905 on the border between Norway and Sweden.

These equations further imply that mutual disarmament without satisfaction is not permanent. Assume that x and y vanish simultaneously at some time $t=t_0$. At this time, $dx/dt=g$ and $dy/dt=h$. Thus, x and y will not remain zero if g and h are positive. Instead, both nations will rearm.

Unilateral disarmament corresponds to setting $y=0$ at a certain instant of time. At this time, $dy/dt=lx+h$. This implies that y will not remain zero if either h or x is positive. Thus, unilateral disarmament is never permanent. This accords with the historical fact that Germany, whose army was reduced by the Treaty of Versailles to 100,000 men, a level far below that of several of her neighbors, insisting on rearming during the years 1933–36.

A race in armaments occurs when the "defense" terms predominate in (1). In this case,

$$\frac{dx}{dt}=ky, \qquad \frac{dy}{dt}=lx. \tag{2}$$

Every solution of (2) is of the form

$$x(t)=Ae^{\sqrt{kl}\,t}+Be^{-\sqrt{kl}\,t}, \qquad y(t)=\sqrt{\frac{l}{k}}\left[Ae^{\sqrt{kl}\,t}-Be^{-\sqrt{kl}\,t}\right].$$

Therefore, both $x(t)$ and $y(t)$ approach infinity if A is positive. This infinity can be interpreted as war.

Now, the system of equations (1) is not quite correct, since it does not take into effect the cooperation, or trade, between Andersland and Jedesland. As we see today, mutual cooperation between nations tends to decrease their fears and suspicions. We correct our model by changing the meaning of $x(t)$ and $y(t)$; we let the variables $x(t)$ and $y(t)$ stand for "threats" minus "cooperation." Specifically, we set $x=U-U_0$ and $y=V-V_0$, where U is the defense budget of Jedesland, V is the defense budget of Andersland, U_0 is the amount of goods exported by Jedesland to

Andersland and V_0 is the amount of goods exported by Andersland to Jedesland. Observe that cooperation evokes reciprocal cooperation, just as armaments provoke more armaments. In addition, nations have a tendency to reduce cooperation on account of the expense which it involves. Thus, the system of equations (1) still describes this more general state of affairs.

The system of equations (1) has a single equilibrium solution

$$x = x_0 = \frac{kh + \beta g}{\alpha\beta - kl}, \qquad y = y_0 = \frac{lg + \alpha h}{\alpha\beta - kl} \tag{3}$$

if $\alpha\beta - kl \neq 0$. We are interested in determining whether this equilibrium solution is stable or unstable. To this end, we write (1) in the form $\dot{\mathbf{w}} = \mathbf{A}\mathbf{w} + \mathbf{f}$, where

$$\mathbf{w}(t) = \begin{pmatrix} x(t) \\ y(t) \end{pmatrix}, \qquad \mathbf{f} = \begin{pmatrix} g \\ h \end{pmatrix}, \quad \text{and} \quad \mathbf{A} = \begin{pmatrix} -\alpha & k \\ l & -\beta \end{pmatrix}.$$

The equilibrium solution is

$$\mathbf{w} = \mathbf{w}_0 = \begin{pmatrix} x_0 \\ y_0 \end{pmatrix},$$

where $\mathbf{A}\mathbf{w}_0 + \mathbf{f} = \mathbf{0}$. Setting $\mathbf{z} = \mathbf{w} - \mathbf{w}_0$, we obtain that

$$\dot{\mathbf{z}} = \dot{\mathbf{w}} = \mathbf{A}\mathbf{w} + \mathbf{f} = \mathbf{A}(\mathbf{z} + \mathbf{w}_0) + \mathbf{f} = \mathbf{A}\mathbf{z} + \mathbf{A}\mathbf{w}_0 + \mathbf{f} = \mathbf{A}\mathbf{z}.$$

Clearly, the equilibrium solution $\mathbf{w}(t) = \mathbf{w}_0$ of $\dot{\mathbf{w}} = \mathbf{A}\mathbf{w} + \mathbf{f}$ is stable if, and only if, $\mathbf{z} = \mathbf{0}$ is a stable solution of $\dot{\mathbf{z}} = \mathbf{A}\mathbf{z}$. To determine the stability of $\mathbf{z} = \mathbf{0}$ we compute

$$p(\lambda) = \det\begin{pmatrix} -\alpha - \lambda & k \\ l & -\beta - \lambda \end{pmatrix} = \lambda^2 + (\alpha + \beta)\lambda + \alpha\beta - kl.$$

The roots of $p(\lambda)$ are

$$\lambda = \frac{-(\alpha+\beta) \pm \left[(\alpha+\beta)^2 - 4(\alpha\beta - kl)\right]^{1/2}}{2}$$

$$= \frac{-(\alpha+\beta) \pm \left[(\alpha-\beta)^2 + 4kl\right]^{1/2}}{2}.$$

Notice that both roots are real and unequal to zero. Moreover, both roots are negative if $\alpha\beta - kl > 0$, and one root is positive if $\alpha\beta - kl < 0$. Thus, $\mathbf{z}(t) \equiv \mathbf{0}$, and consequently the equilibrium solution $x(t) \equiv x_0$, $y(t) \equiv y_0$ is stable if $\alpha\beta - kl > 0$ and unstable if $\alpha\beta - kl < 0$.

Let us now tackle the difficult problem of estimating the coefficients α, β, k, l, g, and h. There is no way, obviously, of measuring g and h. However, it is possible to obtain reasonable estimates for α, β, k, and l. Observe that the units of these coefficients are reciprocal times. Physicists and engineers would call α^{-1} and β^{-1} relaxation times, for if y and g were identically zero, then $x(t) = e^{-\alpha(t-t_0)}x(t_0)$. This implies that $x(t_0 + \alpha^{-1}) =$

$x(t_0)/e$. Hence, α^{-1} is the time required for Jedesland's armaments to be reduced in the ratio 2.718 if that nation has no grievances and no other nation has any armaments. Richardson estimates α^{-1} to be the lifetime of Jedesland's parliament. Thus, $\alpha = 0.2$ for Great Britain, since the lifetime of Britain's parliament is five years.

To estimate k and l we take a hypothetical case in which $g = 0$ and $y = y_1$, so that $dx/dt = ky_1 - \alpha x$. When $x = 0$, $1/k = y_1/(dx/dt)$. Thus, $1/k$ is the time required for Jedesland to catch up to Andersland provided that (i) Andersland's armaments remain constant, (ii) there are no grievances, and (iii) the cost of armaments doesn't slow Jedesland down. Consider now the German rearmament during 1933–36. Germany started with nearly zero armaments and caught up with her neighbors in about three years. Assuming that the slowing effect of α nearly balanced the Germans' very strong grievances g, we take $k = 0.3$ (year)$^{-1}$ for Germany. Further, we observe that k is obviously proportional to the amount of industry that a nation has. Thus, $k = 0.15$ for a nation which has only half the industrial capacity of Germany, and $k = 0.9$ for a nation which has three times the industrial capacity of Germany.

Let us now check our model against the European arms race of 1909–1914. France was allied with Russia, and Germany was allied with Austria–Hungary. Neither Italy or Britain was in a definite alliance with either party. Thus, let Jedesland represent the alliance of France with Russia, and let Andersland represent the alliance of Germany with Austria–Hungary. Since these two alliances were roughly equal in size we take $k = l$, and since each alliance was roughly three times the size of Germany, we take $k = l = 0.9$. We also assume that $\alpha = \beta = 0.2$. Then,

$$\frac{dx}{dt} = -\alpha x + ky + g, \qquad \frac{dy}{dt} = kx - \alpha y + h. \tag{4}$$

Equation (4) has a unique equilibrium point

$$x_0 = \frac{kh + \alpha g}{\alpha^2 - k^2}, \qquad y_0 = \frac{kg + \alpha h}{\alpha^2 - k^2}.$$

This equilibrium is unstable since

$$\alpha\beta - kl = \alpha^2 - k^2 = 0.04 - 0.81 = -0.77.$$

This, of course, is in agreement with the historical fact that these two alliances went to war with each other.

Now, the model we have constructed is very crude since it assumes that the grievances g and h are constant in time. This is obviously not true. The grievances g and h are not even continuous functions of time since they jump instantaneously by large amounts. (It's safe to assume, though, that g and h are relatively constant over long periods of time.) In spite of this, the system of equations (4) still provides a very accurate description of the arms race preceding World War I. To demonstrate this, we add the two equations of (4) together, to obtain that

$$\frac{d}{dt}(x+y)=(k-\alpha)(x+y)+g+h. \tag{5}$$

Recall that $x=U-U_0$ and $y=V-V_0$, where U and V are the defense budgets of the two alliances, and U_0 and V_0 are the amount of goods exported from each alliance to the other. Hence,

$$\frac{d}{dt}(U+V)=(k-\alpha)\left\{U+V-\left[U_0+V_0-\frac{g+h}{k-\alpha}-\frac{1}{k-\alpha}\frac{d}{dt}(U_0+V_0)\right]\right\}. \tag{6}$$

The defense budgets for the two alliances are set out in Table I.

Table 1. Defense budgets expressed in millions of £ sterling

	1909		1910		1911		1912		1913
France	48.6		50.9		57.1		63.2		74.7
Russia	66.7		68.5		70.7		81.8		92.0
Germany	63.1		62.0		62.5		68.2		95.4
Austria–Hungary	20.8		23.4		24.6		25.5		26.9
Total $U+V$	199.2		204.8		214.9		238.7		289.0
$\Delta(U+V)$		5.6		10.1		23.8		50.3	
$U+V$ at same date		202.0		209.8		226.8		263.8	

In Figure 1 we plot the annual increment of $U+V$ against the average of $U+V$ for the two years used in forming the increment. Notice how close these four points, denoted by $\circ$, are to the straight line

$$\Delta(U+V)=0.73(U+V-194). \tag{7}$$

Thus, foreign politics does indeed have a machine-like predictability. Equations (6) and (7) imply that

$$g+h=(k-\alpha)(U_0+V_0)-\Delta(U_0+V_0)-194$$

and $k-\alpha=0.73$. This is in excellent agreement with Richardson's estimates of 0.9 for k and 0.2 for α. Finally, observe from (7) that the total defense budgets of the two alliances will increase if $U+V$ is greater than 194 million, and will decrease otherwise. In actual fact, the defense expenditures of the two alliances was 199.2 million in 1909 while the trade between the two alliances amounted to only 171.8 million. Thus began an arms race which led eventually to World War I.

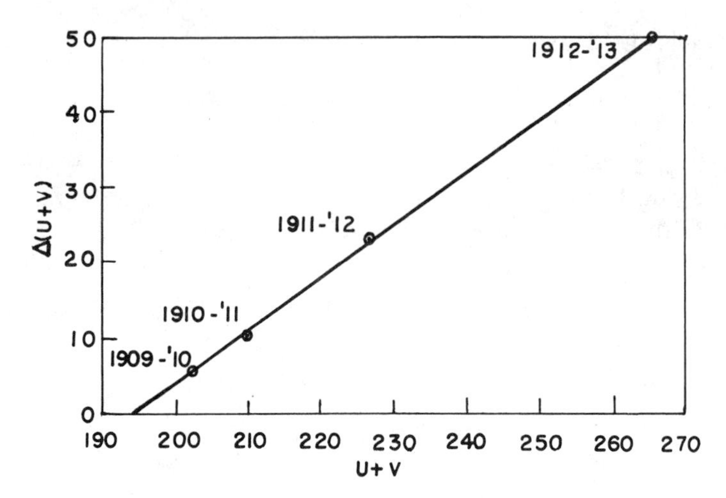

Figure 1. Graph of $\Delta(U+V)$ versus $U+V$

Reference

Richardson, L. F., "Generalized foreign politics," The British Journal of Psychology, monograph supplement #23, 1939.

Exercises

1. Suppose that what moves a government to arm is not the magnitude of other nations' armaments, but the difference between its own and theirs. Then,

$$\frac{dx}{dt}=k(y-x)-\alpha x+g, \qquad \frac{dy}{dt}=l(x-y)-\beta y+h.$$

Show that every solution of this system of equations is stable if $k_1l_1<(\alpha_1+k_1)(\beta_1+l_1)$ and unstable if $k_1l_1>(\alpha_1+k_1)(\beta_1+l_1)$.

2. Consider the case of three nations, each having the same defense coefficient k and the same restraint coefficient α. Then,

$$\frac{dx}{dt}=-\alpha x+ky+kz+g_1$$
$$\frac{dy}{dt}=kx-\alpha y+kz+g_2$$
$$\frac{dz}{dt}=kx+ky-\alpha z+g_3.$$

Setting

$$\mathbf{u}=\begin{pmatrix}x\\y\\z\end{pmatrix},\quad \mathbf{A}=\begin{pmatrix}-\alpha & k & k\\ k & -\alpha & k\\ k & k & -\alpha\end{pmatrix},\quad\text{and}\quad \mathbf{g}=\begin{pmatrix}g_1\\g_2\\g_3\end{pmatrix},$$

we see that $\dot{\mathbf{u}}=\mathbf{A}\mathbf{u}+\mathbf{g}$.

(a) Show that $p(\lambda)\equiv\det(\mathbf{A}-\lambda\mathbf{I})=-(\alpha+\lambda)^3+3k^2(\alpha+\lambda)+2k^3$.

(b) Show that $p(\lambda)=0$ when $\lambda=-\alpha-k$. Use this information to find the remaining two roots of $p(\lambda)$.

(c) Show that every solution $\mathbf{u}(t)$ is stable if $2k<\alpha$, and unstable if $2k>\alpha$.

3. Suppose in Problem 2 that the z nation is a pacifist nation, while x and y are pugnacious nations. Then,

$$\frac{dx}{dt} = -\alpha x + ky + kz + g_1$$
$$\frac{dy}{dt} = kx - \alpha y + kz + g_2 \qquad (*)$$
$$\frac{dz}{dt} = 0\cdot x + 0\cdot y - \alpha z + g_3.$$

Show that every solution $x(t), y(t), z(t)$ of (*) is stable if $k<\alpha$, and unstable if $k>\alpha$.

4.5.2 Lanchester's combat models and the battle of Iwo Jima

During the first World War, F. W. Lanchester [4] pointed out the importance of the concentration of troops in modern combat. He constructed mathematical models from which the expected results of an engagement could be obtained. In this section we will derive two of these models, that of a conventional force versus a conventional force, and that of a conventional force versus a guerilla force. We will then solve these models, or equations, and derive "Lanchester's square law," which states that the *strength* of a combat force is proportional to the square of the number of combatants entering the engagement. Finally, we will fit one of these models, with astonishing accuracy, to the battle of Iwo Jima in World War II.

(a) *Construction of the models*

Suppose that an "x-force" and a "y-force" are engaged in combat. For simplicity, we define the strengths of these two forces as their number of combatants. (See Howes and Thrall [3] for another definition of combat strength.) Thus let $x(t)$ and $y(t)$ denote the number of combatants of the x and y forces, where t is measured in days from the start of the combat. Clearly, the rate of change of each of these quantities equals its *reinforcement rate* minus its *operational loss rate* minus its *combat loss rate*.

The operational loss rate. The operational loss rate of a combat force is its loss rate due to non-combat mishaps; i.e., desertions, diseases, etc. Lanchester proposed that the operational loss rate of a combat force is proportional to its strength. However, this does not appear to be very realistic. For example, the desertion rate in a combat force depends on a host of psychological and other intangible factors which are difficult even to describe, let alone quantify. We will take the easy way out here and consider only those engagements in which the operational loss rates are negligible.

The combat loss rate. Suppose that the x-force is a conventional force which operates in the open, comparatively speaking, and that every member of this force is within "kill range" of the enemy y. We also assume that as soon as the conventional force suffers a loss, fire is concentrated on the remaining combatants. Under these "ideal" conditions, the combat loss rate of a conventional force x equals $ay(t)$, for some positive constant a. This constant is called the *combat effectiveness coefficient* of the y-force.

The situation is very different if x is a guerilla force, invisible to its opponent y and occupying a region R. The y-force fires into R but cannot know when a kill has been made. It is certainly plausible that the combat loss rate for a guerilla force x should be proportional to $x(t)$, for the larger $x(t)$, the greater the probability that an opponent's shot will kill. On the other hand, the combat loss rate for x is also proportional to $y(t)$, for the larger y, the greater the number of x-casualties. Thus, the combat loss rate for a guerilla force x equals $cx(t)y(t)$, where the constant c is called the *combat effectiveness coefficient* of the opponent y.

The reinforcement rate. The reinforcement rate of a combat force is the rate at which new combatants enter (or are withdrawn from) the battle. We denote the reinforcement rates of the x- and y-forces by $f(t)$ and $g(t)$ respectively.

Under the assumptions listed above, we can now write down the following two Lanchestrian models for conventional–guerilla combat.

$$\text{Conventional combat:} \quad \begin{cases} \dfrac{dx}{dt} = -ay + f(t) \\[2mm] \dfrac{dy}{dt} = -bx + g(t) \end{cases} \tag{1a}$$

$$\begin{matrix}\text{Conventional–guerilla combat:} \\ (x = \text{guerilla})\end{matrix} \quad \begin{cases} \dfrac{dx}{dt} = -cxy + f(t) \\[2mm] \dfrac{dy}{dt} = -dx + g(t) \end{cases} \tag{1b}$$

The system of equations (1a) is a linear system and can be solved explicitly once $a, b, f(t)$, and $g(t)$ are known. On the other hand, the system of equations (1b) is nonlinear, and its solution is much more difficult. (Indeed, it can only be obtained with the aid of a digital computer.)

It is very instructive to consider the special case where the reinforcement rates are zero. This situation occurs when the two forces are "isolated." In this case (1a) and (1b) reduce to the simpler systems

$$\frac{dx}{dt} = -ay, \qquad \frac{dy}{dt} = -bx \tag{2a}$$

and

$$\frac{dx}{dt} = -cxy, \qquad \frac{dy}{dt} = -dx. \tag{2b}$$

Conventional combat: The square law. The orbits of system (2a) are the solution curves of the equation

$$\frac{dy}{dx}=\frac{bx}{ay} \quad \text{or} \quad ay\frac{dy}{dx}=bx.$$

Integrating this equation gives

$$ay^2-bx^2=ay_0^2-bx_0^2=K. \tag{3}$$

The curves (3) define a family of hyperbolas in the x–y plane and we have indicated their graphs in Figure 1. The arrowheads on the curves indicate the direction of changing strengths as time passes.

Let us adopt the criterion that one force wins the battle if the other force vanishes first. Then, y wins if $K>0$ since the x-force has been annihilated by the time $y(t)$ has decreased to $\sqrt{K/a}$. Similarly, x wins if $K<0$.

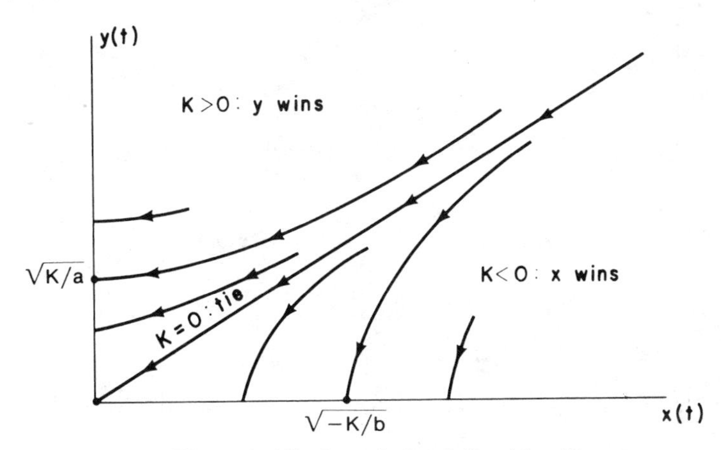

Figure 1. The hyperbolas defined by (3)

Remark 1. Equation (3) is often referred to as "Lanchester's square law," and the system (2a) is often called the square law model, since the strengths of the opposing forces appear *quadratically* in (3). This terminology is rather anomolous since the system (2a) is actually a linear system.

Remark 2. The y-force always seeks to establish a setting in which $K>0$. That is to say, the y-force wants the inequality

$$ay_0^2>bx_0^2$$

to hold. This can be accomplished by increasing a; i.e. by using stronger and more accurate weapons, or by increasing the initial force y_0. Notice though that a doubling of a results in a doubling of ay_0^2 while a doubling of y_0 results in a *four-fold* increase of ay_0^2. This is the essence of Lanchester's square law of conventional combat.

Conventional–guerilla combat. The orbits of system (2b) are the solution curves of the equation

$$\frac{dy}{dx} = \frac{dx}{cxy} = \frac{d}{cy}. \tag{4}$$

Multiplying both sides of (4) by cy and integrating gives

$$cy^2 - 2dx = cy_0^2 - 2dx_0 = M. \tag{5}$$

The curves (5) define a family of parabolas in the x–y plane, and we have indicated their graphs in Figure 2. The y-force wins if $M > 0$, since the x-force has been annihilated by the time $y(t)$ has decreased to $\sqrt{M/c}$. Similarly, x wins if $M < 0$.

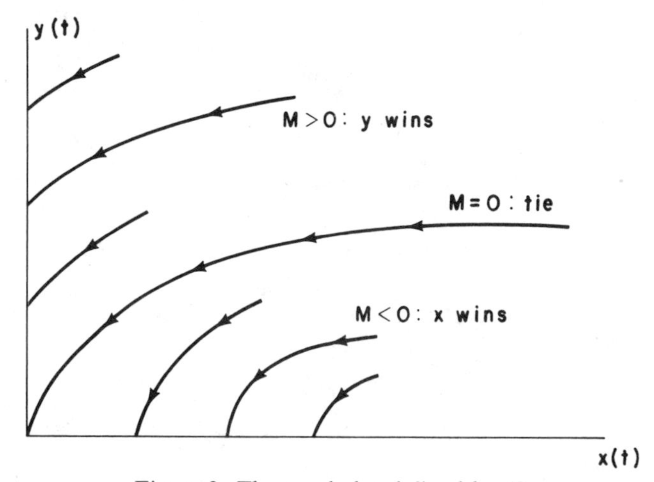

Figure 2. The parabolas defined by (5)

Remark. It is usually impossible to determine, a priori, the numerical value of the combat coefficients a, b, c, and d. Thus, it would appear that Lanchester's combat models have little or no applicability to real-life engagements. However, this is not so. As we shall soon see, it is often possible to determine suitable values of a and b (or c and d) using data from the battle itself. Once these values are established for one engagement, they are known for all other engagements which are fought under similar conditions.

(b) *The battle of Iwo Jima*

One of the fiercest battles of World War II was fought on the island of Iwo Jima, 660 miles south of Tokyo. Our forces coveted Iwo Jima as a bomber base close to the Japanese mainland, while the Japanese needed the island as a base for fighter planes attacking our aircraft on their way to bombing missions over Tokyo and other major Japanese cities. The American invasion of Iwo Jima began on February 19, 1945, and the fighting was intense throughout the month long combat. Both sides suffered heavy casualties (see Table 1). The Japanese had been ordered to fight to the last man, and this is exactly what they did. The island was declared "secure" by the American forces on the 28th day of the battle, and all active combat ceased on the 36th day. (The last two Japanese holdouts surrendered in 1951!)

Table 1. Casualties at Iwo Jima

Total United States casualties at Iwo Jima				
	Killed, missing or died of wounds	Wounded	Combat Fatigue	Total
Marines	5,931	17,272	2,648	25,851
Navy units:				
Ships and air units	633	1,158		1,791
Medical corpsmen	195	529		724
Seabees	51	218		269
Doctors and dentists	2	12		14
Army units in battle	9	28		37
Grand totals	6,821	19,217	2,648	28,686
Japanese casualties at Iwo Jima				
Defense forces (Estimated)		Prisoners		Killed
21,000		Marine 216		20,000
		Army 867		
		Total 1,083		

(Newcomb [6], page 296)

The following data is available to us from the battle of Iwo Jima.

1. *Reinforcement rates.* During the conflict Japanese troops were neither withdrawn nor reinforced. The Americans, on the other hand, landed 54,000 troops on the first day of the battle, none on the second, 6,000 on the third, none on the fourth and fifth, 13,000 on the sixth day, and none thereafter. There were no American troops on Iwo Jima prior to the start of the engagement.

2. *Combat losses.* Captain Clifford Morehouse of the United States Marine Corps (see Morehouse [5]) kept a daily count of all American combat losses. Unfortunately, no such records are available for the Japanese forces. Most probably, the casualty lists kept by General Kuribayashi (commander of the Japanese forces on Iwo Jima) were destroyed in the battle itself, while whatever records were kept in Tokyo were consumed in the fire bombings of the remaining five months of the war. However, we can infer from Table 1 that approximately 21,500 Japanese forces were on Iwo Jima at the start of the battle. (Actually, Newcomb arrived at the fig-

ure of 21,000 for the Japanese forces, but this is a little low since he apparently did not include some of the living and dead found in the caves in the final days.)

3. *Operational losses*. The operational losses on both sides were negligible.

Now, let $x(t)$ and $y(t)$ denote respectively, the active American and Japanese forces on Iwo Jima t days after the battle began. The data above suggests the following Lanchestrian model for the battle of Iwo Jima:

$$\begin{aligned} \frac{dx}{dt} &= -ay + f(t) \\ \frac{dy}{dt} &= -bx \end{aligned} \tag{6}$$

where a and b are the combat effectiveness coefficients of the Japanese and American forces, respectively, and

$$f(t) = \begin{cases} 54{,}000 & 0 \leqslant t < 1 \\ 0 & 1 \leqslant t < 2 \\ 6{,}000 & 2 \leqslant t < 3 \\ 0 & 3 \leqslant t < 5 \\ 13{,}000 & 5 \leqslant t < 6 \\ 0 & t \geqslant 6 \end{cases}$$

Using the method of variation of parameters developed in Section 3.12 or the method of elimination in Section 2.14, it is easily seen that the solution of (6) which satisfies $x(0)=0$, $y(0)=y_0=21{,}500$ is given by

$$x(t) = -\sqrt{\frac{a}{b}}\, y_0 \cosh\sqrt{ab}\, t + \int_0^t \cosh\sqrt{ab}\,(t-s) f(s)\, ds \tag{7a}$$

and

$$y(t) = y_0 \cosh\sqrt{ab}\, t - \sqrt{\frac{b}{a}} \int_0^t \sinh\sqrt{ab}\,(t-s) f(s)\, ds \tag{7b}$$

where

$$\cosh x \equiv (e^x + e^{-x})/2 \quad \text{and} \quad \sinh x \equiv (e^x - e^{-x})/2.$$

The problem before us now is this: Do there exist constants a and b so that (7a) yields a good fit to the data compiled by Morehouse? This is an extremely important question. An affirmative answer would indicate that Lanchestrian models do indeed describe real life battles, while a negative answer would shed a dubious light on much of Lanchester's work.

As we mentioned previously, it is extremely difficult to compute the combat effectiveness coefficients a and b of two opposing forces. However, it is often possible to determine suitable values of a and b once the data for the battle is known, and such is the case for the battle of Iwo Jima.

The calculation of a and b. Integrating the second equation of (6) between 0 and s gives

$$y(s)-y_0=-b\int_0^s x(t)\,dt$$

so that

$$b=\frac{y_0-y(s)}{\int_0^s x(t)\,dt}. \tag{8}$$

In particular, setting $s=36$ gives

$$b=\frac{y_0-y(36)}{\int_0^{36} x(t)\,dt}=\frac{21{,}500}{\int_0^{36} x(t)\,dt}. \tag{9}$$

Now the integral on the right-hand side of (9) can be approximated by the Riemann sum

$$\int_0^{36} x(t)\,dt \cong \sum_{i=1}^{36} x(i)$$

and for $x(i)$ we enter the number of effective American troops on the ith day of the battle. Using the data available from Morehouse, we compute for b the value

$$b=\frac{21{,}500}{2{,}037{,}000}=0.0106. \tag{10}$$

Remark. We would prefer to set $s=28$ in (8) since that was the day the island was declared secure, and the fighting was only sporadic after this day. However, we don't know $y(28)$. Thus, we are forced here to take $s=36$.

Next, we integrate the first equation of (6) between $t=0$ and $t=28$ and obtain that

$$\begin{aligned} x(28) &= -a\int_0^{28} y(t)\,dt+\int_0^{28} f(t)\,dt \\ &= -a\int_0^{28} y(t)\,dt+73{,}000. \end{aligned}$$

There were 52,735 effective American troops on the 28th day of the battle. Thus

$$a=\frac{73{,}000-52{,}735}{\int_0^{28} y(t)\,dt}=\frac{20{,}265}{\int_0^{28} y(t)\,dt}. \tag{11}$$

Finally, we approximate the integral on the right-hand side of (11) by the Riemann sum

$$\int_0^{28} y(t)\,dt \cong \sum_{j=1}^{28} y(j)$$

and we approximate $y(j)$ by

$$\begin{aligned} y(j) &= y_0 - b\int_0^j x(t)\,dt \\ &\cong 21{,}500 - b\sum_{i=i}^{j} x(i). \end{aligned}$$

Again, we replace $x(i)$ by the number of effective American troops on the ith day of the battle. The result of this calculation is (see Engel [2])

$$a = \frac{20{,}265}{372{,}500} = 0.0544. \tag{12}$$

Figure 3 below compares the actual American troop strength with the values predicted by Equation (7a) (with $a=0.0544$ and $b=0.0106$). The fit is remarkably good. Thus, it appears that a Lanchestrian model does indeed describe real life engagements.

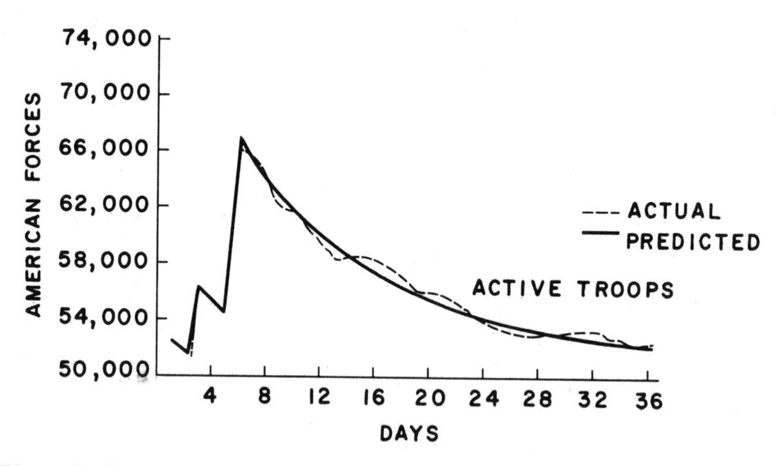

Figure 3. Comparison of actual troop strength with predicted troop strength

Remark. The figures we have used for American reinforcements include *all* the personnel put ashore, both combat troops and support troops. Thus the numbers a and b that we have computed should be interpreted as the *average* effectiveness per man ashore.

References

1. Coleman, C. S., Combat Models, MAA Workshop on Modules in Applied Math, Cornell University, Aug. 1976.
2. Engel, J. H., A verification of Lanchester's law, *Operations Research*, **2**, (1954), 163–171.
3. Howes, D. R., and Thrall, R. M., A theory of ideal linear weights for heterogeneous combat forces, *Naval Research Logistics Quarterly*, vol. 20, 1973, pp. 645–659.
4. Lanchester, F. W., *Aircraft in Warfare, the Dawn of the Fourth Arm*. Tiptree, Constable and Co., Ltd., 1916.
5. Morehouse, C. P., *The Iwo Jima Operation*, USMCR, Historical Division, Hdqr. USMC, 1946.
6. Newcomb, R. F., *Iwo Jima*. New York: Holt, Rinehart, and Winston, 1965.

EXERCISES

1. Derive Equations (7a) and (7b).

2. The system of equations

$$\begin{aligned}\dot{x} &= -ay\\ \dot{y} &= -by - cxy\end{aligned}\tag{13}$$

is a Lanchestrian model for conventional–guerilla combat, in which the operational loss rate of the guerilla force y is proportional to $y(t)$.
(a) Find the orbits of (13).
(b) Who wins the battle?

3. The system of equations

$$\begin{aligned}\dot{x} &= -ay\\ \dot{y} &= -bx - cxy\end{aligned}\tag{14}$$

is a Lanchestrian model for conventional–guerilla combat, in which the operational loss rate of the guerilla force y is proportional to the strength of the conventional force x. Find the orbits of (14).

4. The system of equations

$$\begin{aligned}\dot{x} &= -cxy\\ \dot{y} &= -dxy\end{aligned}\tag{15}$$

is a Lanchestrian model for guerilla–guerilla combat in which the operational loss rates are negligible.
(a) Find the orbits of (15).
(b) Who wins the battle?

5. The system of equations

$$\begin{aligned}\dot{x} &= -ay - cxy\\ \dot{y} &= -bx - dxy\end{aligned}\tag{16}$$

is a Lanchestrian model for guerilla–guerilla combat in which the operational loss rate of each force is proportional to the strength of its opponent. Find the orbits of (16).

6. The system of equations

$$\begin{aligned} \dot{x} &= -ax - cxy \\ \dot{y} &= -by - dxy \end{aligned} \tag{17}$$

is a Lanchestrian model for guerilla–guerilla combat in which the operational loss rate of each force is proportional to its strength.
(a) Find the orbits of (17).
(b) Show that the x and y axes are both orbits of (17).
(c) Using the fact (to be proved in Section 4.6) that two orbits of (17) cannot intersect, show that there is no clear-cut winner in this battle. *Hint*: Show that $x(t)$ and $y(t)$ can never become zero in finite time. (Using lemmas 1 and 2 of Section 4.8, it is easy to show that both $x(t)$ and $y(t)$ approach zero as $t \to \infty$.)

4.6 Qualitative properties of orbits

In this section we will derive two very important properties of the solutions and orbits of the system of differential equations

$$\dot{\mathbf{x}} = \mathbf{f}(\mathbf{x}), \qquad \mathbf{x} = \begin{bmatrix} x_1 \\ \vdots \\ x_n \end{bmatrix}, \qquad \mathbf{f}(\mathbf{x}) = \begin{bmatrix} f_1(x_1, \ldots, x_n) \\ \vdots \\ f_n(x_1, \ldots, x_n) \end{bmatrix}. \tag{1}$$

The first property deals with the existence and uniqueness of orbits, and the second property deals with the existence of periodic solutions of (1). We begin with the following existence–uniqueness theorem for the solutions of (1).

Theorem 3. *Let each of the functions $f_1(x_1, \ldots, x_n), \ldots, f_n(x_1, \ldots, x_n)$ have continuous partial derivatives with respect to $x_1, \ldots, x_n$. Then, the initial-value problem $\dot{\mathbf{x}} = \mathbf{f}(\mathbf{x})$, $\mathbf{x}(t_0) = \mathbf{x}^0$ has one, and only one solution $\mathbf{x} = \mathbf{x}(t)$, for every $\mathbf{x}^0$ in R^n.*

We prove Theorem 3 in exactly the same manner as we proved the existence–uniqueness theorem for the scalar differential equation $\dot{x} = f(t,x)$. Indeed, the proof given in Section 1.10 carries over here word for word. We need only interpret the quantity $|\mathbf{x}(t) - \mathbf{y}(t)|$, where $\mathbf{x}(t)$ and $\mathbf{y}(t)$ are vector-valued functions, as the length of the vector $\mathbf{x}(t) - \mathbf{y}(t)$. That is to say, if we interpret $|\mathbf{x}(t) - \mathbf{y}(t)|$ as

$$|\mathbf{x}(t) - \mathbf{y}(t)| = \max\{|x_1(t) - y_1(t)|, \ldots, |x_n(t) - y_n(t)|\},$$

then the proof of Theorem 2, Section 1.10 is valid even for vector-valued functions $\mathbf{f}(t, \mathbf{x})$ (see Exercises 13–14).

Next, we require the following simple but extremely useful lemma.

Lemma 1. *If $\mathbf{x} = \boldsymbol{\phi}(t)$ is a solution of* (1), *then $\mathbf{x} = \boldsymbol{\phi}(t + c)$ is again a solution of* (1).

The meaning of Lemma 1 is the following. Let $\mathbf{x}=\boldsymbol{\phi}(t)$ be a solution of (1) and let us replace every t in the formula for $\boldsymbol{\phi}(t)$ by $t+c$. In this manner we obtain a new function $\hat{\mathbf{x}}(t)=\boldsymbol{\phi}(t+c)$. Lemma 1 states that $\hat{\mathbf{x}}(t)$ is again a solution of (1). For example, $x_1=\tan t$, $x_2=\sec^2 t$ is a solution of the system of differential equations $dx_1/dt=x_2$, $dx_2/dt=2x_1x_2$. Hence, $x_1=\tan(t+c)$, $x_2=\sec^2(t+c)$ is again a solution, for any constant c.

PROOF OF LEMMA 1. If $\mathbf{x}=\boldsymbol{\phi}(t)$ is a solution of (1), then $d\boldsymbol{\phi}(t)/dt=\mathbf{f}(\boldsymbol{\phi}(t))$; that is, the two functions $d\boldsymbol{\phi}(t)/dt$ and $\mathbf{h}(t)\equiv\mathbf{f}(\boldsymbol{\phi}(t))$ agree at every single time. Fix a time t and a constant c. Since $d\boldsymbol{\phi}/dt$ and $\mathbf{h}$ agree at every time, they must agree at time $t+c$. Hence,

$$\frac{d\boldsymbol{\phi}}{dt}(t+c)=\mathbf{h}(t+c)=\mathbf{f}(\boldsymbol{\phi}(t+c)).$$

But, $d\boldsymbol{\phi}/dt$ evaluated at $t+c$ equals the derivative of $\hat{\mathbf{x}}(t)\equiv\boldsymbol{\phi}(t+c)$, evaluated at t. Therefore,

$$\frac{d}{dt}\boldsymbol{\phi}(t+c)=\mathbf{f}(\boldsymbol{\phi}(t+c)).$$ □

Remark 1. Lemma 1 can be verified explicitly for the linear equation $\dot{\mathbf{x}}=\mathbf{A}\mathbf{x}$. Every solution $\mathbf{x}(t)$ of this equation is of the form $\mathbf{x}(t)=e^{\mathbf{A}t}\mathbf{v}$, for some constant vector $\mathbf{v}$. Hence,

$$\mathbf{x}(t+c)=e^{\mathbf{A}(t+c)}\mathbf{v}=e^{\mathbf{A}t}e^{\mathbf{A}c}\mathbf{v}$$

since $(\mathbf{A}t)\mathbf{A}c=\mathbf{A}c(\mathbf{A}t)$ for all values of t and c. Therefore, $\mathbf{x}(t+c)$ is again a solution of $\dot{\mathbf{x}}=\mathbf{A}\mathbf{x}$ since it is of the form $e^{\mathbf{A}t}$ times the constant vector $e^{\mathbf{A}c}\mathbf{v}$.

Remark 2. Lemma 1 is not true if the function $\mathbf{f}$ in (1) depends explicitly on t. To see this, suppose that $\mathbf{x}=\boldsymbol{\phi}(t)$ is a solution of the nonautonomous differential equation $\dot{\mathbf{x}}=\mathbf{f}(t,\mathbf{x})$. Then, $\dot{\boldsymbol{\phi}}(t+c)=\mathbf{f}(t+c,\boldsymbol{\phi}(t+c))$. Consequently, the function $\mathbf{x}=\boldsymbol{\phi}(t+c)$ satisfies the differential equation

$$\dot{\mathbf{x}}=\mathbf{f}(t+c,\mathbf{x}),$$

and this equation is different from (1) if $\mathbf{f}$ depends explicitly on t.

We are now in a position to derive the following extremely important properties of the solutions and orbits of (1).

Property 1. (Existence and uniqueness of orbits.) Let each of the functions $f_1(x_1,\ldots,x_n),\ldots,f_n(x_1,\ldots,x_n)$ have continuous partial derivatives with respect to $x_1,\ldots,x_n$. Then, there exists one, and only one, orbit through every point $\mathbf{x}^0$ in R^n. In particular, if the orbits of two solutions $\mathbf{x}=\boldsymbol{\phi}(t)$ and $\mathbf{x}=\boldsymbol{\psi}(t)$ of (1) have one point in common, then they must be identical.

Property 2. Let $\mathbf{x}=\boldsymbol{\phi}(t)$ be a solution of (1). If $\boldsymbol{\phi}(t_0+T)=\boldsymbol{\phi}(t_0)$ for some t_0 and $T>0$, then $\boldsymbol{\phi}(t+T)$ is identically equal to $\boldsymbol{\phi}(t)$. In other words, if a solution $\mathbf{x}(t)$ of (1) returns to its starting value after a time $T>0$, then it must be periodic, with period T (i.e. it must repeat itself over every time interval of length T.)

PROOF OF PROPERTY 1. Let $\mathbf{x}^0$ be any point in the n-dimensional phase space $x_1,\ldots,x_n$, and let $\mathbf{x}=\boldsymbol{\phi}(t)$ be the solution of the initial-value problem $\dot{\mathbf{x}}=\mathbf{f}(\mathbf{x})$, $\mathbf{x}(0)=\mathbf{x}^0$. The orbit of this solution obviously passes through $\mathbf{x}^0$. Hence, there exists at least one orbit through every point $\mathbf{x}^0$. Now, suppose that the orbit of another solution $\mathbf{x}=\boldsymbol{\psi}(t)$ also passes through $\mathbf{x}^0$. This means that there exists $t_0(\neq 0)$ such that $\boldsymbol{\psi}(t_0)=\mathbf{x}^0$. By Lemma 1,

$$\mathbf{x}=\boldsymbol{\psi}(t+t_0)$$

is also a solution of (1). Observe that $\boldsymbol{\psi}(t+t_0)$ and $\boldsymbol{\phi}(t)$ have the same value at $t=0$. Hence, by Theorem 3, $\boldsymbol{\psi}(t+t_0)$ equals $\boldsymbol{\phi}(t)$ for all time t. This implies that the orbits of $\boldsymbol{\phi}(t)$ and $\boldsymbol{\psi}(t)$ are identical. To wit, if $\boldsymbol{\xi}$ is a point on the orbit of $\boldsymbol{\phi}(t)$; that is, $\boldsymbol{\xi}=\boldsymbol{\phi}(t_1)$ for some t_1, then $\boldsymbol{\xi}$ is also on the orbit of $\boldsymbol{\psi}(t)$, since $\boldsymbol{\xi}=\boldsymbol{\phi}(t_1)=\boldsymbol{\psi}(t_1+t_0)$. Conversely, if $\boldsymbol{\xi}$ is a point on the orbit of $\boldsymbol{\psi}(t)$; that is, there exists t_2 such that $\boldsymbol{\psi}(t_2)=\boldsymbol{\xi}$, then $\boldsymbol{\xi}$ is also on the orbit of $\boldsymbol{\phi}(t)$ since $\boldsymbol{\xi}=\boldsymbol{\psi}(t_2)=\boldsymbol{\phi}(t_2-t_0)$. □

PROOF OF PROPERTY 2. Let $\mathbf{x}=\boldsymbol{\phi}(t)$ be a solution of (1) and suppose that $\boldsymbol{\phi}(t_0+T)=\boldsymbol{\phi}(t_0)$ for some numbers t_0 and T. Then, the function $\boldsymbol{\psi}(t)=\boldsymbol{\phi}(t+T)$ is also a solution of (1) which agrees with $\boldsymbol{\phi}(t)$ at time $t=t_0$. By Theorem 3, therefore, $\boldsymbol{\psi}(t)=\boldsymbol{\phi}(t+T)$ is identically equal to $\boldsymbol{\phi}(t)$. □

Property 2 is extremely useful in applications, especially when $n=2$. Let $x=x(t)$, $y=y(t)$ be a periodic solution of the system of differential equations

$$\frac{dx}{dt}=f(x,y), \qquad \frac{dy}{dt}=g(x,y). \tag{2}$$

If $x(t+T)=x(t)$ and $y(t+T)=y(t)$, then the orbit of this solution is a closed curve C in the x–y plane. In every time interval $t_0\leqslant t\leqslant t_0+T$, the solution moves once around C. Conversely, suppose that the orbit of a solution $x=x(t), y=y(t)$ of (2) is a closed curve containing no equilibrium points of (2). Then, the solution $x=x(t), y=y(t)$ is periodic. To prove this, recall that a solution $x=x(t)$, $y=y(t)$ of (2) moves along its orbit with velocity $[f^2(x,y)+g^2(x,y)]^{1/2}$. If its orbit C is a closed curve containing no equilibrium points of (2), then the function $[f^2(x,y)+g^2(x,y)]^{1/2}$ has a positive minimum for (x,y) on C. Hence, the orbit of $x=x(t)$, $y=y(t)$ must return to its starting point $x_0=x(t_0), y_0=y(t_0)$ in some finite time T. But this implies that $x(t+T)=x(t)$ and $y(t+T)=y(t)$ for all t.

Example 1. Prove that every solution $z(t)$ of the second-order differential equation $(d^2z/dt^2)+z+z^3=0$ is periodic.

PROOF. We convert this second-order equation into a system of two first-order equations by setting $x = z$, $y = dz/dt$. Then,

$$\frac{dx}{dt} = y, \qquad \frac{dy}{dt} = -x - x^3. \tag{3}$$

The orbits of (3) are the solution curves

$$\frac{y^2}{2} + \frac{x^2}{2} + \frac{x^4}{4} = c^2 \tag{4}$$

of the scalar equation $dy/dx = -(x + x^3)/y$. Equation (4) defines a closed curve in the x–y plane (see Exercise 7). Moreover, the only equilibrium point of (3) is $x = 0, y = 0$. Consequently, every solution $x = z(t), y = z'(t)$ of (3) is a periodic function of time. Notice, however, that we cannot compute the period of any particular solution. □

Example 2. Prove that every solution of the system of differential equations

$$\frac{dx}{dt} = ye^{1+x^2+y^2}, \qquad \frac{dy}{dt} = -xe^{1+x^2+y^2} \tag{5}$$

is periodic.

Solution. The orbits of (5) are the solution curves $x^2 + y^2 = c^2$ of the first-order scalar equation $dy/dx = -x/y$. Moreover, $x = 0$, $y = 0$ is the only equilibrium point of (5). Consequently, every solution $x = x(t)$, $y = y(t)$ of (5) is a periodic function of time.

EXERCISES

1. Show that all solutions $x(t), y(t)$ of

$$\frac{dx}{dt} = x^2 + y\sin x, \qquad \frac{dy}{dt} = -1 + xy + \cos y$$

which start in the first quadrant ($x > 0$, $y > 0$) must remain there for all time (both backwards and forwards).

2. Show that all solutions $x(t), y(t)$ of

$$\frac{dx}{dt} = y(e^x - 1), \qquad \frac{dy}{dt} = x + e^y$$

which start in the right half plane ($x > 0$) must remain there for all time.

3. Show that all solutions $x(t), y(t)$ of

$$\frac{dx}{dt} = 1 + x^2 + y^2, \qquad \frac{dy}{dt} = xy + \tan y$$

which start in the upper half plane ($y > 0$) must remain there for all time.

4. Show that all solutions $x(t), y(t)$ of

$$\frac{dx}{dt} = -1 - y + x^2, \qquad \frac{dy}{dt} = x + xy$$

which start inside the unit circle $x^2 + y^2 = 1$ must remain there for all time. *Hint*: Compute $d(x^2 + y^2)/dt$.

5. Let $x(t)$, $y(t)$ be a solution of

$$\frac{dx}{dt}=y+x^2, \qquad \frac{dy}{dt}=x+y^2$$

with $x(t_0)\neq y(t_0)$. Show that $x(t)$ can never equal $y(t)$.

6. Can a figure 8 ever be an orbit of

$$\frac{dx}{dt}=f(x,y), \qquad \frac{dy}{dt}=g(x,y)$$

where f and g have continuous partial derivatives with respect to x and y?

7. Show that the curve $y^2+x^2+x^4/2=2c^2$ is closed. *Hint*: Show that there exist two points $y=0$, $x=\pm\alpha$ which lie on this curve.

Prove that all solutions of the following second-order equations are periodic.

8. $\dfrac{d^2z}{dt^2}+z^3=0$

9. $\dfrac{d^2z}{dt^2}+z+z^5=0$

10. $\dfrac{d^2z}{dt^2}+e^{z^2}=1$

11. $\dfrac{d^2z}{dt^2}+\dfrac{z}{1+z^2}=0$

12. Show that all solutions $z(t)$ of

$$\frac{d^2z}{dt^2}+z-2z^3=0$$

are periodic if $\dot{z}^2(0)+z^2(0)-z^4(0)<\frac{1}{4}$, and unbounded if

$$\dot{z}^2(0)+z^2(0)-z^4(0)>\tfrac{1}{4}.$$

13. (a) Let

$$L=n\times\max_{i,j=1,\dots,n}|\partial f_i/\partial x_j|, \quad \text{for} \quad |\mathbf{x}-\mathbf{x}^0|\leqslant b.$$

Show that $|\mathbf{f}(\mathbf{x})-\mathbf{f}(\mathbf{y})|\leqslant L|\mathbf{x}-\mathbf{y}|$ if $|\mathbf{x}-\mathbf{x}^0|\leqslant b$ and $|\mathbf{y}-\mathbf{x}^0|\leqslant b$.

(b) Let $M=\max|\mathbf{f}(\mathbf{x})|$ for $|\mathbf{x}-\mathbf{x}^0|\leqslant b$. Show that the Picard iterates

$$\mathbf{x}_{j+1}(t)=\mathbf{x}^0+\int_{t_0}^{t}\mathbf{f}(\mathbf{x}_j(s))\,ds, \qquad \mathbf{x}_0(t)=\mathbf{x}^0$$

converge to a solution $\mathbf{x}(t)$ of the initial-value problem $\dot{\mathbf{x}}=\mathbf{f}(\mathbf{x})$, $\mathbf{x}(t_0)=\mathbf{x}^0$ on the interval $|t-t_0|\leqslant b/M$. *Hint*: The proof of Theorem 2, Section 1.10 carries over here word for word.

14. Compute the Picard iterates $\mathbf{x}_j(t)$ of the initial-value problem $\dot{\mathbf{x}}=\mathbf{A}\mathbf{x}$, $\mathbf{x}(0)=\mathbf{x}^0$, and verify that they approach $e^{\mathbf{A}t}\mathbf{x}^0$ as j approaches infinity.

4.7 Phase portraits of linear systems

In this section we present a complete picture of all orbits of the linear differential equation

$$\dot{\mathbf{x}}=\mathbf{A}\mathbf{x}, \qquad \mathbf{x}=\begin{pmatrix}x_1\\x_2\end{pmatrix}, \qquad \mathbf{A}=\begin{pmatrix}a & b\\ c & d\end{pmatrix}. \tag{1}$$

This picture is called a phase portrait, and it depends almost completely on the eigenvalues of the matrix $\mathbf{A}$. It also changes drastically as the eigenvalues of $\mathbf{A}$ change sign or become imaginary.

When analyzing Equation (1), it is often helpful to visualize a vector

$$\mathbf{x}=\begin{pmatrix} x_1 \\ x_2 \end{pmatrix}$$

in R^2 as a direction, or directed line segment, in the plane. Let

$$\mathbf{x}=\begin{pmatrix} x_1 \\ x_2 \end{pmatrix}$$

be a vector in R^2 and draw the directed line segment $\vec{x}$ from the point $(0,0)$ to the point (x_1,x_2), as in Figure 1a. This directed line segment is parallel to the line through $(0,0)$ with direction numbers x_1,x_2 respectively. If we visualize the vector $\mathbf{x}$ as being this directed line segment $\vec{x}$, then we see that the vectors $\mathbf{x}$ and $c\mathbf{x}$ are parallel if c is positive, and antiparallel if c is negative. We can also give a nice geometric interpretation of vector addition. Let $\mathbf{x}$ and $\mathbf{y}$ be two vectors in R^2. Draw the directed line segment $\vec{x}$, and place the vector $\vec{y}$ at the tip of $\vec{x}$. The vector $\vec{x}+\vec{y}$ is then the composi-

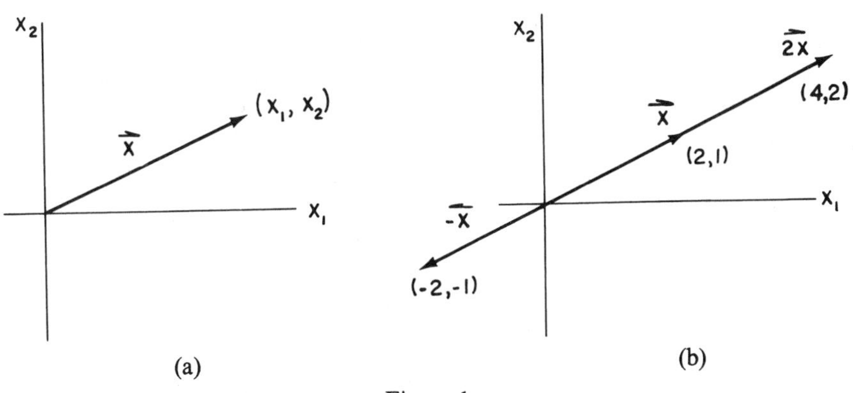

Figure 1

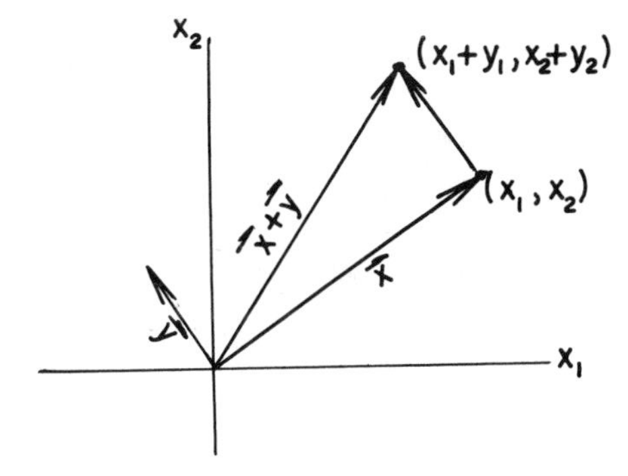

Figure 2

tion of these two directed line segments (see Figure 2). This construction is known as the parallelogram law of vector addition.

We are now in a position to derive the phase portraits of (1). Let λ_1 and λ_2 denote the two eigenvalues of **A**. We distinguish the following cases.

1. $\lambda_2<\lambda_1<0$. Let $\mathbf{v}^1$ and $\mathbf{v}^2$ be eigenvectors of **A** with eigenvalues λ_1 and λ_2 respectively. In the x_1-x_2 plane we draw the four half-lines l_1, l_1', l_2, and l_2', as shown in Figure 3. The rays l_1 and l_2 are parallel to $\mathbf{v}^1$ and $\mathbf{v}^2$, while the rays l_1' and l_2' are parallel to $-\mathbf{v}^1$ and $-\mathbf{v}^2$. Observe first that $\mathbf{x}(t)=ce^{\lambda_1 t}\mathbf{v}^1$ is a solution of (1) for any constant c. This solution is always proportional to $\mathbf{v}^1$, and the constant of proportionality, $ce^{\lambda_1 t}$, runs from $\pm\infty$ to 0, depending as to whether c is positive or negative. Hence, the orbit of this solution is the half-line l_1 for $c>0$, and the half-line l_1' for $c<0$. Similarly, the orbit of the solution $\mathbf{x}(t)=ce^{\lambda_2 t}\mathbf{v}^2$ is the half-line l_2 for $c>0$, and the half-line l_2' for $c<0$. The arrows on these four lines in Figure 3 indicate in what direction $\mathbf{x}(t)$ moves along its orbit.

Next, recall that every solution $\mathbf{x}(t)$ of (1) can be written in the form

$$\mathbf{x}(t)=c_1e^{\lambda_1 t}\mathbf{v}^1+c_2e^{\lambda_2 t}\mathbf{v}^2 \tag{2}$$

for some choice of constants c_1 and c_2. Obviously, every solution $\mathbf{x}(t)$ of (1) approaches $\binom{0}{0}$ as t approaches infinity. Hence, every orbit of (1) approaches the origin $x_1=x_2=0$ as t approaches infinity. We can make an even stronger statement by observing that $e^{\lambda_2 t}\mathbf{v}^2$ is very small compared to $e^{\lambda_1 t}\mathbf{v}^1$ when t is very large. Therefore, $\mathbf{x}(t)$, for $c_1\neq 0$, comes closer and closer to $c_1e^{\lambda_1 t}\mathbf{v}^1$ as t approaches infinity. This implies that the tangent to the orbit of $\mathbf{x}(t)$ approaches l_1 if c_1 is positive, and l_1' if c_1 is negative. Thus,

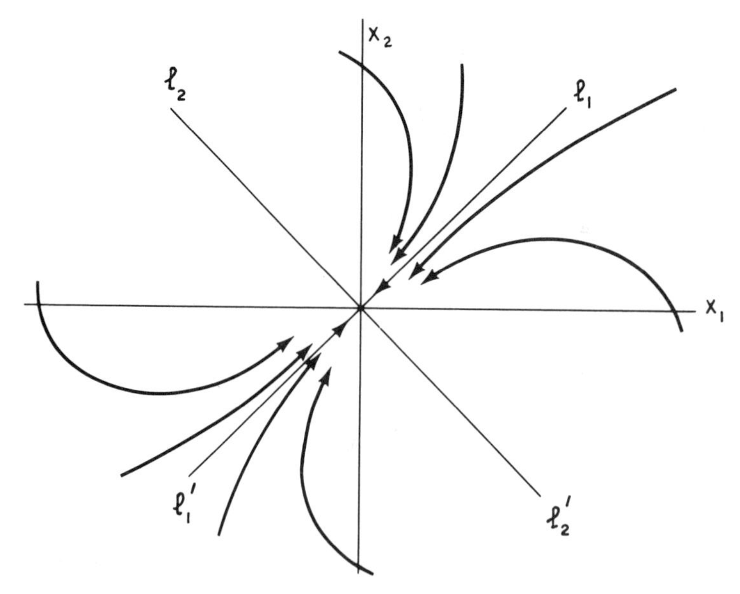

Figure 3. Phase portrait of a stable node

the phase portrait of (1) has the form described in Figure 3. The distinguishing feature of this phase portrait is that every orbit, with the exception of a single line, approaches the origin in a fixed direction (if we consider the directions $\mathbf{v}^1$ and $-\mathbf{v}^1$ equivalent). In this case we say that the equilibrium solution $\mathbf{x}=\mathbf{0}$ of (1) is a stable node.

Remark. The orbit of every solution $\mathbf{x}(t)$ of (1) approaches the origin $x_1=x_2=0$ as t approaches infinity. However, this point does not belong to the orbit of any nontrivial solution $\mathbf{x}(t)$.

1′. $0<\lambda_1<\lambda_2$. The phase portrait of (1) in this case is exactly the same as Figure 3, except that the direction of the arrows is reversed. Hence, the equilibrium solution $\mathbf{x}(t)=\mathbf{0}$ of (1) is an unstable node if both eigenvalues of $\mathbf{A}$ are positive.

2. $\lambda_1=\lambda_2<0$. In this case, the phase portrait of (1) depends on whether $\mathbf{A}$ has one or two linearly independent eigenvectors. (a) Suppose that $\mathbf{A}$ has two linearly independent eigenvectors $\mathbf{v}^1$ and $\mathbf{v}^2$ with eigenvalue $\lambda<0$. In this case, every solution $\mathbf{x}(t)$ of (1) can be written in the form

$$\mathbf{x}(t)=c_1e^{\lambda t}\mathbf{v}^1+c_2e^{\lambda t}\mathbf{v}^2=e^{\lambda t}\left(c_1\mathbf{v}^1+c_2\mathbf{v}^2\right) \tag{2}$$

for some choice of constants c_1 and c_2. Now, the vector $e^{\lambda t}(c_1\mathbf{v}^1+c_2\mathbf{v}^2)$ is parallel to $c_1\mathbf{v}^1+c_2\mathbf{v}^2$ for all t. Hence, the orbit of every solution $\mathbf{x}(t)$ of (1) is a half-line. Moreover, the set of vectors $\{c_1\mathbf{v}^1+c_2\mathbf{v}^2\}$, for all choices of c_1 and c_2, cover every direction in the x_1-x_2 plane, since $\mathbf{v}^1$ and $\mathbf{v}^2$ are linearly independent. Hence, the phase portrait of (1) has the form described in Figure 4a. (b) Suppose that $\mathbf{A}$ has only one linearly independent eigenvector $\mathbf{v}$, with eigenvalue λ. In this case, $\mathbf{x}^1(t)=e^{\lambda t}\mathbf{v}$ is one solution of (1). To find a second solution of (1) which is independent of $\mathbf{x}^1$, we observe that $(\mathbf{A}-\lambda\mathbf{I})^2\mathbf{u}=\mathbf{0}$ for every vector $\mathbf{u}$. Hence,

$$\mathbf{x}(t)=e^{\mathbf{A}t}\mathbf{u}=e^{\lambda t}e^{(\mathbf{A}-\lambda\mathbf{I})t}\mathbf{u}=e^{\lambda t}\left[\mathbf{u}+t(\mathbf{A}-\lambda\mathbf{I})\mathbf{u}\right] \tag{3}$$

is a solution of (1) for any choice of $\mathbf{u}$. Equation (3) can be simplified by observing that $(\mathbf{A}-\lambda\mathbf{I})\mathbf{u}$ must be a multiple k of $\mathbf{v}$. This follows immediately from the equation $(\mathbf{A}-\lambda\mathbf{I})[(\mathbf{A}-\lambda\mathbf{I})\mathbf{u}]=\mathbf{0}$, and the fact that $\mathbf{A}$ has only one linearly independent eigenvector $\mathbf{v}$. Choosing $\mathbf{u}$ independent of $\mathbf{v}$, we see that every solution $\mathbf{x}(t)$ of (1) can be written in the form

$$\mathbf{x}(t)=c_1e^{\lambda t}\mathbf{v}+c_2e^{\lambda t}\left(\mathbf{u}+kt\mathbf{v}\right)=e^{\lambda t}\left(c_1\mathbf{v}+c_2\mathbf{u}+c_2kt\mathbf{v}\right), \tag{4}$$

for some choice of constants c_1 and c_2. Obviously, every solution $\mathbf{x}(t)$ of (1) approaches $\binom{0}{0}$ as t approaches infinity. In addition, observe that $c_1\mathbf{v}+c_2\mathbf{u}$ is very small compared to $c_2kt\mathbf{v}$ if c_2 is unequal to zero and t is very large. Hence, the tangent to the orbit of $\mathbf{x}(t)$ approaches $\pm\mathbf{v}$ (depending on the sign of c_2) as t approaches infinity, and the phase portrait of (1) has the form described in Figure 4b.

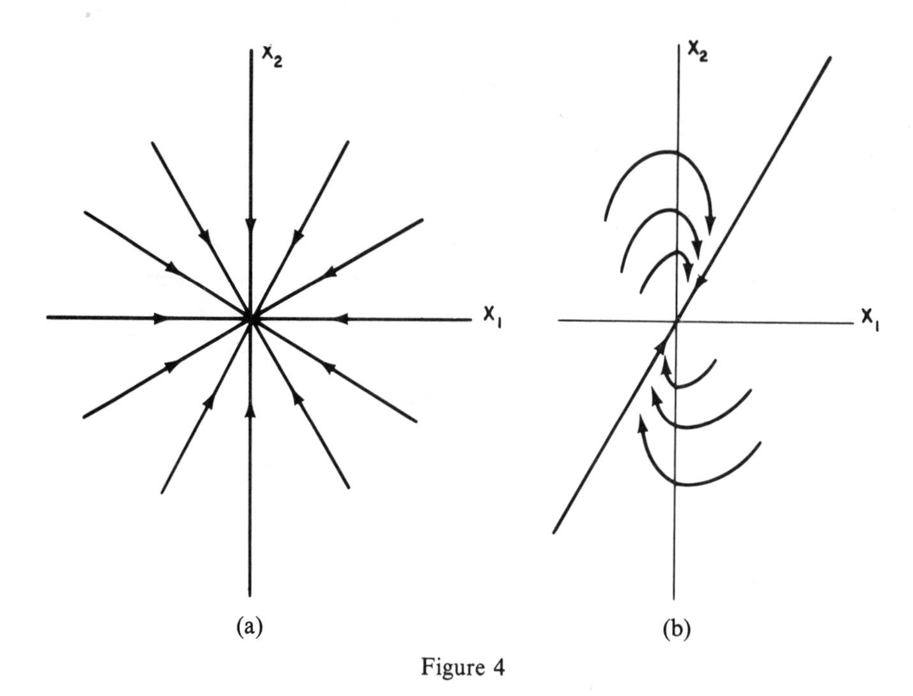

Figure 4

2′. $\lambda_1 = \lambda_2 > 0$. The phase portraits of (1) in the cases (2a)′ and (2b)′ are exactly the same as Figures 4a and 4b, except that the direction of the arrows is reversed.

3. $\lambda_1 < 0 < \lambda_2$. Let $\mathbf{v}^1$ and $\mathbf{v}^2$ be eigenvectors of $\mathbf{A}$ with eigenvalues λ_1 and λ_2 respectively. In the $x_1 - x_2$ plane we draw the four half-lines l_1, l_1', l_2, and l_2'; the half-lines l_1 and l_2 are parallel to $\mathbf{v}^1$ and $\mathbf{v}^2$, while the half-lines l_1' and l_2' are parallel to $-\mathbf{v}^1$ and $-\mathbf{v}^2$. Observe first that every solution $\mathbf{x}(t)$ of (1) is of the form

$$\mathbf{x}(t) = c_1 e^{\lambda_1 t}\mathbf{v}^1 + c_2 e^{\lambda_2 t}\mathbf{v}^2 \tag{5}$$

for some choice of constants c_1 and c_2. The orbit of the solution $\mathbf{x}(t) = c_1 e^{\lambda_1 t}\mathbf{v}^1$ is l_1 for $c_1 > 0$ and l_1' for $c_1 < 0$, while the orbit of the solution $\mathbf{x}(t) = c_2 e^{\lambda_2 t}\mathbf{v}^2$ is l_2 for $c_2 > 0$ and l_2' for $c_2 < 0$. Note, too, the direction of the arrows on l_1, l_1', l_2, and l_2'; the solution $\mathbf{x}(t) = c_1 e^{\lambda_1 t}\mathbf{v}^1$ approaches $\binom{0}{0}$ as t approaches infinity, whereas the solution $\mathbf{x}(t) = c_2 e^{\lambda_2 t}\mathbf{v}^2$ becomes unbounded (for $c_2 \neq 0$) as t approaches infinity. Next, observe that $e^{\lambda_1 t}\mathbf{v}^1$ is very small compared to $e^{\lambda_2 t}\mathbf{v}^2$ when t is very large. Hence, every solution $\mathbf{x}(t)$ of (1) with $c_2 \neq 0$ becomes unbounded as t approaches infinity, and its orbit approaches either l_2 or l_2'. Finally, observe that $e^{\lambda_2 t}\mathbf{v}^2$ is very small compared to $e^{\lambda_1 t}\mathbf{v}^1$ when t is very large negative. Hence, the orbit of any solution $\mathbf{x}(t)$ of (1), with $c_1 \neq 0$, approaches either l_1 or l_1' as t approaches minus infinity. Consequently, the phase portrait of (1) has the form described in Figure 5. This phase portrait resembles a "saddle" near $x_1 = x_2 =$

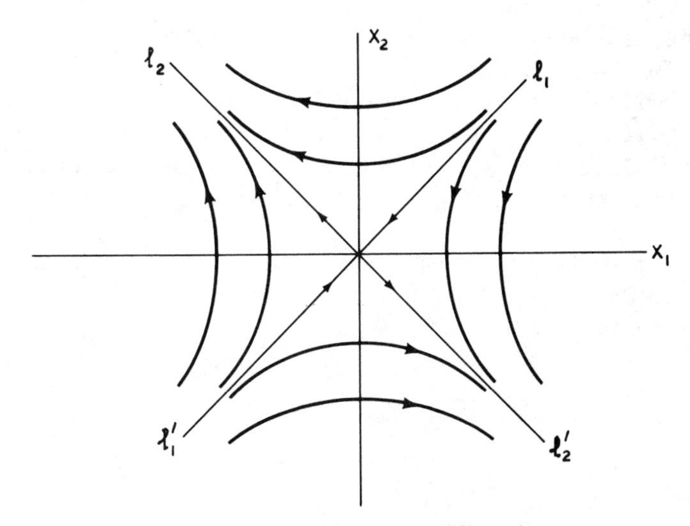

Figure 5. Phase portrait of a saddle point

0. For this reason, we say that the equilibrium solution $\mathbf{x}(t)=\mathbf{0}$ of (1) is a saddle point if the eigenvalues of $\mathbf{A}$ have opposite sign.

4. $\lambda_1=\alpha+i\beta, \lambda_2=\alpha-i\beta, \beta\neq 0$. Our first step in deriving the phase portrait of (1) is to find the general solution of (1). Let $\mathbf{z}=\mathbf{u}+i\mathbf{v}$ be an eigenvector of $\mathbf{A}$ with eigenvalue $\alpha+i\beta$. Then,

$$\begin{aligned}\mathbf{x}(t)&=e^{(\alpha+i\beta)t}(\mathbf{u}+i\mathbf{v})=e^{\alpha t}(\cos\beta t+i\sin\beta t)(\mathbf{u}+i\mathbf{v})\\&=e^{\alpha t}[\mathbf{u}\cos\beta t-\mathbf{v}\sin\beta t]+ie^{\alpha t}[\mathbf{u}\sin\beta t+\mathbf{v}\cos\beta t]\end{aligned}$$

is a complex-valued solution of (1). Therefore,

$$\mathbf{x}^1(t)=e^{\alpha t}[\mathbf{u}\cos\beta t-\mathbf{v}\sin\beta t]$$

and

$$\mathbf{x}^2(t)=e^{\alpha t}[\mathbf{u}\sin\beta t+\mathbf{v}\cos\beta t]$$

are two real-valued linearly independent solutions of (1), and every solution $\mathbf{x}(t)$ of (1) is of the form $\mathbf{x}(t)=c_1\mathbf{x}^1(t)+c_2\mathbf{x}^2(t)$. This expression can be written in the form (see Exercise 15)

$$\mathbf{x}(t)=e^{\alpha t}\begin{pmatrix}R_1\cos(\beta t-\delta_1)\\R_2\cos(\beta t-\delta_2)\end{pmatrix}\tag{6}$$

for some choice of constants $R_1\geqslant 0$, $R_2\geqslant 0$, δ_1, and δ_2. We distinguish the following cases.

(a) $\alpha=0$: Observe that both

$$x_1(t)=R_1\cos(\beta t-\delta_1)\quad\text{and}\quad x_2(t)=R_2\cos(\beta t-\delta_2)$$

are periodic functions of time with period $2\pi/\beta$. The function $x_1(t)$ varies between $-R_1$ and $+R_1$, while $x_2(t)$ varies between $-R_2$ and $+R_2$. Consequently, the orbit of any solution $\mathbf{x}(t)$ of (1) is a closed curve surrounding the origin $x_1 = x_2 = 0$, and the phase portrait of (1) has the form described in Figure 6a. For this reason, we say that the equilibrium solution $\mathbf{x}(t) = \mathbf{0}$ of (1) is a center when the eigenvalues of $\mathbf{A}$ are pure imaginary.

The direction of the arrows in Figure 6a must be determined from the differential equation (1). The simplest way of doing this is to check the sign of $\dot{x}_2$ when $x_2 = 0$. If $\dot{x}_2$ is greater than zero for $x_2 = 0$ and $x_1 > 0$, then all solutions $\mathbf{x}(t)$ of (1) move in the counterclockwise direction; if $\dot{x}_2$ is less than zero for $x_2 = 0$ and $x_1 > 0$, then all solutions $\mathbf{x}(t)$ of (1) move in the clockwise direction.

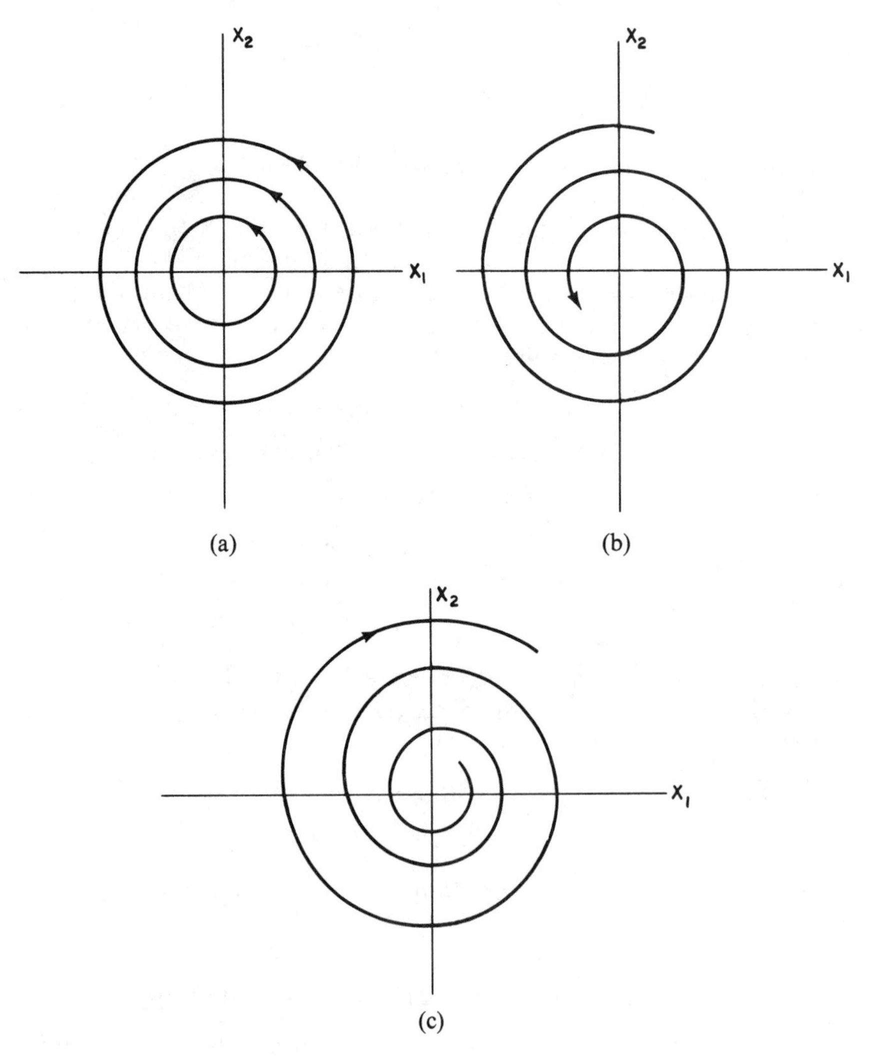

Figure 6. (a) $\alpha = 0$; (b) $\alpha < 0$; (c) $\alpha > 0$

(b) $\alpha<0$: In this case, the effect of the factor $e^{\alpha t}$ in Equation (6) is to change the simple closed curves of Figure 6a into the spirals of Figure 6b. This is because the point $\mathbf{x}(2\pi/\beta)=e^{2\pi\alpha/\beta}\mathbf{x}(0)$ is closer to the origin than $\mathbf{x}(0)$. Again, the direction of the arrows in Figure 6b must be determined directly from the differential equation (1). In this case, we say that the equilibrium solution $\mathbf{x}(t)=\mathbf{0}$ of (1) is a stable focus.

(c) $\alpha>0$: In this case, all orbits of (1) spiral away from the origin as t approaches infinity (see Figure 6c), and the equilibrium solution $\mathbf{x}(t)=\mathbf{0}$ of (1) is called an unstable focus.

Finally, we mention that the phase portraits of nonlinear systems, in the neighborhood of an equilibrium point, are often very similar to the phase portraits of linear systems. More precisely, let $\mathbf{x}=\mathbf{x}^0$ be an equilibrium solution of the nonlinear equation $\dot{\mathbf{x}}=\mathbf{f}(\mathbf{x})$, and set $\mathbf{u}=\mathbf{x}-\mathbf{x}^0$. Then, (see Section 4.3) we can write the differential equation $\dot{\mathbf{x}}=\mathbf{f}(\mathbf{x})$ in the form

$$\dot{\mathbf{u}}=\mathbf{A}\mathbf{u}+\mathbf{g}(\mathbf{u}) \tag{7}$$

where $\mathbf{A}$ is a constant matrix and $\mathbf{g}(\mathbf{u})$ is very small compared to $\mathbf{u}$. We state without proof the following theorem.

Theorem 4. *Suppose that* $\mathbf{u}=\mathbf{0}$ *is either a node, saddle, or focus point of the differential equation* $\dot{\mathbf{u}}=\mathbf{A}\mathbf{u}$. *Then, the phase portrait of the differential equation* $\dot{\mathbf{x}}=\mathbf{f}(\mathbf{x})$, *in a neighborhood of* $\mathbf{x}=\mathbf{x}^0$, *has one of the forms described in Figures* 3, 5, *and* 6 (b *and* c), *depending as to whether* $\mathbf{u}=\mathbf{0}$ *is a node, saddle, or focus.*

Example 1. Draw the phase portrait of the linear equation

$$\dot{\mathbf{x}}=\mathbf{A}\mathbf{x}=\begin{pmatrix} -2 & -1 \\ 4 & -7 \end{pmatrix}\mathbf{x}. \tag{8}$$

Solution. It is easily verified that

$$\mathbf{v}^1=\begin{pmatrix} 1 \\ 1 \end{pmatrix} \quad \text{and} \quad \mathbf{v}^2=\begin{pmatrix} 1 \\ 4 \end{pmatrix}$$

are eigenvectors of $\mathbf{A}$ with eigenvalues -3 and -6, respectively. Therefore, $\mathbf{x}=\mathbf{0}$ is a stable node of (8), and the phase portrait of (8) has the form described in Figure 7. The half-line l_1 makes an angle of 45° with the x_1 axis, while the half-line l_2 makes an angle of θ degrees with the x_1-axis, where $\tan\theta=4$.

Example 2. Draw the phase portrait of the linear equation

$$\dot{\mathbf{x}}=\mathbf{A}\mathbf{x}=\begin{pmatrix} 1 & -3 \\ -3 & 1 \end{pmatrix}\mathbf{x}. \tag{9}$$

Solution. It is easily verified that

$$\mathbf{v}^1=\begin{pmatrix} 1 \\ 1 \end{pmatrix} \quad \text{and} \quad \mathbf{v}^2=\begin{pmatrix} -1 \\ 1 \end{pmatrix}$$

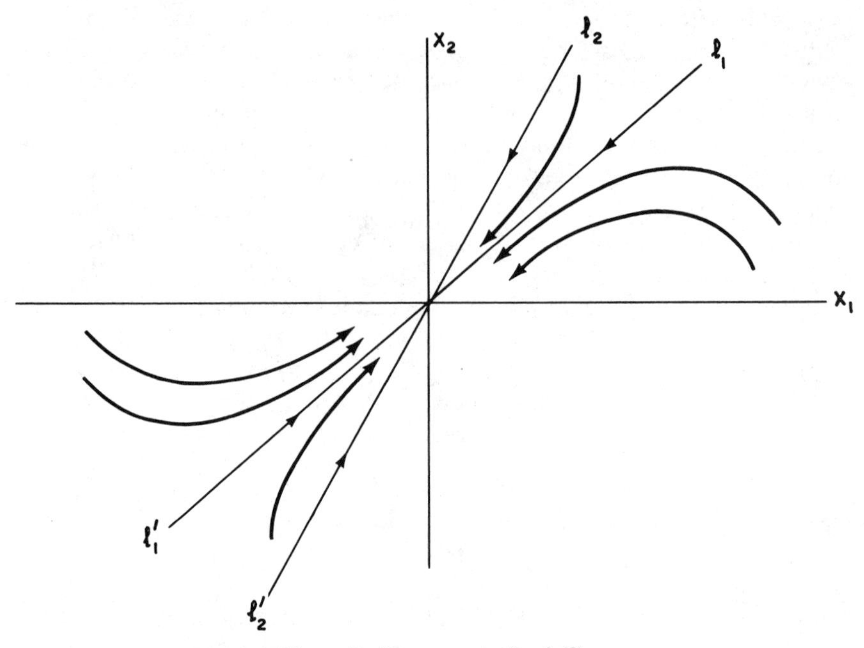

Figure 7. Phase portrait of (8)

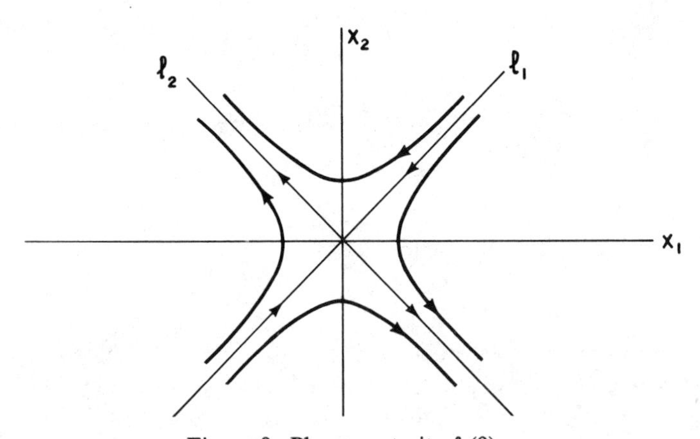

Figure 8. Phase portrait of (9)

are eigenvectors of $\mathbf{A}$ with eigenvalues -2 and 4, respectively. Therefore, $\mathbf{x}=\mathbf{0}$ is a saddle point of (9), and its phase portrait has the form described in Figure 8. The half-line l_1 makes an angle of 45° with the x_1-axis, and the half-line l_2 is at right angles to l_1.

Example 3. Draw the phase portrait of the linear equation

$$\dot{\mathbf{x}}=\mathbf{A}\mathbf{x}=\begin{pmatrix} -1 & 1 \\ -1 & -1 \end{pmatrix}\mathbf{x}. \tag{10}$$

Solution. The eigenvalues of $\mathbf{A}$ are $-1\pm i$. Hence, $\mathbf{x}=\mathbf{0}$ is a stable focus of (10) and every nontrivial orbit of (10) spirals into the origin as t approaches infinity. To determine the direction of rotation of the spiral, we observe that $\dot{x}_2=-x_1$ when $x_2=0$. Thus, $\dot{x}_2$ is negative for $x_1>0$ and $x_2=0$. Consequently, all nontrivial orbits of (10) spiral into the origin in the clockwise direction, as shown in Figure 9.

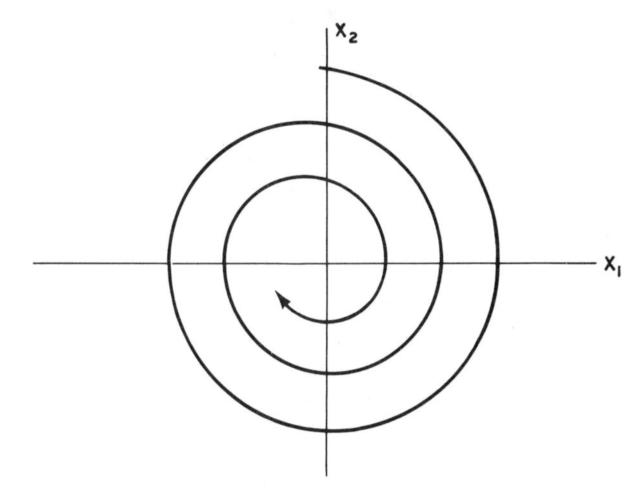

Figure 9. Phase portrait of (10)

EXERCISES

Draw the phase portraits of each of the following systems of differential equations.

1. $\dot{\mathbf{x}}=\begin{pmatrix}-5 & 1\\ 1 & -5\end{pmatrix}\mathbf{x}$ 2. $\dot{\mathbf{x}}=\begin{pmatrix}0 & -1\\ 8 & -6\end{pmatrix}\mathbf{x}$ 3. $\dot{\mathbf{x}}=\begin{pmatrix}4 & -1\\ -2 & 5\end{pmatrix}\mathbf{x}$

4. $\dot{\mathbf{x}}=\begin{pmatrix}-4 & -1\\ 1 & -6\end{pmatrix}\mathbf{x}$ 5. $\dot{\mathbf{x}}=\begin{pmatrix}1 & -4\\ -8 & 4\end{pmatrix}\mathbf{x}$ 6. $\dot{\mathbf{x}}=\begin{pmatrix}3 & -1\\ 5 & -3\end{pmatrix}\mathbf{x}$

7. $\dot{\mathbf{x}}=\begin{pmatrix}0 & 2\\ -2 & -1\end{pmatrix}\mathbf{x}$ 8. $\dot{\mathbf{x}}=\begin{pmatrix}1 & -1\\ 5 & -3\end{pmatrix}\mathbf{x}$ 9. $\dot{\mathbf{x}}=\begin{pmatrix}2 & 1\\ -5 & -2\end{pmatrix}\mathbf{x}$

10. Show that every orbit of

$$\dot{\mathbf{x}}=\begin{pmatrix}0 & 4\\ -9 & 0\end{pmatrix}\mathbf{x}$$

is an ellipse.

11. The equation of motion of a spring-mass system with damping (see Section 2.6) is $m\ddot{z}+c\dot{z}+kz=0$, where m, c, and k are positive numbers. Convert this equation to a system of first-order equations for $x=z$, $y=\dot{z}$, and draw the phase portrait of this system. Distinguish the overdamped, critically damped, and underdamped cases.

12. Suppose that a 2×2 matrix $\mathbf{A}$ has 2 linearly independent eigenvectors with eigenvalue λ. Show that $\mathbf{A}=\lambda\mathbf{I}$.

13. This problem illustrates Theorem 4. Consider the system

$$\frac{dx}{dt}=y, \qquad \frac{dy}{dt}=x+2x^3. \qquad (*)$$

(a) Show that the equilibrium solution $x=0, y=0$ of the linearized system $\dot{x}=y, \dot{y}=x$ is a saddle, and draw the phase portrait of the linearized system.
(b) Find the orbits of (*), and then draw its phase portrait.
(c) Show that there are exactly two orbits of (*) (one for $x>0$ and one for $x<0$) on which $x\to 0, y\to 0$ as $t\to\infty$. Similarly, there are exactly two orbits of (*) on which $x\to 0, y\to 0$ as $t\to-\infty$. Thus, observe that the phase portraits of (*) and the linearized system look the same near the origin.

14. Verify Equation (6). *Hint*: The expression $a\cos\omega t+b\sin\omega t$ can always be written in the form $R\cos(\omega t-\delta)$ for suitable choices of R and δ.

4.8 Long time behavior of solutions; the Poincaré–Bendixson Theorem

We consider now the problem of determining the long time behavior of all solutions of the differential equation

$$\dot{\mathbf{x}}=\mathbf{f}(\mathbf{x}), \qquad \mathbf{x}=\begin{bmatrix} x_1\\ \vdots \\ x_n\end{bmatrix}, \qquad \mathbf{f}(\mathbf{x})=\begin{bmatrix} f_1(x_1,\dots,x_n)\\ \vdots \\ f_n(x_1,\dots,x_n)\end{bmatrix}. \qquad (1)$$

This problem has been solved completely in the special case that $\mathbf{f}(\mathbf{x})=\mathbf{A}\mathbf{x}$. As we have seen in Sections 4.2 and 4.7, all solutions $\mathbf{x}(t)$ of $\dot{\mathbf{x}}=\mathbf{A}\mathbf{x}$ must exhibit one of the following four types of behavior: (i) $\mathbf{x}(t)$ is constant in time; (ii) $\mathbf{x}(t)$ is a periodic function of time; (iii) $\mathbf{x}(t)$ is unbounded as t approaches infinity; and (iv) $\mathbf{x}(t)$ approaches an equilibrium point as t approaches infinity.

A partial solution to this problem, in the case of nonlinear $\mathbf{f}(\mathbf{x})$, was given in Section 4.3. In that section we provided sufficient conditions that every solution $\mathbf{x}(t)$ of (1), whose initial value $\mathbf{x}(0)$ is sufficiently close to an equilibrium point $\boldsymbol{\xi}$, must ultimately approach $\boldsymbol{\xi}$ as t approaches infinity. In many applications it is often possible to go much further and prove that every physically (biologically) realistic solution approaches a single equilibrium point as time evolves. In this context, the following two lemmas play an extremely important role.

Lemma 1. *Let $g(t)$ be a monotonic increasing (decreasing) function of time for $t\geqslant t_0$, with $g(t)\leqslant c(\geqslant c)$ for some constant c. Then, $g(t)$ has a limit as t approaches infinity.*

PROOF. Suppose that $g(t)$ is monotonic increasing for $t\geqslant t_0$, and $g(t)$ is bounded from above. Let l be the least upper bound of g; that is, l is the smallest number which is not exceeded by the values of $g(t)$, for $t\geqslant t_0$. This number must be the limit of $g(t)$ as t approaches infinity. To prove this, let $\varepsilon>0$ be given, and observe that there exists a time $t_\varepsilon \geqslant t_0$ such that $l-g(t_\varepsilon) < \varepsilon$. (If no such time t_ε exists, then l is not the least upper bound of g.) Since $g(t)$ is monotonic, we see that $l-g(t)<\varepsilon$ for $t \geqslant t_\varepsilon$. This shows that $l=\lim_{t\to\infty} g(t)$. □

Lemma 2. *Suppose that a solution* $\mathbf{x}(t)$ *of* (1) *approaches a vector* $\boldsymbol{\xi}$ *as* t *approaches infinity. Then,* $\boldsymbol{\xi}$ *is an equilibrium point of* (1).

PROOF. Suppose that $\mathbf{x}(t)$ approaches $\boldsymbol{\xi}$ as t approaches infinity. Then, $x_j(t)$ approaches ξ_j, where ξ_j is the jth component of $\boldsymbol{\xi}$. This implies that $|x_j(t_1)-x_j(t_2)|$ approaches zero as both t_1 and t_2 approach infinity, since

$$\begin{aligned} |x_j(t_1)-x_j(t_2)| &= |(x_j(t_1)-\xi_j)+(\xi_j-x_j(t_2))| \\ &\leqslant |x_j(t_1)-\xi_j|+|x_j(t_2)-\xi_j|. \end{aligned}$$

In particular, let $t_1=t$ and $t_2=t_1+h$, for some fixed positive number h. Then, $|x_j(t+h)-x_j(t)|$ approaches zero as t approaches infinity. But

$$x_j(t+h)-x_j(t)=h\frac{dx_j(\tau)}{dt}=hf_j(x_1(\tau),\dots,x_n(\tau)),$$

where τ is some number between t and $t+h$. Finally, observe that $f_j(x_1(\tau),\dots,x_n(\tau))$ must approach $f_j(\xi_1,\dots,\xi_n)$ as t approaches infinity. Hence, $f_j(\xi_1,\dots,\xi_n)=0, j=1,2,\dots,n$, and this proves Lemma 1. □

Example 1. Consider the system of differential equations

$$\frac{dx}{dt}=ax-bxy-ex^2, \qquad \frac{dy}{dt}=-cy+dxy-fy^2 \tag{2}$$

where a, b, c, d, e, and f are positive constants. This system (see Section 4.10) describes the population growth of two species x and y, where species y is dependent upon species x for its survival. Suppose that $c/d>a/e$. Prove that every solution $x(t)$, $y(t)$ of (2), with $x(0)$ and $y(0)>0$, approaches the equilibrium solution $x=a/e$, $y=0$, as t approaches infinity.

Solution. Our first step is to show that every solution $x(t), y(t)$ of (2) which starts in the first quadrant ($x>0$, $y>0$) at $t=0$ must remain in the first quadrant for all future time. (If this were not so, then the model (2) could not correspond to reality.) To this end, recall from Section 1.5 that

$$x(t)=\frac{ax_0}{ex_0+(a-ex_0)e^{-at}}, \qquad y(t)=0$$

is a solution of (2) for any choice of x_0. The orbit of this solution is the point $(0,0)$ for $x_0=0$; the line $0<x<a/e$ for $0<x_0<a/e$; the point $(a/e,0)$ for $x_0=a/e$; and the line $a/e<x<\infty$ for $x_0>a/e$. Thus, the x-axis, for $x\geqslant 0$, is the union of four disjoint orbits of (2). Similarly, (see Exercise 14), the positive y-axis is a single orbit of (2). Thus, if a solution

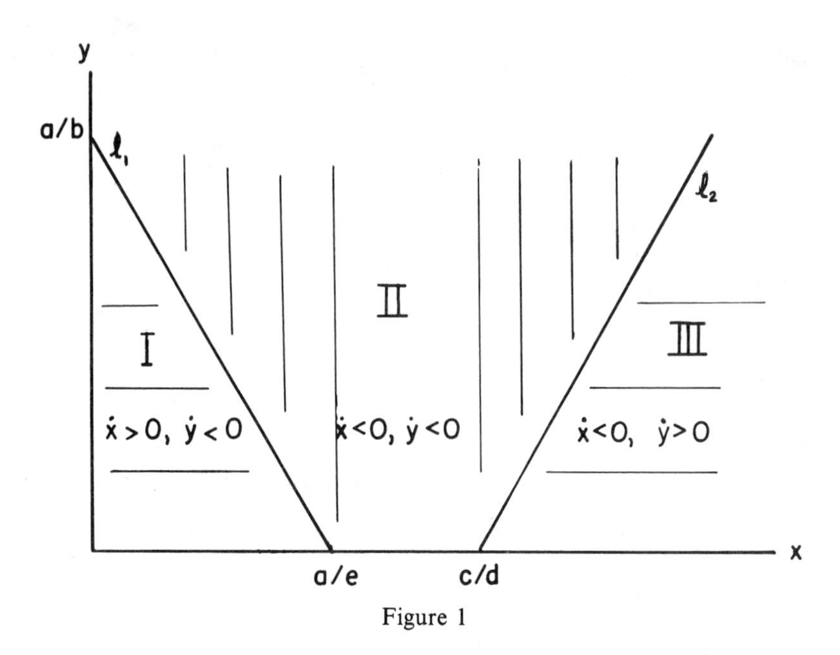

Figure 1

$x(t), y(t)$ of (2) leaves the first quadrant, its orbit must cross another orbit, and this is precluded by the uniqueness of orbits (Property 1, Section 4.6).

Our next step is to divide the first quadrant into regions where dx/dt and dy/dt have fixed signs. This is accomplished by drawing the lines $l_1: a-by-ex=0$, and $l_2: -c+dx-fy=0$, in the x–y plane. These lines divide the first quadrant into three regions I, II, and III as shown in Figure 1. (The lines l_1 and l_2 do not intersect in the first quadrant if $c/d>a/e$.) Now, observe that $ex+by$ is less than a in region I, while $ex+by$ is greater than a in regions II and III. Consequently, dx/dt is positive in region I and negative in regions II and III. Similarly, dy/dt is negative in regions I and II and positive in region III.

Next, we prove the following four simple lemmas.

Lemma 3. *Any solution $x(t), y(t)$ of (2) which starts in region I at time $t=t_0$ will remain in this region for all future time $t\geq t_0$ and ultimately approach the equilibrium solution $x=a/e, y=0$.*

PROOF. Suppose that a solution $x(t), y(t)$ of (2) leaves region I at time $t=t^*$. Then, $\dot{x}(t^*)=0$, since the only way a solution can leave region I is by crossing the line l_1. Differentiating both sides of the first equation of (2) with respect to t and setting $t=t^*$ gives

$$\frac{d^2x(t^*)}{dt^2}=-bx(t^*)\frac{dy(t^*)}{dt}.$$

This quantity is positive. Hence, $x(t)$ has a minimum at $t=t^*$. But this is impossible, since $x(t)$ is always increasing whenever $x(t), y(t)$ is in region

I. Thus, any solution $x(t)$, $y(t)$ of (2) which starts in region I at time $t=t_0$ will remain in region I for all future time $t \geqslant t_0$. This implies that $x(t)$ is a monotonic increasing function of time, and $y(t)$ is a monotonic decreasing function of time for $t \geqslant t_0$, with $x(t) < a/e$ and $y(t) > 0$. Consequently, by Lemma 1, both $x(t)$ and $y(t)$ have limits ξ, η respectively, as t approaches infinity. Lemma 2 implies that (ξ, η) is an equilibrium point of (2). Now, it is easily verified that the only equilibrium points of (2) in the region $x \geqslant 0$, $y \geqslant 0$ are $x=0$, $y=0$, and $x=a/e$, $y=0$. Clearly, ξ cannot equal zero since $x(t)$ is increasing in region I. Therefore, $\xi = a/e$ and $\eta = 0$. □

Lemma 4. *Any solution $x(t)$, $y(t)$ of* (2) *which starts in region* III *at time $t=t_0$ must leave this region at some later time.*

PROOF. Suppose that a solution $x(t)$, $y(t)$ of (2) remains in region III for all time $t \geqslant t_0$. Then, $x(t)$ is a monotonic decreasing function of time, and $y(t)$ is a monotonic increasing function of time, for $t \geqslant t_0$. Moreover, $x(t)$ is greater than c/d and $y(t)$ is less than $(dx(t_0)-c)/f$. Consequently, both $x(t)$ and $y(t)$ have limits ξ, η respectively, as t approaches infinity. Lemma 2 implies that (ξ, η) is an equilibrium point of (2). But (ξ, η) cannot equal $(0,0)$ or $(a/e, 0)$ if $x(t)$, $y(t)$ is in region III for $t \geqslant t_0$. This contradiction establishes Lemma 4. □

Lemma 5. *Any solution $x(t)$, $y(t)$ of* (2) *which starts in region* II *at time $t=t_0$ and remains in region II for all future time $t \geqslant t_0$ must approach the equilibrium solution $x=a/e$, $y=0$.*

PROOF. Suppose that a solution $x(t)$, $y(t)$ of (2) remains in region II for all time $t \geqslant t_0$. Then, both $x(t)$ and $y(t)$ are monotonic decreasing functions of time for $t \geqslant t_0$, with $x(t) > 0$ and $y(t) > 0$. Consequently, by Lemma 1, both $x(t)$ and $y(t)$ have limits ξ, η respectively, as t approaches infinity. Lemma 2 implies that (ξ, η) is an equilibrium point of (2). Now, (ξ, η) cannot equal $(0,0)$. Therefore, $\xi = a/e, \eta = 0$. □

Lemma 6. *A solution $x(t)$,$y(t)$ of* (2) *cannot enter region* III *from region* II.

PROOF. Suppose that a solution $x(t)$, $y(t)$ of (2) leaves region II at time $t=t^*$ and enters region III. Then, $\dot{y}(t^*)=0$. Differentiating both sides of the second equation of (2) with respect to t and setting $t=t^*$ gives

$$\frac{d^2y(t^*)}{dt^2} = dy(t^*)\frac{dx(t^*)}{dt}.$$

This quantity is negative. Hence, $y(t)$ has a maximum at $t=t^*$. But this is impossible, since $y(t)$ is decreasing whenever $x(t)$, $y(t)$ is in region II. □

Finally, observe that a solution $x(t)$, $y(t)$ of (2) which starts on l_1 must immediately enter region I, and that a solution which starts on l_2 must immediately enter region II. It now follows immediately from Lemmas 3–6 that every solution $x(t)$, $y(t)$ of (2), with $x(0)>0$ and $y(0)>0$, approaches the equilibrium solution $x=a/e$, $y=0$ as t approaches infinity.

Up to now, the solutions and orbits of the nonlinear equations that we have studied behaved very much like the solutions and orbits of linear equations. In actual fact, though, the situation is very different. The solutions and orbits of nonlinear equations, in general, exhibit a completely different behavior than the solutions and orbits of linear equations. A standard example is the system of equations

$$\frac{dx}{dt}=-y+x(1-x^2-y^2), \qquad \frac{dy}{dt}=x+y(1-x^2-y^2). \tag{3}$$

Since the term x^2+y^2 appears prominently in both equations, it suggests itself to introduce polar coordinates r,θ, where $x=r\cos\theta$, $y=r\sin\theta$, and to rewrite (3) in terms of r and θ. To this end, we compute

$$\begin{aligned}\frac{d}{dt}r^2=2r\frac{dr}{dt}&=2x\frac{dx}{dt}+2y\frac{dy}{dt}\\&=2(x^2+y^2)-2(x^2+y^2)^2=2r^2(1-r^2).\end{aligned}$$

Similarly, we compute

$$\frac{d\theta}{dt}=\frac{d}{dt}\arctan\frac{y}{x}=\frac{1}{x^2}\frac{x\dfrac{dy}{dt}-y\dfrac{dx}{dt}}{1+(y/x)^2}=\frac{x^2+y^2}{x^2+y^2}=1.$$

Consequently, the system of equations (3) is equivalent to the system of equations

$$\frac{dr}{dt}=r(1-r^2), \qquad \frac{d\theta}{dt}=1. \tag{4}$$

The general solution of (4) is easily seen to be

$$r(t)=\frac{r_0}{\left[r_0^2+(1-r_0^2)e^{-2t}\right]^{1/2}}, \qquad \theta=t+\theta_0 \tag{5}$$

where $r_0=r(0)$ and $\theta_0=\theta(0)$. Hence,

$$x(t)=\frac{r_0}{\left[r_0^2+(1-r_0^2)e^{-2t}\right]^{1/2}}\cos(t+\theta_0),$$

$$y(t)=\frac{r_0}{\left[r_0^2+(1-r_0^2)e^{-2t}\right]^{1/2}}\sin(t+\theta_0).$$

Now, observe first that $x=0$, $y=0$ is the only equilibrium solution of (3). Second, observe that

$$x(t)=\cos(t+\theta_0), \qquad y(t)=\sin(t+\theta_0)$$

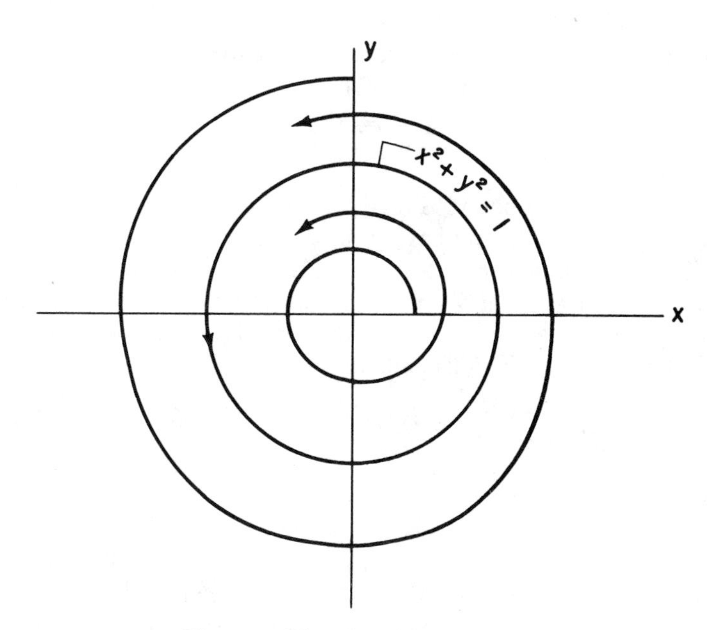

Figure 2. The phase portrait of (3)

when $r_0=1$. This solution is periodic with period 2π, and its orbit is the unit circle $x^2+y^2=1$. Finally, observe from (5) that $r(t)$ approaches one as t approaches infinity, for $r_0\neq 0$. Hence, all the orbits of (3), with the exception of the equilibrium point $x=0$, $y=0$, spiral into the unit circle. This situation is depicted in Figure 2.

The system of equations (3) shows that the orbits of a nonlinear system of equations may spiral into a simple closed curve. This, of course, is not possible for linear systems. Moreover, it is often possible to prove that orbits of a nonlinear system spiral into a closed curve even when we cannot explicitly solve the system of equations, or even find its orbits. This is the content of the following celebrated theorem.

Theorem 5. (Poincaré–Bendixson.) *Suppose that a solution $x=x(t)$, $y=y(t)$ of the system of differential equations*

$$\frac{dx}{dt}=f(x,y),\qquad \frac{dy}{dt}=g(x,y) \tag{6}$$

remains in a bounded region of the plane which contains no equilibrium points of (6). *Then, its orbit must spiral into a simple closed curve, which is itself the orbit of a periodic solution of* (6).

Example 2. Prove that the second-order differential equation

$$\ddot{z}+(z^2+2\dot{z}^2-1)\dot{z}+z=0 \tag{7}$$

has a nontrivial periodic solution.

Solution. First, we convert Equation (7) to a system of two first-order equations by setting $x=z$ and $y=\dot{z}$. Then,

$$\frac{dx}{dt}=y, \qquad \frac{dy}{dt}=-x+(1-x^2-2y^2)y. \tag{8}$$

Next, we try and find a bounded region R in the $x-y$ plane, containing no equilibrium points of (8), and having the property that every solution $x(t)$, $y(t)$ of (8) which starts in R at time $t=t_0$, remains there for all future time $t \geqslant t_0$. It can be shown that a simply connected region such as a square or disc will never work. Therefore, we try and take R to be an annulus surrounding the origin. To this end, compute

$$\frac{d}{dt}\left(\frac{x^2+y^2}{2}\right)=x\frac{dx}{dt}+y\frac{dy}{dt}=(1-x^2-2y^2)y^2,$$

and observe that $1-x^2-2y^2$ is positive for $x^2+y^2<\frac{1}{2}$ and negative for $x^2+y^2>1$. Hence, $x^2(t)+y^2(t)$ is increasing along any solution $x(t)$, $y(t)$ of (8) when $x^2+y^2<\frac{1}{2}$ and decreasing when $x^2+y^2>1$. This implies that any solution $x(t)$, $y(t)$ of (8) which starts in the annulus $\frac{1}{2}<x^2+y^2<1$ at time $t=t_0$ will remain in this annulus for all future time $t \geqslant t_0$. Now, this annulus contains no equilibrium points of (8). Consequently, by the Poincaré–Bendixson Theorem, there exists at least one periodic solution $x(t)$, $y(t)$ of (8) lying entirely in this annulus, and then $z=x(t)$ is a nontrivial periodic solution of (7).

Exercises

1. *What Really Happened at the Paris Peace Talks*

The original plan developed by Henry Kissinger and Le Duc Tho to settle the Vietnamese war is described below. It was agreed that 1 million South Vietnamese ants and 1 million North Vietnamese ants would be placed in the backyard of the Presidential palace in Paris and be allowed to fight it out for a long period of time. If the South Vietnamese ants destroyed nearly all the North Vietnamese ants, then South Vietnam would retain control of all of its land. If the North Vietnamese ants were victorious, then North Vietnam would take over all of South Vietnam. If they appeared to be fighting to a standoff, then South Vietnam would be partitioned according to the proportion of ants remaining. Now, the South Vietnamese ants, denoted by S, and the North Vietnamese ants, denoted by N, compete against each other according to the following differential equations:

$$\begin{aligned}\frac{dS}{dt}&=\frac{1}{10}S-\frac{1}{20}S\times N\\ \frac{dN}{dt}&=\frac{1}{100}N-\frac{1}{100}N^2-\frac{1}{100}S\times N.\end{aligned} \tag{*}$$

Note that these equations correspond to reality since the South Vietnamese ants multiply much more rapidly than the North Vietnamese ants, but the North Vietnamese ants are much better fighters.

The battle began at 10:00 sharp on the morning of May 19, 1972, and was supervised by a representative of Poland and a representative of Canada. At 2:43 p.m. on the afternoon of May 21, the representative of Poland, being unhappy with the progress of the battle, slipped a bag of North Vietnamese ants into the backyard, but he was spotted by the eagle eyes of the representative of Canada. The South Vietnamese immediately claimed a foul and called off the agreement, thus setting the stage for the protracted talks that followed in Paris. The representative of Poland was hauled before a judge in Paris for sentencing. The judge, after making some remarks about the stupidity of the South Vietnamese, gave the Polish representative a very light sentence. Justify mathematically the judge's decision. *Hint*:

(a) Show that the lines $N=2$ and $N+S=1$ divide the first quadrant into three regions (see Figure 3) in which dS/dt and dN/dt have fixed signs.

(b) Show that every solution $S(t), N(t)$ of (*) which starts in either region I or region III must eventually enter region II.

(c) Show that every solution $S(t), N(t)$ of (*) which starts in region II must remain there for all future time.

(d) Conclude from (c) that $S(t)\to\infty$ for all solutions $S(t), N(t)$ of (*) with $S(t_0)$ and $N(t_0)$ positive. Conclude too that $N(t)$ has a finite limit ($\leqslant 2$) as $t\to\infty$.

(e) To prove that $N(t)\to 0$, observe that there exists t_0 such that $dN/dt \leqslant -N$ for $t \geqslant t_0$. Conclude from this inequality that $N(t)\to 0$ as $t\to\infty$.

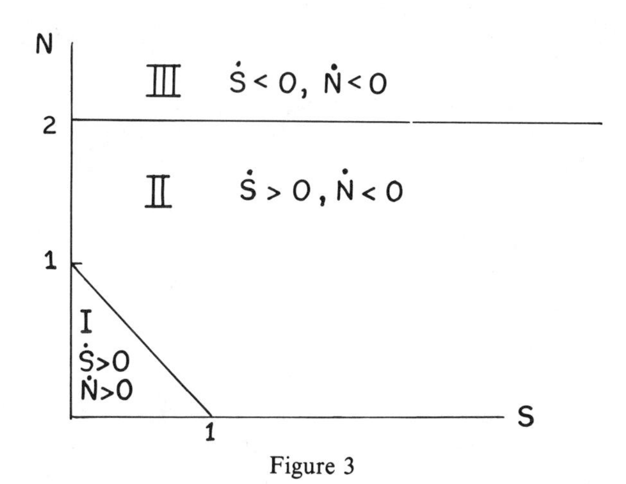

Figure 3

2. Consider the system of differential equations

$$\frac{dx}{dt} = ax - bxy, \qquad \frac{dy}{dt} = cy - dxy - ey^2 \tag{*}$$

with $a/b > c/e$. Prove that $y(t)\to 0$ as $t\to\infty$, for every solution $x(t), y(t)$ of (*) with $x(t_0)$ and $y(t_0)$ positive. *Hint*: Follow the outline in Exercise 1.

3. (a) Without computing the eigenvalues of the matrix

$$\begin{pmatrix} -3 & 1 \\ 1 & -3 \end{pmatrix},$$

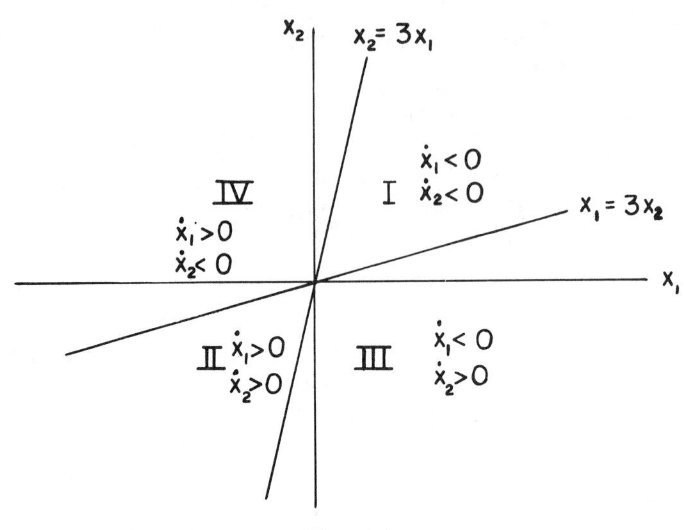

Figure 4

prove that every solution $\mathbf{x}(t)$ of

$$\dot{\mathbf{x}} = \begin{pmatrix} -3 & 1 \\ 1 & -3 \end{pmatrix} \mathbf{x}$$

approaches zero as t approaches infinity. *Hint*: (a) Show that the lines $x_2 = 3x_1$ and $x_1 = 3x_2$ divide the $x_1 - x_2$ plane into four regions (see Figure 4) in which $\dot{x}_1$ and $\dot{x}_2$ have fixed signs.

(b) Show that every solution $\mathbf{x}(t)$ which starts in either region I or II must remain there for all future time and ultimately approach the equilibrium solution $\mathbf{x} = \mathbf{0}$.

(c) Show that every solution $\mathbf{x}(t)$ which remains exclusively in region III or IV must ultimately approach the equilibrium solution $\mathbf{x} = \mathbf{0}$.

A closed curve C is said to be a limit cycle of

$$\dot{x} = f(x,y), \qquad \dot{y} = g(x,y) \tag{*}$$

if orbits of (*) spiral into it, or away from it. It is stable if all orbits of (*) passing sufficiently close to it must ultimately spiral into it, and unstable otherwise. Find all limit cycles of each of the following systems of differential equations. (*Hint*: Compute $d(x^2+y^2)/dt$. Observe too, that C must be the orbit of a periodic solution of (*) if it contains no equilibrium points of (*).)

4. $\dot{x} = -y - \dfrac{x(x^2+y^2-2)}{\sqrt{x^2+y^2}}$

$\dot{y} = x - \dfrac{y(x^2+y^2-2)}{\sqrt{x^2+y^2}}$

5. $\dot{x} = x - x^3 - xy^2$

$\dot{y} = y - y^3 - yx^2$

6. $\dot{x}=y+x(x^2+y^2-1)(x^2+y^2-2)$
$\dot{y}=-x+y(x^2+y^2-1)(x^2+y^2-2)$

7. $\dot{x}=xy+x\cos(x^2+y^2)$
$\dot{y}=-x^2+y\cos(x^2+y^2)$

8. (a) Show that the system

$$\dot{x}=y+xf(r)/r, \qquad \dot{y}=-x+yf(r)/r \qquad (r^2=x^2+y^2) \tag{*}$$

has limit cycles corresponding to the zeros of $f(r)$. What is the direction of motion on these curves?

(b) Determine all limit cycles of (*) and discuss their stability if $f(r)=(r-3)^2(r^2-5r+4)$.

Use the Poincaré–Bendixson Theorem to prove the existence of a nontrivial periodic solution of each of the following differential equations.

9. $\ddot{z}+(z^2+\dot{z}^4-2)z=0$

10. $\ddot{z}+[\ln(z^2+4\dot{z}^2)]\dot{z}+z=0$

11. (a) According to Green's theorem in the plane, if C is a closed curve which is sufficiently "smooth," and if f and g are continuous and have continuous first partial derivatives, then

$$\oint_C [f(x,y)\,dy-g(x,y)\,dx]=\iint_R \big[f_x(x,y)+g_y(x,y)\big]\,dx\,dy$$

where R is the region enclosed by C. Assume that $x(t)$, $y(t)$ is a periodic solution of $\dot{x}=f(x,y)$, $\dot{y}=g(x,y)$, and let C be the orbit of this solution. Show that for this curve, the line integral above is zero.

(b) Suppose that f_x+g_y has the same sign throughout a simply connected region D in the $x-y$ plane. Show that the system of equations $\dot{x}=f(x,y)$, $\dot{y}=g(x,y)$ can have no periodic solution which is entirely in D.

12. Show that the system of differential equations

$$\dot{x}=x+y^2+x^3, \qquad \dot{y}=-x+y+yx^2$$

has no nontrivial periodic solution.

13. Show that the system of differential equations

$$\dot{x}=x-xy^2+y^3, \qquad \dot{y}=3y-yx^2+x^3$$

has no nontrivial periodic solution which lies inside the circle $x^2+y^2=4$.

14. (a) Show that $x=0, y=\psi(t)$ is a solution of (2) for any function $\psi(t)$ satisfying $\dot{\psi}=-c\psi-f\psi^2$.

(b) Choose $\psi(t_0)>0$. Show that the orbit of $x=0, y=\psi(t)$ (for all t for which ψ exists) is the positive y axis.

4.9 Introduction to bifurcation theory

Consider the system of equations

$$\dot{\mathbf{x}}=\mathbf{f}(\mathbf{x},\varepsilon) \tag{1}$$

where

$$\mathbf{x}=\begin{pmatrix} x_1 \\ x_2 \end{pmatrix},$$

and ε is a scalar. Intuitively speaking a bifurcation point of (1) is a value of ε at which the solutions of (1) change their behavior. More precisely, we say that $\varepsilon=\varepsilon_0$ is a bifurcation point of (1) if the phase portraits of (1) for $\varepsilon<\varepsilon_0$ and $\varepsilon>\varepsilon_0$ are different.

Remark. In the examples that follow we will appeal to our intuition in deciding whether two phase portraits are the same or are different. In more advanced courses we define two phase portraits to be the same, or topologically equivalent, if there exists a continuous transformation of the plane onto itself which maps one phase portrait onto the other.

Example 1. Find the bifurcation points of the system

$$\dot{\mathbf{x}}=\mathbf{A}\mathbf{x}=\begin{pmatrix} 1 & \varepsilon \\ 1 & -1 \end{pmatrix}\mathbf{x} \tag{2}$$

Solution. The characteristic polynomial of the matrix $\mathbf{A}$ is

$$\begin{aligned} p(\lambda) &= \det(\mathbf{A}-\lambda\mathbf{I}) \\ &= \det\begin{pmatrix} 1-\lambda & \varepsilon \\ 1 & -1-\lambda \end{pmatrix} \\ &= (\lambda-1)(\lambda+1)-\varepsilon \\ &= \lambda^2-(1+\varepsilon). \end{aligned}$$

The roots of $p(\lambda)$ are $\pm\sqrt{1+\varepsilon}$ for $\varepsilon>-1$, and $\pm\sqrt{-\varepsilon-1}\,i$ for $\varepsilon<-1$. This

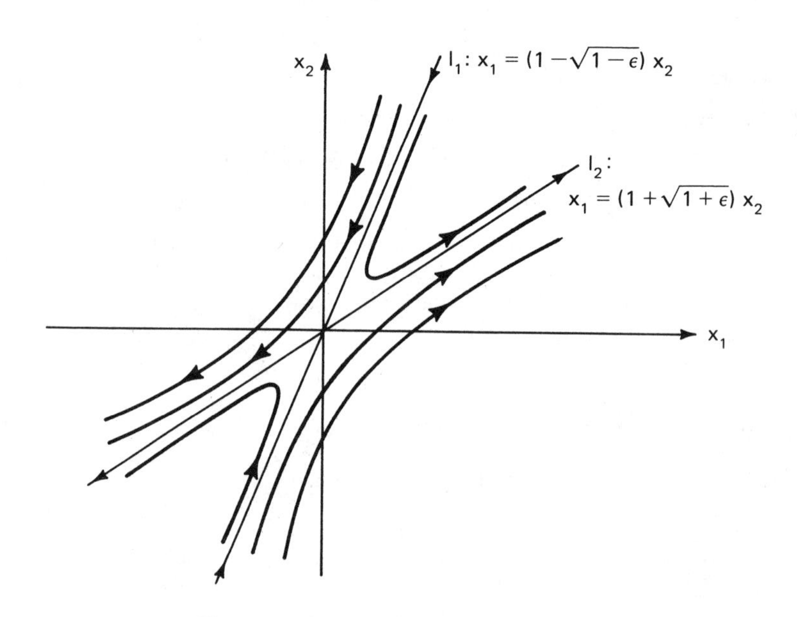

Figure 1. Phase portrait of (2) for $\varepsilon>-1$

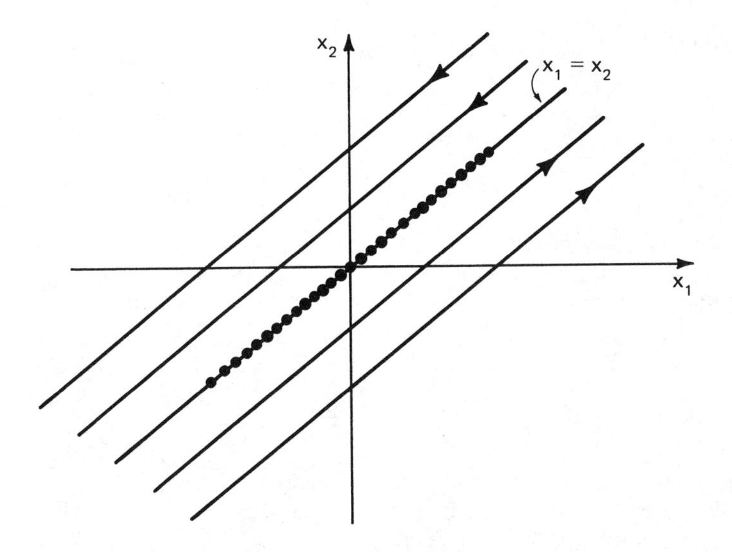

Figure 2. Phase portrait of (2) for $\varepsilon=-1$

implies that $\mathbf{x}=\mathbf{0}$ is a saddle for $\varepsilon>-1$, and a center for $\varepsilon<-1$. We conclude, therefore, that $\varepsilon=-1$ is a bifurcation point of (2). It is also clear that Eq. (2) has no other bifurcation points.

It is instructive to see how the solutions of (2) change as ε passes through the bifurcation value -1. For $\varepsilon>-1$, the eigenvalues of $\mathbf{A}$ are

$$\lambda_1=\sqrt{1+\varepsilon}\,,\quad \lambda_2=-\sqrt{1+\varepsilon}\,.$$

It is easily verified (see Exercise 10) that

$$\mathbf{x}^1=\begin{pmatrix}1+\sqrt{1+\varepsilon}\\ 1\end{pmatrix}$$

is an eigenvector of $\mathbf{A}$ with eigenvalue $\sqrt{1+\varepsilon}$, while

$$\mathbf{x}^2=\begin{pmatrix}1-\sqrt{1+\varepsilon}\\ 1\end{pmatrix}$$

is an eigenvector with eigenvalue $-\sqrt{1+\varepsilon}$. Hence, the phase portrait of (2) has the form shown in Figure 1. As $\varepsilon\to-1$ from the left, the lines l_1 and l_2 both approach the line $x_1=x_2$. This line is a line of equilibrium points of (2) when $\varepsilon=-1$, while each line $x_1-x_2=c$ $(c\neq 0)$ is an orbit of (2) for $\varepsilon=-1$. The phase portrait of (2) for $\varepsilon=-1$ is given in Figure 2.

Example 2. Find the bifurcation points of the system

$$\dot{\mathbf{x}}=\mathbf{A}\mathbf{x}=\begin{pmatrix}0 & -1\\ \varepsilon & -1\end{pmatrix}\mathbf{x}. \tag{3}$$

Solution. The characteristic polynomial of the matrix $\mathbf{A}$ is

$$p(\lambda)=\det(\mathbf{A}-\lambda\mathbf{I})=\det\begin{pmatrix}-\lambda & -1\\ \varepsilon & -1-\lambda\end{pmatrix}$$
$$=\lambda(1+\lambda)+\varepsilon$$
$$=\lambda^2+\lambda+\varepsilon$$

and the roots of $p(\lambda)$ are

$$\lambda_1=\frac{-1+\sqrt{1-4\varepsilon}}{2},\ \lambda_2=\frac{-1-\sqrt{1-4\varepsilon}}{2}.$$

Observe that λ_1 is positive and λ_2 is negative for $\varepsilon<0$. Hence, $\mathbf{x}=\mathbf{0}$ is a saddle for $\varepsilon<0$. For $0<\varepsilon<1/4$, both λ_1 and λ_2 are negative. Hence $\mathbf{x}=\mathbf{0}$ is a stable node for $0<\varepsilon<1/4$. Both λ_1 and λ_2 are complex, with negative real part, for $\varepsilon>1/4$. Hence $\mathbf{x}=\mathbf{0}$ is a stable focus for $\varepsilon>1/4$. Note that the phase portrait of (3) changes as ε passes through 0 and $1/4$. We conclude, therefore, that $\varepsilon=0$ and $\varepsilon=1/4$ are bifurcation points of (3).

Example 3. Find the bifurcation points of the system of equations

$$\frac{dx_1}{dt}=x_2$$
$$\frac{dx_2}{dt}=x_1^2-x_2-\varepsilon. \tag{4}$$

Solution. (i) We first find the equilibrium points of (4). Setting $dx_1/dt=0$ gives $x_2=0$, and then setting $dx_2/dt=0$ gives $x_1^2-\varepsilon=0$, so that $x_1=\pm\sqrt{\varepsilon}$, $\varepsilon>0$. Hence, $(\sqrt{\varepsilon},0)$ and $(-\sqrt{\varepsilon},0)$ are two equilibrium points of (4) for $\varepsilon>0$. The system (4) has no equilibrium points when $\varepsilon<0$. We conclude, therefore, that $\varepsilon=0$ is a bifurcation point of (4). (ii) We now analyze the behavior of the solutions of (4) near the equilibrium points $(\pm\sqrt{\varepsilon},0)$ to determine whether this system has any additional bifurcation points. Setting

$$u=x_1\mp\sqrt{\varepsilon},\quad v=x_2$$

gives

$$\frac{du}{dt}=v$$
$$\frac{dv}{dt}=\left(u\pm\sqrt{\varepsilon}\right)^2-v-\varepsilon=\pm2\sqrt{\varepsilon}\,u-v+u^2. \tag{5}$$

The system (5) can be written in the form

$$\frac{d}{dt}\begin{pmatrix} u \\ v \end{pmatrix} = \begin{pmatrix} 0 & 1 \\ \pm 2\sqrt{\varepsilon} & -1 \end{pmatrix}\begin{pmatrix} u \\ v \end{pmatrix} + \begin{pmatrix} 0 \\ u^2 \end{pmatrix}.$$

By Theorem 4, the phase portrait of (4) near the equilibrium solution

$$\begin{pmatrix} x_1 \\ x_2 \end{pmatrix} = \begin{pmatrix} \pm\sqrt{\varepsilon} \\ 0 \end{pmatrix}$$

is determined by the phase portrait of the linearized system

$$\frac{d}{dt}\begin{pmatrix} u \\ v \end{pmatrix} = \mathbf{A}\begin{pmatrix} u \\ v \end{pmatrix} = \begin{pmatrix} 0 & 1 \\ \pm 2\sqrt{\varepsilon} & -1 \end{pmatrix}\begin{pmatrix} u \\ v \end{pmatrix}.$$

To find the eigenvalues of $\mathbf{A}$ we compute

$$\begin{aligned} p(\lambda) &= \det(\mathbf{A} - \lambda\mathbf{I}) \\ &= \det\begin{pmatrix} -\lambda & 1 \\ \pm 2\sqrt{\varepsilon} & -1-\lambda \end{pmatrix} \\ &= \lambda^2 + \lambda \mp 2\sqrt{\varepsilon}\,. \end{aligned}$$

Hence, the eigenvalues of $\mathbf{A}$ when $u = x_1 - \sqrt{\varepsilon}$ are

$$\lambda_1 = \frac{-1+\sqrt{1+8\sqrt{\varepsilon}}}{2}, \quad \lambda_2 = \frac{-1-\sqrt{1+8\sqrt{\varepsilon}}}{2} \tag{6}$$

while the eigenvalues of $\mathbf{A}$ when $u = x_1 + \sqrt{\varepsilon}$ are

$$\lambda_1 = \frac{-1+\sqrt{1-8\sqrt{\varepsilon}}}{2}, \quad \lambda_2 = \frac{-1-\sqrt{1-8\sqrt{\varepsilon}}}{2}. \tag{7}$$

Observe from (6) that $\lambda_1 > 0$, while $\lambda_2 < 0$. Thus, the system (4) behaves like a saddle near the equilibrium points $\begin{pmatrix} \sqrt{\varepsilon} \\ 0 \end{pmatrix}$. On the other hand, we see from (7) that both λ_1 and λ_2 are negative for $0 < \varepsilon < 1/64$, and complex for $\varepsilon > 1/64$. Consequently, the system (4) near the equilibrium solution $\begin{pmatrix} -\sqrt{\varepsilon} \\ 0 \end{pmatrix}$ behaves like a stable node for $0 < \varepsilon < 1/64$, and a stable focus for $\varepsilon > 1/64$. It can be shown that the phase portraits of a stable node and a stable focus are equivalent. Consequently, $\varepsilon = 1/64$ is *not* a bifurcation point of (4).

Another situation which is included in the context of bifurcation theory, and which is of much current research interest now, is when the system (1) has a certain number of equilibrium, or periodic, solutions for $\varepsilon=\varepsilon_0$, and a different number for $\varepsilon\neq\varepsilon_0$. Suppose, for example, that $\mathbf{x}(t)$ is an equilibrium, or periodic, solution of (1) for $\varepsilon=0$, and $\mathbf{x}^1(t),\mathbf{x}^2(t),\ldots,\mathbf{x}^k(t)$ are equilibria, or periodic solutions of (1) for $\varepsilon\neq 0$ which approach $\mathbf{x}(t)$ as $\varepsilon\to 0$. In this case we say that the solutions $\mathbf{x}^1(t),\mathbf{x}^2(t),\ldots,\mathbf{x}^k(t)$ *bifurcate* from $\mathbf{x}(t)$. We illustrate this situation with the following example.

Example 4. Find all equilibrium solutions of the system of equations

$$\begin{aligned}\frac{dx_1}{dt}&=3\varepsilon x_1-3\varepsilon x_2-x_1^2-x_2^2\\ \frac{dx}{dt}&=\varepsilon x_1-x_1x_2=x_1(\varepsilon-x_2).\end{aligned}\tag{8}$$

Solution. Let $\begin{pmatrix}x_1\\x_2\end{pmatrix}$ be an equilibrium solution of the system (8). The second equation of (8) implies that $x_1=0$ or $x_2=\varepsilon$.

$x_1=0$. In this case, the first equation of (8) implies that

$$0=3\varepsilon x_2+x_2^2=x_2(3\varepsilon+x_2)$$

so that $x_2=0$ or $x_2=-3\varepsilon$. Thus $\begin{pmatrix}0\\0\end{pmatrix}$ and $\begin{pmatrix}0\\-3\varepsilon\end{pmatrix}$ are two equilibrium points of (8).

$x_2=\varepsilon$. In this case, the first equation of (8) implies that

$$3\varepsilon x_1-3\varepsilon^2-x_1^2-\varepsilon^2=0$$

or

$$x_1^2-3\varepsilon x_1+4\varepsilon^2=0.\tag{9}$$

The solutions

$$x_1=\frac{3\varepsilon\pm\sqrt{9\varepsilon^2-16\varepsilon^2}}{2}$$

of (9) are complex. Thus,

$$\mathbf{x}^1=\begin{pmatrix}0\\0\end{pmatrix}\quad\text{and}\quad\mathbf{x}^2=\begin{pmatrix}0\\-3\varepsilon\end{pmatrix}$$

are two equilibrium points of (8), for $\varepsilon\neq 0$, which bifurcate from the single equilibrium point $\mathbf{x}=\begin{pmatrix}0\\0\end{pmatrix}$ when $\varepsilon=0$.

Exercises

Find the bifurcation points of each of the following systems of equations.

1. $\dot{\mathbf{x}}=\begin{pmatrix}1 & \varepsilon\\ \varepsilon & 1\end{pmatrix}\mathbf{x}$

2. $\dot{\mathbf{x}}=\begin{pmatrix}1 & 1\\ \varepsilon & 1\end{pmatrix}\mathbf{x}$

3. $\dot{\mathbf{x}}=\begin{pmatrix}0 & 2\\ -2 & \varepsilon\end{pmatrix}\mathbf{x}$

4. $\dot{\mathbf{x}}=\begin{pmatrix}0 & 2\\ 2 & \varepsilon\end{pmatrix}\mathbf{x}$

5. $\dot{\mathbf{x}}=\begin{pmatrix}0 & -\varepsilon\\ \varepsilon & 0\end{pmatrix}\mathbf{x}$

In each of Problems 6–8, show that more than one equilibrium solutions bifurcate from the equilibrium solution $\mathbf{x}=\mathbf{0}$ when $\varepsilon=0$.

6. $\dot{x}_1=\varepsilon x_1-\varepsilon x_2-x_1^2+x_2^2$
$\dot{x}_2=\varepsilon x_2+x_1x_2$

7. $\dot{x}_1=\varepsilon x_1-x_1^2-x_1x_2$
$\dot{x}_2=-2\varepsilon x_1+2\varepsilon x_2+x_1x_2-x_2^2$

8. $\dot{x}_1=\varepsilon x_2+x_1x_2$
$\dot{x}_2=-\varepsilon x_1+\varepsilon x_2+x_1^2+x_2^2$

9. Consider the system of equations

$$\begin{aligned}\dot{x}_1&=3\varepsilon x_1-5\varepsilon x_2-x_1^2+x_2^2\\ \dot{x}_2&=2\varepsilon x_1-\varepsilon x_2.\end{aligned}\qquad (*)$$

(a) Show that each point on the lines $x_2=x_1$ and $x_2=-x_1$ are equilibrium points of $(*)$ for $\varepsilon=0$.

(b) Show that

$$\begin{pmatrix}x_1\\ x_2\end{pmatrix}=\begin{pmatrix}0\\ 0\end{pmatrix}\quad\text{and}\quad\begin{pmatrix}x_1\\ x_2\end{pmatrix}=\tfrac{7}{3}\varepsilon\begin{pmatrix}1\\ 2\end{pmatrix}.$$

are the only equilibrium points of $(*)$ for $\varepsilon\neq 0$.

10. Show that

$$\mathbf{x}^1=\begin{pmatrix}1+\sqrt{1+\varepsilon}\\ 1\end{pmatrix}\quad\text{and}\quad\mathbf{x}^2=\begin{pmatrix}1-\sqrt{1+\varepsilon}\\ 1\end{pmatrix}$$

are eigenvectors of the matrix $\begin{pmatrix}1 & \varepsilon\\ 1 & -1\end{pmatrix}$ with eigenvalues $\sqrt{1+\varepsilon}$ and $-\sqrt{1+\varepsilon}$ respectively.

4.10 Predator–prey problems; or why the percentage of sharks caught in the Mediterranean Sea rose dramatically during World War I

In the mid 1920's the Italian biologist Umberto D'Ancona was studying the population variations of various species of fish that interact with each other. In the course of his research, he came across some data on percentages-of-total-catch of several species of fish that were brought into different Mediterranean ports in the years that spanned World War I. In particular, the data gave the percentage-of-total-catch of selachians, (sharks, skates, rays, etc.) which are not very desirable as food fish. The data for the port of Fiume, Italy, during the years 1914–1923 is given below.

1914	1915	1916	1917	1918
11.9%	21.4%	22.1%	21.2%	36.4%

1919	1920	1921	1922	1923
27.3%	16.0%	15.9%	14.8%	10.7%

D'Ancona was puzzled by the very large increase in the percentage of selachians during the period of the war. Obviously, he reasoned, the increase in the percentage of selachians was due to the greatly reduced level of fishing during this period. But how does the intensity of fishing affect the fish populations? The answer to this question was of great concern to D'Ancona in his research on the struggle for existence between competing species. It was also of concern to the fishing industry, since it would have obvious implications for the way fishing should be done.

Now, what distinguishes the selachians from the food fish is that the selachians are predators, while the food fish are their prey; the selachians depend on the food fish for their survival. At first, D'Ancona thought that this accounted for the large increase of selachians during the war. Since the level of fishing was greatly reduced during this period, there were more prey available to the selachians, who therefore thrived and multiplied rapidly. However, this explanation does not hold any water since there were also more food fish during this period. D'Ancona's theory only shows that there are more selachians when the level of fishing is reduced; it does not explain why a reduced level of fishing is *more* beneficial to the predators than to their prey.

After exhausting all possible biological explanations of this phenomenon, D'Ancona turned to his colleague, the famous Italian mathematician Vito Volterra. Hopefully, Volterra would formulate a mathematical model of the growth of the selachians and their prey, the food fish, and this model would provide the answer to D'Ancona's question. Volterra began his analysis of this problem by separating all the fish into the prey population $x(t)$ and the predator population $y(t)$. Then, he reasoned that the food fish do not compete very intensively among themselves for their food supply since this is very abundant, and the fish population is not very dense. Hence, in the absence of the selachians, the food fish would grow according to the Malthusian law of population growth $\dot{x} = ax$, for some positive constant a. Next, reasoned Volterra, the number of contacts per unit time between predators and prey is bxy, for some positive constant b. Hence, $\dot{x} = ax - bxy$. Similarly, Volterra concluded that the predators have a natural rate of decrease $-cy$ proportional to their present number, and that they also increase at a rate dxy proportional to their present number y and their food supply x. Thus,

$$\frac{dx}{dt} = ax - bxy, \qquad \frac{dy}{dt} = -cy + dxy. \tag{1}$$

The system of equations (1) governs the interaction of the selachians and food fish in the absence of fishing. We will carefully analyze this system and derive several interesting properties of its solutions. Then, we will include the effect of fishing in our model, and show why a reduced level of fishing is more beneficial to the selachians than to the food fish. In fact, we will derive the surprising result that a reduced level of fishing is actually harmful to the food fish.

Observe first that (1) has two equilibrium solutions $x(t)=0, y(t)=0$ and $x(t)=c/d, y(t)=a/b$. The first equilibrium solution, of course, is of no interest to us. This system also has the family of solutions $x(t)=x_0e^{at}, y(t)=0$ and $x(t)=0, y(t)=y_0e^{-ct}$. Thus, both the x and y axes are orbits of (1). This implies that every solution $x(t)$, $y(t)$ of (1) which starts in the first quadrant $x>0, y>0$ at time $t=t_0$ will remain there for all future time $t \geqslant t_0$.

The orbits of (1), for $x,y \neq 0$ are the solution curves of the first-order equation

$$\frac{dy}{dx} = \frac{-cy+dxy}{ax-bxy} = \frac{y(-c+dx)}{x(a-by)}. \tag{2}$$

This equation is separable, since we can write it in the form

$$\frac{a-by}{y}\frac{dy}{dx} = \frac{-c+dx}{x}.$$

Consequently, $a\ln y - by + c\ln x - dx = k_1$, for some constant k_1. Taking exponentials of both sides of this equation gives

$$\frac{y^a}{e^{by}}\frac{x^c}{e^{dx}} = K \tag{3}$$

for some constant K. Thus, the orbits of (1) are the family of curves defined by (3), and these curves are *closed* as we now show.

Lemma 1. *Equation* (3) *defines a family of closed curves for* $x, y>0$.

PROOF. Our first step is to determine the behavior of the functions $f(y)=y^a/e^{by}$ and $g(x)=x^c/e^{dx}$ for x and y positive. To this end, observe that $f(0)=0, f(\infty)=0$, and $f(y)$ is positive for $y>0$. Computing

$$f'(y) = \frac{ay^{a-1}-by^a}{e^{by}} = \frac{y^{a-1}(a-by)}{e^{by}},$$

we see that $f(y)$ has a single critical point at $y=a/b$. Consequently, $f(y)$ achieves its maximum value $M_y=(a/b)^a/e^a$ at $y=a/b$, and the graph of

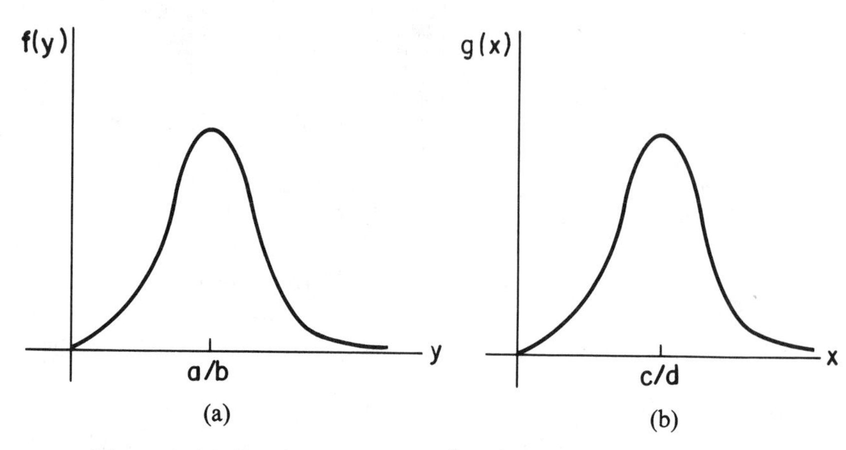

Figure 1. (a) Graph of $f(y)=y^a e^{-by}$; (b) Graph of $g(x)=x^c e^{-dx}$

$f(y)$ has the form described in Figure 1a. Similarly, $g(x)$ achieves its maximum value $M_x=(c/d)^c/e^c$ at $x=c/d$, and the graph of $g(x)$ has the form described in Figure 1b.

From the preceding analysis, we conclude that Equation (3) has no solution $x,y>0$ for $K>M_x M_y$, and the single solution $x=c/d$, $y=a/b$ for $K=M_x M_y$. Thus, we need only consider the case $K=\lambda M_y$, where λ is a positive number less than M_x. Observe first that the equation $x^c/e^{dx}=\lambda$ has one solution $x=x_m<c/d$, and one solution $x=x_M>c/d$. Hence, the equation

$$f(y)=y^a e^{-by}=\left(\frac{\lambda}{x^c e^{-dx}}\right)M_y$$

has no solution y when x is less than x_m or greater than x_M. It has the single solution $y=a/b$ when $x=x_m$ or x_M, and it has two solutions $y_1(x)$ and $y_2(x)$ for each x between x_m and x_M. The smaller solution $y_1(x)$ is always less than a/b, while the larger solution $y_2(x)$ is always greater than a/b. As x approaches either x_m or x_M, both $y_1(x)$ and $y_2(x)$ approach a/b. Consequently, the curves defined by (3) are closed for x and y positive, and have the form described in Figure 2. Moreover, none of these closed curves (with the exception of $x=c/d$, $y=a/b$) contain any equilibrium points of (1). Therefore, all solutions $x(t)$, $y(t)$ of (1), with $x(0)$ and $y(0)$ positive, are *periodic* functions of time. That is to say, each solution $x(t)$, $y(t)$ of (1), with $x(0)$ and $y(0)$ positive, has the property that $x(t+T)=x(t)$ and $y(t+T)=y(t)$ for some positive T. □

Now, the data of D'Ancona is really an *average* over each one year period of the proportion of predators. Thus, in order to compare this data with the predictions of (1), we must compute the "average values" of $x(t)$ and $y(t)$, for any solution $x(t)$, $y(t)$ of (1). Remarkably, we can find these average values even though we cannot compute $x(t)$ and $y(t)$ exactly. This is the content of Lemma 2.

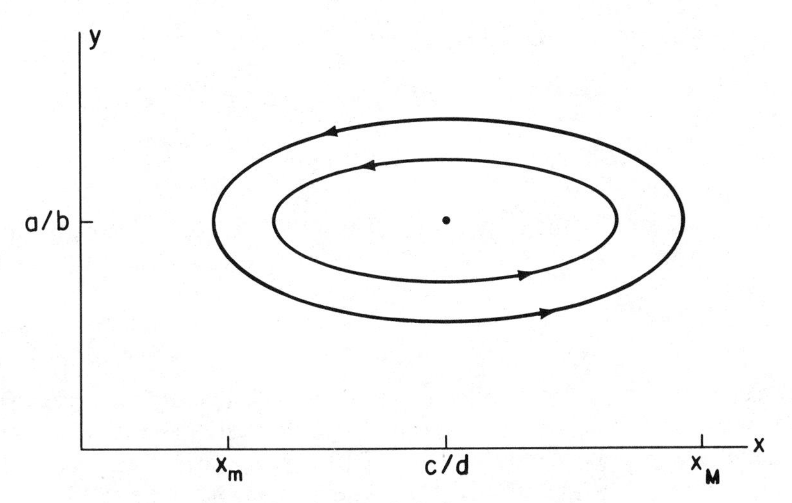

Figure 2. Orbits of (1) for x,y positive

Lemma 2. *Let $x(t), y(t)$ be a periodic solution of (1), with period $T>0$. Define the average values of x and y as*

$$\bar{x}=\frac{1}{T}\int_0^T x(t)\,dt, \qquad \bar{y}=\frac{1}{T}\int_0^T y(t)\,dt.$$

Then, $\bar{x}=c/d$ and $\bar{y}=a/b$. In other words, the average values of $x(t)$ and $y(t)$ are the equilibrium values.

PROOF. Dividing both sides of the first equation of (1) by x gives $\dot{x}/x = a-by$, so that

$$\frac{1}{T}\int_0^T \frac{\dot{x}(t)}{x(t)}\,dt=\frac{1}{T}\int_0^T [a-by(t)]\,dt.$$

Now, $\int_0^T \dot{x}(t)/x(t)\,dt=\ln x(T)-\ln x(0)$, and this equals zero since $x(T)= x(0)$. Consequently,

$$\frac{1}{T}\int_0^T by(t)\,dt=\frac{1}{T}\int_0^T a\,dt=a,$$

so that $\bar{y}=a/b$. Similarly, by dividing both sides of the second equation of (1) by $Ty(t)$ and integrating from 0 to T, we obtain that $\bar{x}=c/d$. □

We are now ready to include the effects of fishing in our model. Observe that fishing decreases the population of food fish at a rate $\varepsilon x(t)$, and decreases the population of selachians at a rate $\varepsilon y(t)$. The constant ε reflects the intensity of fishing; i.e., the number of boats at sea and the number of nets in the water. Thus, the true state of affairs is described by the modified system of differential equations

$$\begin{aligned}\frac{dx}{dt} &= ax - bxy - \varepsilon x = (a-\varepsilon)x - bxy\\ \frac{dy}{dt} &= -cy + dxy - \varepsilon y = -(c+\varepsilon)y + dxy.\end{aligned} \tag{4}$$

This system is exactly the same as (1) (for $a-\varepsilon>0$), with a replaced by $a-\varepsilon$, and c replaced by $c+\varepsilon$. Hence, the average values of $x(t)$ and $y(t)$ are now

$$\bar{x} = \frac{c+\varepsilon}{d}, \qquad \bar{y} = \frac{a-\varepsilon}{b}. \tag{5}$$

Consequently, a moderate amount of fishing ($\varepsilon < a$) actually increases the number of food fish, on the average, and decreases the number of selachians. Conversely, a reduced level of fishing increases the number of selachians, on the average, and *decreases* the number of food fish. This remarkable result, which is known as Volterra's principle, explains the data of D'Ancona, and completely solves our problem.

Volterra's principle has spectacular applications to insecticide treatments, which destroy both insect predators and their insect prey. It implies that the application of insecticides will actually increase the population of those insects which are kept in control by other predatory insects. A remarkable confirmation comes from the cottony cushion scale insect (Icerya purchasi), which, when accidentally introduced from Australia in 1868, threatened to destroy the American citrus industry. Thereupon, its natural Australian predator, a ladybird beetle (Novius Cardinalis) was introduced, and the beetles reduced the scale insects to a low level. When DDT was discovered to kill scale insects, it was applied by the orchardists in the hope of further reducing the scale insects. However, in agreement with Volterra's principle, the effect was an increase of the scale insect!

Oddly enough, many ecologists and biologists refused to accept Volterra's model as accurate. They pointed to the fact that the oscillatory behavior predicted by Volterra's model is not observed in most predator–prey systems. Rather, most predator–prey systems tend to equilibrium states as time evolves. Our answer to these critics is that the system of differential equations (1) is not intended as a model of the general predator–prey interaction. This is because the food fish and selachians do not compete intensively among themselves for their available resources. A more general model of predator–prey interactions is the system of differential equations

$$\dot{x} = ax - bxy - ex^2, \qquad \dot{y} = -cy + dxy - fy^2. \tag{6}$$

Here, the term ex^2 reflects the internal competition of the prey x for their limited external resources, and the term fy^2 reflects the competition among the predators for the limited number of prey. The solutions of (6) are not, in general, periodic. Indeed, we have already shown in Example 1 of Section 4.8 that all solutions $x(t)$, $y(t)$ of (6), with $x(0)$ and $y(0)$ positive, ultimately approach the equilibrium solution $x=a/e$, $y=0$ if c/d is greater than a/e. In this situation, the predators die out, since their available food supply is inadequate for their needs.

Surprisingly, some ecologists and biologists even refuse to accept the more general model (6) as accurate. As a counterexample, they cite the experiments of the mathematical biologist G. F. Gause. In these experiments, the population was composed of two species of protozoa, one of which, Didinium nasatum, feeds on the other, Paramecium caudatum. In all of Gause's experiments, the Didinium quickly destroyed the Paramecium and then died of starvation. This situation cannot be modeled by the system of equations (6), since no solution of (6) with $x(0)y(0)\neq 0$ can reach $x=0$ or $y=0$ in finite time.

Our answer to these critics is that the Didinium are a special, and atypical type of predator. On the one hand, they are ferocious attackers and require a tremendous amount of food; a Didinium demands a fresh Paramecium every three hours. On the other hand, the Didinium don't perish from an insufficient supply of Paramecium. They continue to multiply, but give birth to smaller offspring. Thus, the system of equations (6) does not accurately model the interaction of Paramecium and Didinium. A better model, in this case, is the system of differential equations

$$\frac{dx}{dt}=ax-b\sqrt{x}\,y,\qquad \frac{dy}{dt}=\begin{cases} d\sqrt{x}\,y, & x\neq 0\\ -cy, & x=0\end{cases}. \tag{7}$$

It can be shown (see Exercise 6) that every solution $x(t)$, $y(t)$ of (7) with $x(0)$ and $y(0)$ positive reaches $x=0$ in finite time. This does not contradict the existence–uniqueness theorem, since the function

$$g(x,y)=\begin{cases} d\sqrt{x}\,y, & x\neq 0\\ -cy, & x=0\end{cases}$$

does not have a partial derivative with respect to x or y, at $x=0$.

Finally, we mention that there are several predator–prey interactions in nature which cannot be modeled by any system of ordinary differential equations. These situations occur when the prey are provided with a refuge that is inaccessible to the predators. In these situations, it is impossible to make any definitive statements about the future number of predators and prey, since we cannot predict how many prey will be stupid enough to leave their refuge. In other words, this process is now *random*, rather than *deterministic*, and therefore cannot be modeled by a system of ordinary differential equations. This was verified directly in a famous experiment of Gause. He placed five Paramecium and three Didinium in each of thirty identical test tubes, and provided the Paramecium with a refuge from the Didinium. Two days later, he found the predators dead in four tubes, and a mixed population containing from two to thirty-eight Paramecium in the remaining twenty-six tubes.

Reference

Volterra, V: "Leçons sur la théorie mathématique de la lutte pour la vie." Paris, 1931.

Exercises

1. Find all biologically realistic equilibrium points of (6) and determine their stability.

2. We showed in Section 4.8 that $y(t)$ ultimately approaches zero for all solutions $x(t), y(t)$ of (6), if $c/d > a/e$. Show that there exist solutions $x(t), y(t)$ of (6) for which $y(t)$ increases at first to a maximum value, and then decreases to zero. (To an observer who sees only the predators without noticing the prey, such a case of a population passing through a maximum to total extinction would be very difficult to explain.)

3. In many instances, it is the adult members of the prey who are chiefly attacked by the predators, while the young members are better protected, either by their smaller size, or by their living in a different station. Let x_1 be the number of adult prey, x_2 the number of young prey, and y the number of predators. Then,

$$\dot{x}_1 = -a_1x_1 + a_2x_2 - bx_1y$$
$$\dot{x}_2 = nx_1 - (a_1 + a_2)x_2$$
$$\dot{y} = -cy + dx_1y$$

where a_2x_2 represents the number of young (per unit time) growing into adults, and n represents the birth rate proportional to the number of adults. Find all equilibrium solutions of this system.

4. There are several situations in nature where species 1 preys on species 2 which in turn preys on species 3. One case of this kind of population is the Island of Komodo in Malaya which is inhabited by giant carnivorous reptiles, and by mammals—their food—which feed on the rich vegetation of the island. We assume that the reptiles have no direct influence on the vegetation, and that only the plants compete among themselves for their available resources. A system of differential equations governing this interaction is

$$\dot{x}_1 = -a_1x_1 - b_{12}x_1x_2 + c_{13}x_1x_3$$
$$\dot{x}_2 = -a_2x_2 + b_{21}x_1x_2$$
$$\dot{x}_3 = a_3x_3 - a_4x_3^2 - c_{31}x_1x_3$$

Find all equilibrium solutions of this system.

5. Consider a predator–prey system where the predator has alternate means of support. This system can be modelled by the differential equations

$$\dot{x}_1 = \alpha_1x_1(\beta_1 - x_1) + \gamma_1x_1x_2$$
$$\dot{x}_2 = \alpha_2x_2(\beta_2 - x_2) - \gamma_2x_1x_2$$

where $x_1(t)$ and $x_2(t)$ are the predators and prey populations, respectively, at time t.

(a) Show that the change of coordinates $\beta_i y_i(t) = x_i(t/\alpha_i\beta_i)$ reduces this system of equations to

$$\dot{y}_1 = y_1(1-y_1) + a_1 y_1 y_2, \qquad \dot{y}_2 = y_2(1-y_2) - a_2 y_1 y_2$$

where $a_1 = \gamma_1\beta_2/\alpha_1\beta_1$ and $a_2 = \gamma_2\beta_1/\alpha_2\beta_2$.

(b) What are the stable equilibrium populations when (i) $0 < a_2 < 1$, (ii) $a_2 > 1$?

(c) It is observed that $a_1 = 3a_2$ (a_2 is a measure of the aggressiveness of the predator). What is the value of a_2 if the predator's instinct is to maximize its stable equilibrium population?

6. (a) Let $x(t)$ be a solution of $\dot{x} = ax - M\sqrt{x}$, with $M > a\sqrt{x(t_0)}$. Show that

$$a\sqrt{x} = M - \left(M - a\sqrt{x(t_0)}\right)e^{a(t-t_0)/2}.$$

(b) Conclude from (a) that $x(t)$ approaches zero in finite time.

(c) Let $x(t)$, $y(t)$ be a solution of (7), with $by(t_0) > a\sqrt{x(t_0)}$. Show that $x(t)$ reaches zero in finite time. *Hint*: Observe that $y(t)$ is increasing for $t \geqslant t_0$.

(d) It can be shown that $by(t)$ will eventually exceed $a\sqrt{x(t)}$ for every solution $x(t)$, $y(t)$ of (7) with $x(t_0)$ and $y(t_0)$ positive. Conclude, therefore, that all solutions $x(t)$, $y(t)$ of (7) achieve $x = 0$ in finite time.

4.11 The principle of competitive exclusion in population biology

It is often observed, in nature, that the struggle for existence between two similar species competing for the same limited food supply and living space nearly always ends in the complete extinction of one of the species. This phenomenon is known as the "principle of competitive exclusion." It was first enunciated, in a slightly different form, by Darwin in 1859. In his paper 'The origin of species by natural selection' he writes: "As the species of the same genus usually have, though by no means invariably, much similarity in habits and constitutions and always in structure, the struggle will generally be more severe between them, if they come into competition with each other, than between the species of distinct genera."

There is a very interesting biological explanation of the principle of competitive exclusion. The cornerstone of this theory is the idea of a "niche." A niche indicates what place a given species occupies in a community; i.e., what are its habits, food and mode of life. It has been observed that as a result of competition two similar species rarely occupy the same niche. Rather, each species takes possession of those kinds of food and modes of life in which it has an advantage over its competitor. If the two species tend to occupy the same niche then the struggle for existence between them will be very intense and result in the extinction of the weaker species.

An excellent illustration of this theory is the colony of terns inhabiting the island of Jorilgatch in the Black Sea. This colony consists of four different species of terns: sandwich-tern, common-tern, blackbeak-tern, and little-tern. These four species band together to chase away predators from the colony. However, there is a sharp difference between them as regards the procuring of food. The sandwich-tern flies far out into the open sea to hunt certain species, while the blackbeak-tern feeds exclusively on land. On the other hand, common-tern and little-tern catch fish close to the shore. They sight the fish while flying and dive into the water after them. The little-tern seizes his fish in shallow swampy places, whereas the common-tern hunts somewhat further from shore. In this manner, these four similar species of tern living side by side upon a single small island differ sharply in all their modes of feeding and procuring food. Each has a niche in which it has a distinct advantage over its competitors.

In this section we present a rigorous mathematical proof of the law of competitive exclusion. This will be accomplished by deriving a system of differential equations which govern the interaction between two similar species, and then showing that every solution of the system approaches an equilibrium state in which one of the species is extinct.

In constructing a mathematical model of the struggle for existence between two competing species, it is instructive to look again at the logistic law of population growth

$$\frac{dN}{dt} = aN - bN^2. \tag{1}$$

This equation governs the growth of the population $N(t)$ of a single species whose members compete among themselves for a limited amount of food and living space. Recall (see Section 1.5) that $N(t)$ approaches the limiting population $K = a/b$, as t approaches infinity. This limiting population can be thought of as the maximum population of the species which the microcosm can support. In terms of K, the logistic law (1) can be rewritten in the form

$$\frac{dN}{dt} = aN\left(1 - \frac{b}{a}N\right) = aN\left(1 - \frac{N}{K}\right) = aN\left(\frac{K-N}{K}\right). \tag{2}$$

Equation (2) has the following interesting interpretation. When the population N is very low, it grows according to the Malthusian law $dN/dt = aN$. The term aN is called the "biotic potential" of the species. It is the potential rate of increase of the species under ideal conditions, and it is realized if there are no restrictions on food and living space, and if the individual members of the species do not excrete any toxic waste products. As the population increases though, the biotic potential is reduced by the factor $(K-N)/K$, which is the relative number of still vacant places in the microcosm. Ecologists call this factor the environmental resistance to growth.

Now, let $N_1(t)$ and $N_2(t)$ be the population at time t of species 1 and 2 respectively. Further, let K_1 and K_2 be the maximum population of species 1 and 2 which the microcosm can support, and let a_1N_1 and a_2N_2 be the biotic potentials of species 1 and 2. Then, $N_1(t)$ and $N_2(t)$ satisfy the system of differential equations

$$\frac{dN_1}{dt}=a_1N_1\left(\frac{K_1-N_1-m_2}{K_1}\right),\qquad \frac{dN_2}{dt}=a_2N_2\left(\frac{K_2-N_2-m_1}{K_2}\right), \tag{3}$$

where m_2 is the total number of places of the first species which are taken up by members of the second species, and m_1 is the total number of places of the second species which are taken up by members of the first species. At first glance it would appear that $m_2=N_2$ and $m_1=N_1$. However, this is not generally the case, for it is highly unlikely that two species utilize the environment in identical ways. Equal numbers of individuals of species 1 and 2 do not, on the average, consume equal quantities of food, take up equal amounts of living space and excrete equal amounts of waste products of the same chemical composition. In general, we must set $m_2=\alpha N_2$ and $m_1=\beta N_1$, for some constants α and β. The constants α and β indicate the degree of influence of one species upon the other. If the interests of the two species do not clash, and they occupy separate niches, then both α and β are zero. If the two species lay claim to the same niche and are very similar, then α and β are very close to one. On the other hand, if one of the species, say species 2, utilizes the environment very unproductively; i.e., it consumes a great deal of food or excretes very poisonous waste products, then one individual of species 2 takes up the place of many individuals of species 1. In this case, then, the coefficient α is very large.

We restrict ourselves now to the case where the two species are nearly identical, and lay claim to the same niche. Then, $\alpha=\beta=1$, and $N_1(t)$ and $N_2(t)$ satisfy the system of differential equations

$$\frac{dN_1}{dt}=a_1N_1\left(\frac{K_1-N_1-N_2}{K_1}\right),\qquad \frac{dN_2}{dt}=a_2N_2\left(\frac{K_2-N_1-N_2}{K_2}\right). \tag{4}$$

In this instance, we expect the struggle for existence between species 1 and 2 to be very intense, and to result in the extinction of one of the species. This is indeed the case as we now show.

Theorem 6 (Principle of competitive exclusion). *Suppose that K_1 is greater than K_2. Then, every solution $N_1(t)$, $N_2(t)$ of (4) approaches the equilibrium solution $N_1=K_1$, $N_2=0$ as t approaches infinity. In other words, if species 1 and 2 are very nearly identical, and the microcosm can support more members of species 1 than of species 2, then species 2 will ultimately become extinct.*

Our first step in proving Theorem 6 is to show that $N_1(t)$ and $N_2(t)$ can never become negative. To this end, recall from Section 1.5 that

$$N_1(t)=\frac{K_1N_1(0)}{N_1(0)+(K_1-N_1(0))e^{-a_1t}},\qquad N_2(t)=0$$

is a solution of (4) for any choice of $N_1(0)$. The orbit of this solution in the N_1–N_2 plane is the point (0,0) for $N_1(0)=0$; the line $0<N_1<K_1$, $N_2=0$ for $0<N_1(0)<K_1$; the point $(K_1,0)$ for $N_1(0)=K_1$; and the line $K_1<N_1<\infty$, $N_2=0$ for $N_1(0)>K_1$. Thus, the N_1 axis, for $N_1 \geqslant 0$, is the union of four distinct orbits . Similarly, the N_2 axis, for $N_2 \geqslant 0$, is the union of four distinct orbits of (4). This implies that all solutions $N_1(t)$, $N_2(t)$ of (4) which start in the first quadrant $(N_1>0, N_2>0)$ of the N_1–N_2 plane must remain there for all future time.

Our second step in proving Theorem 6 is to split the first quadrant into regions in which both dN_1/dt and dN_2/dt have fixed signs. This is accomplished in the following manner. Let l_1 and l_2 be the lines $K_1-N_1-N_2=0$ and $K_2-N_1-N_2=0$, respectively. Observe that dN_1/dt is negative if (N_1,N_2) lies above l_1, and positive if (N_1,N_2) lies below l_1. Similarly, dN_2/dt is negative if (N_1,N_2) lies above l_2, and positive if (N_1,N_2) lies below l_2. Thus, the two parallel lines l_1 and l_2 split the first quadrant of the N_1–N_2 plane into three regions (see Figure 1) in which both dN_1/dt and dN_2/dt have fixed signs. Both $N_1(t)$ and $N_2(t)$ increase with time (along any solution of (4)) in region I; $N_1(t)$ increases, and $N_2(t)$ decreases, with time in region II; and both $N_1(t)$ and $N_2(t)$ decrease with time in region III.

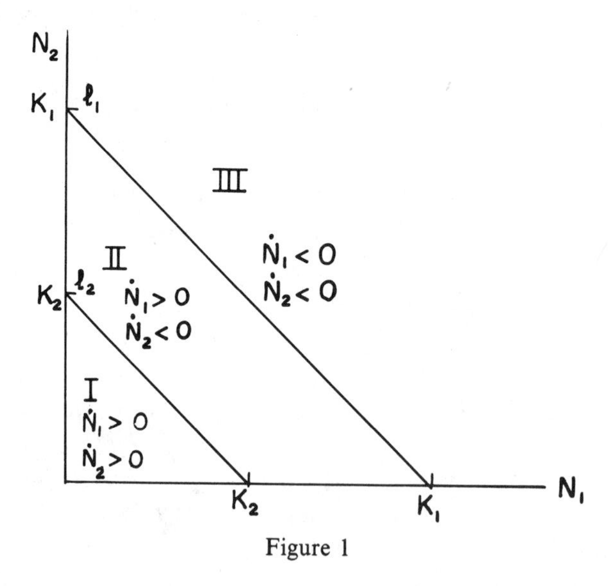

Figure 1

Lemma 1. *Any solution $N_1(t)$, $N_2(t)$ of* (4) *which starts in region* I *at* $t=t_0$ *must leave this region at some later time.*

PROOF. Suppose that a solution $N_1(t)$, $N_2(t)$ of (4) remains in region I for all time $t \geqslant t_0$. This implies that both $N_1(t)$ and $N_2(t)$ are monotonic increasing functions of time for $t \geqslant t_0$, with $N_1(t)$ and $N_2(t)$ less than K_2. Consequently, by Lemma 1 of Section 4.8, both $N_1(t)$ and $N_2(t)$ have limits

ξ,η respectively, as t approaches infinity. Lemma 2 of Section 4.8 implies that (ξ,η) is an equilibrium point of (4). Now, the only equilibrium points of (4) are $(0,0)$, $(K_1,0)$, and $(0,K_2)$, and (ξ,η) obviously cannot equal any of these three points. We conclude therefore, that any solution $N_1(t)$, $N_2(t)$ of (4) which starts in region I must leave this region at a later time. □

Lemma 2. *Any solution $N_1(t)$, $N_2(t)$ of* (4) *which starts in region* II *at time $t=t_0$ will remain in this region for all future time $t\geqslant t_0$, and ultimately approach the equilibrium solution $N_1=K_1$, $N_2=0$.*

PROOF. Suppose that a solution $N_1(t)$, $N_2(t)$ of (4) leaves region II at time $t=t^*$. Then, either $\dot{N}_1(t^*)$ or $\dot{N}_2(t^*)$ is zero, since the only way a solution of (4) can leave region II is by crossing l_1 or l_2. Assume that $\dot{N}_1(t^*)=0$. Differentiating both sides of the first equation of (4) with respect to t and setting $t=t^*$ gives

$$\frac{d^2N_1(t^*)}{dt^2}=\frac{-a_1N_1(t^*)}{K_1}\frac{dN_2(t^*)}{dt}.$$

This quantity is positive. Hence, $N_1(t)$ has a minimum at $t=t^*$. But this is impossible, since $N_1(t)$ is increasing whenever a solution $N_1(t)$, $N_2(t)$ of (4) is in region II. Similarly, if $\dot{N}_2(t^*)=0$, then

$$\frac{d^2N_2(t^*)}{dt^2}=\frac{-a_2N_2(t^*)}{K_2}\frac{dN_1(t^*)}{dt}.$$

This quantity is negative, implying that $N_2(t)$ has a maximum at $t=t^*$. But this is impossible, since $N_2(t)$ is decreasing whenever a solution $N_1(t)$, $N_2(t)$ of (4) is in region II.

The previous argument shows that any solution $N_1(t)$, $N_2(t)$ of (4) which starts in region II at time $t=t_0$ will remain in region II for all future time $t\geqslant t_0$. This implies that $N_1(t)$ is monotonic increasing and $N_2(t)$ is monotonic decreasing for $t\geqslant t_0$, with $N_1(t)<K_1$ and $N_2(t)>K_2$. Consequently, by Lemma 1 of Section 4.8, both $N_1(t)$ and $N_2(t)$ have limits ξ,η respectively, as t approaches infinity. Lemma 2 of Section 4.8 implies that (ξ,η) is an equilibrium point of (4). Now, (ξ,η) obviously cannot equal $(0,0)$ or $(0,K_2)$. Consequently, $(\xi,\eta)=(K_1,0)$, and this proves Lemma 2. □

Lemma 3. *Any solution $N_1(t)$, $N_2(t)$ of* (4) *which starts in region* III *at time $t=t_0$ and remains there for all future time must approach the equilibrium solution $N_1(t)=K_1$, $N_2(t)=0$ as t approaches infinity.*

PROOF. If a solution $N_1(t)$, $N_2(t)$ of (4) remains in region III for $t\geqslant t_0$, then both $N_1(t)$ and $N_2(t)$ are monotonic decreasing functions of time for $t\geqslant t_0$, with $N_1(t)>0$ and $N_2(t)>0$. Consequently, by Lemma 1 of Section 4.8, both $N_1(t)$ and $N_2(t)$ have limits ξ,η respectively, as t approaches infinity. Lemma 2 of Section 4.8 implies that (ξ,η) is an equilibrium point of (4). Now, (ξ,η) obviously cannot equal $(0,0)$ or $(0,K_2)$. Consequently, $(\xi,\eta)=(K_1,0)$. □

PROOF OF THEOREM 6. Lemmas 1 and 2 above state that every solution $N_1(t)$, $N_2(t)$ of (4) which starts in regions I or II at time $t=t_0$ must approach the equilibrium solution $N_1=K_1$, $N_2=0$ as t approaches infinity. Similarly, Lemma 3 shows that every solution $N_1(t)$, $N_2(t)$ of (4) which starts in region III at time $t=t_0$ and remains there for all future time must also approach the equilibrium solution $N_1=K_1$, $N_2=0$. Next, observe that any solution $N_1(t)$, $N_2(t)$ of (4) which starts on l_1 or l_2 must immediately afterwards enter region II. Finally, if a solution $N_1(t), N_2(t)$ of (4) leaves region III, then it must cross the line l_1 and immediately afterwards enter region II. Lemma 2 then forces this solution to approach the equilibrium solution $N_1=K_1$, $N_2=0$. □

Theorem 6 deals with the case of identical species; i.e., $\alpha=\beta=1$. By a similar analysis (see Exercises 4–6) we can predict the outcome of the struggle for existence for all values of α and β.

Reference

Gause, G. F., 'The Struggle for Existence,' Dover Publications, New York, 1964.

EXERCISES

1. Rewrite the system of equations (4) in the form

$$\frac{K_1}{a_1N_1}\frac{dN_1}{dt}=K_1-N_1-N_2, \qquad \frac{K_2}{a_2N_2}\frac{dN_2}{dt}=K_2-N_1-N_2.$$

Then, subtract these two equations and integrate to obtain directly that $N_2(t)$ approaches zero for all solutions $N_1(t)$, $N_2(t)$ of (4) with $N_1(t_0)>0$.

2. The system of differential equations

$$\begin{aligned}\frac{dN_1}{dt}&=N_1[-a_1+c_1(1-b_1N_1-b_2N_2)]\\ \frac{dN_2}{dt}&=N_2[-a_2+c_2(1-b_1N_1-b_2N_2)]\end{aligned} \tag{*}$$

is a model of two species competing for the same limited resource. Suppose that $c_1>a_1$ and $c_2>a_2$. Deduce from Theorem 6 that $N_1(t)$ ultimately approaches zero if $a_1c_2>a_2c_1$, and $N_2(t)$ ultimately approaches zero if $a_1c_2<a_2c_1$.

3. In 1926, Volterra presented the following model of two species competing for the same limited food supply:

$$\begin{aligned}\frac{dN_1}{dt}&=[b_1-\lambda_1(h_1N_1+h_2N_2)]N_1\\ \frac{dN_2}{dt}&=[b_2-\lambda_2(h_1N_1+h_2N_2)]N_2.\end{aligned}$$

Suppose that $b_1/\lambda_1>b_2/\lambda_2$. (The coefficient b_i/λ_i is called the susceptibility of species i to food shortages.) Prove that species 2 will ultimately become extinct if $N_1(t_0)>0$.

Problems 4–6 are concerned with the system of equations

$$\frac{dN_1}{dt}=\frac{a_1N_1}{K_1}(K_1-N_1-\alpha N_2),\qquad \frac{dN_2}{dt}=\frac{a_2N_2}{K_2}(K_2-N_2-\beta N_1).\quad (*)$$

4. (a) Assume that $K_1/\alpha > K_2$ and $K_2/\beta < K_1$. Show that $N_2(t)$ approaches zero as t approaches infinity for every solution $N_1(t)$, $N_2(t)$ of (*) with $N_1(t_0)>0$.

(b) Assume that $K_1/\alpha < K_2$ and $K_2/\beta > K_1$. Show that $N_1(t)$ approaches zero as t approaches infinity for every solution $N_1(t)$, $N_2(t)$ of (*) with $N_1N_2(t_0)>0$. *Hint*: Draw the lines $l_1: N_1+\alpha N_2=K_1$ and $l_2: N_2+\beta N_1=K_2$, and follow the proof of Theorem 6.

5. Assume that $K_1/\alpha > K_2$ and $K_2/\beta > K_1$. Prove that all solutions $N_1(t)$, $N_2(t)$ of (*), with both $N_1(t_0)$ and $N_2(t_0)$ positive, ultimately approach the equilibrium solution

$$N_1=N_1^0=\frac{K_1-\alpha K_2}{1-\alpha\beta},\qquad N_2=N_2^0=\frac{K_2-\beta K_1}{1-\alpha\beta}.$$

Hint:

(a) Draw the lines $l_1: N_1+\alpha N_2=K_1$ and $l_2: N_2+\beta N_1=K_2$. The two lines divide the first quadrant into four regions (see Figure 2) in which both $\dot{N}_1$ and $\dot{N}_2$ have fixed signs.

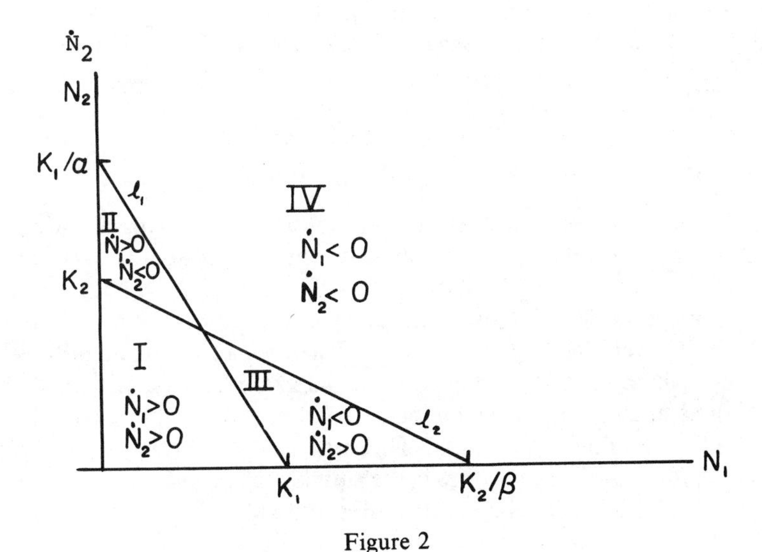

Figure 2

(b) Show that all solutions $N_1(t)$, $N_2(t)$ of (*) which start in either region II or III must remain in these regions and ultimately approach the equilibrium solution $N_1=N_1^0$, $N_2=N_2^0$.

(c) Show that all solutions $N_1(t)$, $N_2(t)$ of (*) which remain exclusively in region I or region IV for all time $t\geqslant t_0$ must ultimately approach the equilibrium solution $N_1=N_1^0$, $N_2=N_2^0$.

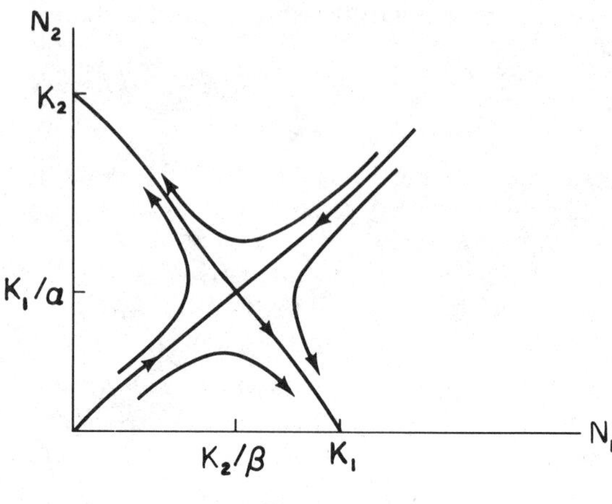

Figure 3

6. Assume that $K_1/\alpha < K_2$ and $K_2/\beta < K_1$.
  (a) Show that the equilibrium solution $N_1=0$, $N_2=0$ of (*) is unstable.
  (b) Show that the equilibrium solutions $N_1=K_1$, $N_2=0$ and $N_1=0$, $N_2=K_2$ of (*) are asymptotically stable.
  (c) Show that the equilibrium solution $N_1=N_1^0$, $N_2=N_2^0$ (see Exercise 5) of (*) is a saddle point. (This calculation is very cumbersome.)
  (d) It is not too difficult to see that the phase portrait of (*) must have the form described in Figure 3.

4.12 The Threshold Theorem of epidemiology

Consider the situation where a small group of people having an infectious disease is inserted into a large population which is capable of catching the disease. What happens as time evolves? Will the disease die out rapidly, or will an epidemic occur? How many people will ultimately catch the disease? To answer these questions we will derive a system of differential equations which govern the spread of an infectious disease within a population, and analyze the behavior of its solutions. This approach will also lead us to the famous Threshold Theorem of epidemiology which states that an epidemic will occur only if the number of people who are susceptible to the disease exceeds a certain threshold value.

We begin with the assumptions that the disease under consideration confers permanent immunity upon any individual who has completely recovered from it, and that it has a negligibly short incubation period. This latter assumption implies that an individual who contracts the disease becomes infective immediately afterwards. In this case we can divide the population into three classes of individuals: the infective class (I), the susceptible class (S) and the removed class (R). The infective class consists of those individuals who are capable of transmitting the disease to others.

The susceptible class consists of those individuals who are not infective, but who are capable of catching the disease and becoming infective. The removed class consists of those individuals who have had the disease and are dead, or have recovered and are permanently immune, or are isolated until recovery and permanent immunity occur.

The spread of the disease is presumed to be governed by the following rules.

Rule 1: The population remains at a fixed level N in the time interval under consideration. This means, of course, that we neglect births, deaths from causes unrelated to the disease under consideration, immigration and emigration.

Rule 2: The rate of change of the susceptible population is proportional to the product of the number of members of (S) and the number of members of (I).

Rule 3: Individuals are removed from the infectious class (I) at a rate proportional to the size of (I).

Let $S(t), I(t)$, and $R(t)$ denote the number of individuals in classes (S), (I), and (R), respectively, at time t. It follows immediately from Rules 1–3 that $S(t), I(t), R(t)$ satisfies the system of differential equations

$$\begin{aligned} \frac{dS}{dt} &= -rSI \\ \frac{dI}{dt} &= rSI - \gamma I \\ \frac{dR}{dt} &= \gamma I \end{aligned} \tag{1}$$

for some positive constants r and γ. The proportionality constant r is called the infection rate, and the proportionality constant γ is called the removal rate.

The first two equations of (1) do not depend on R. Thus, we need only consider the system of equations

$$\frac{dS}{dt} = -rSI, \qquad \frac{dI}{dt} = rSI - \gamma I \tag{2}$$

for the two unknown functions $S(t)$ and $I(t)$. Once $S(t)$ and $I(t)$ are known, we can solve for $R(t)$ from the third equation of (1). Alternately, observe that $d(S+I+R)/dt = 0$. Thus,

$$S(t) + I(t) + R(t) = \text{constant} = N$$

so that $R(t) = N - S(t) - I(t)$.

The orbits of (2) are the solution curves of the first-order equation

$$\frac{dI}{dS} = \frac{rSI - \gamma I}{-rSI} = -1 + \frac{\gamma}{rS}. \tag{3}$$

Integrating this differential equation gives

$$I(S) = I_0 + S_0 - S + \rho \ln \frac{S}{S_0}, \tag{4}$$

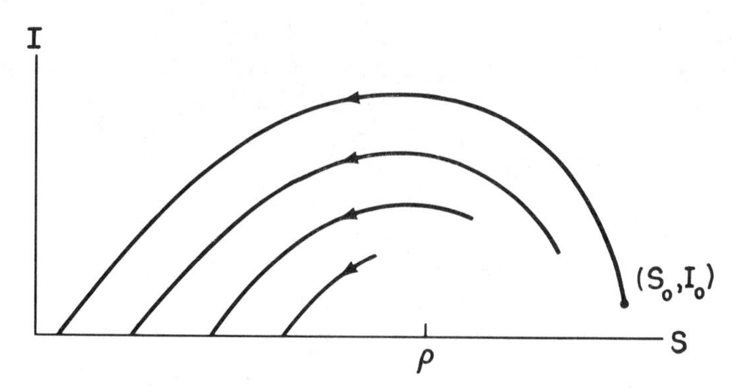

Figure 1. The orbits of (2)

where S_0 and I_0 are the number of susceptibles and infectives at the initial time $t=t_0$, and $\rho=\gamma/r$. To analyze the behavior of the curves (4), we compute $I'(S)=-1+\rho/S$. The quantity $-1+\rho/S$ is negative for $S>\rho$, and positive for $S<\rho$. Hence, $I(S)$ is an increasing function of S for $S<\rho$, and a decreasing function of S for $S>\rho$.

Next, observe that $I(0)=-\infty$ and $I(S_0)=I_0>0$. Consequently, there exists a unique point S_∞, with $0<S_\infty<S_0$, such that $I(S_\infty)=0$, and $I(S)>0$ for $S_\infty<S\leqslant S_0$. The point $(S_\infty,0)$ is an equilibrium point of (2) since both dS/dt and dI/dt vanish when $I=0$. Thus, the orbits of (2), for $t_0\leqslant t<\infty$, have the form described in Figure 1.

Let us see what all this implies about the spread of the disease within the population. As t runs from t_0 to ∞, the point $(S(t),I(t))$ travels along the curve (4), and it moves along the curve in the direction of decreasing S, since $S(t)$ decreases monotonically with time. Consequently, if S_0 is less than ρ, then $I(t)$ decreases monotonically to zero, and $S(t)$ decreases monotonically to S_∞. Thus, if a small group of infectives I_0 is inserted into a group of susceptibles S_0, with $S_0<\rho$, then the disease will die out rapidly. On the other hand, if S_0 is greater than ρ, then $I(t)$ increases as $S(t)$ decreases to ρ, and it achieves a maximum value when $S=\rho$. It only starts decreasing when the number of susceptibles falls below the threshold value ρ. From these results we may draw the following conclusions.

Conclusion 1: An epidemic will occur only if the number of susceptibles in a population exceeds the threshold value $\rho=\gamma/r$.

Conclusion 2: The spread of the disease does not stop for lack of a susceptible population; it stops only for lack of infectives. In particular, some individuals will escape the disease altogether.

Conclusion 1 corresponds to the general observation that epidemics tend to build up more rapidly when the density of susceptibles is high due to overcrowding, and the removal rate is low because of ignorance, inadequate isolation and inadequate medical care. On the other hand, outbreaks tend to be of only limited extent when good social conditions entail lower densities of susceptibles, and when removal rates are high because of good public health vigilance and control.

If the number of susceptibles S_0 is initially greater than, but close to, the threshold value ρ, then we can estimate the number of individuals who ultimately contract the disease. Specifically, if $S_0-\rho$ is small compared to ρ, then the number of individuals who ultimately contract the disease is approximately $2(S_0-\rho)$. This is the famous Threshold Theorem of epidemiology, which was first proven in 1927 by the mathematical biologists Kermack and McKendrick.

Theorem 7 (Threshold Theorem of epidemiology). *Let $S_0=\rho+\nu$ and assume that ν/ρ is very small compared to one. Assume moreover, that the number of initial infectives I_0 is very small. Then, the number of individuals who ultimately contract the disease is 2ν. In other words, the level of susceptibles is reduced to a point as far below the threshold as it originally was above it.*

PROOF. Letting t approach infinity in (4) gives

$$0=I_0+S_0-S_\infty+\rho\ln\frac{S_\infty}{S_0}.$$

If I_0 is very small compared to S_0, then we can neglect it, and write

$$\begin{aligned}0&=S_0-S_\infty+\rho\ln\frac{S_\infty}{S_0}\\&=S_0-S_\infty+\rho\ln\left[\frac{S_0-(S_0-S_\infty)}{S_0}\right]\\&=S_0-S_\infty+\rho\ln\left[1-\left(\frac{S_0-S_\infty}{S_0}\right)\right].\end{aligned}$$

Now, if $S_0-\rho$ is small compared to ρ, then S_0-S_∞ will be small compared to S_0. Consequently, we can truncate the Taylor series

$$\ln\left[1-\left(\frac{S_0-S_\infty}{S_0}\right)\right]=-\left(\frac{S_0-S_\infty}{S_0}\right)-\frac{1}{2}\left(\frac{S_0-S_\infty}{S_0}\right)^2+\ldots$$

after two terms. Then,

$$\begin{aligned}0&=S_0-S_\infty-\rho\left(\frac{S_0-S_\infty}{S_0}\right)-\frac{\rho}{2}\left(\frac{S_0-S_\infty}{S_0}\right)^2\\&=(S_0-S_\infty)\left[1-\frac{\rho}{S_0}-\frac{\rho}{2S_0^2}(S_0-S_\infty)\right].\end{aligned}$$

Solving for $S_0 - S_\infty$, we see that

$$S_0 - S_\infty = 2S_0\left(\frac{S_0}{\rho} - 1\right) = 2(\rho+\nu)\left[\frac{\rho+\nu}{\rho} - 1\right]$$

$$= 2(\rho+\nu)\frac{\nu}{\rho} = 2\rho\left(1+\frac{\nu}{\rho}\right)\frac{\nu}{\rho} \cong 2\nu. \qquad \square$$

During the course of an epidemic it is impossible to accurately ascertain the number of new infectives each day or week, since the only infectives who can be recognized and removed from circulation are those who seek medical aid. Public health statistics thus record only the number of new removals each day or week, not the number of new infectives. Therefore, in order to compare the results predicted by our model with data from actual epidemics, we must find the quantity dR/dt as a function of time. This is accomplished in the following manner. Observe first that

$$\frac{dR}{dt} = \gamma I = \gamma(N - R - S).$$

Second, observe that

$$\frac{dS}{dR} = \frac{dS/dt}{dR/dt} = \frac{-rSI}{\gamma I} = \frac{-S}{\rho}.$$

Hence, $S(R) = S_0 e^{-R/\rho}$ and

$$\frac{dR}{dt} = \gamma\left(N - R - S_0 e^{-R/\rho}\right). \tag{5}$$

Equation (5) is separable, but cannot be solved explicitly. However, if the epidemic is not very large, then R/ρ is small and we can truncate the Taylor series

$$e^{-R/\rho} = 1 - \frac{R}{\rho} + \frac{1}{2}\left(\frac{R}{\rho}\right)^2 + \dots$$

after three terms. With this approximation,

$$\frac{dR}{dt} = \gamma\left[N - R - S_0\left[1 - R/\rho + \frac{1}{2}(R/\rho)^2\right]\right]$$

$$= \gamma\left[N - S_0 + \left(\frac{S_0}{\rho} - 1\right)R - \frac{S_0}{2}\left(\frac{R}{\rho}\right)^2\right].$$

The solution of this equation is

$$R(t) = \frac{\rho^2}{S_0}\left[\frac{S_0}{\rho} - 1 + \alpha\tanh\left(\frac{1}{2}\alpha\gamma t - \phi\right)\right] \tag{6}$$

where

$$\alpha=\left[\left(\frac{S_0}{\rho}-1\right)^2+\frac{2S_0(N-S_0)}{\rho^2}\right]^{1/2}, \qquad \phi=\tanh^{-1}\frac{1}{\alpha}\left(\frac{S_0}{\rho}-1\right)$$

and the hyperbolic tangent function $\tanh z$ is defined by

$$\tanh z=\frac{e^z-e^{-z}}{e^z+e^{-z}}.$$

It is easily verified that

$$\frac{d}{dz}\tanh z=\operatorname{sech}^2 z=\frac{4}{(e^z+e^{-z})^2}.$$

Hence,

$$\frac{dR}{dt}=\frac{\gamma\alpha^2\rho^2}{2S_0}\operatorname{sech}^2\left(\frac{1}{2}\alpha\gamma t-\phi\right). \tag{7}$$

Equation (7) defines a symmetric bell shaped curve in the t–dR/dt plane (see Figure 2). This curve is called the epidemic curve of the disease. It illustrates very well the common observation that in many actual epidemics, the number of new cases reported each day climbs to a peak value and then dies away again.

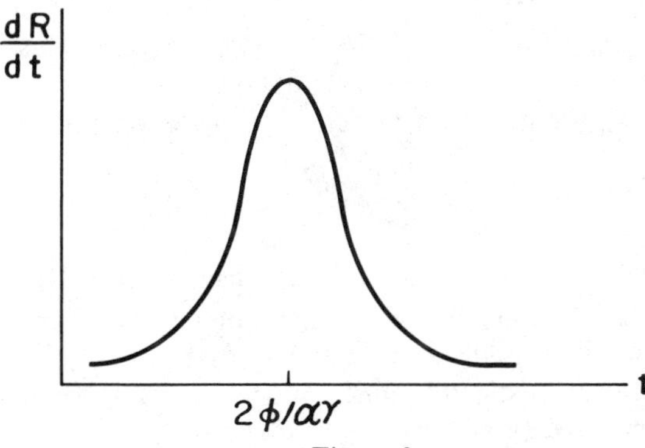

Figure 2

Kermack and McKendrick compared the values predicted for dR/dt from (7) with data from an actual plague in Bombay which spanned the last half of 1905 and the first half of 1906. They set

$$\frac{dR}{dt}=890\operatorname{sech}^2(0.2t-3.4)$$

with t measured in weeks, and compared these values with the number of deaths per week from the plague. This quantity is a very good approximation of dR/dt, since almost all cases terminated fatally. As can be seen from Figure 3, there is excellent agreement between the actual values of

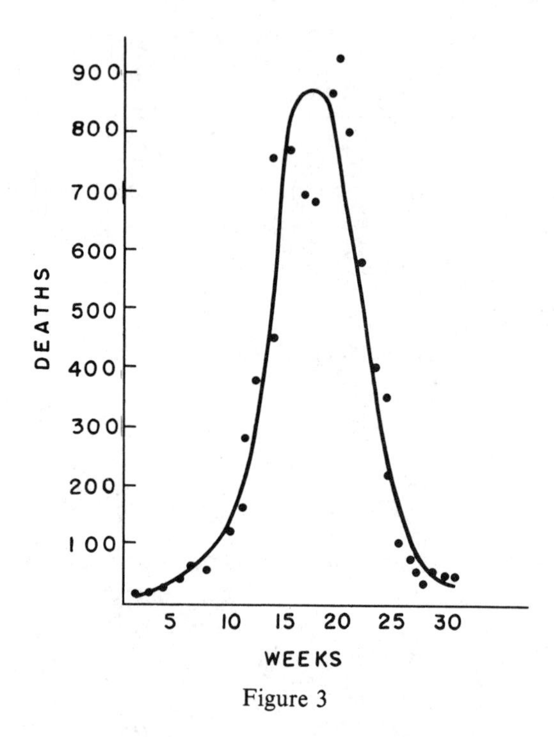

Figure 3

dR/dt, denoted by •, and the values predicted by (7). This indicates, of course, that the system of differential equations (1) is an accurate and reliable model of the spread of an infectious disease within a population of fixed size.

References

Bailey, N. T. J., 'The mathematical theory of epidemics,' 1957, New York.

Kermack, W. O. and McKendrick, A. G., Contributions to the mathematical theory of epidemics, Proceedings Roy. Stat. Soc., A, 115, 700–721, 1927.

Waltman, P., 'Deterministic threshold models in the theory of epidemics,' Springer-Verlag, New York, 1974.

Exercises

1. Derive Equation (6).

2. Suppose that the members of (S) are vaccinated against the disease at a rate λ proportional to their number. Then,

$$\frac{dS}{dt} = -rSI - \lambda S, \qquad \frac{dI}{dt} = rSI - \gamma I. \tag{*}$$

(a) Find the orbits of (*).

(b) Conclude from (a) that $S(t)$ approaches zero as t approaches infinity, for every solution $S(t)$, $I(t)$ of (*).

3. Suppose that the members of (S) are vaccinated against the disease at a rate λ proportional to the product of their numbers and the square of the members of (I). Then,

$$\frac{dS}{dt} = -rSI - \lambda SI^2, \qquad \frac{dI}{dt} = I(rS-\gamma). \qquad (*)$$

(a) Find the orbits of (*).
(b) Will any susceptibles remain after the disease dies out?

4. The *intensity* i of an epidemic is the proportion of the total number of susceptibles that finally contracts the disease. Show that

$$i = \frac{I_0 + S_0 - S_\infty}{S_0}$$

where S_∞ is a root of the equation

$$S = S_0 e^{(S - S_0 - I_0)/\rho}.$$

5. Compute the intensity of the epidemic if $\rho = 1000$, $I_0 = 10$, and (a) $S_0 = 1100$, (b) $S_0 = 1200$, (c) $S_0 = 1300$, (d) $S_0 = 1500$, (e) $S_0 = 1800$, (f) $S_0 = 1900$. (This cannot be done analytically.)

6. Let R_∞ denote the total number of individuals who contract the disease.
(a) Show that $R_\infty = I_0 + S_0 - S_\infty$.
(b) Let R_1 denote the members of (R) who are removed from the population prior to the peak of the epidemic. Compute R_1/R_∞ for each of the values of S_0 in 5a–5f. Notice that most of the removals occur after the peak. This type of asymmetry is often found in actual notifications of infectious diseases.

7. It was observed in London during the early 1900's, that large outbreaks of measles epidemics recurred about once every two years. The mathematical biologist H. E. Soper tried to explain this phenomenon by assuming that the stock of susceptibles is constantly replenished by new recruits to the population. Thus, he assumed that

$$\frac{dS}{dt} = -rSI + \mu, \qquad \frac{dI}{dt} = rSI - \gamma I \qquad (*)$$

for some positive constants r, γ, and μ.
(a) Show that $S = \gamma/r$, $I = \mu/\gamma$ is the only equilibrium solution of (*).
(b) Show that every solution $S(t)$, $I(t)$ of (*) which starts sufficiently close to this equilibrium point must ultimately approach it as t approaches infinity.
(c) It can be shown that every solution $S(t)$, $I(t)$ of (*) approaches the equilibrium solution $S = \gamma/r$, $I = \mu/\gamma$ as t approaches infinity. Conclude, therefore, that the system (*) does not predict recurrent outbreaks of measles epidemics. Rather, it predicts that the disease will ultimately approach a steady state.

4.13 A model for the spread of gonorrhea

Gonorrhea ranks first today among reportable communicable diseases in the United States. There are more reported cases of gonorrhea every year than the combined totals for syphilis, measles, mumps, and infectious hepatitis. Public health officials estimate that more than 2,500,000 Americans contract gonorrhea every year. This painful and dangerous disease, which is caused by the gonococcus germ, is spread from person to person by sexual contact. A few days after the infection there is usually itching and burning of the genital area, particularly while urinating. About the same time a discharge develops which males will notice, but which females may not notice. Infected women may have no easily recognizable symptoms, even while the disease does substantial internal damage. Gonorrhea can only be cured by antibiotics (usually penicillin). However, treatment must be given early if the disease is to be stopped from doing serious damage to the body. If untreated, gonorrhea can result in blindness, sterility, arthritis, heart failure, and ultimately, death.

In this section we construct a mathematical model of the spread of gonorrhea. Our work is greatly simplified by the fact that the incubation period of gonorrhea is very short (3–7 days) compared to the often quite long period of active infectiousness. Thus, we will assume in our model that an individual becomes infective immediately after contracting gonorrhea. In addition, gonorrhea does not confer even partial immunity to those individuals who have recovered from it. Immediately after recovery, an individual is again susceptible. Thus, we can split the sexually active and promiscuous portion of the population into two groups, susceptibles and infectives. Let $c_1(t)$ be the total number of promiscuous males, $c_2(t)$ the total number of promiscuous females, $x(t)$ the total number of infective males, and $y(t)$ the total number of infective females, at time t. Then, the total numbers of susceptible males and susceptible females are $c_1(t)-x(t)$ and $c_2(t)-y(t)$ respectively. The spread of gonorrhea is presumed to be governed by the following rules:

1. Male infectives are cured at a rate a_1 proportional to their total number, and female infectives are cured at a rate a_2 proportional to their total number. The constant a_1 is larger than a_2 since infective males quickly develop painful symptoms and therefore seek prompt medical attention. Female infectives, on the other hand, are usually asymptomatic, and therefore are infectious for much longer periods.

2. New infectives are added to the male population at a rate b_1 proportional to the total number of male susceptibles and female infectives. Similarly, new infectives are added to the female population at a rate b_2 proportional to the total number of female susceptibles and male infectives.

3. The total numbers of promiscuous males and promiscuous females remain at constant levels c_1 and c_2, respectively.

It follows immediately from rules 1–3 that

$$\frac{dx}{dt} = -a_1x + b_1(c_1-x)y$$
$$\frac{dy}{dt} = -a_2y + b_2(c_2-y)x. \tag{1}$$

Remark. The system of equations (1) treats only those cases of gonorrhea which arise from heterosexual contacts; the case of homosexual contacts (assuming no interaction between heterosexuals and homosexuals) is treated in Exercises 5 and 6. The number of cases of gonorrhea which arise from homosexual encounters is a small percentage of the total number of incidents of gonorrhea. Interestingly enough, this situation is completely reversed in the case of syphilis. Indeed, more than 90% of all cases of syphilis reported in the state of Rhode Island during 1973 resulted from homosexual encounters. (This statistic is not as startling as it first appears. Within ten to ninety days after being infected with syphilis, an individual usually develops a chancre sore at the spot where the germs entered the body. A homosexual who contracts syphilis as a result of anal intercourse with an infective will develop a chancre sore on his rectum. This individual, naturally, will be reluctant to seek medical attention, since he will then have to reveal his identity as a homosexual. Moreover, he feels no sense of urgency, since the chancre sore is usually painless and disappears after several days. With gonorrhea, on the other hand, the symptoms are so painful and unmistakable that a homosexual will seek prompt medical attention. Moreover, he need not reveal his identity as a homosexual since the symptoms of gonorrhea appear in the genital area.)

Our first step in analyzing the system of differential equations (1) is to show that they are realistic. Specifically, we must show that $x(t)$ and $y(t)$ can never become negative, and can never exceed c_1 and c_2, respectively. This is the content of Lemmas 1 and 2.

Lemma 1. *If $x(t_0)$ and $y(t_0)$ are positive, then $x(t)$ and $y(t)$ are positive for all $t \geqslant t_0$.*

Lemma 2. *If $x(t_0)$ is less than c_1 and $y(t_0)$ is less than c_2, then $x(t)$ is less than c_1 and $y(t)$ is less than c_2 for all $t \geqslant t_0$.*

PROOF OF LEMMA 1. Suppose that Lemma 1 is false. Let $t^* > t_0$ be the first time at which either x or y is zero. Assume that x is zero first. Then, evaluating the first equation of (1) at $t = t^*$ gives $\dot{x}(t^*) = b_1 c_1 y(t^*)$. This quantity is positive. (Note that $y(t^*)$ cannot equal zero since $x = 0$, $y = 0$ is an equilibrium solution of (1).) Hence, $x(t)$ is less than zero for t close to, and less than t^*. But this contradicts our assumption that t^* is the first time at which $x(t)$ equals zero. We run into the same contradiction if $y(t^*) = 0$. Thus, both $x(t)$ and $y(t)$ are positive for $t \geqslant t_0$. □

PROOF OF LEMMA 2. Suppose that Lemma 2 is false. Let $t^* > t_0$ be the first time at which either $x = c_1$, or $y = c_2$. Suppose that $x(t^*) = c_1$. Evaluating the first equation of (1) at $t = t^*$ gives $\dot{x}(t^*) = -a_1 c_1$. This quantity is negative. Hence, $x(t)$ is greater than c_1 for t close to, and less than t^*. But this contradicts our assumption that t^* is the first time at which $x(t)$ equals c_1. We run into the same contradiction if $y(t^*)=c_2$. Thus, $x(t)$ is less than c_1 and $y(t)$ is less than c_2 for $t \geqslant t_0$. □

Having shown that the system of equations (1) is a realistic model of gonorrhea, we now see what predictions it makes concerning the future course of this disease. Will gonorrhea continue to spread rapidly and uncontrollably as the data in Figure 1 seems to suggest, or will it level off eventually? The following extremely important theorem of epidemiology provides the answer to this question.

Theorem 8.

(a) *Suppose that a_1a_2 is less than $b_1b_2c_1c_2$. Then, every solution $x(t)$, $y(t)$ of (1) with $0<x(t_0)<c_1$ and $0<y(t_0)<c_2$, approaches the equilibrium solution*

$$x=\frac{b_1b_2c_1c_2-a_1a_2}{a_1b_2+b_1b_2c_2}, \qquad y=\frac{b_1b_2c_1c_2-a_1a_2}{a_2b_1+b_1b_2c_1}$$

as t approaches infinity. In other words, the total numbers of infective males and infective females will ultimately level off.

(b) *Suppose that a_1a_2 is greater than $b_1b_2c_1c_2$. Then every solution $x(t)$, $y(t)$ of (1) with $0<x(t_0)<c_1$ and $0<y(t_0)<c_2$, approaches zero as t approaches infinity. In other words, gonorrhea will ultimately die out.*

Our first step in proving part (a) of Theorem 8 is to split the rectangle $0<x<c_1$, $0<y<c_2$ into regions in which both dx/dt and dy/dt have fixed signs. This is accomplished in the following manner. Setting $dx/dt=0$ in (1), and solving for y as a function of x gives

$$y=\frac{a_1x}{b_1(c_1-x)}\equiv\phi_1(x).$$

Similarly, setting $dy/dt=0$ in (1) gives

$$x=\frac{a_2y}{b_2(c_2-y)}, \quad \text{or} \quad y=\frac{b_2c_2x}{a_2+b_2x}\equiv\phi_2(x).$$

Observe first that $\phi_1(x)$ and $\phi_2(x)$ are monotonic increasing functions of x; $\phi_1(x)$ approaches infinity as x approaches c_1, and $\phi_2(x)$ approaches c_2 as x approaches infinity. Second, observe that the curves $y=\phi_1(x)$ and $y=\phi_2(x)$ intersect at $(0,0)$ and at (x_0,y_0) where

$$x_0=\frac{b_1b_2c_1c_2-a_1a_2}{a_1b_2+b_1b_2c_2}, \qquad y_0=\frac{b_1b_2c_1c_2-a_1a_2}{a_2b_1+b_1b_2c_1}.$$

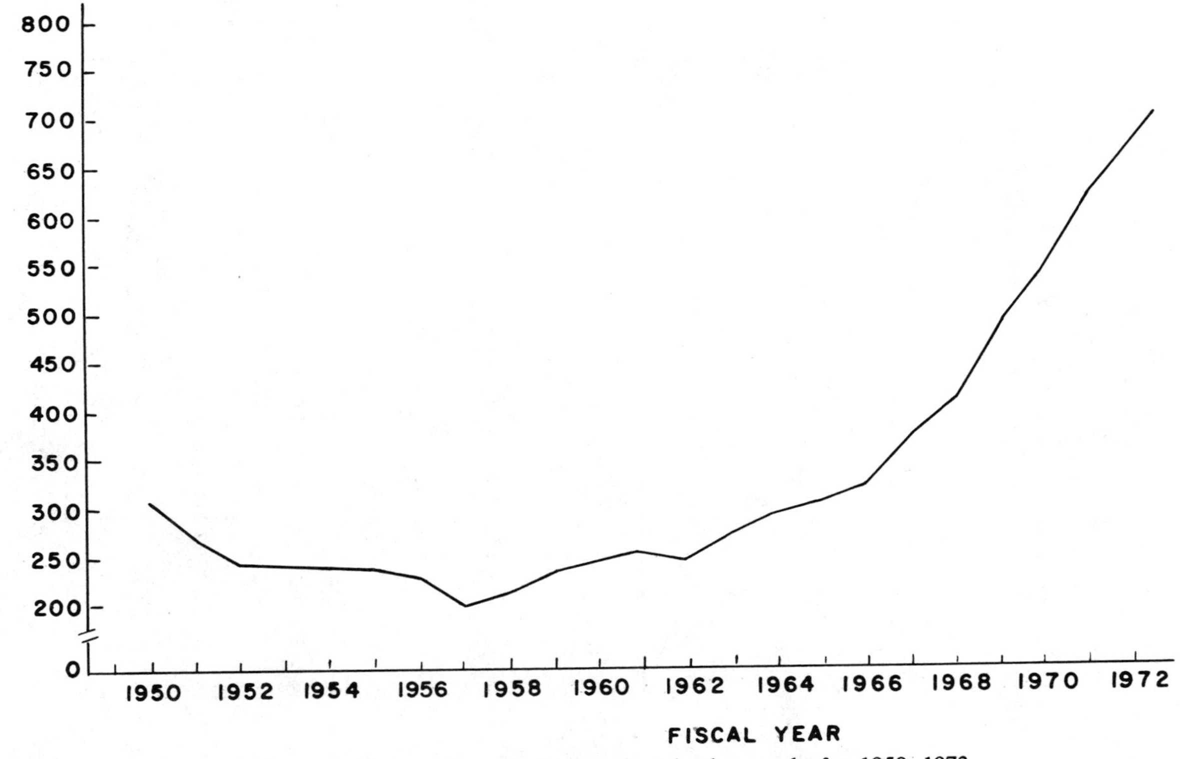

Figure 1. Reported cases of gonorrhea, in thousands, for 1950–1973.

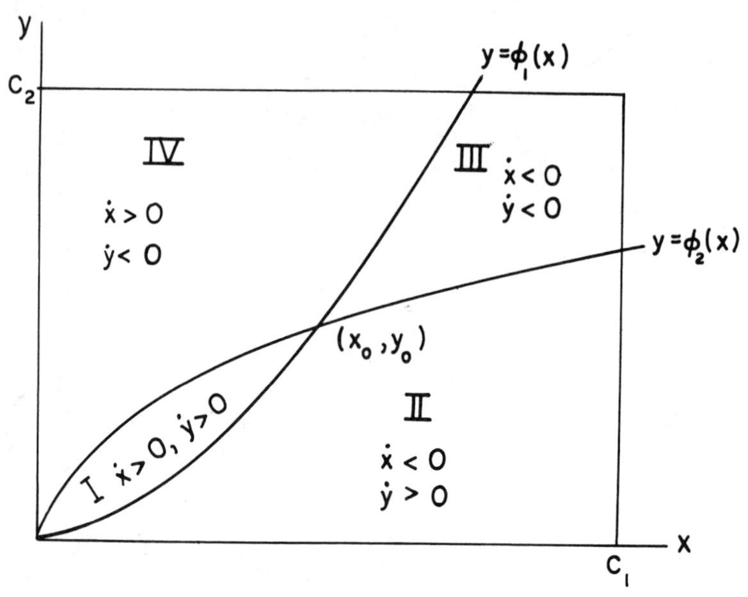

Figure 2

Third, observe that $\phi_2(x)$ is increasing faster than $\phi_1(x)$ at $x=0$, since

$$\phi_2'(0)=\frac{b_2c_2}{a_2}>\frac{a_1}{b_1c_1}=\phi_1'(0).$$

Hence, $\phi_2(x)$ lies above $\phi_1(x)$ for $0<x<x_0$, and $\phi_2(x)$ lies below $\phi_1(x)$ for $x_0<x<c_1$, as shown in Figure 2. The point (x_0, y_0) is an equilibrium point of (1) since both dx/dt and dy/dt are zero when $x=x_0$ and $y=y_0$.

Finally, observe that dx/dt is positive at any point (x,y) above the curve $y=\phi_1(x)$, and negative at any point (x,y) below this curve. Similarly, dy/dt is positive at any point (x,y) below the curve $y=\phi_2(x)$, and negative at any point (x,y) above this curve. Thus, the curves $y=\phi_1(x)$ and $y=\phi_2(x)$ split the rectangle $0<x<c_1$, $0<y<c_2$ into four regions in which dx/dt and dy/dt have fixed signs (see Figure 2).

Next, we require the following four simple lemmas.

Lemma 3. *Any solution $x(t), y(t)$ of (1) which starts in region I at time $t=t_0$ will remain in this region for all future time $t\geqslant t_0$ and approach the equilibrium solution $x=x_0$, $y=y_0$ as t approaches infinity.*

PROOF. Suppose that a solution $x(t), y(t)$ of (1) leaves region I at time $t=t^*$. Then, either $\dot{x}(t^*)$ or $\dot{y}(t^*)$ is zero, since the only way a solution of (1) can leave region I is by crossing the curve $y=\phi_1(x)$ or $y=\phi_2(x)$. Assume that $\dot{x}(t^*)=0$. Differentiating both sides of the first equation of (1) with respect to t and setting $t=t^*$ gives

$$\frac{d^2x(t^*)}{dt^2}=b_1(c_1-x(t^*))\frac{dy(t^*)}{dt}.$$

This quantity is positive, since $x(t^*)$ is less than c_1, and dy/dt is positive on the curve $y=\phi_1(x)$, $0<x<x_0$. Hence, $x(t)$ has a minimum at $t=t^*$. But this is impossible, since $x(t)$ is increasing whenever the solution $x(t), y(t)$ is in region I. Similarly, if $\dot{y}(t^*)=0$, then

$$\frac{d^2y(t^*)}{dt^2}=b_2(c_2-y(t^*))\frac{dx(t^*)}{dt}.$$

This quantity is positive, since $y(t^*)$ is less than c_2, and dx/dt is positive on the curve $y=\phi_2(x)$, $0<x<x_0$. Hence, $y(t)$ has a minimum at $t=t^*$. But this is impossible, since $y(t)$ is increasing whenever the solution $x(t), y(t)$ is in region I.

The previous argument shows that any solution $x(t)$,$y(t)$ of (1) which starts in region I at time $t=t_0$ will remain in region I for all future time $t\geqslant t_0$. This implies that $x(t)$ and $y(t)$ are monotonic increasing functions of time for $t\geqslant t_0$, with $x(t)<x_0$ and $y(t)<y_0$. Consequently, by Lemma 1 of Section 4.8, both $x(t)$ and $y(t)$ have limits ξ,η, respectively, as t approaches infinity. Lemma 2 of Section 4.8 implies that (ξ,η) is an equilibrium point of (1). Now, it is easily seen from Figure 2 that the only equilibrium points of (1) are $(0,0)$ and (x_0, y_0). But (ξ,η) cannot equal $(0,0)$ since both $x(t)$ and $y(t)$ are increasing functions of time. Hence, $(\xi,\eta)=(x_0, y_0)$, and this proves Lemma 3. □

Lemma 4. *Any solution $x(t), y(t)$ of (1) which starts in region III at time $t=t_0$ will remain in this region for all future time and ultimately approach the equilibrium solution $x=x_0, y=y_0$.*

PROOF. Exactly the same as Lemma 3 (see Exercise 1). □

Lemma 5. *Any solution $x(t), y(t)$ of (1) which starts in region II at time $t=t_0$, and remains in region II for all future time, must approach the equilibrium solution $x=x_0, y=y_0$ as t approaches infinity.*

PROOF. If a solution $x(t)$, $y(t)$ of (1) remains in region II for $t\geqslant t_0$, then $x(t)$ is monotonic decreasing and $y(t)$ is monotonic increasing for $t\geqslant t_0$. Moreover, $x(t)$ is positive and $y(t)$ is less than c_2, for $t\geqslant t_0$. Consequently, by Lemma 1 of Section 4.8, both $x(t)$ and $y(t)$ have limits ξ,η respectively, as t approaches infinity. Lemma 2 of Section 4.8 implies that (ξ,η) is an equilibrium point of (1). Now, (ξ,η) cannot equal $(0,0)$ since $y(t)$ is increasing for $t\geqslant t_0$. Therefore, $(\xi,\eta)=(x_0,y_0)$, and this proves Lemma 5. □

Lemma 6. *Any solution $x(t), y(t)$ of* (1) *which starts in region* IV *at time $t = t_0$ and remains in region* IV *for all future time, must approach the equilibrium solution $x = x_0$, $y = y_0$ as t approaches infinity.*

PROOF. Exactly the same as Lemma 5 (see Exercise 2). □

We are now in a position to prove Theorem 8.

PROOF OF THEOREM 8. (a) Lemmas 3 and 4 state that every solution $x(t)$, $y(t)$ of (1) which starts in region I or III at time $t = t_0$ must approach the equilibrium solution $x = x_0$, $y = y_0$ as t approaches infinity. Similarly, Lemmas 5 and 6 state that every solution $x(t), y(t)$ of (1) which starts in region II or IV and which remains in these regions for all future time, must also approach the equilibrium solution $x = x_0$, $y = y_0$. Now, observe that if a solution $x(t), y(t)$ of (1) leaves region II or IV, then it must cross the curve $y = \phi_1(x)$ or $y = \phi_2(x)$, and immediately afterwards enter region I or region III. Consequently, all solutions $x(t), y(t)$ of (1) which start in regions II and IV or on the curves $y = \phi_1(x)$ and $y = \phi_2(x)$, must also approach the equilibrium solution $x(t) = x_0$, $y(t) = y_0$. □

(b) PROOF #1. If a_1a_2 is greater than $b_1b_2c_1c_2$, then the curves $y = \phi_1(x)$ and $y = \phi_2(x)$ have the form described in Figure 3 below. In region I, dx/dt is positive and dy/dt is negative; in region II, both dx/dt and dy/dt are negative; and in region III, dx/dt is negative and dy/dt is positive. It is a simple matter to show (see Exercise 3) that every solution $x(t), y(t)$ of (1) which starts in region II at time $t = t_0$ must remain in this region for all

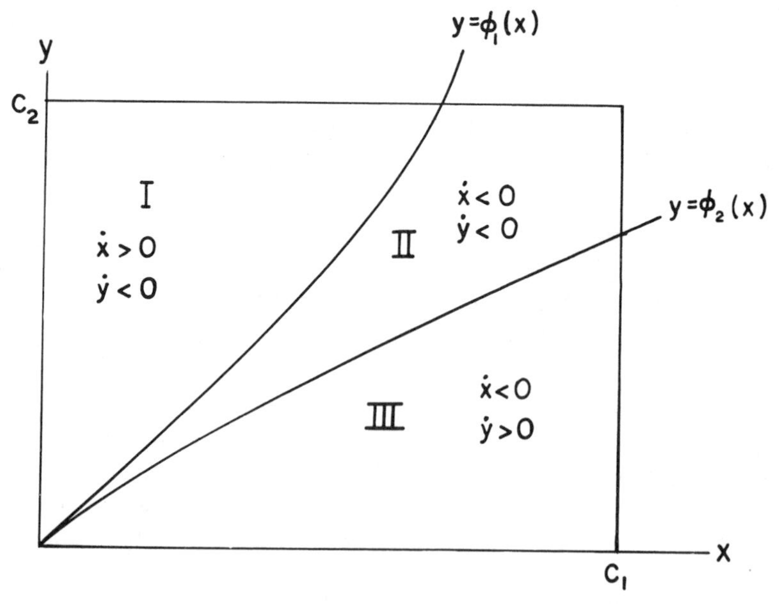

Figure 3

future time, and approach the equilibrium solution $x=0$, $y=0$ as t approaches infinity. It is also trivial to show that every solution $x(t)$, $y(t)$ of (1) which starts in region I or region III at time $t=t_0$ must cross the curve $y=\phi_1(x)$ or $y=\phi_2(x)$, and immediately afterwards enter region II (see Exercise 4). Consequently, every solution $x(t)$, $y(t)$ of (1), with $0<x(t_0)<c_1$ and $0<y(t_0)<c_2$, approaches the equilibrium solution $x=0$, $y=0$ as t approaches infinity. □

PROOF #2. We would now like to show how we can use the Poincaré–Bendixson theorem to give an elegant proof of part (b) of Theorem 8. Observe that the system of differential equations (1) can be written in the form

$$\frac{d}{dt}\begin{pmatrix} x \\ y \end{pmatrix}=\begin{pmatrix} -a_1 & b_1c_1 \\ b_2c_2 & -a_2 \end{pmatrix}\begin{pmatrix} x \\ y \end{pmatrix}-\begin{pmatrix} b_1xy \\ b_2xy \end{pmatrix}. \tag{2}$$

Thus, by Theorem 2 of Section 4.3, the stability of the solution $x=0$, $y=0$ of (2) is determined by the stability of the equilibrium solution $x=0$, $y=0$ of the linearized system

$$\frac{d}{dt}\begin{pmatrix} x \\ y \end{pmatrix}=\mathbf{A}\begin{pmatrix} x \\ y \end{pmatrix}=\begin{pmatrix} -a_1 & b_1c_1 \\ b_2c_2 & -a_2 \end{pmatrix}\begin{pmatrix} x \\ y \end{pmatrix}.$$

The characteristic polynomial of the matrix $\mathbf{A}$ is

$$\lambda^2+(a_1+a_2)\lambda+a_1a_2-b_1b_2c_1c_2$$

whose roots are

$$\lambda=\frac{-(a_1+a_2)\pm\left[(a_1+a_2)^2-4(a_1a_2-b_1b_2c_1c_2)\right]^{1/2}}{2}.$$

It is easily verified that both these roots are real and negative. Hence, the equilibrium solution $x=0$, $y=0$ of (2) is asymptotically stable. This implies that any solution $x(t)$, $y(t)$ of (1) which starts sufficiently close to the origin $x=y=0$ will approach the origin as t approaches infinity. Now, suppose that a solution $x(t)$, $y(t)$ of (1), with $0<x(t_0)<c_1$ and $0<y(t_0)<c_2$, does not approach the origin as t approaches infinity. By the previous remark, this solution must always remain a minimum distance from the origin. Consequently, its orbit for $t\geqslant t_0$ lies in a bounded region in the $x-y$ plane which contains no equilibrium points of (1). By the Poincaré–Bendixson Theorem, therefore, its orbit must spiral into the orbit of a periodic solution of (1). But the system of differential equations (1) has no periodic solution in the first quadrant $x\geqslant 0$, $y\geqslant 0$. This follows immediately from Exercise 11, Section 4.8, and the fact that

$$\frac{\partial}{\partial x}\left[-a_1x+b_1(c_1-x)y\right]+\frac{\partial}{\partial y}\left[-a_2y+b_2(c_2-y)x\right]$$
$$=-(a_1+a_2+b_1y+b_2x)$$

is strictly negative if both x and y are nonnegative. Consequently, every solution $x(t)$, $y(t)$ of (1), with $0<x(t_0)<c_1$ and $0<y(t_0)<c_2$ approaches the equilibrium solution $x=0, y=0$ as t approaches infinity. □

Now, it is quite difficult to evaluate the coefficients a_1, a_2, b_1, b_2, c_1, and c_2. Indeed, it is impossible to obtain even a crude estimate of a_2, which should be interpreted as the average amount of time that a female remains infective. (Similarly, a_1 should be interpreted as the average amount of time that a male remains infective.) This is because most females do not exhibit symptoms. Thus, a female can be infective for an amount of time varying from just one day to well over a year. Nevertheless, it is still possible to ascertain from public health data that a_1a_2 is less than $b_1b_2c_1c_2$, as we now show. Observe that the condition $a_1a_2<b_1b_2c_1c_2$ is equivalent to

$$1<\left(\frac{b_1c_1}{a_2}\right)\left(\frac{b_2c_2}{a_1}\right).$$

The quantity b_1c_1/a_2 can be interpreted as the average number of males that one female infective contacts during her infectious period, if every male is susceptible. Similarly, the quantity b_2c_2/a_1 can be interpreted as the average number of females that one male infective contacts during his infectious period, if every female is susceptible. The quantities b_1c_1/a_2 and b_2c_2/a_1 are called the maximal female and male contact rates, respectively. Theorem 8 can now be interpreted in the following manner.

(a) If the product of the maximal male and female contact rates is greater than one, then gonorrhea will approach a nonzero steady state.
(b) If the product of the maximal male and female contact rates is less than one, then gonorrhea will die out eventually.

In 1973, the average number of female contacts named by a male infective during his period of infectiousness was 0.98, while the average number of male contacts named by a female infective during her period of infectiousness was 1.15. These numbers are very good approximations of the maximal male and female contact rates, respectively, and their product does not exceed the product of the maximal male and female contact rates. (The number of contacts of a male or female infective during their period of infectiousness is slightly less than the maximal male or female contact rates. However, the *actual* number of contacts is often greater than the number of contacts named by an infective.) The product of 1.15 with 0.98 is 1.0682. Thus, gonorrhea will ultimately approach a nonzero steady state.

Remark. Our model of gonorrhea is rather crude since it lumps all promiscuous males and all promiscuous females together, regardless of age. A more accurate model can be obtained by separating the male and female populations into different age groups and then computing the rate of change of infectives in each age group. This has been done recently, but the analysis is too difficult to present here. We just mention that a result completely analogous to Theorem 8 is obtained: either gonorrhea dies out in each age group, or it approaches a constant, positive level in each age group.

Exercises

In Problems 1 and 2, we assume that $a_1a_2 < b_1b_2c_1c_2$.

1. (a) Suppose that a solution $x(t), y(t)$ of (1) leaves region III of Figure 2 at time $t = t^*$ by crossing the curve $y = \phi_1(x)$ or $y = \phi_2(x)$. Conclude that either $x(t)$ or $y(t)$ has a maximum at $t = t^*$. Then, show that this is impossible. Conclude, therefore, that any solution $x(t), y(t)$ of (1) which starts in region III at time $t = t_0$ must remain in region III for all future time $t > t_0$.
(b) Conclude from (a) that any solution $x(t), y(t)$ of (1) which starts in region III has a limit ξ, η as t approaches infinity. Then, show that (ξ, η) must equal (x_0, y_0).

2. Suppose that a solution $x(t), y(t)$ of (1) remains in region IV of Figure 2 for all time $t \geqslant t_0$. Prove that $x(t)$ and $y(t)$ have limits ξ, η respectively, as t approaches infinity. Then conclude that (ξ, η) must equal (x_0, y_0).

In Problems 3 and 4, we assume that $a_1a_2 > b_1b_2c_1c_2$.

3. Suppose that a solution $x(t), y(t)$ of (1) leaves region II of Figure 3 at time $t = t^*$ by crossing the curve $y = \phi_1(x)$ or $y = \phi_2(x)$. Show that either $x(t)$ or $y(t)$ has a maximum at $t = t^*$. Then, show that this is impossible. Conclude, therefore, that every solution $x(t), y(t)$ of (1) which starts in region II at time $t = t_0$ must remain in region II for all future time $t \geqslant t_0$.

4. (a) Suppose that a solution $x(t), y(t)$ of (1) remains in either region I or III of Figure 3 for all time $t \geqslant t_0$. Show that $x(t)$ and $y(t)$ have limits ξ, η respectively, as t approaches infinity.
(b) Conclude from Lemma 1 of Section 4.8 that $(\xi, \eta) = (0, 0)$.
(c) Show that (ξ, η) cannot equal $(0, 0)$ if $x(t), y(t)$ remains in region I or region III for all time $t \geqslant t_0$.
(d) Show that any solution $x(t), y(t)$ of (1) which starts on either $y = \phi_1(x)$ or $y = \phi_2(x)$ will immediately afterwards enter region II.

5. Assume that $a_1a_2 < b_1b_2c_1c_2$. Prove directly, using Theorem 2 of Section 4.3, that the equilibrium solution $x = x_0$, $y = y_0$ of (1) is asymptotically stable. *Warning*: The calculations are extremely tedious.

6. Assume that the number of homosexuals remains constant in time. Call this constant c. Let $x(t)$ denote the number of homosexuals who have gonorrhea at time t. Assume that homosexuals are cured of gonorrhea at a rate α_1, and that new infectives are added at a rate $\beta_1(c - x)x$.
(a) Show that $\dot{x} = -\alpha_1 x + \beta_1 x(c - x)$.
(b) What happens to $x(t)$ as t approaches infinity?

7. Suppose that the number of homosexuals $c(t)$ grows according to the logistic law $\dot{c} = c(a - bc)$, for some positive constants a and b. Let $x(t)$ denote the number of homosexuals who have gonorrhea at time t, and assume (see Problem 6) that $\dot{x} = -\alpha_1 x + \beta_1 x(c - x)$. What happens to $x(t)$ as t approaches infinity?

5 Separation of variables and Fourier series

5.1 Two point boundary-value problems

In the applications which we will study in this chapter, we will be confronted with the following problem.

Problem: For which values of λ can we find nontrivial functions $y(x)$ which satisfy

$$\frac{d^2y}{dx^2}+\lambda y=0; \qquad ay(0)+by'(0)=0, \quad cy(l)+dy'(l)=0? \tag{1}$$

Equation (1) is called a boundary-value problem, since we prescribe information about the solution $y(x)$ and its derivative $y'(x)$ at two distinct points, $x=0$ and $x=l$. In an initial-value problem, on the other hand, we prescribe the value of y and its derivative at a single point $x=x_0$.

Our intuitive feeling, at this point, is that the boundary-value problem (1) has nontrivial solutions $y(x)$ only for certain exceptional values λ. To wit, $y(x)=0$ is certainly one solution of (1), and the existence–uniqueness theorem for second-order linear equations would seem to imply that a solution $y(x)$ of $y''+\lambda y=0$ is determined uniquely once we prescribe two additional pieces of information. Let us test our intuition on the following simple, but extremely important example.

Example 1. For which values of λ does the boundary-value problem

$$\frac{d^2y}{dx^2}+\lambda y=0; \qquad y(0)=0, \quad y(l)=0 \tag{2}$$

have nontrivial solutions?

Solution.

(i) $\lambda=0$. Every solution $y(x)$ of the differential equation $y''=0$ is of the form $y(x)=c_1x+c_2$, for some choice of constants c_1 and c_2. The condition $y(0)=0$ implies that $c_2=0$, and the condition $y(l)=0$ then implies that $c_1=0$. Thus, $y(x)=0$ is the only solution of the boundary-value problem (2), for $\lambda=0$.

(ii) $\lambda<0$: In this case, every solution $y(x)$ of $y''+\lambda y=0$ is of the form $y(x)=c_1e^{\sqrt{-\lambda}\,x}+c_2e^{-\sqrt{-\lambda}\,x}$, for some choice of constants c_1 and c_2. The boundary conditions $y(0)=y(l)=0$ imply that

$$c_1+c_2=0, \qquad e^{\sqrt{-\lambda}\,l}c_1+e^{-\sqrt{-\lambda}\,l}c_2=0. \tag{3}$$

The system of equations (3) has a nonzero solution c_1,c_2 if, and only if,

$$\det\begin{pmatrix} 1 & 1 \\ e^{\sqrt{-\lambda}\,l} & e^{-\sqrt{-\lambda}\,l} \end{pmatrix}=e^{-\sqrt{-\lambda}\,l}-e^{\sqrt{-\lambda}\,l}=0.$$

This implies that $e^{\sqrt{-\lambda}\,l}=e^{-\sqrt{-\lambda}\,l}$, or $e^{2\sqrt{-\lambda}\,l}=1$. But this is impossible, since e^z is greater than one for $z>0$. Hence, $c_1=c_2=0$ and the boundary-value problem (2) has no nontrivial solutions $y(x)$ when λ is negative.

(iii) $\lambda>0$: In this case, every solution $y(x)$ of $y''+\lambda y=0$ is of the form $y(x)=c_1\cos\sqrt{\lambda}\,x+c_2\sin\sqrt{\lambda}\,x$, for some choice of constants c_1 and c_2. The condition $y(0)=0$ implies that $c_1=0$, and the condition $y(l)=0$ then implies that $c_2\sin\sqrt{\lambda}\,l=0$. This equation is satisfied, for any choice of c_2, if $\sqrt{\lambda}\,l=n\pi$, or $\lambda=n^2\pi^2/l^2$, for some positive integer n. Hence, the boundary-value problem (2) has nontrivial solutions $y(x)=c\sin n\pi x/l$ for $\lambda=n^2\pi^2/l^2$, $n=1,2,\ldots$.

Remark. Our calculations for the case $\lambda<0$ can be simplified if we write every solution $y(x)$ in the form $y=c_1\cosh\sqrt{-\lambda}\,x+c_2\sinh\sqrt{-\lambda}\,x$, where

$$\cosh\sqrt{-\lambda}\,x=\frac{e^{\sqrt{-\lambda}\,x}+e^{-\sqrt{-\lambda}\,x}}{2}$$

and

$$\sinh\sqrt{-\lambda}\,x=\frac{e^{\sqrt{-\lambda}\,x}-e^{-\sqrt{-\lambda}\,x}}{2}.$$

The condition $y(0)=0$ implies that $c_1=0$, and the condition $y(l)=0$ then implies that $c_2\sinh\sqrt{-\lambda}\,l=0$. But $\sinh z$ is positive for $z>0$. Hence, $c_2=0$, and $y(x)=0$.

Example 1 is indicative of the general boundary-value problem (1). Indeed, we have the following remarkable theorem which we state, but do not prove.

Theorem 1. *The boundary-value problem* (1) *has nontrivial solutions* $y(x)$ *only for a denumerable set of values* $\lambda_1,\lambda_2,\ldots$, *where* $\lambda_1 \leqslant \lambda_2\ldots$, *and* λ_n *approaches infinity as* n *approaches infinity. These special values of* λ *are called eigenvalues of* (1), *and the nontrivial solutions* $y(x)$ *are called eigenfunctions of* (1). *In this terminology, the eigenvalues of* (2) *are* $\pi^2/l^2, 4\pi^2/l^2, 9\pi^2/l^2,\ldots$, *and the eigenfunctions of* (2) *are all constant multiples of* $\sin\pi x/l, \sin 2\pi x/l,\ldots$.

There is a very natural explanation of why we use the terms eigenvalue and eigenfunction in this context. Let $\mathbf{V}$ be the set of all functions $y(x)$ which have two continuous derivatives and which satisfy $ay(0)+by'(0)=0, cy(l)+dy'(l)=0$. Clearly, $\mathbf{V}$ is a vector space, of infinite dimension. Consider now the linear operator, or transformation L, defined by the equation

$$[Ly](x) = -\frac{d^2y}{dx^2}(x). \tag{4}$$

The solutions $y(x)$ of (1) are those functions y in $\mathbf{V}$ for which $Ly=\lambda y$. That is to say, the solutions $y(x)$ of (1) are exactly those functions y in $\mathbf{V}$ which are transformed by L into multiples λ of themselves.

Example 2. Find the eigenvalues and eigenfunctions of the boundary-value problem

$$\frac{d^2y}{dx^2}+\lambda y=0; \qquad y(0)+y'(0)=0, \quad y(1)=0. \tag{5}$$

Solution.

(i) $\lambda=0$. Every solution $y(x)$ of $y''=0$ is of the form $y(x)=c_1x+c_2$, for some choice of constants c_1 and c_2. The conditions $y(0)+y'(0)=0$ and $y(1)=0$ both imply that $c_2=-c_1$. Hence, $y(x)=c(x-1)$, $c\neq 0$, is a nontrivial solution of (5) when $\lambda=0$; i.e., $y(x)=c(x-1)$, $c\neq 0$, is an eigenfunction of (5) with eigenvalue zero.

(ii) $\lambda<0$. In this case, every solution $y(x)$ of $y''+\lambda y=0$ is of the form $y(x)=c_1\cosh\sqrt{-\lambda}\,x+c_2\sinh\sqrt{-\lambda}\,x$, for some choice of constants c_1 and c_2. The boundary conditions $y(0)+y'(0)=0$ and $y(1)=0$ imply that

$$c_1+\sqrt{-\lambda}\,c_2=0, \qquad \cosh\sqrt{-\lambda}\,c_1+\sinh\sqrt{-\lambda}\,c_2=0. \tag{6}$$

(Observe that $(\cosh x)'=\sinh x$ and $(\sinh x)'=\cosh x$.) The system of equations (6) has a nontrivial solution c_1,c_2 if, and only if,

$$\det\begin{pmatrix} 1 & \sqrt{-\lambda} \\ \cosh\sqrt{-\lambda} & \sinh\sqrt{-\lambda} \end{pmatrix} = \sinh\sqrt{-\lambda} - \sqrt{-\lambda}\cosh\sqrt{-\lambda} = 0.$$

This implies that

$$\sinh\sqrt{-\lambda} = \sqrt{-\lambda}\cosh\sqrt{-\lambda}\,. \tag{7}$$

But Equation (7) has no solution $\lambda<0$. To see this, let $z=\sqrt{-\lambda}$, and consider the function $h(z)=z\cosh z-\sinh z$. This function is zero for $z=0$ and is positive for $z>0$, since its derivative

$$h'(z)=\cosh z+z\sinh z-\cosh z=z\sinh z$$

is strictly positive for $z>0$. Hence, no negative number λ can satisfy (7).

(iii) $\lambda>0$. In this case, every solution $y(x)$ of $y''+\lambda y=0$ is of the form $y(x)=c_1\cos\sqrt{\lambda}\,x+c_2\sin\sqrt{\lambda}\,x$, for some choice of constants c_1 and c_2. The boundary conditions imply that

$$c_1+\sqrt{\lambda}\,c_2=0,\qquad \cos\sqrt{\lambda}\,c_1+\sin\sqrt{\lambda}\,c_2=0. \tag{8}$$

The system of equations (8) has a nontrivial solution c_1,c_2 if, and only if,

$$\det\begin{pmatrix} 1 & \sqrt{\lambda} \\ \cos\sqrt{\lambda} & \sin\sqrt{\lambda} \end{pmatrix}=\sin\sqrt{\lambda}-\sqrt{\lambda}\cos\sqrt{\lambda}=0.$$

This implies that

$$\tan\sqrt{\lambda}=\sqrt{\lambda}\,. \tag{9}$$

To find those values of λ which satisfy (9), we set $\xi=\sqrt{\lambda}$ and draw the graphs of the functions $\eta=\xi$ and $\eta=\tan\xi$ in the $\xi-\eta$ plane (see Figure 1); the ξ coordinate of each point of intersection of these curves is then a root of the equation $\xi=\tan\xi$. It is clear that these curves intersect exactly once in the interval $\pi/2<\xi<3\pi/2$, and this occurs at a point $\xi_1>\pi$. Similarly, these two curves intersect exactly once in the interval $3\pi/2<\xi<5\pi/2$, and this occurs at a point $\xi_2>2\pi$. More generally, the curves $\eta=\xi$ and $\eta=\tan\xi$ intersect exactly once in the interval

$$\frac{(2n-1)\pi}{2}<\xi<\frac{(2n+1)\pi}{2}$$

and this occurs at a point $\xi_n>n\pi$.

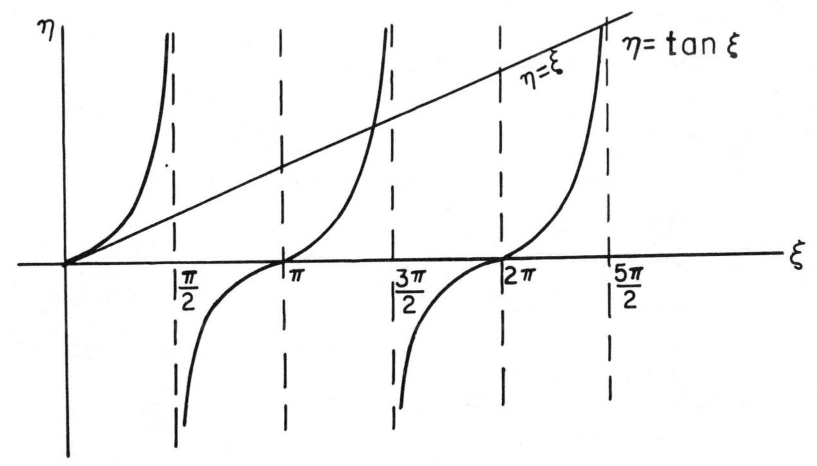

Figure 1. Graphs of $\eta=\xi$ and $\eta=\tan\xi$

Finally, the curves $\eta=\xi$ and $\eta=\tan\xi$ do not intersect in the interval $0<\xi<\pi/2$. To prove this, set $h(\xi)=\tan\xi-\xi$ and compute

$$h'(\xi)=\sec^2\xi-1=\tan^2\xi.$$

This quantity is positive for $0<\xi<\pi/2$. Consequently, the eigenvalues of (5) are $\lambda_1=\xi_1^2, \lambda_2=\xi_2^2,\ldots$, and the eigenfunction of (5) are all constant multiples of the functions $-\sqrt{\lambda_1}\cos\sqrt{\lambda_1}\,x+\sin\sqrt{\lambda_1}\,x, -\sqrt{\lambda_2}\cos\sqrt{\lambda_2}\,x+\sin\sqrt{\lambda_2}\,x,\ldots$. We cannot compute λ_n exactly. Nevertheless, we know that

$$n^2\pi^2<\lambda_n<(2n+1)^2\pi^2/4.$$

In addition, it is clear that λ_n approaches $(2n+1)^2\pi^2/4$ as n approaches infinity.

Exercises

Find the eigenvalues and eigenfunctions of each of the following boundary-value problems.

1. $y''+\lambda y=0;\quad y(0)=0,\ \ y'(l)=0$

2. $y''+\lambda y=0;\quad y'(0)=0,\ \ y'(l)=0$

3. $y''-\lambda y=0;\quad y'(0)=0,\ \ y'(l)=0$

4. $y''+\lambda y=0;\quad y'(0)=0,\ \ y(l)=0$

5. $y''+\lambda y=0;\quad y(0)=0,\ \ y(\pi)-y'(\pi)=0$

6. $y''+\lambda y=0;\quad y(0)-y'(0)=0,\ \ y(1)=0$

7. $y''+\lambda y=0;\quad y(0)-y'(0)=0,\ \ y(\pi)-y'(\pi)=0$

8. For which values of λ does the boundary-value problem

$$y''-2y'+(1+\lambda)y=0;\qquad y(0)=0,\ \ y(1)=0$$

have a nontrivial solution?

9. For which values of λ does the boundary-value problem

$$y''+\lambda y=0;\qquad y(0)=y(2\pi),\ \ y'(0)=y'(2\pi)$$

have a nontrivial solution?

10. Consider the boundary-value problem

$$y''+\lambda y=f(t);\qquad y(0)=0,\ \ y(1)=0 \qquad (*)$$

(a) Show that (*) has a unique solution $y(t)$ if λ is not an eigenvalue of the homogeneous problem.

(b) Show that (*) may have no solution $y(t)$ if λ is an eigenvalue of the homogeneous problem.

(c) Let λ be an eigenvalue of the homogeneous problem. Determine conditions on f so that (*) has a solution $y(t)$. Is this solution unique?

5.2 Introduction to partial differential equations

Up to this point, the differential equations that we have studied have all been relations involving one or more functions of a single variable, and their derivatives. In this sense, these differential equations are *ordinary* differential equations. On the other hand, many important problems in applied mathematics give rise to *partial* differential equations. A partial differential equation is a relation involving one or more functions of *several* variables, and their partial derivatives. For example, the equation

$$\frac{\partial^3 u}{\partial x^3}+\left(\frac{\partial u}{\partial t}\right)^2=\frac{\partial^2 u}{\partial x^2}$$

is a partial differential equation for the function $u(x,t)$, and the equations

$$\frac{\partial u}{\partial x}=\frac{\partial v}{\partial y}, \qquad \frac{\partial u}{\partial y}=-\frac{\partial v}{\partial x}$$

are a system of partial differential equations for the two functions $u(x,y)$ and $v(x,y)$. The order of a partial differential equation is the order of the highest partial derivative that appears in the equation. For example, the order of the partial differential equation

$$\frac{\partial^2 u}{\partial t^2}=2\frac{\partial^2 u}{\partial x \partial t}+u$$

is two, since the order of the highest partial derivative that appears in this equation is two.

There are three classical partial differential equations of order two which appear quite often in applications, and which dominate the theory of partial differential equations. These equations are

$$\frac{\partial u}{\partial t}=\alpha^2\frac{\partial^2 u}{\partial x^2}, \tag{1}$$

$$\frac{\partial^2 u}{\partial t^2}=c^2\frac{\partial^2 u}{\partial x^2} \tag{2}$$

and

$$\frac{\partial^2 u}{\partial x^2}+\frac{\partial^2 u}{\partial y^2}=0. \tag{3}$$

Equation (1) is known as the heat equation, and it appears in the study of heat conduction and other diffusion processes. For example, consider a thin metal bar of length l whose surface is insulated. Let $u(x,t)$ denote the temperature in the bar at the point x at time t. This function satisfies the partial differential equation (1) for $0<x<l$. The constant α^2 is known as the thermal diffusivity of the bar, and it depends solely on the material from which the bar is made.

Equation (2) is known as the wave equation, and it appears in the study of acoustic waves, water waves and electromagnetic waves. Some form of this equation, or a generalization of it, almost invariably arises in any mathematical analysis of phenomena involving the propagation of waves in a continuous medium. (We will gain some insight into why this is so in Section 5.7.) The wave equation also appears in the study of mechanical vibrations. Suppose, for example, that an elastic string of length l, such as a violin string or guy wire, is set in motion so that it vibrates in a vertical plane. Let $u(x,t)$ denote the vertical displacement of the string at the point x at time t (see Figure 1). If all damping effects, such as air resistance, are negligible, and if the amplitude of the motion is not too large, then $u(x,t)$ will satisfy the partial differential equation (2) on the interval $0 \leqslant x \leqslant l$. In this case, the constant c^2 is H/ρ, where H is the horizontal component of the tension in the string, and ρ is the mass per unit length of the string.

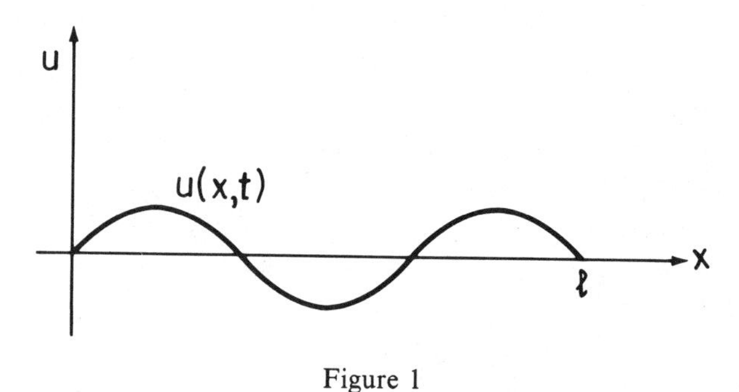

Figure 1

Equation (3) is known as Laplace's equation, and is the most famous of all partial differential equations. It arises in the study of such diverse applications as steady state heat flow, vibrating membranes, and electric and gravitational potentials. For this reason, Laplace's equation is often referred to as the potential equation.

In addition to the differential equation (1), (2), or (3), we will often impose initial and boundary conditions on the function u. These conditions will be dictated to us by the physical and biological problems themselves; they will be chosen so as to guarantee that our equation has a unique solution.

As a model case for the heat equation (1), we consider a thin metal bar of length l whose sides are insulated, and we let $u(x,t)$ denote the temperature in the bar at the point x at time t. In order to dermine the temperature in the bar at any time t we need to know (i) the initial temperature distribution in the bar, and (ii) what is happening at the ends of the bar. Are they held at constant temperatures, say 0°C, or are they insulated, so that no heat can pass through them? (This latter condition implies that $u_x(0,t) = u_x(l,t) = 0$.) Thus, a "well posed" problem for diffusion processes is the heat equation (1), together with the initial condition $u(x,0)=f(x)$, $0<x<l$, and the boundary conditions $u(0,t)=u(l,t)=0$, or $u_x(0,t)=u_x(l,t)=0$.

As a model case for the wave equation, we consider an elastic string of length l, whose ends are fixed, and which is set in motion in a vertical plane. In order to determine the position $u(x,t)$ of the string at any time t we need to know (i) the initial position of the string, and (ii) the initial velocity of the string. It is also implicit that $u(0,t)=u(l,t)=0$. Thus, a well posed problem for wave propagation is the differential equation (2) together with the initial conditions $u(x,0)=f(x)$, $u_t(x,0)=g(x)$, and the boundary conditions $u(0,t)=u(l,t)=0$.

The partial differential equation (3) does not contain the time t, so that we do not expect any "initial conditions" to be imposed here. In the problems that arise in applications, we are given u, or its normal derivative, on the boundary of a given region R, and we seek to determine $u(x,y)$ inside R. The problem of finding a solution of Laplace's equation which takes on given boundary values is known as a Dirichlet problem, while the problem of finding a solution of Laplace's equation whose normal derivative takes on given boundary values is known as a Neumann problem.

In Section 5.3 we will develop a very powerful method, known as the method of separation of variables, for solving the boundary-value problem (strictly speaking, we should say "initial boundary-value problem")

$$\frac{\partial u}{\partial t}=\alpha^2\frac{\partial^2 u}{\partial x^2};\qquad u(x,0)=f(x),\quad 0<x<l;\quad u(0,t)=u(l,t)=0.$$

After developing the theory of Fourier series in Sections 5.4 and 5.5, we will show that the method of separation of variables can also be used to solve more general problems of heat conduction, and several important problems of wave propagation and potential theory.

5.3 The heat equation; separation of variables

Consider the boundary-value problem

$$\frac{\partial u}{\partial t}=\alpha^2\frac{\partial^2 u}{\partial x^2};\qquad u(x,0)=f(x),\quad 0<x<l;\quad u(0,t)=u(l,t)=0.\tag{1}$$

Our goal is to find the solution $u(x,t)$ of (1). To this end, it is helpful to recall how we solved the initial-value problem

$$\frac{d^2y}{dt^2}+p(t)\frac{dy}{dt}+q(t)y=0;\qquad y(0)=y_0,\quad y'(0)=y_0'.\tag{2}$$

First we showed that the differential equation

$$\frac{d^2y}{dt^2}+p(t)\frac{dy}{dt}+q(t)y=0\tag{3}$$

is linear; that is, any linear combination of solutions of (3) is again a solution of (3). And then, we found the solution $y(t)$ of (2) by taking an appropriate linear combination $c_1y_1(t)+c_2y_2(t)$ of two linearly independent solutions $y_1(t)$ and $y_2(t)$ of (3). Now, it is easily verified that any linear combination $c_1u_1(x,t)+\ldots+c_nu_n(x,t)$ of solutions $u_1(x,t),\ldots,u_n(x,t)$ of

$$\frac{\partial u}{\partial t}=\alpha^2\frac{\partial^2 u}{\partial x^2} \tag{4}$$

is again a solution of (4). In addition, if $u_1(x,t),\ldots,u_n(x,t)$ satisfy the boundary conditions $u(0,t)=u(l,t)=0$, then the linear combination $c_1u_1+\ldots+c_nu_n$ also satisfies these boundary conditions. This suggests the following "game plan" for solving the boundary-value problem (1):

(a) Find as many solutions $u_1(x,t), u_2(x,t),\ldots$ as we can of the boundary-value problem

$$\frac{\partial u}{\partial t}=\alpha^2\frac{\partial^2 u}{\partial x^2}; \qquad u(0,t)=u(l,t)=0. \tag{5}$$

(b) Find the solution $u(x,t)$ of (1) by taking an appropriate linear combination of the functions $u_n(x,t)$, $n=1,2,\ldots$.

(a) Since we don't know, as yet, how to solve any partial differential equations, we must reduce the problem of solving (5) to that of solving one or more ordinary differential equations. This is accomplished by setting $u(x,t)=X(x)T(t)$ (hence the name separation of variables). Computing

$$\frac{\partial u}{\partial t}=XT' \quad\text{and}\quad \frac{\partial^2 u}{\partial x^2}=X''T$$

we see that $u(x,t)=X(x)T(t)$ is a solution of the equation $u_t=\alpha^2u_{xx}$ ($u_t=\partial u/\partial t$ and $u_{xx}=\partial^2u/\partial x^2$) if

$$XT'=\alpha^2X''T. \tag{6}$$

Dividing both sides of (6) by α^2XT gives

$$\frac{X''}{X}=\frac{T'}{\alpha^2T}. \tag{7}$$

Now, observe that the left-hand side of (7) is a function of x alone, while the right-hand side of (7) is a function of t alone. This implies that

$$\frac{X''}{X}=-\lambda, \quad\text{and}\quad \frac{T'}{\alpha^2T}=-\lambda \tag{8}$$

for some constant λ. (The only way that a function of x can equal a function of t is if both are constant. To convince yourself of this, let $f(x)=g(t)$ and fix t_0. Then, $f(x)=g(t_0)$ for all x, so that $f(x)=\text{constant}=c_1$, and this immediately implies that $g(t)$ also equals c_1.) In addition, the boundary conditions

$$0=u(0,t)=X(0)T(t),$$

and

$$0=u(l,t)=X(l)T(t)$$

imply that $X(0)=0$ and $X(l)=0$ (otherwise, u must be identically zero). Thus, $u(x,t)=X(x)T(t)$ is a solution of (5) if

$$X''+\lambda X=0; \qquad X(0)=0, \quad X(l)=0 \tag{9}$$

and

$$T'+\lambda\alpha^2 T=0. \tag{10}$$

At this point, the constant λ is arbitrary. However, we know from Example 1 of Section 5.1 that the boundary-value problem (9) has a nontrivial solution $X(x)$ only if $\lambda=\lambda_n=n^2\pi^2/l^2$, $n=1,2,\dots$; and in this case,

$$X(x)=X_n(x)=\sin\frac{n\pi x}{l}.$$

Equation (10), in turn, implies that

$$T(t)=T_n(t)=e^{-\alpha^2 n^2\pi^2 t/l^2}.$$

(Actually, we should multiply both $X_n(x)$ and $T_n(t)$ by constants; however, we omit these constants here since we will soon be taking linear combinations of the functions $X_n(x)T_n(t)$.) Hence,

$$u_n(x,t)=\sin\frac{n\pi x}{l}e^{-\alpha^2 n^2\pi^2 t/l^2}$$

is a nontrivial solution of (5) for every positive integer n.

(b) Suppose that $f(x)$ is a finite linear combination of the functions $\sin n\pi x/l$; that is,

$$f(x)=\sum_{n=1}^{N} c_n \sin\frac{n\pi x}{l}.$$

Then,

$$u(x,t)=\sum_{n=1}^{N} c_n \sin\frac{n\pi x}{l}e^{-\alpha^2 n^2\pi^2 t/l^2}$$

is the desired solution of (1), since it is a linear combination of solutions of (5), and it satisfies the initial condition

$$u(x,0)=\sum_{n=1}^{N} c_n \sin\frac{n\pi x}{l}=f(x), \qquad 0<x<l.$$

Unfortunately, though, most functions $f(x)$ cannot be expanded as a finite linear combination of the functions $\sin n\pi x/l$, $n=1,2,\dots$, on the interval $0<x<l$. This leads us to ask the following question.

Question: Can an arbitrary function $f(x)$ be written as an *infinite* linear combination of the functions $\sin n\pi x/l$, $n=1,2,\dots$, on the interval $0<x<$

l? In other words, given an arbitrary function f, can we find constants $c_1, c_2, \ldots$, such that

$$f(x) = c_1 \sin\frac{\pi x}{l} + c_2 \sin\frac{2\pi x}{l} + \ldots = \sum_{n=1}^{\infty} c_n \sin\frac{n\pi x}{l}; \qquad 0<x<l?$$

Remarkably, the answer to this question is yes, as we show in Section 5.5.

Example 1. At time $t=0$, the temperature $u(x,0)$ in a thin copper rod ($\alpha^2 = 1.14$) of length one is $2\sin 3\pi x + 5\sin 8\pi x$, $0 \leqslant x \leqslant 1$. The ends of the rod are packed in ice, so as to maintain them at 0°C. Find the temperature $u(x,t)$ in the rod at any time $t>0$.

Solution. The temperature $u(x,t)$ satisfies the boundary-value problem

$$\frac{\partial u}{\partial t} = 1.14\frac{\partial^2 u}{\partial x^2}; \qquad \begin{cases} u(x,0) = 2\sin 3\pi x + 5\sin 8\pi x, & 0<x<1 \\ u(0,t) = u(1,t) = 0 \end{cases}$$

and this implies that

$$u(x,t) = 2\sin 3\pi x\, e^{-9(1.14)\pi^2 t} + 5\sin 8\pi x\, e^{-64(1.14)\pi^2 t}.$$

Exercises

Find a solution $u(x,t)$ of the following problems.

1. $\dfrac{\partial u}{\partial t} = 1.71\dfrac{\partial^2 u}{\partial x^2}$; $\begin{cases} u(x,0) = \sin \pi x/2 + 3\sin 5\pi x/2, & 0<x<2 \\ u(0,t) = u(2,t) = 0 \end{cases}$

2. $\dfrac{\partial u}{\partial t} = 1.14\dfrac{\partial^2 u}{\partial x^2}$; $\begin{cases} u(x,0) = \sin \pi x/2 - 3\sin 2\pi x, & 0<x<2 \\ u(0,t) = u(2,t) = 0 \end{cases}$

3. Use the method of separation of variables to solve the boundary-value problem

$$\frac{\partial u}{\partial t} = \frac{\partial^2 u}{\partial x^2} + u; \qquad \begin{cases} u(x,0) = 3\sin 2\pi x - 7\sin 4\pi x, & 0<x<10 \\ u(0,t) = u(10,t) = 0 \end{cases}$$

Use the method of separation of variables to solve each of the following boundary-value problems.

4. $\dfrac{\partial u}{\partial t} = \dfrac{\partial u}{\partial y}$; $u(0,y) = e^{y} + e^{-2y}$

5. $\dfrac{\partial u}{\partial t} = \dfrac{\partial u}{\partial y}$; $u(t,0) = e^{-3t} + e^{2t}$

6. $\dfrac{\partial u}{\partial t} = \dfrac{\partial u}{\partial y} + u$; $u(0,y) = 2e^{-y} - e^{2y}$

7. $\dfrac{\partial u}{\partial t} = \dfrac{\partial u}{\partial y} - u$; $u(t,0) = e^{-5t} + 2e^{-7t} - 14e^{13t}$

8. Determine whether the method of separation of variables can be used to replace each of the following partial differential equations by pairs of ordinary differential equations. If so, find the equations.

(a) $tu_{tt}+u_x=0$ (b) $tu_{xx}+xu_t=0$

(c) $u_{xx}+(x-y)u_{yy}=0$ (d) $u_{xx}+2u_{xt}+u_t=0$

9. The heat equation in two space dimensions is

$$u_t=\alpha^2(u_{xx}+u_{yy}). \qquad (*)$$

(a) Assuming that $u(x,y,t)=X(x)Y(y)T(t)$, find ordinary differential equations satisfied by X, Y, and T.

(b) Find solutions $u(x,y,t)$ of (*) which satisfy the boundary conditions $u(0,y,t)=0$, $u(a,y,t)=0$, $u(x,0,t)=0$, and $u(x,b,t)=0$.

10. The heat equation in two space dimensions may be expressed in terms of polar coordinates as

$$u_t=\alpha^2\left[u_{rr}+\frac{1}{r}u_r+\frac{1}{r^2}u_{\theta\theta}\right].$$

Assuming that $u(r,\theta,t)=R(r)\Theta(\theta)T(t)$, find ordinary differential equations satisfied by R, Θ, and T.

5.4 Fourier series

On December 21, 1807, an engineer named Joseph Fourier announced to the prestigious French Academy of Sciences that an arbitrary function $f(x)$ could be expanded in an infinite series of sines and cosines. Specifically, let $f(x)$ be defined on the interval $-l\leqslant x\leqslant l$, and compute the numbers

$$a_n=\frac{1}{l}\int_{-l}^{l}f(x)\cos\frac{n\pi x}{l}\,dx, \qquad n=0,1,2,\ldots \qquad (1)$$

and

$$b_n=\frac{1}{l}\int_{-l}^{l}f(x)\sin\frac{n\pi x}{l}\,dx, \qquad n=1,2,\ldots \qquad (2)$$

Then, the infinite series

$$\frac{a_0}{2}+a_1\cos\frac{\pi x}{l}+b_1\sin\frac{\pi x}{l}+\ldots=\frac{a_0}{2}+\sum_{n=1}^{\infty}\left[a_n\cos\frac{n\pi x}{l}+b_n\sin\frac{n\pi x}{l}\right] \qquad (3)$$

converges to $f(x)$. Fourier's announcement caused a loud furor in the Academy. Many of its prominent members, including the famous mathematician Lagrange, thought this result to be pure nonsense, since at that time it could not be placed on a rigorous foundation. However, mathematicians have now developed the theory of "Fourier series" to such an extent that whole volumes have been written on it. (Just recently, in fact, they have succeeded in establishing exceedingly sharp conditions for the Fourier series (3) to converge. This result ranks as one of the great mathematical theorems of the twentieth century.) The following theorem, while not the most general theorem possible, covers most of the situations that arise in applications.

Theorem 2. *Let f and f' be piecewise continuous on the interval $-l \leqslant x \leqslant l$. (This means that f and f' have only a finite number of discontinuities on this interval, and both f and f' have right- and left-hand limits at each point of discontinuity.) Compute the numbers a_n and b_n from* (1) *and* (2) *and form the infinite series* (3). *This series, which is called the Fourier series for f on the interval $-l \leqslant x \leqslant l$, converges to $f(x)$ if f is continuous at x, and to $\frac{1}{2}[f(x+0)+f(x-0)]$* if f is discontinuous at x. At $x = \pm l$, the Fourier series* (3) *converges to $\frac{1}{2}[f(l)+f(-l)]$, where $f(\pm l)$ is the limit of $f(x)$ as x approaches $\pm l$.*

Remark. The quantity $\frac{1}{2}[f(x+0)+f(x-0)]$ is the average of the right- and left-hand limits of f at the point x. If we define $f(x)$ to be the average of the right- and left-hand limits of f at any point of discontinuity x, then the Fourier series (3) converges to $f(x)$ for all points x in the interval $-l < x < l$.

Example 1. Let f be the function which is 0 for $-1 \leqslant x < 0$ and 1 for $0 \leqslant x \leqslant 1$. Compute the Fourier series for f on the interval $-1 \leqslant x \leqslant 1$.

Solution. In this problem, $l = 1$. Hence, from (1) and (2),

$$a_0 = \int_{-1}^{1} f(x)\,dx = \int_0^1 dx = 1,$$

$$a_n = \int_{-1}^{1} f(x)\cos n\pi x\,dx = \int_0^1 \cos n\pi x\,dx = 0, \qquad n \geqslant 1$$

and

$$b_n = \int_{-1}^{1} f(x)\sin n\pi x\,dx = \int_0^1 \sin n\pi x\,dx$$

$$= \frac{1}{n\pi}(1-\cos n\pi) = \frac{1-(-1)^n}{n\pi}, \qquad n \geqslant 1.$$

Notice that $b_n = 0$ for n even, and $b_n = 2/n\pi$ for n odd. Hence, the Fourier series for f on the interval $-1 \leqslant x \leqslant 1$ is

$$\frac{1}{2} + \frac{2\sin\pi x}{\pi} + \frac{2\sin 3\pi x}{3\pi} + \frac{2\sin 5\pi x}{5\pi} + \dots.$$

By Theorem 2, this series converges to 0 if $-1 < x < 0$, and to 1 if $0 < x < 1$. At $x = -1$, 0, and $+1$, this series reduces to the single number $\frac{1}{2}$, which is the value predicted for it by Theorem 2.

*The quantity $f(x+0)$ denotes the limit from the right of f at the point x. Similarly, $f(x-0)$ denotes the limit of f from the left.

Example 2. Let f be the function which is 1 for $-2 \leqslant x < 0$ and x for $0 \leqslant x \leqslant 2$. Compute the Fourier series for f on the interval $-2 \leqslant x \leqslant 2$.
Solution. In this problem $l = 2$. Hence from (1) and (2),

$$a_0 = \frac{1}{2}\int_{-2}^{2} f(x)\,dx = \frac{1}{2}\int_{-2}^{0} dx + \frac{1}{2}\int_{0}^{2} x\,dx = 2$$

$$a_n = \frac{1}{2}\int_{-2}^{2} f(x)\cos\frac{n\pi x}{2}\,dx = \frac{1}{2}\int_{-2}^{0}\cos\frac{n\pi x}{2}\,dx + \frac{1}{2}\int_{0}^{2} x\cos\frac{n\pi x}{2}\,dx$$

$$= \frac{2}{(n\pi)^2}(\cos n\pi - 1), \qquad n \geqslant 1$$

and

$$b_n = \frac{1}{2}\int_{-2}^{2} f(x)\sin\frac{n\pi x}{2}\,dx = \frac{1}{2}\int_{-2}^{0}\sin\frac{n\pi x}{2}\,dx + \frac{1}{2}\int_{0}^{2} x\sin\frac{n\pi x}{2}\,dx$$

$$= -\frac{1}{n\pi}(1 + \cos n\pi), \qquad n \geqslant 1.$$

Notice that $a_n = 0$ if n is even; $a_n = -4/n^2\pi^2$ if n is odd; $b_n = 0$ if n is odd; and $b_n = -2/n\pi$ if n is even. Hence, the Fourier series for f on the interval $-2 \leqslant x \leqslant 2$ is

$$1 - \frac{4}{\pi^2}\cos\frac{\pi x}{2} - \frac{1}{\pi}\sin\pi x - \frac{4}{9\pi^2}\cos\frac{3\pi x}{2} - \frac{1}{2\pi}\sin 2\pi x + \dots$$

$$= 1 - \frac{4}{\pi^2}\sum_{n=0}^{\infty}\frac{\cos(2n+1)\pi x/2}{(2n+1)^2} - \frac{1}{\pi}\sum_{n=1}^{\infty}\frac{\sin n\pi x}{n}. \qquad (4)$$

By Theorem 2, this series converges to 1 if $-2 < x < 0$; to x, if $0 < x < 2$; to $\frac{1}{2}$ if $x = 0$; and to $\frac{3}{2}$ if $x \pm 2$. Now, at $x = 0$, the Fourier series (4) is

$$1 - \frac{4}{\pi^2}\left[\frac{1}{1^2} + \frac{1}{3^2} + \frac{1}{5^2} + \frac{1}{7^2} + \dots\right].$$

Thus, we deduce the remarkable identity

$$\frac{1}{2} = 1 - \frac{4}{\pi^2}\left[\frac{1}{1^2} + \frac{1}{3^2} + \frac{1}{5^2} + \frac{1}{7^2} + \dots\right]$$

or

$$\frac{1}{1^2} + \frac{1}{3^2} + \frac{1}{5^2} + \frac{1}{7^2} + \dots = \frac{\pi^2}{8}.$$

The Fourier coefficients a_n and b_n defined by (1) and (2) can be derived in a simple manner. Indeed, if a piecewise continuous function f can be expanded in a series of sines and cosines on the interval $-l \leqslant x \leqslant l$, then, of necessity, this series must be the Fourier series (3). We prove this in the following manner. Suppose that f is piecewise continuous, and that

$$f(x)=\frac{c_0}{2}+\sum_{k=1}^{\infty}\left[c_k\cos\frac{k\pi x}{l}+d_k\sin\frac{k\pi x}{l}\right] \tag{5}$$

for some numbers c_k and d_k. Equation (5) is assumed to hold at all but a finite number of points in the interval $-l\leqslant x\leqslant l$. Integrating both sides of (5) between $-l$ and l gives $c_0 l=\int_{-l}^{l}f(x)dx$, since

$$\int_{-l}^{l}\cos\frac{k\pi x}{l}\,dx=\int_{-l}^{l}\sin\frac{k\pi x}{l}\,dx=0;\qquad k=1,2,\ldots^*$$

Similarly, multiplying both sides of (5) by $\cos n\pi x/l$ and integrating between $-l$ and l gives

$$lc_n=\int_{-l}^{l}f(x)\cos\frac{n\pi x}{l}\,dx$$

while multiplying both sides of (5) by $\sin n\pi x/l$ and integrating between $-l$ and l gives

$$ld_n=\int_{-l}^{l}f(x)\sin\frac{n\pi x}{l}\,dx.$$

This follows immediately from the relations (see Exercise 19)

$$\int_{-l}^{l}\cos\frac{n\pi x}{l}\cos\frac{k\pi x}{l}\,dx=\begin{cases}0, & k\neq n\\ l, & k=n\end{cases} \tag{6}$$

$$\int_{-l}^{l}\cos\frac{n\pi x}{l}\sin\frac{k\pi x}{l}\,dx=0 \tag{7}$$

and

$$\int_{-l}^{l}\sin\frac{n\pi x}{l}\sin\frac{k\pi x}{l}\,dx=\begin{cases}0, & k\neq n\\ l, & k=n\end{cases}. \tag{8}$$

Hence, the coefficients c_n and d_n must equal the Fourier coefficients a_n and b_n. In particular, therefore, a function f can be expanded in one, and only one, Fourier series on the interval $-l\leqslant x\leqslant l$.

Example 3. Find the Fourier series for the function $f(x)=\cos^2 x$ on the interval $-\pi\leqslant x\leqslant\pi$.

Solution. By the preceding remark, the function $f(x)=\cos^2 x$ has a unique Fourier series

$$\frac{a_0}{2}+\sum_{n=1}^{\infty}\left[a_n\cos\frac{n\pi x}{\pi}+b_n\sin\frac{n\pi x}{\pi}\right]=\frac{a_0}{2}+\sum_{n=1}^{\infty}\left[a_n\cos nx+b_n\sin nx\right]$$

*It can be shown that it is permissible to integrate the series (5) term by term.

on the interval $-\pi \leqslant x \leqslant \pi$. But we already know that

$$\cos^2 x = \frac{1}{2} + \frac{\cos 2x}{2}.$$

Hence, the Fourier series for $\cos^2 x$ on the interval $-\pi \leqslant x \leqslant \pi$ must be $\frac{1}{2} + \frac{1}{2}\cos 2x$.

The functions $\cos n\pi x/l$ and $\sin n\pi x/l$, $n = 1,2,\ldots$ all have the interesting property that they are periodic with period $2l$; that is, they repeat themselves over every interval of length $2l$. This follows trivially from the identities

$$\cos\frac{n\pi}{l}(x+2l) = \cos\left(\frac{n\pi x}{l} + 2n\pi\right) = \cos\frac{n\pi x}{l}$$

and

$$\sin\frac{n\pi}{l}(x+2l) = \sin\left(\frac{n\pi x}{l} + 2n\pi\right) = \sin\frac{n\pi x}{l}.$$

Hence, the Fourier series (3) converges for all x to a periodic function $F(x)$. This function is called the periodic extension of $f(x)$. It is defined by the equations

$$\begin{cases} F(x) = f(x), & -l < x < l \\ F(x) = \frac{1}{2}[f(l) + f(-l)], & x = \pm l \\ F(x+2l) = F(x). & \end{cases}$$

For example, the periodic extension of the function $f(x) = x$ is described in Figure 1, and the periodic extension of the function $f(x) = |x|$ is the saw-toothed function described in Figure 2.

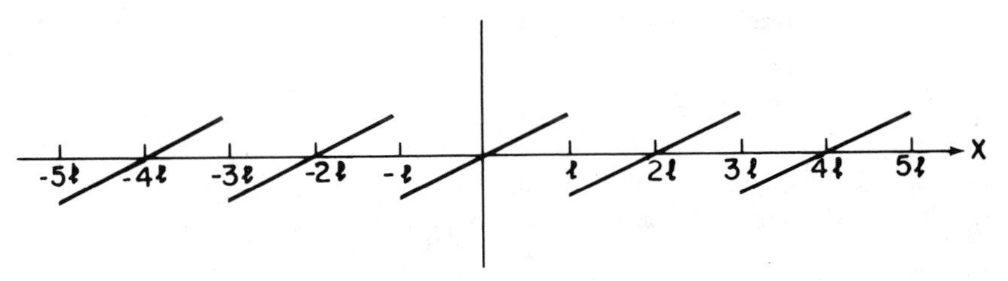

Figure 1. Periodic extension of $f(x) = x$

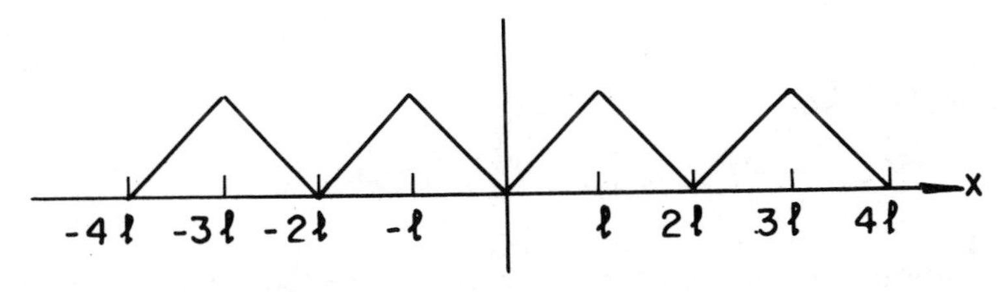

Figure 2. Periodic extension of $f(x) = |x|$

Exercises

In each of Problems 1–13, find the Fourier series for the given function f on the prescribed interval.

1. $f(x)=\begin{cases} -1, & -1\leqslant x<0 \\ 1, & 0\leqslant x\leqslant 1\end{cases}; \quad |x|\leqslant 1$

2. $f(x)=\begin{cases} x, & -2\leqslant x<0 \\ 0, & 0\leqslant x\leqslant 2\end{cases}; \quad |x|\leqslant 2$

3. $f(x)=x; \quad -1\leqslant x\leqslant 1$

4. $f(x)=\begin{cases} -x, & -1\leqslant x<0 \\ x, & 0\leqslant x\leqslant 1\end{cases}; \quad |x|\leqslant 1$

5. $f(x)=\begin{cases} 1, & -2\leqslant x<0 \\ 0, & 0\leqslant x<1; \\ 1, & 1\leqslant x\leqslant 2\end{cases} \quad |x|\leqslant 2$

6. $f(x)=\begin{cases} 0, & -2\leqslant x<1 \\ 3, & 1\leqslant x\leqslant 2\end{cases}; \quad |x|\leqslant 2$

7. $f(x)=\begin{cases} 0, & -l\leqslant x<0 \\ e^x, & 0\leqslant x\leqslant l\end{cases}; \quad |x|\leqslant l$

8. $f(x)=\begin{cases} e^x, & -l\leqslant x<0 \\ 0, & 0\leqslant x\leqslant l\end{cases}; \quad |x|\leqslant l$

9. $f(x)=\begin{cases} e^{-x}, & -l\leqslant x<0 \\ e^x, & 0\leqslant x\leqslant l\end{cases}; \quad -l\leqslant x\leqslant l$

10. $f(x)=e^x; \quad |x|\leqslant l$

11. $f(x)=e^{-x}; \quad |x|\leqslant l$

12. $f(x)=\sin^2 x; \quad |x|\leqslant \pi$

13. $f(x)=\sin^3 x; \quad |x|\leqslant \pi$

14. Let $f(x)=(\pi\cos ax)/2a\sin a\pi$, a not an integer.
(a) Find the Fourier series for f on the interval $-\pi\leqslant x\leqslant \pi$.
(b) Show that this series converges at $x=\pi$ to the value $(\pi/2a)\cot\pi a$.
(c) Use this result to sum the series

$$\frac{1}{1^2-a^2}+\frac{1}{2^2-a^2}+\frac{1}{3^2-a^2}+\dots.$$

15. Suppose that f and f' are piecewise continuous on the interval $-l\leqslant x\leqslant l$. Show that the Fourier coefficients a_n and b_n approach zero as n approaches infinity.

16. Let

$$f(x)=\frac{a_0}{2}+\sum_{n=1}^{\infty}\left[a_n\cos\frac{n\pi x}{l}+b_n\sin\frac{n\pi x}{l}\right].$$

Show that

$$\frac{1}{l}\int_{-l}^{l} f^2(x)\,dx=\frac{a_0^2}{2}+\sum_{n=1}^{\infty}(a_n^2+b_n^2).$$

This relation is known as Parseval's identity. *Hint*: Square the Fourier series for f and integrate term by term.

17. (a) Find the Fourier series for the function $f(x)=x^2$ on the interval $-\pi \leqslant x \leqslant \pi$.

(b) Use Parseval's identity to show that

$$\frac{1}{1^4}+\frac{1}{2^4}+\frac{1}{3^4}+\ldots=\frac{\pi^4}{90}.$$

18. If the Dirac delta function $\delta(x)$ had a Fourier series on the interval $-l \leqslant x \leqslant l$, what would it be?

19. Derive Equations (6)–(8). *Hint*: Use the trigonometric identities

$$\sin A \cos B=\tfrac{1}{2}[\sin(A+B)+\sin(A-B)]$$

$$\sin A \sin B=\tfrac{1}{2}[\cos(A-B)-\cos(A+B)]$$

$$\cos A \cos B=\tfrac{1}{2}[\cos(A+B)+\cos(A-B)].$$

5.5 Even and odd functions

There are certain special cases when the Fourier series of a function f reduces to a pure cosine or a pure sine series. These special cases occur when f is even or odd.

Definition. A function f is said to be *even* if $f(-x)=f(x)$.

Example 1. The function $f(x)=x^2$ is even since

$$f(-x)=(-x)^2=x^2=f(x).$$

Example 2. The function $f(x)=\cos n\pi x/l$ is even since

$$f(-x)=\cos\frac{-n\pi x}{l}=\cos\frac{n\pi x}{l}=f(x).$$

Definition. A function f is said to be *odd* if $f(-x)=-f(x)$.

Example 3. The function $f(x)=x$ is odd since

$$f(-x)=-x=-f(x).$$

Example 4. The function $f(x)=\sin n\pi x/l$ is odd since

$$f(-x)=\sin\frac{-n\pi x}{l}=-\sin\frac{n\pi x}{l}=-f(x).$$

Even and odd functions satisfy the following elementary properties.

1. The product of two even functions is even.
2. The product of two odd functions is even.
3. The product of an odd function with an even function is odd.

The proofs of these assertions are trivial and follow immediately from the definitions. For example, let f and g be odd and let $h(x)=f(x)g(x)$. This function h is even since

$$h(-x)=f(-x)g(-x)=[-f(x)][-g(x)]=f(x)g(x)=h(x).$$

In addition to the multiplicative properties 1–3, even and odd functions satisfy the following integral properties.

4. The integral of an odd function f over a symmetric interval $[-l,l]$ is zero; that is, $\int_{-l}^{l} f(x)\,dx=0$ if f is odd.
5. The integral of an even function f over the interval $[-l,l]$ is twice the integral of f over the interval $[0,l]$; that is,

$$\int_{-l}^{l} f(x)\,dx=2\int_{0}^{l} f(x)\,dx$$

if f is even.

PROOF OF PROPERTY 4. If f is odd, then the area under the curve of f between $-l$ and 0 is the negative of the area under the curve of f between 0 and l. Hence, $\int_{-l}^{l} f(x)\,dx=0$ if f is odd. □

PROOF OF PROPERTY 5. If f is even, then the area under the curve of f between $-l$ and 0 equals the area under the curve of f between 0 and l. Hence,

$$\int_{-l}^{l} f(x)\,dx=\int_{-l}^{0} f(x)\,dx+\int_{0}^{l} f(x)\,dx=2\int_{0}^{l} f(x)\,dx$$

if f is even. □

Concerning even and odd functions, we have the following important lemma.

Lemma 1.

(a) *The Fourier series for an even function is a pure cosine series; that is, it contains no terms of the form* $\sin n\pi x/l$.

(b) *The Fourier series for an odd function is a pure sine series; that is, it contains no terms of the form* $\cos n\pi x/l$.

PROOF. (a) If f is even, then the function $f(x)\sin n\pi x/l$ is odd. Thus, by Property 4, the coefficients

$$b_n=\frac{1}{l}\int_{-l}^{l} f(x)\sin\frac{n\pi x}{l}\,dx, \qquad n=1,2,3,\ldots$$

in the Fourier series for f are all zero.

(b) If f is odd, then the function $f(x)\cos n\pi x/l$ is also odd. Consequently, by Property 4, the coefficients

$$a_n=\frac{1}{l}\int_{-l}^{l} f(x)\cos\frac{n\pi x}{l}\,dx, \qquad n=0,1,2,\ldots$$

in the Fourier series for f are all zero. □

We are now in a position to prove the following extremely important extension of Theorem 2. This theorem will enable us to solve the heat conduction problem of Section 5.3 and many other boundary-value problems that arise in applications.

Theorem 3. *Let f and f' be piecewise continuous on the interval $0 \leqslant x \leqslant l$. Then, on this interval, $f(x)$ can be expanded in either a pure cosine series*

$$\frac{a_0}{2} + \sum_{n=1}^{\infty} a_n \cos\frac{n\pi x}{l},$$

or a pure sine series

$$\sum_{n=1}^{\infty} b_n \sin\frac{n\pi x}{l}.$$

In the former case, the coefficients a_n are given by the formula

$$a_n = \frac{2}{l}\int_0^l f(x)\cos\frac{n\pi x}{l}\,dx, \qquad n=0,1,2,\dots \tag{1}$$

while in the latter case, the coefficients b_n are given by the formula

$$b_n = \frac{2}{l}\int_0^l f(x)\sin\frac{n\pi x}{l}\,dx. \tag{2}$$

PROOF. Consider first the function

$$F(x) = \begin{cases} f(x), & 0 \leqslant x \leqslant l \\ f(-x), & -l \leqslant x < 0 \end{cases}.$$

The graph of $F(x)$ is described in Figure 1, and it is easily seen that F is even. (For this reason, F is called the even extension of f.) Hence, by

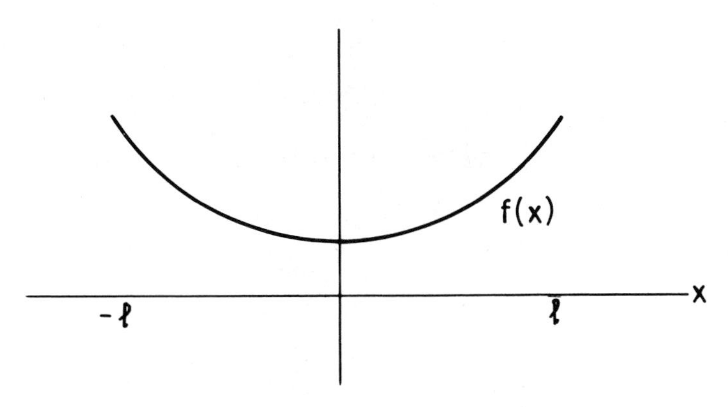

Figure 1. Graph of $F(x)$

Lemma 1, the Fourier series for F on the interval $-l \leqslant x \leqslant l$ is

$$F(x) = \frac{a_0}{2} + \sum_{n=1}^{\infty} a_n \cos\frac{n\pi x}{l}; \qquad a_n = \frac{1}{l}\int_{-l}^{l} F(x)\cos\frac{n\pi x}{l}\,dx. \tag{3}$$

Now, observe that the function $F(x)\cos n\pi x/l$ is even. Thus, by Property 5

$$a_n = \frac{2}{l}\int_0^l F(x)\cos\frac{n\pi x}{l}\,dx = \frac{2}{l}\int_0^l f(x)\cos\frac{n\pi x}{l}\,dx.$$

Finally, since $F(x) = f(x)$, $0 \leqslant x \leqslant l$, we conclude from (3) that

$$f(x) = \frac{a_0}{2} + \sum_{n=1}^{\infty} a_n \cos\frac{n\pi x}{l}, \qquad 0 \leqslant x \leqslant l.$$

Observe too, that the series (3) converges to $f(x)$ for $x = 0$ and $x = l$.

To show that $f(x)$ can also be expanded in a pure sine series, we consider the function

$$G(x) = \begin{cases} f(x), & 0 < x < l \\ -f(-x), & -l < x < 0 \\ 0, & x = 0, \pm l. \end{cases}$$

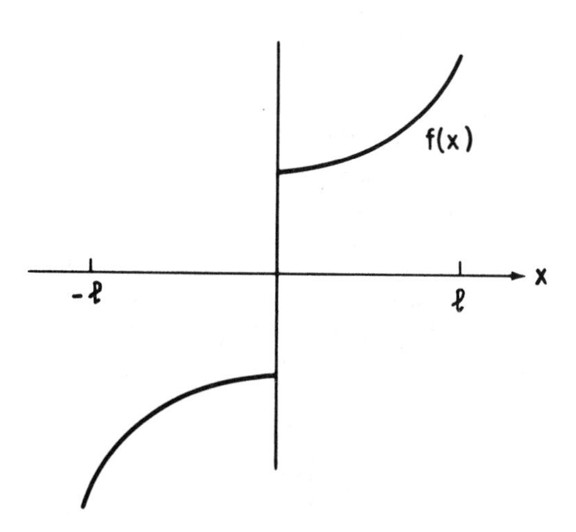

Figure 2. Graph of $G(x)$

The graph of $G(x)$ is described in Figure 2, and it is easily seen that G is odd. (For this reason, G is called the odd extension of f.) Hence, by Lemma 1, the Fourier series for G on the interval $-l \leqslant x \leqslant l$ is

$$G(x) = \sum_{n=1}^{\infty} b_n \sin\frac{n\pi x}{l}; \qquad b_n = \frac{1}{l}\int_{-l}^{l} G(x)\sin\frac{n\pi x}{l}\,dx. \tag{4}$$

Now, observe that the function $G(x)\sin n\pi x/l$ is even. Thus, by Property 5,

$$b_n = \frac{2}{l}\int_0^l G(x)\sin\frac{n\pi x}{l}\,dx = \frac{2}{l}\int_0^l f(x)\sin\frac{n\pi x}{l}\,dx.$$

Finally, since $G(x)=f(x), 0<x<l$, we conclude from (4) that

$$f(x) = \sum_{n=1}^{\infty} b_n \sin\frac{n\pi x}{l}, \qquad 0<x<l.$$

Observe too, that the series (4) is zero for $x=0$ and $x=l$. □

Example 5. Expand the function $f(x)=1$ in a pure sine series on the interval $0<x<\pi$.
Solution. By Theorem 3, $f(x)=\sum_{n=1}^{\infty} b_n \sin nx$, where

$$b_n = \frac{2}{\pi}\int_0^{\pi}\sin nx\,dx = \frac{2}{n\pi}(1-\cos n\pi) = \begin{cases} 0, & n \text{ even} \\ \dfrac{4}{n\pi}, & n \text{ odd}. \end{cases}$$

Hence,

$$1 = \frac{4}{\pi}\left[\frac{\sin x}{1} + \frac{\sin 3x}{3} + \frac{\sin 5x}{5} + \dots\right], \qquad 0<x<\pi.$$

Example 6. Expand the function $f(x)=e^x$ in a pure cosine series on the interval $0 \leqslant x \leqslant 1$.
Solution. By Theorem 3, $f(x)=a_0/2+\sum_{n=1}^{\infty} a_n \cos n\pi x$, where

$$a_0 = 2\int_0^1 e^x\,dx = 2(e-1)$$

and

$$\begin{aligned} a_n &= 2\int_0^1 e^x \cos n\pi x\,dx = 2\,\text{Re}\int_0^1 e^x e^{in\pi x}\,dx \\ &= 2\,\text{Re}\int_0^1 e^{(1+in\pi)x}\,dx = 2\,\text{Re}\left\{\frac{e^{1+in\pi}-1}{1+in\pi}\right\} \\ &= 2\,\text{Re}\left\{\frac{(e\cos n\pi - 1)(1-in\pi)}{1+n^2\pi^2}\right\} = \frac{2(e\cos n\pi - 1)}{1+n^2\pi^2}. \end{aligned}$$

Hence,

$$e^x = e-1+2\sum_{n=1}^{\infty}\frac{(e\cos n\pi - 1)}{1+n^2\pi^2}\cos n\pi x, \qquad 0\leqslant x\leqslant 1.$$

EXERCISES

Expand each of the following functions in a Fourier cosine series on the prescribed interval.

1. $f(x)=e^{-x}; \quad 0\leqslant x\leqslant 1$ **2.** $f(x)=\begin{cases}0, & 0\leqslant x\leqslant 1\\ 1, & 1<x\leqslant 2\end{cases}; \quad 0\leqslant x\leqslant 2$

3. $f(x)=\begin{cases}x, & 0\leqslant x<a\\ a, & a\leqslant x\leqslant 2a\end{cases}; \quad 0\leqslant x\leqslant 2a$

4. $f(x)=\cos^2 x; \quad 0\leqslant x\leqslant \pi$

5. $f(x)=\begin{cases}x, & 0\leqslant x\leqslant l/2\\ l-x, & l/2\leqslant x\leqslant l\end{cases}; \quad 0\leqslant x\leqslant l$

Expand each of the following functions in a Fourier sine series on the prescribed interval.

6. $f(x)=e^{-x}; \quad 0<x<1$ **7.** $f(x)=\begin{cases}0, & 0<x\leqslant 1\\ 1, & 1<x<2\end{cases}; \quad 0<x<2$

8. $f(x)=\begin{cases}x, & 0<x<a\\ a, & a\leqslant x<2a\end{cases}; \quad 0<x<2a$

9. $f(x)=2\sin x\cos x; \quad 0<x<\pi$

10. $f(x)=\begin{cases}x, & 0<x<l/2\\ l-x, & l/2\leqslant x<l\end{cases}; \quad 0<x<l$

11. (a) Expand the function $f(x)=\sin x$ in a Fourier cosine series on the interval $0\leqslant x\leqslant\pi$.

(b) Expand the function $f(x)=\cos x$ in a Fourier sine series on the interval $0<x<\pi$.

(c) Can you expand the function $f(x)=\sin x$ in a Fourier cosine series on the interval $-\pi\leqslant x\leqslant\pi$? Explain.

5.6 Return to the heat equation

We return now to the boundary-value problem

$$\frac{\partial u}{\partial t}=\alpha^2\frac{\partial^2 u}{\partial x^2}; \qquad \begin{cases}u(x,0)=f(x)\\ u(0,t)=u(l,t)=0\end{cases}. \tag{1}$$

We showed in Section 5.3 that the function

$$u(x,t)=\sum_{n=1}^{\infty} c_n \sin\frac{n\pi x}{l}\, e^{-\alpha^2 n^2\pi^2 t/l^2}$$

is (formally) a solution of the boundary-value problem

$$\frac{\partial u}{\partial t}=\alpha^2\frac{\partial^2 u}{\partial x^2}; \qquad u(0,t)=u(l,t)=0 \tag{2}$$

for any choice of constants $c_1,c_2,\ldots$. This led us to ask whether we can find constants $c_1, c_2, \ldots$ such that

$$u(x,0) = \sum_{n=1}^{\infty} c_n \sin\frac{n\pi x}{l} = f(x), \qquad 0 < x < l. \tag{3}$$

As we showed in Section 5.5, the answer to this question is yes; if we choose

$$c_n = \frac{2}{l}\int_0^l f(x)\sin\frac{n\pi x}{l}\,dx,$$

then the Fourier series $\sum_{n=1}^{\infty} c_n \sin n\pi x/l$ converges to $f(x)$ if f is continuous at the point x. Thus,

$$u(x,t) = \frac{2}{l}\sum_{n=1}^{\infty}\left[\int_0^l f(x)\,\sin\frac{n\pi x}{l}\,dx\right]\sin\frac{n\pi x}{l}\,e^{-\alpha^2 n^2\pi^2 t/l^2} \tag{4}$$

is the desired solution of (1).

Remark. Strictly speaking, the solution (4) cannot be regarded as the solution of (1) until we rigorously justify all the limiting processes involved. Specifically, we must verify that the function $u(x,t)$ defined by (4) actually has partial derivatives with respect to x and t, and that $u(x,t)$ satisfies the heat equation $u_t = \alpha^2 u_{xx}$. (It is not true, necessarily, that an infinite sum of solutions of a linear differential equation is again a solution. Indeed, an infinite sum of solutions of a given differential equation need not even be differentiable.) However, in the case of (4) it is possible to show (see Exercise 3) that $u(x,t)$ has partial derivatives with respect to x and t of all orders, and that $u(x,t)$ satisfies the boundary-value problem (1). The argument rests heavily upon the fact that the infinite series (4) converges very rapidly, due to the presence of the factor $e^{-\alpha^2 n^2 \pi^2 t/l^2}$. Indeed, the function $u(x,t)$, for fixed $t>0$, is even analytic for $0<x<l$. Thus, heat conduction is a diffusive process which instantly smooths out any discontinuities that may be present in the initial temperature distribution in the rod. Finally, we observe that $\lim_{t\to\infty} u(x,t) = 0$, for all x, regardless of the initial temperature in the rod. This is in accord with our physical intuition that the heat distribution in the rod should ultimately approach a "steady state"; that is, a state in which the temperature does not change with time.

Example 1. A thin aluminum bar ($\alpha^2 = 0.86$ cm^2/s) 10 cm long is heated to a uniform temperature of 100°C. At time $t=0$, the ends of the bar are plunged into an ice bath at 0°C, and thereafter they are maintained at this temperature. No heat is allowed to escape through the lateral surface of the bar. Find an expression for the temperature at any point in the bar at any later time t.

Solution. Let $u(x,t)$ denote the temperature in the bar at the point x at time t. This function satisfies the boundary-value problem

$$\frac{\partial u}{\partial t}=0.86\frac{\partial^2 u}{\partial x^2};\qquad \begin{cases} u(x,0)=100, & 0<x<10 \\ u(0,t)=u(10,t)=0 \end{cases} \tag{5}$$

The solution of (5) is

$$u(x,t)=\sum_{n=1}^{\infty} c_n \sin\frac{n\pi x}{10}e^{-0.86n^2\pi^2 t/100}$$

where

$$c_n=\frac{1}{5}\int_0^{10} 100\sin\frac{n\pi x}{10}dx=\frac{200}{n\pi}(1-\cos n\pi).$$

Notice that $c_n=0$ if n is even, and $c_n=400/n\pi$ if n is odd. Hence,

$$u(x,t)=\frac{400}{\pi}\sum_{n=1}^{\infty}\frac{\sin(2n+1)\dfrac{\pi x}{10}}{(2n+1)}e^{-0.86(2n+1)^2\pi^2 t/100}.$$

There are several other problems of heat conduction which can be solved by the method of separation of variables. Example 2 below treats the case where the ends of the bar are also insulated, and Exercise 4 treats the case where the ends of the bar are kept at constant, but nonzero temperatures T_1 and T_2.

Example 2. Consider a thin metal rod of length l and thermal diffusivity α^2, whose sides and ends are insulated so that there is no passage of heat through them. Let the initial temperature distribution in the rod be $f(x)$. Find the temperature distribution in the rod at any later time t.
Solution. Let $u(x,t)$ denote the temperature in the rod at the point x at time t. This function satisfies the boundary-value problem

$$\frac{\partial u}{\partial t}=\alpha^2\frac{\partial^2 u}{\partial x^2};\qquad \begin{cases} u(x,0)=f(x), & 0<x<l \\ u_x(0,t)=u_x(l,t)=0 \end{cases} \tag{6}$$

We solve this problem in two steps. First, we will find infinitely many solutions $u_n(x,t)=X_n(x)T_n(t)$ of the boundary-value problem

$$u_t=\alpha^2 u_{xx};\qquad u_x(0,t)=u_x(l,t)=0, \tag{7}$$

and then we will find constants $c_0,c_1,c_2,\ldots$ such that

$$u(x,t)=\sum_{n=0}^{\infty} c_n u_n(x,t)$$

satisfies the initial condition $u(x,0)=f(x)$.

Step 1: Let $u(x,t)=X(x)T(t)$. Computing

$$\frac{\partial u}{\partial t}=XT' \quad \text{and} \quad \frac{\partial^2 u}{\partial x^2}=X''T$$

we see that $u(x,t)$ is a solution of $u_t=\alpha^2 u_{xx}$ if

$$XT'=\alpha^2 X''T, \quad \text{or} \quad \frac{X''}{X}=\frac{T'}{\alpha^2 T}. \tag{8}$$

As we showed in Section 5.3, Equation (8) implies that

$$X''+\lambda X=0 \quad \text{and} \quad T'+\lambda\alpha^2 T=0$$

for some constant λ. In addition, the boundary conditions

$$0=u_x(0,t)=X'(0)T(t) \quad \text{and} \quad 0=u_x(l,t)=X'(l)T(t)$$

imply that $X'(0)=0$ and $X'(l)=0$. Hence $u(x,t)=X(x)T(t)$ is a solution of (7) if

$$X''+\lambda X=0; \qquad X'(0)=0, \quad X'(l)=0 \tag{9}$$

and

$$T'+\lambda\alpha^2 T=0. \tag{10}$$

At this point, the constant λ is arbitrary. However, the boundary-value problem (9) has a nontrivial solution $X(x)$ (see Exercise 1, Section 5.1) only if $\lambda=n^2\pi^2/l^2$, $n=0,1,2,\ldots$, and in this case

$$X(x)=X_n(x)=\cos\frac{n\pi x}{l}.$$

Equation (10), in turn, implies that $T(t)=e^{-\alpha^2 n^2\pi^2 t/l^2}$. Hence,

$$u_n(x,t)=\cos\frac{n\pi x}{l}\, e^{-\alpha^2 n^2\pi^2 t/l^2}$$

is a solution of (7) for every nonnegative integer n.

Step 2: Observe that the linear combination

$$u(x,t)=\frac{c_0}{2}+\sum_{n=1}^{\infty} c_n \cos\frac{n\pi x}{l}\, e^{-\alpha^2 n^2\pi^2 t/l^2}$$

is a solution (formally) of (7) for every choice of constants $c_0,c_1,c_2,\ldots$. Its initial value is

$$u(x,0)=\frac{c_0}{2}+\sum_{n=1}^{\infty} c_n \cos\frac{n\pi x}{l}.$$

Thus, in order to satisfy the initial condition $u(x,0)=f(x)$ we must choose constants $c_0,c_1,c_2,\ldots$ such that

$$f(x)=\frac{c_0}{2}+\sum_{n=1}^{\infty} c_n \cos\frac{n\pi x}{l}, \qquad 0\leqslant x\leqslant l.$$

In other words, we must expand f in a Fourier cosine series on the interval

$0 \leqslant x \leqslant l$. This is precisely the situation in Theorem 3 of Section 5.5, and we conclude, therefore, that

$$c_n = \frac{2}{l}\int_0^l f(x)\cos\frac{n\pi x}{l}\,dx.$$

Hence,

$$u(x,t) = \frac{1}{l}\int_0^l f(x)\,dx + \frac{2}{l}\sum_{n=1}^{\infty}\left[\int_0^l f(x)\cos\frac{n\pi x}{l}\,dx\right]\cos\frac{n\pi x}{l}\,e^{-\alpha^2 n^2\pi^2 t/l^2} \tag{11}$$

is the desired solution of (6).

Remark. Observe from (11) that the temperature in the rod ultimately approaches the steady state temperature

$$\frac{1}{l}\int_0^l f(x)\,dx.$$

This steady state temperature can be interpreted as the "average" of the initial temperature distribution in the rod.

Exercises

1. The ends $x=0$ and $x=10$ of a thin aluminum bar ($\alpha^2=0.86$) are kept at $0°$ C, while the surface of the bar is insulated. Find an expression for the temperature $u(x,t)$ in the bar if initially

(a) $u(x,0)=70, \quad 0<x<10$

(b) $u(x,0)=70\cos x, \quad 0<x<10$

(c) $u(x,0)=\begin{cases} 10x, & 0<x<5 \\ 10(10-x), & 5\leqslant x<10 \end{cases}$

(d) $u(x,0)=\begin{cases} 0, & 0<x<3 \\ 65, & 3\leqslant x<10 \end{cases}$

2. The ends and sides of a thin copper bar ($\alpha^2=1.14$) of length 2 are insulated so that no heat can pass through them. Find the temperature $u(x,t)$ in the bar if initially

(a) $u(x,0)=65\cos^2\pi x, \quad 0\leqslant x\leqslant 2$

(b) $u(x,0)=70\sin x, \quad 0\leqslant x\leqslant 2$

(c) $u(x,0)=\begin{cases} 60x, & 0\leqslant x<1 \\ 60(2-x), & 1\leqslant x\leqslant 2 \end{cases}$

(d) $u(x,0)=\begin{cases} 0, & 0\leqslant x<1 \\ 75, & 1\leqslant x\leqslant 2 \end{cases}$

3. Verify that the function $u(x,t)$ defined by (4) satisfies the heat equation. *Hint*: Use the Cauchy ratio test to show that the infinite series (4) can be differentiated term by term with respect to x and t.

4. A steady state solution $u(x,t)$ of the heat equation $u_t=\alpha^2 u_{xx}$ is a solution $u(x,t)$ which does not change with time.

(a) Show that all steady state solutions of the heat equation are linear functions of x; i.e., $u(x)=Ax+B$.

(b) Find a steady state solution of the boundary-value problem

$$u_t = \alpha^2 u_{xx}; \qquad u(0,t) = T_1, \quad u(l,t) = T_2.$$

(c) Solve the heat conduction problem

$$u_t = \alpha^2 u_{xx}; \qquad \begin{cases} u(x,0) = 75, & 0 < x < 1 \\ u(0,t) = 20, & u(1,t) = 60 \end{cases}$$

Hint: Let $u(x,t) = v(x) + w(x,t)$ where $v(x)$ is the steady state solution of the boundary-value problem $u_t = \alpha^2 u_{xx}$; $u(0,t) = 20$, $u(1,t) = 60$.

5. (a) The ends of a copper rod ($\alpha^2 = 1.14$) 10 cm long are maintained at 0°C, while the center of the rod is maintained at 100°C by an external heat source. Show that the temperature in the rod will ultimately approach a steady state distribution regardless of the initial temperature in the rod. *Hint*: Split this problem into two boundary-value problems.
(b) Assume that the temperature in the rod is at its steady state distribution. At time $t = 0$, the external heat source is removed from the center of the rod, and placed at the left end of the rod. Find the temperature in the rod at any later time t.

6. Solve the boundary-value problem

$$u_t = u_{xx} + u; \qquad \begin{cases} u(x,0) = \cos x, & 0 < x < 1 \\ u(0,t) = 0, & u(1,t) = 0. \end{cases}$$

5.7 The wave equation

We consider now the boundary-value problem

$$\frac{\partial^2 u}{\partial t^2} = c^2 \frac{\partial^2 u}{\partial x^2}; \qquad \begin{cases} u(x,0) = f(x), & u_t(x,0) = g(x) \\ u(0,t) = u(l,t) = 0 \end{cases} \tag{1}$$

which characterizes the propagation of waves in various media, and the mechanical vibrations of an elastic string. This problem, too, can be solved by the method of separation of variables. Specifically, we will (a) find solutions $u_n(x,t) = X_n(x)T_n(t)$ of the boundary-value problem

$$\frac{\partial^2 u}{\partial t^2} = c^2 \frac{\partial^2 u}{\partial x^2}; \qquad u(0,t) = u(l,t) = 0 \tag{2}$$

and (b) find the solution $u(x,t)$ of (1) by taking a suitable linear combination of the functions $u_n(x,t)$.

(a) Let $u(x,t) = X(x)T(t)$. Computing

$$\frac{\partial^2 u}{\partial t^2} = XT'' \quad \text{and} \quad \frac{\partial^2 u}{\partial x^2} = X''T$$

we see that $u(x,t) = X(x)T(t)$ is a solution of the wave equation $u_{tt} = c^2 u_{xx}$

if $XT''=c^2X''T$, or

$$\frac{T''}{c^2T}=\frac{X''}{X}. \tag{3}$$

Next, we observe that the left-hand side of (3) is a function of t alone, while the right-hand side is a function of x alone. This implies that

$$\frac{T''}{c^2T}=-\lambda=\frac{X''}{X}$$

for some constant λ. In addition, the boundary conditions

$$0=u(0,t)=X(0)T(t), \quad \text{and} \quad 0=u(l,t)=X(l)T(t)$$

imply that $X(0)=0$ and $X(l)=0$. Hence $u(x,t)=X(x)T(t)$ is a solution of (2) if

$$X''+\lambda X=0; \qquad X(0)=X(l)=0 \tag{4}$$

and

$$T''+\lambda c^2T=0. \tag{5}$$

At this point, the constant λ is arbitrary. However, the boundary-value problem (4) has a nontrivial solution $X(x)$ only if $\lambda=\lambda_n=n^2\pi^2/l^2$, and in this case,

$$X(x)=X_n(x)=\sin\frac{n\pi x}{l}.$$

Equation (5), in turn, implies that

$$T(t)=T_n(t)=a_n\cos\frac{n\pi ct}{l}+b_n\sin\frac{n\pi ct}{l}.$$

Hence,

$$u_n(x,t)=\sin\frac{n\pi x}{l}\left[a_n\cos\frac{n\pi ct}{l}+b_n\sin\frac{n\pi ct}{l}\right]$$

is a nontrivial solution of (2) for every positive integer n, and every pair of constants a_n, b_n.

(b) The linear combination

$$u(x,t)=\sum_{n=1}^{\infty}\sin\frac{n\pi x}{l}\left[a_n\cos\frac{n\pi ct}{l}+b_n\sin\frac{n\pi ct}{l}\right]$$

formally satisfies the boundary-value problem (2) and the initial conditions

$$u(x,0)=\sum_{n=1}^{\infty}a_n\sin\frac{n\pi x}{l} \quad \text{and} \quad u_t(x,0)=\sum_{n=1}^{\infty}\frac{n\pi c}{l}b_n\sin\frac{n\pi x}{l}.$$

Thus, to satisfy the initial conditions $u(x,0)=f(x)$ and $u_t(x,0)=g(x)$, we must choose the constants a_n and b_n such that

$$f(x)=\sum_{n=1}^{\infty}a_n\sin\frac{n\pi x}{l} \quad \text{and} \quad g(x)=\sum_{n=1}^{\infty}\frac{n\pi c}{l}b_n\sin\frac{n\pi x}{l}$$

on the interval $0<x<l$. In other words, we must expand the functions $f(x)$ and $g(x)$ in Fourier sine series on the interval $0<x<l$. This is precisely the situation in Theorem 3 of Section 5.5, and we conclude, therefore, that

$$a_n=\frac{2}{l}\int_0^l f(x)\sin\frac{n\pi x}{l}\,dx \quad \text{and} \quad b_n=\frac{2}{n\pi c}\int_0^l g(x)\sin\frac{n\pi x}{l}\,dx.$$

For simplicity, we now restrict ourselves to the case where $g(x)$ is zero; that is, the string is released with zero initial velocity. In this case the displacement $u(x,t)$ of the string at any time $t>0$ is given by the formula

$$u(x,t)=\sum_{n=1}^{\infty} a_n\sin\frac{n\pi x}{l}\cos\frac{n\pi ct}{l}; \qquad a_n=\frac{2}{l}\int_0^l f(x)\sin\frac{n\pi x}{l}\,dx. \tag{6}$$

There is a physical significance to the various terms in (6). Each term represents a particular mode in which the string vibrates. The first term ($n=1$) represents the first mode of vibration in which the string oscillates about its equilibrium position with frequency

$$\omega_1=\frac{1}{2\pi}\frac{\pi c}{l}=\frac{c}{2l} \text{ cycles per second.}$$

This lowest frequency is called the fundamental frequency, or first harmonic of the string. Similarly, the nth mode has a frequency

$$\omega_n=\frac{1}{2\pi}\frac{n\pi c}{l}=n\omega_1 \text{ cycles per second}$$

which is called the nth harmonic of the string.

In the case of the vibrating string, all the harmonic frequencies are integer multiples of the fundamental frequency ω_1. Thus, we have music in this case. Of course, if the tension in the string is not large enough, then the sound produced will be of such very low frequency that it is not in the audible range. As we increase the tension in the string, we increase the frequency, and the result is a musical note that can be heard by the human ear.

Justification of solution. We cannot prove directly, as in the case of the heat equation, that the function $u(x,t)$ defined in (6) is a solution of the wave equation. Indeed, we cannot even prove directly that the infinite series (6) has a partial derivative with respect to t and x. For example, on formally computing u_t, we obtain that

$$u_t=-\sum_{n=1}^{\infty}\frac{n\pi c}{l}a_n\sin\frac{n\pi x}{l}\sin\frac{n\pi ct}{l}$$

and due to the presence of the factor n, this series may not converge. However, there is an alternate way to establish the validity of the solution (6). At the same time, we will gain additional insight into the structure of

$u(x,t)$. Observe first that

$$\sin\frac{n\pi x}{l}\cos\frac{n\pi ct}{l}=\frac{1}{2}\left[\sin\frac{n\pi}{l}(x-ct)+\sin\frac{n\pi}{l}(x+ct)\right].$$

Next, let F be the odd periodic extension of f on the interval $-l<x<l$; that is,

$$F(x)=\begin{cases} f(x), & 0<x<l \\ -f(-x), & -l<x<0 \end{cases} \quad \text{and} \quad F(x+2l)=F(x).$$

It is easily verified (see Exercise 6) that the Fourier series for F is

$$F(x)=\sum_{n=1}^{\infty} c_n \sin\frac{n\pi x}{l}, \qquad c_n=\frac{2}{l}\int_0^l f(x)\sin\frac{n\pi x}{l}\,dx.$$

Therefore, we can write $u(x,t)$ in the form

$$u(x,t)=\tfrac{1}{2}\left[F(x-ct)+F(x+ct)\right] \tag{7}$$

and it is now a trivial matter to show that $u(x,t)$ satisfies the wave equation if $f(x)$ has two continuous derivatives.

Equation (7) has the following interpretation. If we plot the graph of the function $y=F(x-ct)$ for any fixed t, we see that it is the same as the graph of $y=F(x)$, except that it is displaced a distance ct in the positive x direction, as shown in Figures 1a and 1b. Thus, $F(x-ct)$ is a *wave* which travels with velocity c in the positive x direction. Similarly, $F(x+ct)$ is a wave which travels with velocity c in the negative x direction. The number c represents the velocity with which a disturbance propagates along the string. If a disturbance occurs at the point x_0, then it will be felt at the point x after a time $t=(x-x_0)/c$ has elapsed. Thus, the wave equation, or some form of it, characterizes the propagation of waves in a medium where disturbances (or signals) travel with a finite, rather than infinite, velocity.

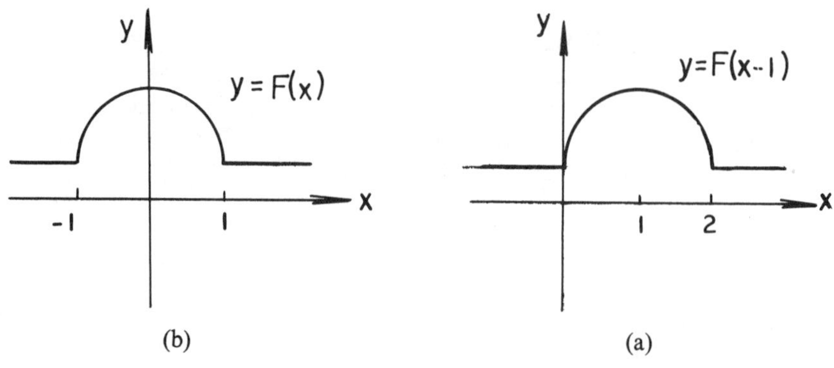

Figure 1

Exercises

Solve each of the following boundary-value problems.

1. $u_{tt}=c^2u_{xx};\quad \begin{cases} u(x,0)=\cos x-1, & u_t(x,0)=0, \quad 0\leqslant x\leqslant 2\pi \\ u(0,t)=0, & u(2\pi,t)=0 \end{cases}$

2. $u_{tt}=c^2u_{xx};\quad \begin{cases} u(x,0)=0, & u_t(x,0)=1, \quad 0\leqslant x\leqslant 1 \\ u(0,t)=0, & u(1,t)=0 \end{cases}$

3. $\begin{matrix} u_{tt}=c^2u_{xx} \\ u(0,t)=u(3,t)=0 \end{matrix};\quad u(x,0)=\begin{cases} x, & 0\leqslant x\leqslant 1, \quad u_t(x,0)=0 \\ 1, & 1\leqslant x\leqslant 2 \\ 3-x, & 2\leqslant x\leqslant 3 \end{cases}$

4. $u_{tt}=c^2u_{xx};\quad \begin{cases} u(x,0)=x\cos\pi x/2, & u_t(x,0)=0, \quad 0\leqslant x\leqslant 1 \\ u(0,t)=0, & u(1,t)=0 \end{cases}$

5. A string of length 10 ft is raised at the middle to a distance of 1 ft, and then released. Describe the motion of the string, assuming that $c^2=1$.

6. Let F be the odd periodic extension of f on the interval $-l<x<l$. Show that the Fourier series

$$\frac{2}{l}\sum_{n=1}^{\infty}\left[\int_0^l f(x)\sin\frac{n\pi x}{l}\,dx\right]\sin\frac{n\pi x}{l}$$

converges to $F(x)$ if F is continuous at x.

7. Show that the transformation $\xi=x-ct$, $\eta=x+ct$ reduces the wave equation to the equation $u_{\xi\eta}=0$. Conclude, therefore, that every solution $u(x,t)$ of the wave equation is of the form $u(x,t)=F(x-ct)+G(x+ct)$ for some functions F and G.

8. Show that the solution of the boundary-value problem

$$u_{tt}=c^2u_{xx};\quad \begin{cases} u(x,0)=f(x), & u_t(x,0)=g(x), \quad -l<x<l \\ u(0,t)=u(l,t)=0 \end{cases}$$

is

$$u(x,t)=\frac{1}{2}[F(x-ct)+F(x+ct)]+\frac{1}{2c}\int_{x-ct}^{x+ct}g(s)\,ds$$

where F is the odd periodic extension of f.

9. The wave equation in two dimensions is $u_{tt}=c^2(u_{xx}+u_{yy})$. Find solutions of this equation by the method of separation of variables.

10. Solve the boundary-value problem

$$u_{tt}=c^2u_{xx}+u;\quad \begin{cases} u(x,0)=f(x), & u_t(x,0)=0, \quad 0<x<l \\ u(0,t)=0, & u(l,t)=0 \end{cases}.$$

5.8 Laplace's equation

We consider now Laplace's equation

$$\frac{\partial^2 u}{\partial x^2}+\frac{\partial^2 u}{\partial y^2}=0. \tag{1}$$

As we mentioned in Section 5.2, two important boundary-value problems that arise in connection with (1) are the Dirichlet problem and the Neumann problem. In a Dirichlet problem we seek a function $u(x,y)$ which satisfies Laplace's equation inside a region R, and which assumes prescribed values on the boundary of R. In a Neumann problem, we seek a function $u(x,y)$ which satisfies Laplace's equation inside a region R, and whose derivative in the direction normal to the boundary of R takes on prescribed values. Both of these problems can be solved by the method of separation of variables if R is a rectangle.

Example 1. Find a function $u(x,y)$ which satisfies Laplace's equation in the rectangle $0<x<a$, $0<y<b$ and which also satisfies the boundary conditions

$$\begin{aligned} u(x,0)&=0, \qquad & u(x,b)&=0\\ u(0,y)&=0, \qquad & u(a,y)&=f(y) \end{aligned}$$

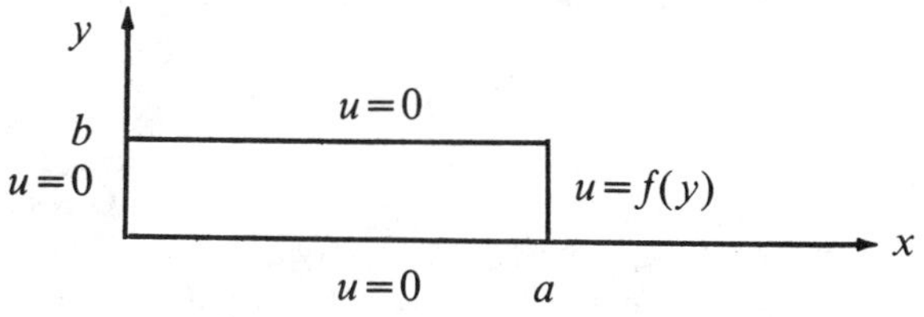

y
b
$u=0$
$u=0$
$u=f(y)$ (2)
x
$u=0$
a

Solution. We solve this problem in two steps. First, we will find functions $u_n(x,y)=X_n(x)Y_n(y)$ which satisfy the boundary-value problem

$$u_{xx}+u_{yy}=0; \qquad u(x,0)=0, \quad u(x,b)=0, \quad u(0,y)=0. \tag{3}$$

Then, we will find constants c_n such that the linear combination

$$u(x,y)=\sum_{n=1}^{\infty} c_n u_n(x,y)$$

satisfies the boundary condition $u(a,y)=f(y)$.

Step 1: Let $u(x,y)=X(x)Y(y)$. Computing $u_{xx}=X''Y$ and $u_{yy}=XY''$, we see that $u(x,y)=X(x)Y(y)$ is a solution of Laplace's equation if $X''Y+XY''=0$, or

$$\frac{Y''}{Y}=-\frac{X''}{X}. \tag{4}$$

Next, we observe that the left-hand side of (4) is a function of y alone, while the right-hand side is a function of x alone. This implies that

$$\frac{Y''}{Y} = -\frac{X''}{X} = -\lambda.$$

for some constant λ. In addition, the boundary conditions

$$0 = u(x,0) = X(x)Y(0), \qquad 0 = u(x,b) = X(x)Y(b),$$
$$0 = u(0,y) = X(0)Y(y)$$

imply that $Y(0)=0$, $Y(b)=0$, and $X(0)=0$. Hence $u(x,y)=XY$ is a solution of (3) if

$$Y'' + \lambda Y = 0; \qquad Y(0)=0, \quad Y(b)=0 \tag{5}$$

and

$$X'' - \lambda X = 0, \qquad X(0) = 0. \tag{6}$$

At this point the constant λ is arbitrary. However, the boundary-value problem (5) has a nontrivial solution $Y(y)$ only if $\lambda = \lambda_n = n^2\pi^2/b^2$, and in this case,

$$Y(y) = Y_n(y) = \sin n\pi y/b.$$

Equation (6), in turn, implies that $X_n(x)$ is proportional to $\sinh n\pi x/b$. (The differential equation $X'' - (n^2\pi^2/b^2)X = 0$ implies that $X(x) = c_1 \cosh n\pi x/b + c_2 \sinh n\pi x/b$ for some choice of constants c_1, c_2, and the initial condition $X(0)=0$ forces c_1 to be zero.) We conclude, therefore, that

$$u_n(x,y) = \sinh\frac{n\pi x}{b}\sin\frac{n\pi y}{b}$$

is a solution of (3) for every positive integer n.

Step 2: The function

$$u(x,y) = \sum_{n=1}^{\infty} c_n \sinh\frac{n\pi x}{b}\sin\frac{n\pi y}{b}$$

is a solution (formally) of (3) for every choice of constants $c_1, c_2, \ldots$. Its value at $x=a$ is

$$u(a,y) = \sum_{n=1}^{\infty} c_n \sinh\frac{n\pi a}{b}\sin\frac{n\pi y}{b}.$$

Therefore, we must choose the constants c_n such that

$$f(y) = \sum_{n=1}^{\infty} c_n \sinh\frac{n\pi a}{b}\sin\frac{n\pi y}{b}, \qquad 0<y<b.$$

In other words, we must expand f in a Fourier sine series on the interval $0<y<b$. This is precisely the situation described in Theorem 3 of Section

5.5, and we conclude, therefore, that

$$c_n=\frac{2}{b\sinh\dfrac{n\pi a}{b}}\int_0^b f(y)\sin\frac{n\pi y}{b}\,dy,\qquad n=1,2,\ldots.$$

Remark. The method of separation of variables can always be used to solve the Dirichlet problem for a rectangle R if u is zero on three sides of R. We can solve an arbitrary Dirichlet problem for a rectangle R by splitting it up into four problems where u is zero on three sides of R (see Exercises 1–4).

Example 2. Find a function $u(x,y)$ which satisfies Laplace's equation in the rectangle $0<x<a$, $0<y<b$, and which also satisfies the boundary conditions

$$\begin{aligned} &u_y(x,0)=0, \quad u_y(x,b)=0\\ &u_x(0,y)=0, \quad u_x(a,y)=f(y) \end{aligned}\tag{7}$$

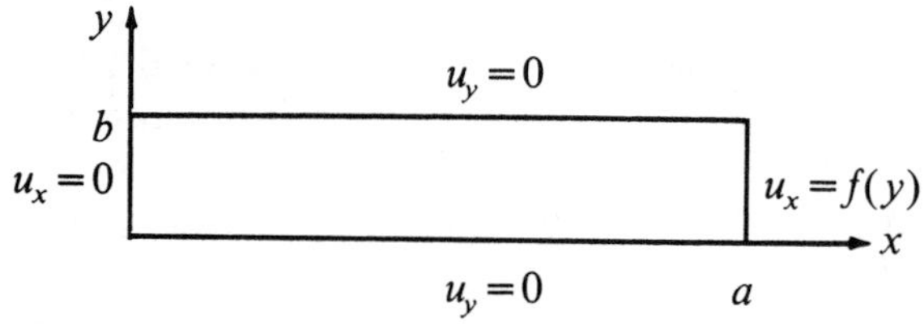

y, b, $u_y=0$, $u_x=0$, $u_x=f(y)$, x, $u_y=0$, a

Solution. We attempt to solve this problem in two steps. First, we will find functions $u_n(x,y)=X_n(x)Y_n(y)$ which satisfy the boundary-value problem

$$u_{xx}+u_{yy}=0;\qquad u_y(x,0)=0,\quad u_y(x,b)=0,\quad\text{and}\quad u_x(0,y)=0.\tag{8}$$

Then, we will try and find constants c_n such that the linear combination $u(x,y)=\sum_{n=0}^{\infty}c_nu_n(x,y)$ satisfies the boundary condition $u_x(a,y)=f(y)$.

Step 1: Set $u(x,y)=X(x)Y(y)$. Then, as in Example 1,

$$\frac{Y''}{Y}=-\frac{X''}{X}=-\lambda$$

for some constant λ. The boundary conditions

$$\begin{aligned} &0=u_y(x,0)=X(x)Y'(0),\quad 0=u_y(x,b)=X(x)Y'(b),\\ &0=u_x(0,y)=X'(0)Y(y) \end{aligned}$$

imply that

$$Y'(0)=0,\quad Y'(b)=0\quad\text{and}\quad X'(0)=0.$$

Hence, $u(x,y)=X(x)Y(y)$ is a solution of (8) if

$$Y''+\lambda Y=0;\qquad Y'(0)=0,\quad Y'(b)=0\tag{9}$$

and

$$X'' - \lambda X = 0; \qquad X'(0) = 0. \tag{10}$$

At this point, the constant λ is arbitrary. However, the boundary-value problem (9) has a nontrivial solution $Y(y)$ only if $\lambda = \lambda_n = n^2\pi^2/b^2, n = 0, 1, 2, \ldots$, and in this case

$$Y(y) = Y_n(y) = \cos n\pi y/b.$$

Equation (10), in turn, implies that $X(x)$ is proportional to $\cosh n\pi x/b$. (The differential equation $X'' - n^2\pi^2 X/b^2 = 0$ implies that $X(x) = c_1 \cosh n\pi x/b + c_2 \sinh n\pi x/b$ for some choice of constants c_1, c_2, and the boundary condition $X'(0) = 0$ forces c_2 to be zero.) We conclude, therefore, that

$$u_n(x,y) = \cosh\frac{n\pi x}{b}\cos\frac{n\pi y}{b}$$

is a solution of (8) for every nonnegative integer n.

Step 2: The function

$$u(x,y) = \frac{c_0}{2} + \sum_{n=1}^{\infty} c_n \cosh\frac{n\pi x}{b}\cos\frac{n\pi y}{b}$$

is a solution (formally) of (8) for every choice of constants $c_0, c_1, c_2, \ldots$. The value of u_x at $x = a$ is

$$u_x(a,y) = \sum_{n=1}^{\infty} \frac{n\pi}{b} c_n \sinh\frac{n\pi a}{b}\cos\frac{n\pi y}{b}.$$

Therefore, we must choose the constants $c_1, c_2, \ldots$, such that

$$f(y) = \sum_{n=1}^{\infty} \frac{n\pi}{b} c_n \sinh\frac{n\pi a}{b}\cos\frac{n\pi y}{b}, \qquad 0 < y < b. \tag{11}$$

Now, Theorem 3 of Section 5.5 states that we can expand $f(y)$ in the cosine series

$$f(y) = \frac{1}{b}\int_0^b f(y)\,dy + \frac{2}{b}\sum_{n=1}^{\infty}\left[\int_0^b f(y)\cos\frac{n\pi y}{b}\,dy\right]\cos\frac{n\pi y}{b} \tag{12}$$

on the interval $0 \leqslant y \leqslant b$. However, we cannot equate coefficients in (11) and (12) since the series (11) has no constant term. Therefore, the condition

$$\int_0^b f(y)\,dy = 0$$

is necessary for this Neumann problem to have a solution. If this is the case, then

$$c_n = \frac{2}{n\pi \sinh \frac{n\pi a}{b}} \int_0^b f(y) \cos \frac{n\pi y}{b} dy, \qquad n \geq 1.$$

Finally, note that c_0 remains arbitrary, and thus the solution $u(x,y)$ is only determined up to an additive constant. This is a property of all Neumann problems.

Exercises

Solve each of the following Dirichlet problems.

1. $\begin{matrix} u_{xx} + u_{yy} = 0 \\ 0 < x < a, \quad 0 < y < b \end{matrix}; \quad \begin{matrix} u(x,0) = 0, & u(x,b) = 0 \\ u(a,y) = 0, & u(0,y) = f(y) \end{matrix}$

2. $\begin{matrix} u_{xx} + u_{yy} = 0 \\ 0 < x < a, \quad 0 < y < b \end{matrix}; \quad \begin{matrix} u(0,y) = 0, & u(a,y) = 0 \\ u(x,0) = 0, & u(x,b) = f(x) \end{matrix}$

Remark. You can do this problem the long way, by separation of variables, or you can try something smart, like interchanging x with y and using the result of Example 1 in the text.

3. $\begin{matrix} u_{xx} + u_{yy} = 0 \\ 0 < x < a, \quad 0 < y < b \end{matrix}; \quad \begin{matrix} u(0,y) = 0, & u(a,y) = 0 \\ u(x,b) = 0, & u(x,0) = f(x) \end{matrix}$

4. $\begin{matrix} u_{xx} + u_{yy} = 0 \\ 0 < x < a, \quad 0 < y < b \end{matrix}; \quad \begin{matrix} u(x,0) = f(x), & u(x,b) = g(x) \\ u(0,y) = h(y), & u(a,y) = k(y) \end{matrix}$

Hint: Write $u(x,y)$ as the sum of 4 functions, each of which is zero on three sides of the rectangle.

5. $\begin{matrix} u_{xx} + u_{yy} = 0 \\ 0 < x < a, \quad 0 < y < b \end{matrix}; \quad \begin{matrix} u(x,0) = 0, & u(x,b) = 1 \\ u(0,y) = 0, & u(a,y) = 1 \end{matrix}$

6. $\begin{matrix} u_{xx} + u_{yy} = 0 \\ 0 < x < a, \quad 0 < y < b \end{matrix}; \quad \begin{matrix} u(x,b) = 0, & u(x,0) = 1 \\ u(0,y) = 0, & u(a,y) = 1 \end{matrix}$

7. $\begin{matrix} u_{xx} + u_{yy} = 0 \\ 0 < x < a, \quad 0 < y < b \end{matrix}; \quad \begin{matrix} u(x,0) = 1, & u(x,b) = 1 \\ u(0,y) = 0, & u(a,y) = 1 \end{matrix}$

8. $\begin{matrix} u_{xx} + u_{yy} = 0 \\ 0 < x < a, \quad 0 < y < b \end{matrix}; \quad \begin{matrix} u(x,0) = 1, & u(x,b) = 1 \\ u(0,y) = 1, & u(a,y) = 1 \end{matrix}$

Remark. *Think*!

9. Solve the boundary-value problem

$$\begin{matrix} u_{xx} + u_{yy} = u \\ 0 < x < 1, \quad 0 < y < 1 \end{matrix}; \quad \begin{matrix} u(x,0) = 0, & u(x,1) = 0 \\ u(0,y) = 0, & u(1,y) = y \end{matrix}$$

10. (a) For which functions $f(y)$ can we find a solution $u(x,y)$ of the Neumann problem

$$\begin{matrix} u_{xx}+u_{yy}=0 \\ 0<x<1, \quad 0<y<1 \end{matrix}; \quad \begin{matrix} u_x(1,y)=0, & u_x(0,y)=f(y) \\ u_y(x,0)=0, & u_y(x,1)=0 \end{matrix}$$

(b) Solve this problem if $f(y)=\sin 2\pi y$.

11. Laplace's equation in three dimensions is

$$u_{xx}+u_{yy}+u_{zz}=0.$$

Assuming that $u=X(x)Y(y)Z(z)$, find 3 ordinary differential equations satisfied by X, Y, and Z.

Appendix A

Some simple facts concerning functions of several variables

1. A function $f(x,y)$ is continuous at the point (x_0,y_0) if for every $\varepsilon>0$ there exists $\delta(\varepsilon)$ such that

$$|f(x,y)-f(x_0,y_0)|<\varepsilon \quad \text{if} \quad |x-x_0|+|y-y_0|<\delta(\varepsilon).$$

2. The partial derivative of $f(x,y)$ with respect to x is the ordinary derivative of f with respect to x, assuming that y is constant. In other words

$$\frac{\partial f(x_0,y_0)}{\partial x}=\lim_{h\to 0}\frac{f(x_0+h,y_0)-f(x_0,y_0)}{h}.$$

3. (a) A function $f(x_1,\ldots,x_n)$ is said to be differentiable if

$$f(x_1+\Delta x_1,\ldots,x_n+\Delta x_n)-f(x_1,\ldots,x_n)=\frac{\partial f}{\partial x_1}\Delta x_1+\ldots+\frac{\partial f}{\partial x_n}\Delta x_n+e$$

where $e/[|\Delta x_1|+\ldots+|\Delta x_n|]$ approaches zero as $|\Delta x_1|+\ldots+|\Delta x_n|$ approaches zero. (b) A function $f(x_1,\ldots,x_n)$ is differentiable in a region R if $\partial f/\partial x_1,\ldots,\partial f/\partial x_n$ are continuous in R.

4. Let $f=f(x_1,\ldots,x_n)$ and $x_j=g_j(y_1,\ldots,y_m)$, $j=1,\ldots,n$. If f and g are differentiable, then

$$\frac{\partial f}{\partial y_k}=\sum_{j=1}^{n}\frac{\partial f}{\partial x_j}\frac{\partial g_j}{\partial y_k}.$$

This is the chain rule of partial differentiation.

5. If all the partial derivatives of order two of f are continuous in a region R, then

$$\frac{\partial^2 f}{\partial x_j \partial x_k} = \frac{\partial^2 f}{\partial x_k \partial x_j}; \qquad j,k=1,\ldots,n.$$

6. The general term in the Taylor series expansion of f about $x_1 = x_1^0, \ldots, x_n = x_n^0$ is

$$\frac{1}{j_1! \ldots j_n!} \frac{\partial^{j_1 + \cdots + j_n} f(x_1^0, \ldots, x_n^0)}{\partial x_1^{j_1} \ldots \partial x_n^{j_n}} (x_1 - x_1^0)^{j_1} \ldots (x_n - x_n^0)^{j_n}.$$

Appendix B

Sequences and series

1. A sequence of numbers a_n, $n=1,2,\ldots$ is said to converge to the limit a if the numbers a_n get closer and closer to a as n approaches infinity. More precisely, the sequence (a_n) converges to a if for every $\varepsilon>0$ there exists $N(\varepsilon)$ such that

$$|a-a_n|<\varepsilon \quad \text{if } n\geqslant N(\varepsilon).$$

2. **Theorem 1.** *If $a_n\to a$ and $b_n\to b$, then $a_n\pm b_n\to a\pm b$.*

PROOF. Let $\varepsilon>0$ be given. Choose $N_1(\varepsilon), N_2(\varepsilon)$ such that

$$|a-a_n|<\frac{\varepsilon}{2} \quad \text{for } n\geqslant N_1(\varepsilon), \quad \text{and} \quad |b-b_n|<\frac{\varepsilon}{2} \quad \text{for } n\geqslant N_2(\varepsilon).$$

Let $N(\varepsilon)=\max\{N_1(\varepsilon),N_2(\varepsilon)\}$. Then, for $n\geqslant N(\varepsilon)$,

$$|a_n\pm b_n-(a\pm b)|\leqslant|a_n-a|+|b_n-b|<\frac{\varepsilon}{2}+\frac{\varepsilon}{2}=\varepsilon.$$ □

3. **Theorem 2.** *Suppose that $a_{n+1}\geqslant a_n$, and there exists a number K such that $|a_n|\leqslant K$, for all n. Then, the sequence (a_n) has a limit.*

PROOF. Exactly the same as the proof of Lemma 1, Section 4.8. □

4. The infinite series $a_1+a_2+\ldots=\sum a_n$ is said to converge if the sequence of partial sums

$$s_1=a_1,\quad s_2=a_1+a_2,\quad \ldots,\quad s_n=a_1+a_2+\ldots+a_n,\ldots$$

has a limit.

5. The sum and difference of two convergent series are also convergent. This follows immediately from Theorem 1.

6. **Theorem 3.** *Let $a_n \geqslant 0$. The series $\sum a_n$ converges if there exists a number K such that $a_1 + \dots + a_n \leqslant K$ for all n.*

PROOF. The sequence of partial sums $s_n = a_1 + \dots + a_n$ satisfies $s_{n+1} \geqslant s_n$. Since $s_n \leqslant K$, we conclude from Theorem 2 that $\sum a_n$ converges. □

7. **Theorem 4.** *The series $\sum a_n$ converges if there exists a number K such that $|a_1| + \dots + |a_n| \leqslant K$ for all n.*

PROOF. From Theorem 3, $\sum |a_n|$ converges. Let $b_n = a_n + |a_n|$. Clearly, $0 \leqslant b_n \leqslant 2|a_n|$. Thus, $\sum b_n$ also converges. This immediately implies that the series

$$\sum a_n = \sum [b_n - |a_n|]$$

also converges. □

8. **Theorem 5** (Cauchy ratio test). *Suppose that the sequence $|a_{n+1}/a_n|$ has a limit λ. Then, the series $\sum a_n$ converges if $\lambda < 1$ and diverges if $\lambda > 1$.*

PROOF. Suppose that $\lambda < 1$. Then, there exists $\rho < 1$, and an index N, such that $|a_{n+1}| \leqslant \rho |a_n|$ for $n \geqslant N$. This implies that $|a_{N+p}| \leqslant \rho^p |a_N|$. Hence,

$$\sum_{n=N}^{N+K} |a_n| \leqslant (1 + \rho + \dots + \rho^K)|a_N| \leqslant \frac{|a_N|}{1-\rho},$$

and $\sum a_n$ converges.

If $\lambda > 1$, then $|a_{N+p}| \geqslant \rho^p |a_N|$, with $\rho > 1$. Thus $|a_n| \to \infty$ as $n \to \infty$. But $|a_n|$ must approach zero if $\sum a_n$ converges to s, since

$$|a_{n+1}| = |s_{n+1} - s_n| \leqslant |s_{n+1} - s| + |s_n - s|$$

and both these quantities approach zero as n approaches infinity. Thus, $\sum a_n$ diverges. □

Appendix C

Introduction to APL

This appendix is a very brief introduction to the programming language APL. It contains all the information needed to solve the computer problems in this text.

In APL we communicate with the computer via a terminal. We type our instructions on an electric typewriter, or keyboard, and these instructions are transmitted (through telephone lines, usually) to the computer. Similarly, the computer replies to us by taking control of the typewriter and causing it to type its reply. A picture of the APL keyboard is given in Figure 1. To enter the upper symbol on a given key, such as the symbol ∇ above the *G*, we depress "shift" while typing the key.

Most programming languages are very simple to learn. This is because a digital computer is essentially very simple in nature. The only mathematical operations which a computer can perform are addition, subtraction, multiplication and division. The real power of the computer lies in its ability to perform millions of such operations in a matter of seconds.

The procedures for signing on and off the computer are very simple but they usually vary from installation to installation. Therefore, we will omit these instructions. We begin with the instructions for addition, subtraction, multiplication and division.

(1) *Addition*. To add the numbers 3 and 4 together we type `3+4` and depress the return key. Depressing the return key tells the computer that we have completed our instruction. While the computer executes this instruction, the keyboard is locked. In a fraction of a second (unless many people are using the computer at the same time) the computer will type out `7`.

(2) *Subtraction*. To subtract the number 4 from 3 we type `3−4`. The computer will respond with `¯1`.

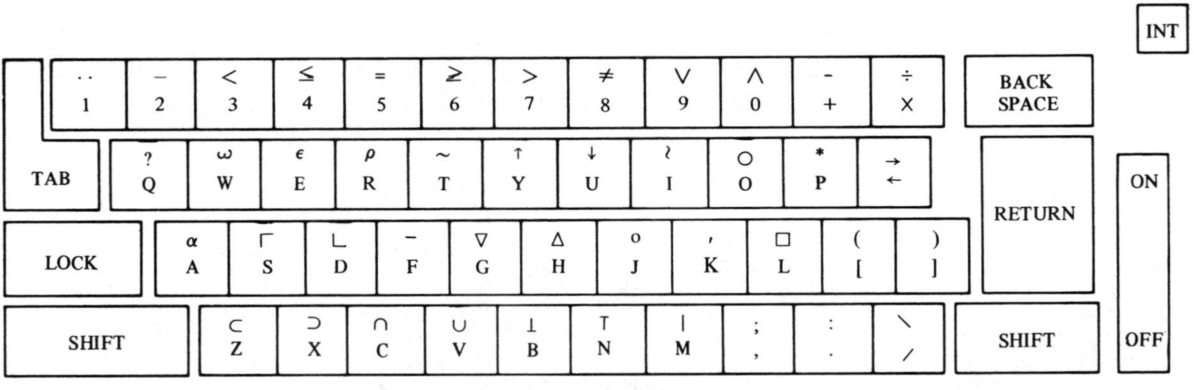

Figure 1. The APL keyboard

Remark. The minus sign that denotes the operation of subtraction is the minus sign above the plus sign on the APL keyboard. The negative sign that the computer types is the negative sign above the 2. This negative sign denotes negative numbers. For example, the instruction ¯2+1 tells the computer to add the numbers -2 and 1 together.

(3) *Multiplication.* To multiply the numbers 12 and 6 together we type 12×6. The computer will respond with 72.

(4) *Division.* To divide the number 6 by 12, we type 6÷12. The computer will respond with 0.5.

Remark. The symbol ÷ can also be used to find the reciprocal of a number. To find the reciprocal of 6 we type ÷6. As another example, suppose that we enter the instruction 3+ ÷6. The computer looks to the immediate left of the division sign. Since it does not see a number there (i.e., it sees + which is an operation) it takes the reciprocal of 6 and adds that to 3. Thus, the result of this instruction is 3.16666667.

Exponentiation. By one simple instruction, the computer can raise any given number to any power. To find the value of $(3.14)^{1/4}$ we type 3.14∗0.25 (or 3.14∗ ÷4).

Remark. The operation of exponentiation is not one of the four mathematical operations that the computer can perform. However, there are very elegant and numerically efficient schemes for computing the power of a given number which involve only the basic four operations (see Sections 1.11 and 1.11.1). A complete set of instructions for one of these schemes is on file in the computer. When we give the instruction for exponentiation, the computer runs through this pre-packaged set of instructions.

Exponential form of numbers. A number can be entered into the computer in one of two ways. The first way is the ordinary decimal form, i.e., 3.14. A second way, which is often used for very large or very small numbers, is the exponential form. In this form, we factor out a power of 10 from a given number. For example, Avogadro's number (the number of molecules in one gram-mole of any substance) is 6.02×10^{23}. This number can be entered into the computer in the form $6.02E23$. More generally, let a and n denote two fixed numbers, with n an integer. The instruction aEn means: multiply a by 10^n. For example,

$$3.2E^{-5}=3.2\times 10^{-5}=0.000032$$

and

$$-1.732E2=-1.732\times 10^{2}=-173.2.$$

The computer prints out the results of its calculations in either decimal or exponential form. (This author has yet to figure out which form the computer will use for a given number.)

Compound expressions. In APL we can include several operations in the same instruction. For example, we can instruct the computer to multiply the numbers 3 and 4 together, divide 6 by 2, and then add the results together. In ordinary arithmetic we would write this as $3\times 4+6\div 2$ $(=15)$. However, we must always remember that the computer is Jewish: it reads from right to left. Thus, when the computer receives the instruction

```
3×4+6÷2
```

it first divides 6 by 2 to give 3. Then it adds 4 to 3 giving 7. Finally, it multiplies 7 by 3 giving 21. The best way of ensuring that the computer does the correct computation is to use parentheses. Thus, the proper way of writing the previous instruction is `(3×4)+6÷2`. This instruction tells the computer to (i) divide 6 by 2; (ii) multiply 3 and 4 together; and (iii) add these two results together. As a second example, suppose we want the computer to evaluate the expression

$$\frac{2}{1+(3.2)^2+6^{1/3}}. \tag{1}$$

This can be accomplished by entering the instruction

```
2÷1+(3.2*2)+6*÷3
```

Upon receiving this instruction the computer, (i) takes the cube root of 6; (ii) squares the number 3.2 and adds that to $6^{1/3}$; (iii) adds 1 to that result; and (iv) divides this number into 2.

Identifiers. There are many instances when we want the computer to store a given number or the result of a calculation. For example, we might need the result of a certain calculation at some later time. However, the computer does not store or save anything unless we specifically tell it to. In APL (and in other programming languages too) we cause the computer to store a given number by the clever device of assigning a name, or *identifier*, to it. This name must begin with a letter of the alphabet and contain only letters and numbers. For example, we can save the result of the calculation $(3.2)^2+6^{1/3}$ by entering the instruction `R←(3.2*2)+6*÷3`. This instruction causes the result of this calculation to be stored in the computer at the address R. In other words, the address R is the computer's way of identifying where it stored this number. Notice too that the compound expression (1) can now be computed by the simple instruction `Z←2÷1+R`. Upon receiving this instruction the computer goes to the address R and sees what number is stored there. Then, it adds one to that number; it divides the resulting number into 2; and it stores this last number at the address Z. If we want to know the result of this calculation, we type `Z`. This instruction tells the computer to print out the number which is stored at the address Z.

Remark 1. It is important that we understand exactly what happens when we assign an identifier to a number. The instruction `R←R+1` is perfectly reasonable in APL, whereas the equation $R = R + 1$ does not make sense mathematically. The instruction R←R+1 tells the computer to add one to the number stored at the address R and to store this sum at the same address. (In this process, of course, the original number stored in R is erased.) If A is an identifier, and we use the phrase "A is four," then we really mean that the number stored at the address A is four.

Remark 2. If we enter the instruction Z←A+3 and we have never defined A, (that is, we have never stored anything at the address A) then the computer will respond with the error message VALUE ERROR, and will type a carat (∧) under the A.

Remark 3. If we enter the instruction 1÷3−A and A is 3, then the computer will respond with the error message DOMAIN ERROR. We will get the same error message if we enter the instruction RAD←B∗0.5 and B is negative.

Functions already stored in the computer. The functions $\sin x$, $\cos x$, $\tan x$, e^x, and $\ln x$ etc. appear so frequently in applications that the instructions for computing them (that is, the instructions for reducing their evaluation to additions, multiplications, subtractions, and divisions) are already stored in the computer.

1. The instructions for computing $\sin x$, $\cos x$, and $\tan x$ are 1○X, 2○X, and 3○X, respectively. The large circle preceding the letter X is the symbol above the letter O on the APL keyboard.

2. The instructions for computing $\arcsin x$, $\arccos x$, and $\arctan x$ are ¯1○X, ¯2○X, and ¯3○X, respectively.

3. The instruction ○X tells the computer to multiply the number X by π. Thus, the instruction A←2○○X∗2 causes the computer to evaluate $\cos \pi x^2$, and to store the result under the name A.

4. The instruction for computing e^x is ∗X. Thus, the instruction 1○∗X tells the computer to evaluate $\sin e^x$.

5. The instruction for computing $\ln x$ is ⍟X. The symbol preceding X is an overstruck symbol. There are two ways of typing it: either we first type ○, backspace and then type ∗, or we first type ∗, backspace and then type ○.

6. The instruction |X tells the computer to take the absolute value of the number x.

7. The instruction !N tells the computer to compute $N!$. The symbol ! is formed by overstriking the quote and the period.

8. The signum function $\operatorname{sgn} x$ in mathematics has the value 1 if x is positive, -1 if x is negative, and 0 if x is zero. APL uses the multiplication sign to denote this function. Thus, if we enter the instruction ×127.6, the computer will respond with 1, and if we enter the instruction ×¯332.9, the computer will respond with ¯1.

Correcting typing errors. If we discover a typing error before an instruction has been entered (that is, before the return key has been depressed) then we backspace to the first incorrect character and depress the ATTN or INT key. (Each APL keyboard has either an ATTN key or an INT key.) The computer's response will be a check mark ($\surd$) below this character to indicate that this and all characters to its right have been erased. We then complete the instruction as it should have been typed.

Vectors. One of the most useful features of APL is the elegant way it handles vectors, or sequences of numbers. The sequence of numbers 1, 3, 9, and 16 can be stored at a single address as a vector with four components by entering the instruction A←1 3 9 16. This instruction tells the computer to store the numbers 1, 3, 9, and 16 at the address A. The number 1 is identified by $A[1]$, the number 3 is identified by $A[2]$, the number 9 is identified by $A[3]$, and the number 16 is identified by $A[4]$. If we enter the instruction A[3], then the computer will respond with 9.

Next, suppose that A and B are two vectors of the same length. The instruction C←A+B tells the computer to add together the corresponding terms of A and B and to store the resulting vector at the address C. The instruction +/A tells the computer to add together all the elements of A. Thus, if we wanted to sum the numbers 3.2, 7, 16, 32.4, and 1.8, we could enter the instruction A←3.2 7 16 32.4 1.8 followed by +/A.

Warning. If the computer does not know that A is a vector (that is, we forgot to tell it) and we enter the instruction A[3], then the computer will respond with an error message, since $A[3]$ is not a valid identifier.

In many instances we want to define a vector of length N where each element is defined recursively. Thus, $A[2]$ is computed in terms of $A[1]$, then $A[3]$ is computed in terms of $A[2]$, and so on. However, if we type either A[1] or A[2], the computer will respond with an error message, since it does not know, as yet, that $A[1],\ldots,A[N]$ are the components of a vector A of length N. We overcome this difficulty by first entering the instruction A←Nρ0. This instruction causes the vector $0\ 0\ldots0$ (N times) to be stored at the address A. We can then define $A[1]$, at which time this new value replaces the value 0 as the first element of A. Then, we can compute $A[2]$, $A[3],\ldots,A[N]$ recursively. For example, the sequence of instructions

```
A←4ρ0
A[1]←1.5
A[2]←A[1]*1.6
A[3]←A[2]*1.7
A[4]←A[3]*1.8
```

causes the computer to store the vector

$$1.5, 1.5^{1.6}, 1.5^{1.6^{1.7}}, 1.5^{1.6^{1.7^{1.8}}}$$

at the address A.

In this context, the instruction A←,X is also very useful. This instruction causes the number X to be identified as a vector with only one component. Similarly, let A and B be two vectors of length l and m respectively. The instruction C←A, B tells the computer to form the $l+m$ dimensional vector C whose first l components are the components of A, and whose remaining m components are the components of B.

Logical expressions. A digital computer can also compare two numbers to see which is larger. To find out whether x is greater than or equal to y, we type X⩾Y. If x is greater than or equal to y, that is, if the expression X⩾Y is true, then the computer will respond with 1. If x is less than y, that is, if the expression X⩾Y is false, then the computer will respond with 0. One of the powerful features of APL is that we can include logical expressions of the form X⩾Y, X<Y, X⩽Y, X=Y, and X≠Y in any APL instruction. For example, suppose that $B=3$ and $C=5$ and we enter the instruction 1+ *B ⩾C. The computer will respond with 2 since the value of $B \geqslant C$ is 0 and $e^0=1$.

Character data. The computer can both print and store messages. One way of accomplishing this is to enclose the message to be printed in quotes and assign it an identifier. For example, the sequence of instructions

```
A←'MARTIN BRAUN IS A GREAT GUY.'
A
```

causes the computer to print the message

```
MARTIN BRAUN IS A GREAT GUY.
```

Next, suppose that T and Y are both vectors of length N which are stored in the computer. For comparison purposes we would like the computer to print out the two vectors T and Y in adjacent columns. This is accomplished by entering the instruction ⍉(2, ⍴Y)⍴T, Y. The symbol ⍉ is typed by overstriking the symbols ○ and \.

Remark. If we enter the instruction T, followed by the instruction Y, then the computer will print out horizontally all the components of T and Y. If necessary, it will use many lines. Needless to say, it will be practically impossible to match up the components of T with the components of Y if N is very large.

APL *Programs*. In many instances, we want the computer to follow a sequence of instructions for a given value of a parameter, and then to repeat that same sequence of instructions for many other values of the parameter. For example, we may want to solve the initial-value problem $dy/dt = f(t,y)$, $y(t_0) = y_0$ for many different values of y_0. Needless to say, we do not want to keep entering these same instructions over and over again every time we change the value of the parameter. Instead, we would like to store the instructions permanently inside the computer. Then, every time we changed the value of the parameter, we would simply inform the computer of the change and tell it to follow the set of instructions we gave it. This is accomplished by defining a *program*.

Definition. A program is a sequence of instructions which are entered and stored in the computer and which can be executed as a whole.

I. *Program definition.*

1. We *open* the definition of a program by typing the symbol ∇ followed by a word which is taken to be the program name. The rules governing program names are the same as those for identifiers. For example, we open the definition of the program Euler by typing ∇ EULER.

2. The computer responds with [1] and waits for the first instruction to be entered. After each instruction the computer responds with the number of the next instruction and waits its entry.

3. We *close* the definition of a program by typing the symbol ∇ right after we type the last instruction.

II. *Program execution.*

1. Typing the program name causes the computer to execute the instructions in that program in sequence, beginning with [1].

2. We can halt the execution of a program at any time by striking the INT (or ATTN) key.

III. *Branching statement.*

1. With one exception, all of the instructions in a program are so-called specification statements. That is to say, they are instructions which either perform mathematical operations, assign identifiers to numbers, or print out data. The one exception is a branching statement of the form →N where N is an integer, or an APL expression that has an integer value. This instruction tells the computer to continue with instruction N. For example, the instruction [15]→4 tells the computer to proceed to instruction 4 when instruction 15 is reached.

2. By convention, program execution terminates if $N = 0$.

3. There are many instances where we want the computer to repeatedly perform a certain operation a specified number of times. We keep track of how many times the operation has been performed with a *counter*, that is, an identifier which is increased by one each time the operation is per-

formed. Such a technique is called *looping*. For example, let the counter be denoted by I and let N be the total number of times that we want to perform the given operation. Suppose, moreover, that the instructions for performing this operation are specified in statements [6]–[8]. Then, our 9th instruction will read [9] I←I+1, and our 10th instruction will read

```
[10] →6×N≥I
```

If $N \geq I$, then $6\times N \geq I$ yields 6, and a branch is made to instruction [6]. If $N<I$, then $6\times N\geq I$ yields 0, and the program terminates. Alternately, our 10th instruction might read [10]→6×ιN≥I. This instruction causes the computer to branch to statement [6] if $N\geq I$, or to the next instruction ([11], in this case) if $N<I$. For example, suppose that we want to evaluate a_{1000}, where a_{1000} is defined recursively through the equations

$$a_{n+1}=a_n-\tfrac{1}{2}a_n^2, \qquad a_1=\tfrac{1}{2}.$$

This can be accomplished by defining the following program.

```
    ∇ EVALUATE
[1] A←1000ρ0
[2] A[1]←0.5
[3] I←1
[4] A[I+1]←A[I]−0.5×A[I]*2
[5] I←I+1
[6] →4×ιI≤999
[7] A[1000]  ∇
```

Typing the program name EVALUATE then causes the computer to run through this sequence of instructions, and to print out the value of a_{1000}.

IV. *Program editing.*

1. The instruction ∇ EULER [□] ∇ tells the computer to print out the program Euler. It is a good programming practice to enter this instruction immediately upon closing a program definition.

2. To change the Nth instruction in a program named Euler after we have closed the program definition, we type ∇ EULER [N] followed by the revised instruction and by ∇.

Extremely important remark. Program execution may stop before completion with an error message. (A frequent cause of this is an undefined identifier.) If this occurs, we immediately type a right-hand arrow →. Otherwise, the computer will keep trying to execute the program and we will be unable to correct it.

Example 1. We want to evaluate the expression $\int_0^x e^{t^2}\,dt$ for many values of x. This can be accomplished in the following manner. Choose a large integer n and set $h=x/n$. Then the Riemann sum

$$h\sum_{j=0}^{n-1} e^{t_j^2}=h\sum_{j=0}^{n-1} e^{(jh)^2}$$

is a good approximation of the integral $\int_0^x e^{t^2}dt$, for h small. The following program will compute this Reimann sum.

```
    ∇ INTEGRAL
[1] H←X÷N
[2] SUM←0
[3] J←0
[4] SUM←SUM+*(J×H)*2
[5] J←J+1
[6] →4×ιJ≤N−1
[7] SUM←H×SUM  ∇.
```

Example 2. A famous mathematician was run over by a hit and run driver while attending a summer meeting of the American Mathematical Society. As the police approached him lying in agony on the street, they asked him if he took down the license plate of the car. "No," said the mathematician, "but the license plate consisted of 3 two digit numbers and these numbers were the lengths of the sides of a right triangle." The policemen quickly procured all the license numbers in the state which were composed of 3 two digit numbers. They fed each of these license numbers into the computer under the variable name A, and then used the following program to identify all possible suspects.

```
    ∇ HIT RUN
[1] S1←(A[1]*2)−(A[2]*2)+A[3]*2
[2] S2←(A[2]*2)−(A[1]*2)+A[3]*2
[3] S3←(A[3]*2)−(A[1]*2)+A[2]*2
[4] →5×(S1×S2×S3)=0
[5] 'THIS IS A POSSIBLE SUSPECT.'  ∇
```

Answers to odd-numbered exercises

Chapter 1

Section 1.2

1. $y(t)=ce^{-\sin t}$ **3.** $y(t)=\dfrac{t+c}{1+t^2}$ **5.** $y(t)=\exp(-\frac{1}{3}t^3)[\int\exp(\frac{1}{3}t^3)\,dt+c]$

7. $y(t)=\dfrac{c+\int(1+t^2)^{1/2}(1+t^4)^{1/4}\,dt}{(1+t^2)^{1/2}(1+t^4)^{1/4}}$ **9.** $y(t)=\exp\left(-\int_0^t\sqrt{1+s^2}\,e^{-s}\,ds\right)$

11. $y(t)=\dfrac{3e^{t^2}-1}{2}$ **13.** $y(t)=e^{-t}\left[2e+\int_1^t\dfrac{e^s}{1+s^2}\,ds\right]$

15. $y(t)=\left(\dfrac{t^5}{5}+\dfrac{2t^3}{3}+t+c\right)(1+t^2)^{-1/2}$ **17.** $y(t)=\begin{cases}2(1-e^{-t}), & 0\leqslant t<1\\ 2(e-1)e^{-t}, & t>1\end{cases}$

21. Each solution approaches a distinct limit as $t\to 0$.

23. All solutions approach zero as $t\to\pi/2$.

Section 1.3

3. 127,328 **5.** (a) $N_{238}(t)=N_{238}(0)2^{-10^{-9}t/4.5}$; (b) $N_{235}(t)=N_{235}(0)2^{-10^{-9}t/.707}$

7. About 13,550 B.C.

Section 1.4

1. $y(t)=\dfrac{t+c}{1-ct}$ **3.** $y(t)=\tan(t-\frac{1}{2}t^2+c)$ **5.** $y(t)=\arcsin(c\sin t)$

7. $y(t)=\left[9+2\ln\left(\dfrac{1+t^2}{5}\right)\right]^{1\ 2}, \quad -\infty<t<\infty$

9. $y(t)=1-[4+2t+2t^2+t^3]^{1/2}, \quad -2<t<\infty$

11. $y(t)=\dfrac{a^2kt}{1+akt}, \quad \dfrac{-1}{ak}<t<\infty$, if $a=b$

$y(t)=\dfrac{ab[1-e^{k(b-a)t}]}{a-be^{k(b-a)t}}, \quad \dfrac{1}{k(b-a)}\ln\dfrac{a}{b}<t<\infty; \quad a\neq b$

13. (b) $y(t)=-t$ and $y(t)=\dfrac{ct^2}{1-ct}$ **15.** $y(t)=\dfrac{t^2-1}{2}$

17. $y=c^2e^{-2\sqrt{t/y}}, \quad t>0;\ y=-c^2e^{2\sqrt{t/y}}, \quad t<0$ **19.** $t+ye^{t/y}=c$

21. (b) $|(b+c)(at+by)+an+bm|=ke^{(b+c)(at-cy)}$

23. $(t+2y)^2-(t+2y)=c-7t$

Section 1.5

3. $a>0.4685$

7. (a) $\dfrac{dp}{dt}=0.003p-0.001p^2-0.002$; (b) $p(t)=\dfrac{1{,}999{,}998-999{,}998e^{-0.001t}}{999{,}999-999{,}998e^{-0.001t}}$,

$p(t)\to 2$ as $t\to\infty$

Section 1.6

1. $p(t)=\dfrac{Ne^{cNt}}{N-1+e^{cNt}}$

Section 1.7

1. $V(t)=\dfrac{W-B}{c}[1-e^{(-cg/W)t}]$

7. $V(t)=\sqrt{322}\left[\dfrac{K_0e^{(64.4)(t-5)/\sqrt{322}}+1}{K_0e^{(64.4)(t-5)/\sqrt{322}}-1}\right], \quad K_0=\dfrac{\sqrt{322}\left(1-\dfrac{1}{\sqrt{e}}\right)+1}{\sqrt{322}\left(1-\dfrac{1}{\sqrt{e}}\right)-1}$

9. (a) $\sqrt{V}-\sqrt{V_0}+\dfrac{mg}{c}\ln\dfrac{mg-c\sqrt{V}}{mg-c\sqrt{V_0}}=\dfrac{-ct}{2m}$; (b) $V_T=\dfrac{m^2g^2}{c^2}$

11. (a) $v\dfrac{dv}{dy}=\dfrac{-gR^3}{(y+R)^2}$; (b) $V_0=\sqrt{2gR}$

Section 1.8

3. $\dfrac{2\ln 5}{\ln 2}$ yrs. **5.** 10^4e^{10} **7.** (a) $c(t)=1-e^{-0.001t}$; (b) $1000\ln\frac{5}{4}$ min.

9. $c(t)=\frac{1}{2}(1-e^{-0.02t})+\dfrac{Q_0}{150}e^{-0.02t}$ **11.** 48% **13.** $xy=c$

15. $x^2+y^2=cy$ **17.** $y^2(\ln y^2-1)=2(c-x^2)$

Section 1.9

3. $t^2 \sin y + y^3 e^t = c$ **5.** $y \tan t + \sec t + y^2 = c$ **7.** $y(t) = t^{-2/3}$

9. $y(t) = -t^2 + \sqrt{t^4 - (t^3 - 1)}$ **11.** $t^3 y + \frac{1}{2}(t^2 y^2) = 10$

13. $a = -2$; $y(t) = \pm\left[\dfrac{t(2t-1)}{2(1+ct)}\right]^{1/2}$ **19.** $a = 1$; $y(t) = \arcsin[(c-t)e^t]$

Section 1.10

1. $y_n(t) = t^2 + \dfrac{t^4}{2!} + \dfrac{t^6}{3!} + \ldots + \dfrac{t^{2n}}{n!}$

3. $y_1(t) = e^t - 1$; $y_2(t) = t - e^t + \dfrac{1+e^{2t}}{2}$

$$y_3(t) = -\frac{107}{48} + \frac{t}{4} + \frac{t^2}{2} + \frac{t^3}{3} + 2(1-t)e^t + \frac{(1+t)e^{2t}}{2} - \frac{e^{3t}}{3} + \frac{e^{4t}}{16}$$

19. $y(t) = \sin\dfrac{t^2}{2}$

Section 1.11

3. (a) $0 < x_0 < 10$ **5.** 1.48982 **7.** (c) 0.73909

Section 1.11.1

5. 0.61547 **7.** 0.22105 **9.** 1.2237

Section 1.12

1. $y_n = (-7)^n - \frac{1}{4}[(-7)^n - 1]$ **3.** (a) $E_n \leqslant \frac{1}{2}(3^n - 1)$; (b) $E_n \leqslant 2(2^n - 1)$

5. $y_n = a_1 \ldots a_{n-1}\left[y_1 + \sum_{j=1}^{n-1} \dfrac{b_j}{a_1 \ldots a_j}\right]$ **7.** $y_{25} = \dfrac{1}{25}\left[1 + \sum_{j=1}^{24} 2^j\right]$

9. (a) $x = \$251.75$; (b) $x = \$289.50$

Section 1.13

1. $h = 0.1$, $y_{10} = 1$; $h = 0.025$, $y_{40} = 1$

3. $h = 0.1$, $y_{10} = 1$; $h = 0.025$, $y_{40} = 1$

5. $h = 0.1$, $y_{10} = 1.80516901$; $h = 0.025$, $y_{40} = 1.94633036$

Section 1.13.1

1. $E_k \leqslant \dfrac{3h}{2}(e^{4/5} - 1)$ **3.** $E_k \leqslant \dfrac{h(1+e+e^2)}{2e}[e^{e/(e+1)} - 1]$ **5.** $h \leqslant \dfrac{4(10^{-5})}{3(e^2 - 1)}$

Section 1.14

1. $h = 0.1$, $y_{10} = 1$; $h = 0.025$, $y_{40} = 1$ **3.** $h = 0.1$, $y_{10} = 1$; $h = 0.025$, $y_{40} = 1$

5. $h = 0.1$, $y_{10} = 2$; $h = 0.025$, $y_{40} = 2$

Section 1.15

1. $h=0.1, y_{10}=1$; $h=0.025, y_{40}=1$ **3.** $h=0.1, y_{10}=1$; $h=0.025, y_{40}=1$

5. $h=0.1, y_{10}=1.98324929$; $h=0.025, y_{40}=1.99884368$

Section 1.16

1. $h=0.1, y_{10}=1$; $h=0.025, y_{40}=1$ **3.** $h=0.1, y_{10}=1$; $h=0.025, y_{40}=1$

5. $h=0.1, y_{10}=1.99997769$; $h=0.025, y_{40}=1.9999999$

Section 1.17

1. 2.4103 **3.** 0.0506 **5.** 0.4388

Chapter 2

Section 2.1

1. (a) $(4-3t)e^t$; (b) $3\sqrt{3}\, t\sin\sqrt{3}\, t$; (c) $2(4-3t)e^t+12\sqrt{3}\, t\sin\sqrt{3}\, t$; (d) $2-3t^2$; (e) $5(2-3t^2)$; (f) 0; (g) $2-3t^2$

5. (b) $W=\dfrac{-3}{2t^{3/2}}$; (d) $y(t)=2\sqrt{t}$

7. (a) $-(b\sin at\sin bt+a\cos at\cos bt)$; (b) 0; (c) $(b-a)e^{(a+b)t}$; (d) e^{2at}; (e) t; (f) $-be^{2at}$

13. $W=\dfrac{1}{t}$

Section 2.2

1. $y(t)=c_1e^t+c_2e^{-t}$ **3.** $y(t)=e^{3t/2}\left[c_1e^{\sqrt{5}\,t/2}+c_2e^{-\sqrt{5}\,t/2}\right]$

5. $y(t)=\frac{1}{5}(e^{4t}+4e^{-t})$ **7.** $y(t)=\dfrac{\sqrt{5}}{3}e^{-t/2}\left[e^{3\sqrt{5}\,t/10}-e^{-3\sqrt{5}\,t/10}\right]$

9. $V\geqslant -3$ **11.** $y(t)=c_1t+c_2/t^5$

Section 2.2.1

1. $y(t)=e^{-t/2}\left[c_1\cos\dfrac{\sqrt{3}\,t}{2}+c_2\sin\dfrac{\sqrt{3}\,t}{2}\right]$

3. $y(t)=e^{-t}\left(c_1\cos\sqrt{2}\,t+c_2\sin\sqrt{2}\,t\right)$

5. $y(t)=e^{-t/2}\left[\cos\dfrac{\sqrt{7}\,t}{2}-\dfrac{3}{\sqrt{7}}\sin\dfrac{\sqrt{7}\,t}{2}\right]$

9. $y(t)=e^{(t-2)/3}\left[\cos\dfrac{\sqrt{11}\,(t-2)}{3}-\dfrac{4}{\sqrt{11}}\sin\dfrac{\sqrt{11}\,(t-2)}{3}\right]$

11. $y_1(t)=\cos\omega t$, $y_2(t)=\sin\omega t$

15. $\sqrt{i}=\dfrac{\pm(1+i)}{\sqrt{2}}$; $\sqrt{1+i}=\dfrac{\pm 1}{\sqrt{2}}\left[\sqrt{\sqrt{2}+1}+i\sqrt{\sqrt{2}-1}\right]$

$\sqrt{-i}=\dfrac{\pm(1-i)}{\sqrt{2}}$; $\sqrt{\sqrt{i}}=\pm\left[\sqrt{\sqrt{2}-1}+i\sqrt{\sqrt{2}+1}\right]$, $(\sqrt{i}=e^{i\pi/4})$

19. $y(t)=\dfrac{1}{\sqrt{t}}\left[c_1\cos\dfrac{\sqrt{7}}{2}\ln t+c_2\sin\dfrac{\sqrt{7}}{2}\ln t\right]$

Section 2.2.2

1. $y(t)=(c_1+c_2t)e^{3t}$ **3.** $y(t)=(1+\frac{1}{3}t)e^{-t/3}$ **7.** $y(t)=2(t-\pi)e^{2(t-\pi)/3}$

11. $y(t)=(c_1+c_2t)e^{t^2}$ **13.** $y(t)=c_1t+c_2(t^2-1)$ **15.** $y(t)=c_1(t+1)+c_2e^{2t}$

17. $y(t)=c_1(1+3t)+c_2e^{3t}$ **19.** $y(t)=\dfrac{c_1+c_2\ln t}{t}$

Section 2.3

1. $y(t)=c_1+c_2e^{2t}+t^2$ **3.** $y(t)=e^{t^2}+2e^t-2e^{-t^3}$

Section 2.4

1. $y(t)=(c_1+\ln\cos t)\cos t+(c_2+t)\sin t$

3. $y(t)=c_1e^{t/2}+[c_2+9t-2t^2+\frac{1}{3}t^3]e^t$

5. $y(t)=\frac{1}{13}(2\cos t-3\sin t)e^{-t}-e^{-t}+\frac{24}{13}e^{-t/3}$

7. $y(t)=\int_0^t\sqrt{s+1}\,[e^{2(t-s)}-e^{(t-s)}]\,ds$ **9.** $y(t)=c_1t^2+\dfrac{c_2}{t}+\dfrac{t^2\ln t}{3}$

11. $y(t)=\sqrt{t}\left[c_1+c_2\ln t+\int_0^t f\sqrt{s}\,\cos s[\ln t-\ln s]\,ds\right]$

Section 2.5

1. $\psi(t)=\dfrac{t^3-2t-1}{3}$ **3.** $\psi(t)=t(\frac{1}{4}-\frac{1}{4}t+\frac{1}{6}t^2)e^t$ **5.** $\psi(t)=\dfrac{t^2}{2}e^{-t}$

7. $\psi(t)=\frac{1}{16}t[\sin 2t-2t\cos 2t]$ **9.** $\psi(t)=\frac{1}{5}+\frac{1}{17}(\cos 2t-4\sin 2t)$

11. $\psi(t)=\dfrac{-1}{50}(\cos t+7\sin t)+\left(\dfrac{t}{2}-\dfrac{1}{5}\right)\dfrac{te^{2t}}{5}$ **13.** $\psi(t)=t(e^{2t}-e^t)$

15. $\psi(t)=-\frac{1}{16}\cos 3t+\frac{1}{4}t\sin t$

17. (b) $\psi(t)=0.005+\dfrac{1}{32{,}000{,}018}(15\sin 30t-20{,}000\cos 30t)+\frac{15}{2}\sin 10t$

Section 2.6

1. Amplitude $=\frac{1}{4}$, period $=\frac{1}{4}\pi$, frequency $=8$

7. $\alpha\geqslant\beta$, where $\left[1+(1-\beta)^2\right]e^{-2\beta}=10^{-6}$

9. $y(t)=\dfrac{e^{-(t-\pi)}}{2}\left[\left(\dfrac{1}{2}+\dfrac{(\pi+1)}{2}e^{-\pi}\right)\cos(t-\pi)+\dfrac{\pi}{2}e^{-\pi}\sin(t-\pi)\right]$

11. $\pi/2$ seconds

Section 2.6.2

3. $Q(1)=\dfrac{12}{10^6}\left[1-\left(\cos 500\sqrt{3}+\dfrac{1}{\sqrt{3}}\sin 500\sqrt{3}\right)e^{-500}\right]$

Steady state charge $=\dfrac{3}{250{,}000}$

7. $\omega=\left(\dfrac{1}{LC}-\dfrac{R^2}{2L^2}\right)^{1/2}$

Section 2.8

1. $y(t)=a_0e^{-t^2/2}+a_1\sum\limits_{n=0}^{\infty}\dfrac{(-1)^n t^{2n+1}}{3\cdot 5\ldots(2n+1)}$

3. $y(t)=a_0\left[1+\dfrac{3t^2}{2^2}+\dfrac{3t^4}{2^4\cdot 2!}-\dfrac{3t^6}{2^6\cdot 3!}+\dfrac{3\cdot 3t^8}{2^8\cdot 4!}-\dfrac{3\cdot 3\cdot 5t^{10}}{2^{10}\cdot 5!}+\ldots\right]+a_1\left(t+\dfrac{t^3}{3}\right)$

5. $y(t)=\sum\limits_{n=0}^{\infty}(n+1)(t-1)^{2n}$

7. $y(t)=-2[t+\dfrac{t^6}{5\cdot 6}+\dfrac{t^{11}}{5\cdot 6\cdot 10\cdot 11}+\dfrac{t^{16}}{5\cdot 6\cdot 10\cdot 11\cdot 15\cdot 16}+\ldots]$

9. (a) $y_1(t)=1-\dfrac{\lambda t^2}{2!}-\dfrac{\lambda(4-\lambda)t^4}{4!}-\dfrac{\lambda(4-\lambda)(8-\lambda)t^6}{6!}+\ldots$

(b) $y_2(t)=t+(2-\lambda)\dfrac{t^3}{3!}+(2-\lambda)(6-\lambda)\dfrac{t^5}{5!}+(2-\lambda)(6-\lambda)(10-\lambda)\dfrac{t^7}{7!}+\ldots$

11. (a) $y_1(t)=1-\alpha^2\dfrac{t^2}{2!}-\alpha^2(2^2-\alpha^2)\dfrac{t^4}{4!}$

$-\alpha^2(2^2-\alpha^2)(4^2-\alpha^2)\dfrac{t^6}{6!}+\ldots$

$y_2(t)=t+(1^2-\alpha^2)\dfrac{t^3}{3!}+(1^2-\alpha^2)(3^2-\alpha^2)\dfrac{t^5}{5!}+\ldots$

13. (a) $y(t)=1-\frac{1}{2}t^2-\frac{1}{6}t^3+\frac{1}{24}t^4+\ldots$ (b) $y(\frac{1}{2})\approx 0.8592436$

15. (a) $y(t)=1-\frac{1}{2}t^2-\frac{1}{6}t^3+\frac{1}{12}t^4+\ldots$ (b) $y(\frac{1}{2})\approx 0.86087198$

17. (a) $y(t)=3+5t-4t^2+t^3-\frac{3}{8}t^4+\ldots$ (b) $y(\frac{1}{2})\approx 5.3409005$

Section 2.8.1

1. $y(t)=c_1t+c_2/t^5$

3. $y(t)=c_1(t-1)+c_2(t-1)^2$

5. $y(t)=t(c_1+c_2\ln t)$

7. $y(t)=c_1\cos(\ln t)+c_2\sin(\ln t)$

9. $y(t)=\frac{t}{2\sqrt{3}}\left[t^{\sqrt{3}}-\frac{1}{t^{\sqrt{3}}}\right]$

Section 2.8.2

1. Yes **3.** Yes **5.** No

7. $y(t)=\frac{c_1}{t}\left(1-t-\frac{t^2}{2!}-\frac{t^3}{3\cdot 3!}-\frac{t^4}{3\cdot 5\cdot 4!}-\cdots\right)$
$+c_2t^{1/2}\left(1+\frac{t}{5}+\frac{t^2}{5\cdot 7\cdot 2!}+\frac{t^3}{5\cdot 7\cdot 9\cdot 3!}+\cdots\right)$

9. $y(t)=c_1\left(1+2t+\frac{t^2}{3}\right)+c_2t^{1/2}\left(1+\frac{t}{2}+\frac{t^2}{2^2\cdot 2!\cdot 5}-\frac{t^3}{2^3\cdot 3!\cdot 5\cdot 7}+\cdots\right)$

11. $y(t)=c_1\left(1+\frac{3t}{1\cdot 3}+\frac{3^2t^2}{3\cdot 7\cdot 2!}+\frac{3^3t^3}{3\cdot 7\cdot 11\cdot 3!}+\cdots\right)$
$+c_2t^{1/4}\left(1+\frac{3t}{5}+\frac{3^2t^2}{5\cdot 9\cdot 2!}+\frac{3^3t^3}{5\cdot 9\cdot 13\cdot 3!}+\cdots\right)$

13. $y_1(t)=t^{5/2}\left(1+\frac{t^2}{2\cdot 5}+\frac{t^4}{2\cdot 4\cdot 5\cdot 7}+\cdots\right)$

$y_2(t)=t^{-1/2}\left(1-\frac{t^2}{2}-\frac{t^4}{2\cdot 4}-\frac{t^6}{2\cdot 4\cdot 6\cdot 8}-\cdots\right)$

15. $y_1(t)=e^{t^2}/2,\ y_2(t)=t^3+\frac{t^5}{5}+\frac{t^7}{5\cdot 7}+\frac{t^9}{5\cdot 7\cdot 9}+\cdots$

17. $y_1(t)=\frac{1}{t}-1,\ y_2(t)=\frac{1}{t}[e^{-t}-(1-t)]$

19. (e) $y_2(t)=t^3e^{-t}\int\frac{e^t}{t^3}\,dt$

21. (b) $y_1(t)=t\left(2-3t+\frac{4t^2}{2!}-\frac{5t^3}{3!}+\frac{6t^4}{4!}\cdots\right)$

(c) $y_2(t)=y_1(t)\int\frac{e^{-t}}{y_1^2(t)}\,dt$

23. (c) $y_2(t)=J_0(t)\int\frac{dt}{tJ_0^2(t)}$

25. (b) $y(t)=1-\frac{\lambda t}{(1!)^2}-\lambda\frac{(1-\lambda)t^2}{(1!)^2(2!)^2}+\cdots$
$+\frac{(-\lambda)(1-\lambda)\cdots(n-1-\lambda)}{(n!)^2}t^n+\cdots$

27. (b) $y_1(t)=1+2t+t^2+\dfrac{4t^3}{15}+\dfrac{t^4}{14}+\cdots$

$$y_2(t)=t^{1/2}\left[1+\frac{5t}{6}+\frac{17t^2}{60}+\frac{89t^3}{(60)(21)}+\frac{941t^4}{(36)(21)(60)}+\cdots\right]$$

Section 2.8.3

1. $y_1(t)=1+4t+4t^2+\cdots$

$y_2(t)=y_1(t)\ln t-\left[8t+12t^2+\dfrac{176}{27}t^3+\cdots\right]$

3. (b) $J_1(t)=\frac{1}{2}t\displaystyle\sum_{n=0}^{\infty}\frac{(-1)^n t^{2n}}{2^{2n}(n+1)!n!}$

(c) $y_2(t)=-J_1(t)\ln t+\dfrac{1}{t}\left[1-\displaystyle\sum_{n=1}^{\infty}\frac{(-1)^n\left(H_n+H_n^{-1}\right)}{2^{2n}n!(n-1)!}t^{2n}\right]$

Section 2.9

1. $\dfrac{1}{s^2}$ **3.** $\dfrac{s-a}{(s-a)^2+b^2}$ **5.** $\dfrac{1}{2}\left[\dfrac{1}{s}+\dfrac{s}{s^2+4a^2}\right]$

7. $\dfrac{1}{2}\left[\dfrac{a+b}{s^2+(a+b)^2}+\dfrac{a-b}{s^2+(a-b)^2}\right]$ **9.** $\sqrt{\dfrac{\pi}{s}}$

15. $y(t)=2e^t-\frac{1}{2}e^{4t}-\frac{1}{2}e^{2t}$ **17.** $Y(s)=\dfrac{s^2+6s+6}{(s+1)^3}$

19. $Y(s)=\dfrac{2s^2+s+2}{(s^2+3s+7)(s^2+1)}$ **23.** $y(t)=-\frac{1}{6}e^t+\frac{1}{2}e^{2t}-\frac{1}{2}e^{3t}+\frac{1}{6}e^{4t}$

Section 2.10

1. $\dfrac{n!}{s^{n+1}}$ **3.** $\dfrac{2as}{(s^2+a^2)^2}$ **5.** $\dfrac{15}{8s^3}\sqrt{\dfrac{\pi}{s}}$

7. (a) $\dfrac{\pi}{2}-\arctan s$; (b) $\ln\dfrac{s}{\sqrt{s^2+a^2}}$; (c) $\ln\dfrac{s-b}{s-a}$ **9.** $-\frac{5}{3}+2e^{-t}+\frac{2}{3}e^{-3t}$

11. $\left[\cosh\dfrac{\sqrt{57}\,t}{2}+\dfrac{3}{\sqrt{57}}\sinh\dfrac{\sqrt{57}\,t}{2}\right]e^{3t/2}$ **13.** $\frac{1}{2}(3-t)t^2e^{-t}$

15. $\frac{1}{2}(\cos t-te^{-t})$ **19.** $y(t)=\cos t+\frac{5}{2}\sin t-\frac{1}{2}t\cos t$ **21.** $y(t)=\frac{1}{6}t^3e^t$

23. $y(t)=1+e^{-t}+\left[\cos\dfrac{\sqrt{3}\,t}{2}-\dfrac{7}{\sqrt{3}}\sin\dfrac{\sqrt{3}\,t}{2}\right]e^{-t/2}$

SECTION 2.11

1. $y(t)=(2+3t)e^{-t}+2H_3(t)\left[(t-5)+(t-1)e^{-(t-3)}\right]$

3. $y(t)=3\cos 2t-\sin 2t+\frac{1}{4}(1-\cos 2t)-\frac{1}{4}(1-\cos 2(t-4))H_4(t)$

5. $y(t)=3\cos t-\sin t+\frac{1}{2}t\sin t+\frac{1}{2}H_{\pi/2}(t)\left[\left(t-\frac{1}{2}\pi\right)\sin t-\cos t\right]$

7. $$y(t)=\frac{1}{49}\left[(7t-1)+\left(\cos\frac{\sqrt{27}\,t}{2}-\frac{13}{\sqrt{27}}\sin\frac{\sqrt{27}\,t}{2}\right)e^{-t/2}-H_2(t)(7t-1)-H_2(t)\left[13\cos\sqrt{27}\,\frac{(t-2)}{2}+\sqrt{27}\sin\frac{\sqrt{27}\,(t-2)}{2}\right]e^{-(t-2)/2}\right]$$

9. $y(t)=te^{t}+H_1(t)\left[2+t+(2t-5)e^{t-1}\right]-H_2(t)\left[1+t+(2t-7)e^{t-2}\right]$

SECTION 2.12

3. (a) $y(t)=(\cosh\frac{1}{2}t-3\sinh\frac{1}{2}t)e^{3t/2}-2H_2(t)\sinh\frac{1}{2}(t-2)e^{3(t-2)/2}$

5. $y(t)=\frac{1}{2}(\sin t-t\cos t)-H_\pi(t)\sin t$

7. $y(t)=3te^{-t}+\frac{1}{2}t^2e^{-t}+3H_3(t)(t-3)e^{-(t-3)}$

SECTION 2.13

1. $\dfrac{e^{bt}-e^{at}}{b-a}$ **3.** $\dfrac{a\sin at-b\sin bt}{a^2-b^2}$ **5.** $\dfrac{\sin at-at\cos at}{2a}$ **7.** $t-\sin t$

9. $\frac{1}{2}t\sin t$ **11.** $(t-2)+(t+2)e^{-t}$ **13.** $y(t)=t+\frac{3}{2}\sin 2t$ **15.** $y(t)=\frac{1}{2}t^2$

17. $y(t)=\frac{1}{2}\sin t+(1-\frac{3}{2}t)e^{-t}$

SECTION 2.14

1. $x(t)=c_1e^{3t}+3c_2e^{4t},\quad y(t)=c_1e^{3t}+2c_2e^{4t}$

3. $x(t)=[(c_1+c_2)\sin t+(c_1-c_2)\cos t]e^{-2t}$, $y(t)=[c_1\cos t+c_2\sin t]e^{-2t}$

5. $x(t)=\frac{7}{4}e^{3t}+\frac{1}{4}e^{-t},\quad y(t)=\frac{7}{8}e^{3t}-\frac{1}{8}e^{-t}$

7. $x(t)=\cos t\; e^{-t},\quad y(t)=(2\cos t+\sin t)e^{-t}$

9. $x(t)=2(t\cos t+3t\sin t+\sin t)e^{t},\quad y(t)=-2t\sin t\; e^{t}$

11. $x(t)=-4\sin t-\frac{1}{2}t\sin t-t\cos t+5\cos t\ln(\sec t+\tan t)$,
$y(t)=-t\sin t-t\cos t-\frac{1}{2}\sin^2 t\cos t-5\sin t\cos t+5\sin^2 t$
$-\frac{1}{2}\sin^3 t+5(\cos t-\sin t)\ln(\sec t+\tan t)$

SECTION 2.15

1. $y(t)=c_1e^{t}+c_2e^{-t}+c_3e^{2t}$ **3.** $y(t)=(c_1+c_2t+c_3t^2)e^{2t}+c_4e^{-t}$ **5.** $y(t)=0$

7. $y(t)=-3-2t-\frac{1}{2}t^2+(3-t)e^{t}$ **9.** $\psi(t)=1-\cos t-\ln\cos t-\sin t\ln(\sec t+\tan t)$

11. $\psi(t)=\frac{1}{\sqrt{2}}\int_0^t\big[\sin(t-s)/\sqrt{2}\,\cosh(t-s)/\sqrt{2}$
$-\cos(t-s)/\sqrt{2}\,\sinh(t-s)/\sqrt{2}\,\big]g(s)\,ds$

13. $\psi(t)=\frac{1}{4}t(e^{-2t}-1)-\frac{1}{5}\sin t$ **15.** $\psi(t)=\frac{1}{4}t^2\big[\big(\frac{1}{2}-\frac{1}{3}t\big)\cos t+\big(\frac{3}{4}+\frac{1}{6}t-\frac{1}{12}t^2\big)\sin t\big]$

17. $\psi(t)=t-1+\frac{1}{2}te^{-t}$

Chapter 3

Section 3.1

1. $\dot{x}_1=x_2$
$\dot{x}_2=x_3$
$\dot{x}_3=-x_2^2$

3. $\dot{x}_1=x_2$
$\dot{x}_2=x_3$
$\dot{x}_3=x_4$
$\dot{x}_4=1-x_3$

7. $\dot{\mathbf{x}}=\begin{pmatrix}5 & 5\\ -1 & 7\end{pmatrix}\mathbf{x},\quad \mathbf{x}(3)=\begin{pmatrix}0\\6\end{pmatrix}$

9. $\dot{\mathbf{x}}=\begin{pmatrix}0 & 0 & -1\\ 1 & 0 & 0\\ 0 & -1 & 0\end{pmatrix}\mathbf{x},\quad \mathbf{x}(-1)=\begin{pmatrix}2\\3\\4\end{pmatrix}$ **11.** (a) $\begin{pmatrix}1\\3\\-1\end{pmatrix}$; (b) $\begin{pmatrix}2\\0\\-1\end{pmatrix}$; (c) $\begin{pmatrix}-1\\4\\2\end{pmatrix}$

13. (a) $\begin{pmatrix}11\\16\\7\end{pmatrix}$; (b) $\begin{pmatrix}-7\\2\\-3\end{pmatrix}$; (c) $\begin{pmatrix}5\\10\\1\end{pmatrix}$; (d) $\begin{pmatrix}-1\\9\\1\end{pmatrix}$ **15.** $\mathbf{A}=\begin{pmatrix}\frac{1}{2} & \frac{7}{2}\\ -2 & 4\end{pmatrix}$

Section 3.2

1. Yes **3.** No **5.** No **7.** Yes **9.** No **11.** Yes

Section 3.3

1. Linearly dependent **3.** Linearly independent

5. (a) Linearly dependent; (b) Linearly dependent

7. (b) $y_1(t)=e^t,\quad y_2(t)=e^{-t}$ **9.** $p_1(t)=t-1,\quad p_2(t)=t^2-6$

11. $\mathbf{x}^1=\begin{pmatrix}-1\\-1\\1\end{pmatrix}$ **17.** $x_1=y_1$
$x_2=\frac{1}{\sqrt{2}}(y_2-y_3)$
$x_3=\frac{1}{\sqrt{2}}(y_2+y_3)$

Section 3.4

1. $\mathbf{x}^1(t)=\begin{bmatrix}\cos\sqrt{3}\,t/2\;e^{-t/2}\\ \big(-\frac{1}{2}\cos\sqrt{3}\,t/2-\frac{1}{2}\sqrt{3}\,\sin\sqrt{3}\,t/2\big)e^{-t/2}\end{bmatrix}$

$\mathbf{x}^2(t)=\begin{bmatrix}\sin\sqrt{3}\,t/2\;e^{-t/2}\\ \big(\frac{3}{2}\cos\sqrt{3}\,t/2-\frac{1}{2}\sin\sqrt{3}\,t/2\big)e^{-t/2}\end{bmatrix}$

3. $\mathbf{x}^1(t)=\begin{pmatrix}0\\e^t\end{pmatrix}$, $\mathbf{x}^2(t)=\begin{pmatrix}e^t\\2te^t\end{pmatrix}$ **5.** Yes **7.** Yes **9.** No **11.** (b) No

SECTION 3.5

3. -48 **5.** 0 **7.** -97 **11.** No solutions

13. $\begin{bmatrix}x_1\\x_2\\x_3\\x_4\end{bmatrix}=\begin{bmatrix}1\\0\\0\\0\end{bmatrix}$ **15.** $\begin{bmatrix}x_1\\x_2\\x_3\\x_4\end{bmatrix}=c\begin{bmatrix}-1\\-2\\1\\4\end{bmatrix}$

SECTION 3.6

1. $\mathbf{AB}=\begin{pmatrix}1&2&25\\1&7&10\\10&6&12\end{pmatrix}$, $\mathbf{BA}=\begin{pmatrix}15&52&0\\5&-1&1\\10&18&6\end{pmatrix}$

3. $\mathbf{AB}=\begin{pmatrix}1&1&2\\1&2&1\\2&1&1\end{pmatrix}$, $\mathbf{BA}=\begin{pmatrix}1&2&1\\2&1&1\\1&1&2\end{pmatrix}$

9. $\frac{1}{72}\begin{pmatrix}-3&5&9\\18&-6&18\\6&14&-18\end{pmatrix}$ **11.** $\frac{i}{2}\begin{pmatrix}2i&0&-2i\\1-2i&-1&-1+i\\1-2i&1&1+i\end{pmatrix}$

13. $\frac{i}{2}\begin{pmatrix}0&-2i&0\\-1&1&1-i\\1&-1&-1-i\end{pmatrix}$

SECTION 3.7

1. $\mathbf{b}=\begin{bmatrix}b_1\\b_2\\b_1+2b_2\end{bmatrix}$ **3.** $\mathbf{b}=\begin{bmatrix}2b_2+3b_3\\b_2\\b_3\end{bmatrix}$ **5.** $\mathbf{x}=c\begin{pmatrix}1\\-1\\1\end{pmatrix}+\begin{bmatrix}-4\\\frac{7}{2}\\0\end{bmatrix}$

7. No solutions **9.** $\mathbf{x}=\begin{pmatrix}1\\0\\0\end{pmatrix}$ **11.** $\lambda=1,-1$

13. (a) $\lambda=-1$; (b) $\mathbf{x}=\frac{1}{6}\begin{pmatrix}5\\-1\\0\end{pmatrix}+c\begin{pmatrix}-1\\-7\\6\end{pmatrix}$

SECTION 3.8

1. $\mathbf{x}(t)=c_1\begin{pmatrix}1\\1\end{pmatrix}e^{3t}+c_2\begin{pmatrix}3\\2\end{pmatrix}e^{4t}$ **3.** $\mathbf{x}(t)=c_1\begin{pmatrix}1\\0\\-1\end{pmatrix}e^{-t}+c_2\begin{pmatrix}0\\2\\-1\end{pmatrix}e^{-t}+c_3\begin{pmatrix}2\\1\\2\end{pmatrix}e^{8t}$

5. $\mathbf{x}(t)=c_1\begin{pmatrix}2\\0\\-1\end{pmatrix}e^{-10t}+\left[c_2\begin{pmatrix}0\\1\\0\end{pmatrix}+c_3\begin{pmatrix}1\\0\\2\end{pmatrix}\right]e^{5t}$

7. $\mathbf{x}(t)=\frac{7}{4}\begin{pmatrix}1\\2\end{pmatrix}e^{3t}+\frac{1}{4}\begin{pmatrix}1\\-2\end{pmatrix}e^{-t}$

9. $\mathbf{x}(t)=\begin{pmatrix}1\\-2\\-1\end{pmatrix}e^{2t}$ **11.** $\mathbf{x}(t)=\begin{pmatrix}-2\\0\\3\end{pmatrix}e^{-2t}$

13. (b) $\mathbf{x}(t)=9\begin{pmatrix}1\\0\\1\end{pmatrix}e^{t-1}+\frac{4}{3}\begin{pmatrix}-7\\2\\-13\end{pmatrix}e^{-(t-1)}+\frac{4}{3}\begin{pmatrix}1\\1\\1\end{pmatrix}e^{2(t-1)}$

Section 3.9

1. $\mathbf{x}(t)=\left[c_1\begin{pmatrix}2\\1-\sin t\end{pmatrix}+c_2\begin{pmatrix}2\sin t\\\cos t+\sin t\end{pmatrix}\right]e^{-2t}$

3. $\mathbf{x}(t)=e^{t}\left[c_1\begin{pmatrix}2\\-2\\3\end{pmatrix}+c_2\begin{pmatrix}0\\\cos 2t\\\sin 2t\end{pmatrix}+c_3\begin{pmatrix}0\\\sin 2t\\-\cos 2t\end{pmatrix}\right]$ **5.** $\mathbf{x}(t)=\begin{pmatrix}\cos t\\2\cos t+\sin t\end{pmatrix}e^{-t}$

7. $\mathbf{x}(t)=\begin{pmatrix}2\\-2\\1\end{pmatrix}e^{-2t}+\begin{bmatrix}-\sqrt{2}\sin\sqrt{2}\,t-\sqrt{2}\cos\sqrt{2}\,t\\\cos\sqrt{2}\,t-\sqrt{2}\sin\sqrt{2}\,t\\-3\cos\sqrt{2}\,t\end{bmatrix}$ **9.** $\mathbf{x}(0)=\begin{pmatrix}x_1\\0\\x_3\end{pmatrix}$

Section 3.10

1. $\mathbf{x}(t)=c_1\begin{pmatrix}0\\1\\1\end{pmatrix}e^{-2t}+\left[c_2\begin{pmatrix}1\\1\\0\end{pmatrix}+c_3\begin{pmatrix}t\\t\\1\end{pmatrix}\right]e^{-t}$

3. $\mathbf{x}(t)=c_1\begin{pmatrix}0\\0\\1\end{pmatrix}e^{-2t}+\left[c_2\begin{pmatrix}1\\0\\0\end{pmatrix}+c_3\begin{pmatrix}-t\\1\\0\end{pmatrix}\right]e^{-t}$

5. $\mathbf{x}(t)=\begin{pmatrix}1\\0\\1\end{pmatrix}e^{t}$ **7.** $\mathbf{x}(t)=\begin{pmatrix}1-t\\t\\t\end{pmatrix}e^{2t}$

13. (b) $\frac{1}{10}\begin{bmatrix}7e^{t}-5e^{2t}+8e^{-t} & -e^{t}+5e^{2t}-4e^{-t} & 3e^{t}-5e^{2t}-2e^{-t}\\21e^{t}-5e^{2t}-16e^{-t} & -3e^{t}+5e^{2t}+8e^{-t} & 9e^{t}-5e^{2t}-4e^{-t}\\14e^{t}+10e^{2t}-24e^{-t} & -2e^{t}-10e^{2t}+12e^{-t} & 6e^{t}+10e^{2t}-6e^{-t}\end{bmatrix}$

15. $\mathbf{I}+\frac{\mathbf{A}}{\alpha}(e^{\alpha t}-1)$

Section 3.11

1. $\frac{1}{5}\begin{bmatrix}e^{-2t}+5e^{2t}-e^{3t} & 5e^{2t}-5e^{3t} & e^{-2t}-e^{3t}\\-e^{-2t}+e^{3t} & 5e^{3t} & -e^{-2t}+e^{3t}\\4e^{-2t}-5e^{2t}+e^{3t} & -5e^{2t}+5e^{3t} & 4e^{-2t}+e^{3t}\end{bmatrix}$

3. $\frac{1}{5}\begin{bmatrix} 6e^{2t}-\cos t-2\sin t & -2e^{2t}+2\cos t+4\sin t & 2e^{2t}-2\cos t+\sin t \\ 3e^{2t}-3\cos t-\sin t & -e^{2t}+6\cos t+2\sin t & e^{2t}-\cos t+3\sin t \\ 5\sin t & -10\sin t & 5\cos t \end{bmatrix}$

5. $\begin{bmatrix} (t+1)e^{-t} & (t+1)e^{-t}-e^{-2t} & e^{-t}-e^{-2t} \\ -te^{-t} & -te^{-t}+e^{-2t} & e^{-2t}-e^{-t} \\ te^{-t} & te^{-t} & e^{-t} \end{bmatrix}$

7. $\mathbf{A}=\begin{pmatrix} 3 & 1 & -1 \\ 1 & 3 & -1 \\ 3 & 3 & -1 \end{pmatrix}$ **9.** $\mathbf{A}=\frac{1}{13}\begin{pmatrix} 16 & -25 & 30 \\ 8 & -6 & -24 \\ 0 & 13 & 26 \end{pmatrix}$ **11.** No

Section 3.12

1. $\mathbf{x}(t)=2e^{t}\begin{pmatrix} t\cos t+3t\sin t+\sin t \\ -t\sin t \end{pmatrix}$

3. $\mathbf{x}(t)=\begin{bmatrix} -4\sin t-\frac{1}{2}t\sin t-t\cos t+5\cos t\ln(\sec t+\tan t) \\ -t\sin t-t\cos t+\sin t\cos^2 t-\frac{1}{2}\sin^2 t\cos t \\ -5\sin t\cos t+5\sin^2 t-\frac{1}{2}\sin^3 t \\ +5(\cos t-\sin t)\ln(\sec t+\tan t) \end{bmatrix}$

5. $\mathbf{x}(t)=\begin{bmatrix} 3e^{3t}-2e^{2t}-te^{2t} \\ e^{2t} \\ 3e^{3t}-2e^{2t} \end{bmatrix}$ **11.** $\psi(t)=\begin{pmatrix} \frac{3}{4}t^2+\frac{1}{2}t+\frac{1}{8} \\ -\frac{1}{4}t^2-t+\frac{1}{8} \end{pmatrix}$

13. $\psi(t)=\begin{pmatrix} 2 \\ -2 \\ -1 \end{pmatrix}e^{t}$ **17.** $\psi(t)=\begin{pmatrix} 1 \\ -1 \\ -1 \end{pmatrix}te^{3t}$

Section 3.13

1. $\mathbf{x}(t)=e^{-t}\begin{pmatrix} 3 \\ 2 \end{pmatrix}+e^{4t}\begin{pmatrix} -3 \\ 3 \end{pmatrix}$

3. $\mathbf{x}(t)=-\frac{4}{3}\begin{pmatrix} 1 \\ 2 \end{pmatrix}e^{-t}+\frac{1}{6}\begin{pmatrix} 2 \\ 1 \end{pmatrix}e^{2t}+\frac{1}{2}\begin{pmatrix} 0 \\ 1 \end{pmatrix}-t\begin{pmatrix} 1 \\ 1 \end{pmatrix}+3\begin{pmatrix} 1 \\ 1 \end{pmatrix}e^{t}$

5. $\mathbf{x}(t)=e^{t}\begin{pmatrix} 1-t-t^2 \\ 1-\frac{3}{2}t^2 \end{pmatrix}$ **7.** $\mathbf{x}(t)=2e^{t}\begin{pmatrix} t\cos t+3t\sin t+\sin t \\ -t\sin t \end{pmatrix}$

9. $\mathbf{x}(t)=\begin{cases} \begin{pmatrix} \cos 2t+\sin 2t \\ 2\sin 2t \end{pmatrix}, & t<\pi \\ \begin{pmatrix} \cos 2t \\ \sin 2t+\cos 2t \end{pmatrix}, & t>\pi \end{cases}$ **11.** $\mathbf{x}(t)=\begin{pmatrix} 1-t \\ t \\ t \end{pmatrix}e^{2t}$

13. $\mathbf{x}(t)=\begin{bmatrix} t-t^2-\frac{1}{6}t^3-\frac{1}{2}t^4+\frac{1}{12}t^5 \\ -\frac{1}{2}t^2 \\ t^2+\frac{1}{6}t^3 \end{bmatrix}$ **15.** $\mathbf{x}(t)=\begin{bmatrix} 1 \\ 1+t \\ 1 \\ 1+2t \end{bmatrix}e^{3t}$

Chapter 4

Section 4.1

1. $x=0, y=0$; $x=0, y=1$; $x=1, y=0$; $x=\frac{1}{2}, y=\frac{1}{4}$

3. $x=0, y=0, z=0$; $x=\frac{c}{d}, y=\frac{a}{b}, z=-\left(\frac{c^2}{d^2}+\frac{a^2}{b^2}\right)$

5. $x=0, y=y_0$; y_0 arb.
$x=x_0, y=1$; x_0 arb.
$x=x_0, y=-1$; x_0 arb.

7. $x=0, y=-2, z=n\pi$

Section 4.2

1. Stable **3.** Unstable **5.** Asymptotically stable

7. Asymptotically stable **9.** Stable

11. $x(t)=1$ is stable; $x(t)=0$ is unstable.

Section 4.3

1. $x=0, y=0$ is unstable; $x=1, y=0$ is unstable;
$x=-1, y=0$ is unstable; $x=0, y=2^{1/4}$ is stable;
$x=0, y=-2^{1/4}$ is stable.

3. $x=0, y=1$ is unstable; $x=0, y=-1$ is unstable;
$x=1, y=0$ is unstable; $x=-1, y=0$ is stable.

5. $x=0, y=n\pi$ is unstable for all integers n.

7. Unstable **9.** Unstable **11.** Stable **13.** Stable

15. Unstable **17.** Unstable

Section 4.4

1. $y=\cos(x-1)^2$ **3.** $y=\tan x$ **5.** $x^2+y^2=c^2$

7. The points $x=x_0, y=y_0$ with $x_0+y_0=-1$, and the circles $x^2+y^2=c^2$, minus these points.

9. The points $x=0, y=y_0$; $x=x_0, y=0$, and the curves $y=ce^{-(2/3)e^{3x}}$, $x>0$; $y=ce^{-(2/3)e^{3x}}$, $x<0$.

11. The points $x=0, y=y_0$, and the curves

$$(by-dx)-\frac{(ad-bc)}{d}\ln|c-dy|=k, \quad x>0$$
$$(by-dx)-\frac{(ad-bc)}{d}\ln|c-dy|=k, \quad x<0$$

13. The point $x=0, y=0$; and the curves $xy^2-\frac{1}{3}x^3=c$.

Section 4.5.1

3. $\frac{y}{c}-\frac{b}{c^2}\ln(b+cy)=\frac{x^2}{2a}+k$ **5.** $\frac{y}{d}-\frac{x}{c}=\frac{b}{d^2}\ln(b+dy)-\frac{a}{c^2}\ln(a+cx)+k$

Section 4.7

13. (b) $y^2 = x^2 + x^4 + c^2$

Section 4.8

5. The circle $x^2 + y^2 = 1$. Note that every point on this circle is an equilibrium point.

7. The circles $x^2 + y^2 = (2n+1)\pi/2$

Section 4.9

1. $\varepsilon = -1, 1$

3. $\varepsilon = -2, 2$

5. No bifurcation points

Section 4.10

1. $x=0, y=0$ is unstable; $x=(a/e), y=0$ is stable if $ad < ec$, and unstable if $ad > ec$; $x=(af+bc)/(ef+bd), y=(ad-ec)/(bd+ef)$ exists only for $ad > ec$ and is stable.

3. $x_1 = 0, x_2 = 0, y = 0$; $x_1 = c/d, x_2 = nc/[d(a_1+a_2)], y = na_2 - a_1^2 - a_1a_2$; assuming that $na_2 > a_1^2 + a_1a_2$.

5. (b) (i) $y_1 = \dfrac{1+a_1}{1+a_1a_2}, y_2 = \dfrac{1-a_2}{1-a_1a_2}$; (ii) $y_1 = 1, y_2 = 0$; (c) $a_2 = \frac{1}{3}$

Section 4.12

3. (a) $rI + \lambda I^2/2 = \gamma \ln S - rS + c$; (b) Yes

5. (a) 0.24705; (b) 0.356; (c) 0.45212; (d) 0.60025; (e) 0.74305; (f) 0.77661

Section 4.13

7. Either $x(t) \to 0$ or $x(t) \to \dfrac{\beta_1 c - \alpha_1}{\beta_1}$ (for $\beta_1 c - \alpha_1 > 0$)

Chapter 5

Section 5.1

1. $\lambda_n = \dfrac{(2n+1)^2\pi^2}{4l^2}, y(x) = c\sin\dfrac{(2n+1)\pi x}{2l}$

3. $\lambda = 0, y = c$; $\lambda = \dfrac{-n^2\pi^2}{l^2}, y(x) = c\cos\dfrac{n\pi x}{l}$

5. $y(x) = c\sinh\sqrt{-\lambda_0}\, x$, where $\sinh\sqrt{-\lambda_0}\,\pi = \sqrt{-\lambda_0}\cosh\sqrt{-\lambda_0}\,\pi$; $y(x) = c\sin\sqrt{\lambda_n}\, x$, where $\tan\sqrt{\lambda_n}\,\pi = \sqrt{\lambda_n}$.

7. $\lambda = -1, y(x) = ce^x$; $\lambda = n^2, y(x) = c[n\cos nx + \sin nx]$

Section 5.3

1. $u(x,t)=\sin\frac{1}{2}\pi x\, e^{-1.71\pi^2 t/4}+3\sin\frac{5}{2}\pi x\, e^{-(1.71)25\pi^2 t/4}$

3. $u(x,t)=3\sin 2\pi x\, e^{(1-4\pi^2)t}-7\sin 4\pi x\, e^{(1-16\pi^2)t}$

5. $u(t,y)=e^{2(t+y)}+e^{-3(t+y)}$

7. $u(t,y)=e^{-5t}e^{-4y}+2e^{-7t}e^{-6y}-14e^{13t}e^{14y}$

9. (a) $X''-\mu X=0;\quad Y''-(\mu+\lambda)Y=0;\quad T'-\lambda\alpha^2 T=0;$
(b) $u(x,y,t)=\sin\dfrac{n\pi x}{a}\sin\dfrac{n\pi y}{b}\, e^{\alpha^2 n^2\pi^2(b^2-a^2)t/a^2b^2}$

Section 5.4

1. $f(x)=\dfrac{4}{\pi}\left[\dfrac{\sin\pi x}{1}+\dfrac{\sin 3\pi x}{3}+\dfrac{\sin 5\pi x}{5}+\dots\right]$

3. $f(x)=\dfrac{2}{\pi}\left[\dfrac{\sin\pi x}{1}-\dfrac{\sin 2\pi x}{2}+\dfrac{\sin 3\pi x}{3}\pm\dots\right]$

5. $f(x)=\dfrac{3}{4}-\dfrac{1}{\pi}\sum_{n=1}^{\infty}\dfrac{1}{n}\left[\sin\dfrac{n\pi}{2}\cos\dfrac{n\pi x}{2}+\left(1-\cos\dfrac{n\pi}{2}\right)\sin\dfrac{n\pi x}{2}\right]$

7. $f(x)=\dfrac{e^l-1}{2l}+\sum_{n=1}^{\infty}\dfrac{e^l(-1)^n-1}{l^2+n^2\pi^2}\left[l\cos\dfrac{n\pi x}{l}-n\pi\sin\dfrac{n\pi x}{l}\right]$

9. $f(x)=\dfrac{e^l}{l}+2\sum_{n=1}^{\infty}\dfrac{l\left(e^l(-1)^n-1\right)}{l^2+n^2\pi^2}\cos\dfrac{n\pi x}{l}$

11. $f(x)=\dfrac{e^l-1}{l}+\sum_{n=1}^{\infty}\dfrac{(-1)^n(e^l-e^{-l})}{l^2+n^2\pi^2}\left[l\cos\dfrac{n\pi x}{l}+n\pi\sin\dfrac{n\pi x}{l}\right]$

13. $f(x)=\frac{3}{4}\sin x-\frac{1}{4}\sin 3x$

17. (a) $f(x)=\dfrac{\pi^2}{3}+4\sum_{n=1}^{\infty}\dfrac{(-1)^n\cos nx}{n^2}$

Section 5.5

1. $f(x)=\dfrac{e-1}{e}+\sum_{n=1}^{\infty}\dfrac{2(1-\cos n\pi/e)}{1+n^2\pi^2}\cos n\pi x$

3. $f(x)=\dfrac{3a}{4}+\sum_{n=1}^{\infty}\dfrac{4a(\cos n\pi/2\,-1)}{n^2\pi^2}\cos\dfrac{n\pi x}{2a}$

5. $f(x)=\dfrac{l}{4}+\dfrac{2l}{\pi^2}\sum_{n=1}^{\infty}\dfrac{1}{n^2}\left[2\cos n\pi/2\,-1-(-1)^n\right]\cos\dfrac{n\pi x}{l}$

7. $f(x)=\dfrac{2}{\pi}\sum_{n=1}^{\infty}\dfrac{1}{n}\left(\cos n\pi/2\,-(-1)^n\right)\sin\dfrac{n\pi x}{2}$

9. $f(x)=\sin 2x$

11. (a) $\sin x=\dfrac{2}{\pi}+\dfrac{4}{\pi}\left[\dfrac{\cos 2x}{1-2^2}+\dfrac{\cos 4x}{1-4^2}+\dots\right];\quad 0<x<1;$
(b) $\cos x=\dfrac{4}{\pi}\left[\dfrac{2\sin 2x}{2^2-1}+\dfrac{4\sin 4x}{4^2-1}+\dots\right];\quad 0<x<1;$ (c) No

Section 5.6

1. (a) $u(x,t)=\frac{280}{\pi}\sum_{n=0}^{\infty}\frac{\sin(2n+1)\pi x/10}{2n+1}e^{-0.86(2n+1)^2\pi^2 t/100}$;

(b) $u(x,t)=\sum_{n=1}^{\infty}\frac{2n\left(1-(-1)^n\cos 10\right)}{n^2\pi^2-100}\sin\frac{n\pi x}{10}e^{-0.86n^2\pi^2 t/100}$;

(c) $u(x,t)=100\sum_{n=1}^{\infty}\left[\frac{4\sin n\pi/2}{n^2\pi^2}-\frac{(-1)^n}{n\pi}\right]\sin\frac{n\pi x}{100}e^{-0.86n^2\pi^2 t/100}$;

(d) $u(x,t)=\frac{-130}{\pi}\sum_{n=1}^{\infty}\frac{\left((-1)^n-\cos 3\pi/10\right)}{n}\sin\frac{n\pi x}{10}e^{-0.86n^2\pi^2 t/100}$

5. (b) $u(x,t)=10x+800\sum_{n=1}^{\infty}\frac{\sin n\pi/2}{n^2\pi^2}\sin\frac{n\pi x}{10}e^{(1-n^2\pi^2)t}$

Section 5.7

1. $u(x,t)=\frac{4}{\pi}\sum_{n=0}^{\infty}\left[\frac{2n+1}{(2n+1)^2-4}-\frac{1}{2n+1}\right]\sin(2n+1)\frac{x}{2}\cos(2n+1)\frac{ct}{2}$

3. $u(x,t)=\frac{12}{\pi^2}\sum_{n=1}^{\infty}\frac{\sin 2n\pi/3}{n^2}\sin\frac{n\pi x}{3}\cos\frac{n\pi ct}{3}$

5. $u(x,t)=\frac{8}{\pi^2}\sum_{n=1}^{\infty}\frac{1}{n^2}\sin\frac{n\pi}{2}\sin\frac{n\pi x}{2}\cos\frac{n\pi t}{10}$

9. $u(x,y,t)=X(x)Y(y)T(t)$; $\quad Y(y)=a_2\cos\sqrt{\lambda^2-\mu^2}\,y+b_2\sin\sqrt{\lambda^2-\mu^2}\,y$
$X(x)=a_1\cos\mu x+b_1\sin\mu x \quad T(t)=a_3\cos\lambda ct+b_3\sin\lambda ct$

Section 5.8

1. $u(x,y)=\sum_{n=1}^{\infty}c_n\sinh\frac{n\pi(x-a)}{b}\sin\frac{n\pi y}{b};\quad c_n=\frac{-2}{b\sinh n\pi a/b}\int_0^b f(y)\sin\frac{n\pi y}{b}\,dy$

3. $u(x,y)=\sum_{n=1}^{\infty}c_n\sin\frac{n\pi x}{a}\sinh\frac{n\pi(y-b)}{a};\quad c_n=\frac{-2}{a\sinh n\pi b/a}\int_0^a f(x)\sin\frac{n\pi x}{a}\,dx$

5. $$u(x,y)=\frac{4}{\pi}\sum_{n=1}^{\infty}\frac{\sin(2n-1)\frac{\pi x}{a}\sinh(2n-1)\frac{\pi y}{b}}{(2n-1)\sinh(2n-1)\pi b/a}+\frac{4}{\pi}\sum_{n=1}^{\infty}\frac{\sinh(2n-1)\frac{\pi x}{b}\sin(2n-1)\frac{\pi y}{b}}{(2n-1)\sinh(2n-1)\pi a/b}$$

7. $$u(x,y)=1+\frac{4}{\pi}\sum_{n=1}^{\infty}\frac{\sinh(2n-1)\pi\frac{x-a}{b}\sin(2n-1)\frac{\pi y}{b}}{(2n-1)\sinh(2n-1)\pi a/b}$$

9. $$u(x,y)=\frac{-2}{\pi}\sum_{n=1}^{\infty}\frac{(-1)^n\sinh\sqrt{1+n^2\pi^2}\,x\sin n\pi y}{n\sinh\sqrt{1+n^2\pi^2}}$$

11. $X''+\lambda X=0;\quad Y''+\mu Y=0;\quad Z''-(\lambda+\mu)Z=0$

Index